Supply Chain Planning for Clinical Trials – A Practical Guide

Supply Chain Planning for Clinical Trials – A Practical Guide

Ryan Mills
San Francisco
CA, United States

WILEY

Published by John Wiley & Sons, Inc., Hoboken, New Jersey.
Published simultaneously in Canada.

Library of Congress Cataloging-in-Publication Data Applied for:
ISBN: 9781394179558 (Paperback)

Cover Design: Wiley
Cover Image: © LeoWolfert/Shutterstock

Set in 9.5/12.5pt STIXTwoText by Straive, Pondicherry, India

Contents

Preface

Welcome to the exciting, dynamic, and complex world of clinical trial supply chain management! If you're reading this book, there is a good chance you're responsible for managing some aspect of the clinical trial supply chain or are interested in learning how to do so. Congratulations, you've made an excellent career choice. This work is deeply impactful as the decisions made by clinical supply chain planners directly impact millions of clinical trial participants around the world who need access to cutting-edge therapies. Your work contributes to extending and saving lives on a grand scale.

The top priority of this job is to ensure that every person who enters a clinical trial receives the drug they're expecting when they expect to receive it. Theoretically, producing the right amount of drug for a clinical trial should be easy. The total amount of drug required to service a clinical trial is known because the clinical protocol prescribes the number of subjects who will enroll and how much drug each subject will take over the duration of the trial. Multiply these two variables and we have our supply requirement for the trial. In a world of unlimited resources, we could simply make double or triple that amount of drug, and we'd have more than enough to supply the trial. I've seen that strategy employed in my past!

But we don't live in a world of unlimited resources. This is particularly true for the burgeoning class of pre-commercial biopharmaceutical and cell therapy companies for whom every dollar is precious. Supply chain management gives us the tools to develop an optimal supply plan that adequately supplies the trial while also minimizing the amount of money it takes to do so. Across the industry, there has been increasing recognition that effective supply chain management is a potential gold mine of savings for a company. One large-molecule drug substance batch costs anywhere between $2 and 5M, enough to fund a Phase I study (or two) on its own! Meanwhile, a 2016 study showed that drug companies waste almost half of the clinical trial supply they produce.[1] More recent data suggests that the number may be as high as a staggering 75%.[2]

Combatting this waste epidemic starts with better supply chain planning. The word planning has a specific definition in the supply chain lexicon. It is the process of optimizing the manufacture and delivery of goods by balancing supply and demand. When supply is balanced with demand, the consumer gets what they expect, when they expect it, and waste and costs are minimized throughout the supply chain. Simply put, better planning leads to better outcomes.

That's why I decided to write this book. I've worked in clinical supply chain management for over 15 years, partnering with hundreds of dedicated, talented colleagues who are committed to eradicating disease and improving lives. Many people who work in the biopharmaceutical sector have received extensive training in their area of expertise. Doctors, chemists, biologists, and pharmacologists all have degrees that underpin their day-to-day work. However, when you look at the backgrounds of people who work in biopharmaceutical supply chain management, you'll often find a hodgepodge of degrees and backgrounds. While some people do have formal business training via business school or have past experience in logistics or a related discipline, many don't. And even amongst those with a specialized degree, their exposure to the nuances of clinical trials is often limited. This means that most people walking into a clinical supply chain role have never been trained or had exposure to a clinical trial supply chain. Even for those of us who do have formal training in business and supply chain management, we've had to relearn much of what we were taught in the classroom because the practical application differs so much from the theoretical concept in the world of clinical trials.

This is because supply chain planning for clinical trials is fundamentally different from managing a commercially available good that can be purchased at your local grocery store. The demand forecast is built on a completely unique set of assumptions. Supply planning requires an understanding of not only the manufacturing steps in the process but also the quality standards and regulatory requirements of the biopharmaceutical space. Throw in relatively short expiry periods, a heaping dose of stakeholder complexity, and enormous costs, and you've developed a recipe where newcomers to clinical trial supply chain management are often lost and must learn their job the hard way – through trial and error.

This book aims to put a stake in the ground for clinical trial supply chain planning. You've probably heard the phrase "proper planning prevents poor performance," and nowhere is that truer than for clinical trials. There are so many variables in play that can impact supply availability; it takes a trained professional to understand what to do, when to do it, how to do it, and who to work with to make sure it's all done properly.

Unfortunately, there isn't one "right" way to do this. Every biopharmaceutical company, every product, and every clinical trial is unique and has its own set of considerations and challenges. But there are a set of guiding principles and concepts that every clinical planner needs to be aware of and understand. This book attempts to lay those out in a way that's approachable for someone who is entirely new to supply chain management or to clinical trials. Please understand that the content here isn't designed to cover every possible approach, strategy, or scenario. The space is too complex and moves too fast for that, and even an encyclopedia wouldn't have enough room to contain it all. The broad strokes are here though, and a good bit of detail, too.

While this book spends a majority of its time on supply chain planning in the traditional context, there is much more to discuss than just demand and supply planning. To reach the broadest audience possible, the book is laid out in three sections:

- *Clinical Supply Chain Foundations:* Chapters 1–6 are intended for the audience who has never been formally trained in supply chain management or worked on a clinical trial before. These chapters introduce the reader to general supply chain management, clinical trial design, the pharmaceutical supply chain, how quality impacts the clinical trial supply chain, regulatory considerations, and the randomization and trial supply management technology. The intent of these chapters is not to cover every potential topic in detail but rather to provide a broad introduction of the concepts that are most relevant and important for supply chain planning.
- *Supply Chain Planning for Clinical Trials*: Chapters 7–19 comprise the majority of the text and offer a deep dive into the world of supply chain planning for clinical trials. Beginning with an overview of the objectives and elements of demand and supply planning, this section walks the reader through the supply chain planning process in a step-by-step fashion. Demand forecasting, finished product supply planning, finished product distribution planning, expiry planning, and

upstream drug substance and drug product planning are all covered in extensive detail. More advanced concepts like aggregate forecasting, attrition, SKU-switching, and low case-high case planning are also given attention. Supply chain planning in this space is about much more than just the mechanics of creating and maintaining a plan though. It's also about effectively using data to monitor and communicate supply chain performance. And effective business process design and implementation are necessary to get information to the clinical planner and to supply outputs to the stakeholders who need to act on the plan. We'll cover all of these points in this section as well. Special note should be made of Chapter 16, which takes the reader through a day in the life of a clinical planner and provides insight into what makes the job both tremendously exciting and equally challenging. Chapters 17–19 introduce nontraditional clinical trial supply chains including comparators, just-in-time supply chains, and direct-to-patient supply chains and talk about supply chain considerations holistically from both a planning perspective and an operational perspective.

- *Additional Clinical Supply Chain Considerations:* Chapters 20–23 conclude our journey by touching on other aspects of clinical trial supply chain management that influence supply chain planning specifically. Contract manufacturing and logistics are two fundamental areas of the clinical trial supply chain, and planners who are familiar with the intricacies of these spaces tend to produce better plans. It's also impossible to conclude a review of supply chain planning without touching on the tools and technology that clinical planners use to create and maintain their plans. We conclude with a review of the final step of the clinical development process, commercialization, and how to integrate the clinical and commercial supply chains within an organization.

There is also an appendix that catalogues the data elements and equations used in the supply chain planning process, and a glossary that catalogues the definition of key terms, of which there are many. Every chapter concludes with a summary of key takeaways for the reader to consider. More experienced readers may want to glance through the key takeaways first, and if they're familiar with the content, then the reader may choose to skip ahead to more advanced content.

For those readers who are grizzled veterans, you will likely find yourself disagreeing with some of what's recommended here. Good! This book also intends to provoke, and it's my sincere hope to have more discussion in the future about the merits of weeks versus months-of-supply or to challenge my assumptions on just-in-time manufacturing's relevance for clinical labeling. Our industry needs more conversation, more debate, and more transparency when it comes to supply chain planning for clinical trials. We can do better as a collective, and I want this book to push all of us toward more innovation and better performance.

Supply chain planning for clinical trials is a difficult job. It requires expertise in supply chain management and clinical trial design, analytical skills, strategic thinking, and people who can successfully collaborate across functions, cultures, and companies. This is also an incredibly fun, stimulating, and rewarding job. Our industry needs good people who are committed to this work, and people who have the fundamentals to do this work well. My sincere hope is for this book to engage those who are interested, to educate those new to the role, and to reinvigorate those who have been at this work a long time.

Thank you for reading.

-Ryan Mills

Notes

1 Lamberti M., Hsia R., Mahon C., Milligan C., Getz K. (2016). Assessing Global Clinical Supply Logistics. *Applied Clinical Trials* October/November 2016: 26–34.

2 https://lifesciences.n-side.com/blog/decrease-clinical-trial-supply-chain-waste-even-more-with-the-n-side-supply-app-refined-algorithm

Acknowledgments

This book would not have been possible without the contributions of so many, beginning with Wiley and my first managing editor, Dr. Julia Squarr, who saw the value of creating a book on this topic. Dr. Squarr encouraged me to think broadly and to make the book as educational and accessible as possible for people who are interested in learning more about our industry and this space in particular.

I recognize the Association for Supply Chain Management (ASCM) for granting permission to reproduce their Supply Chain Operations Reference (SCOR) model in Chapter 1. The SCOR model is foundational to understanding modern supply chains and provides important context for a large amount of content that's covered in later chapters. I also recognize the International Council for Harmonisation (ICH) for allowing content from their guidelines to be reproduced in works like these.

I'm inspired every day by the people I work with, particularly my supply chain colleagues in industry. Krishna, Emily, Patty, Vipul, and Ben have taught me so much about why people matter more than anything else in the supply chain field. Their dedication, willingness to both teach and learn, and their ambition to create a world-class supply chain have provided boundless inspiration.

Countless friends and colleagues have provided ideas and feedback to this book. Thank you to Sarah Lockwood for her seemingly endless knowledge of clinical development and regulatory affairs. Thank you to Angela Kay for her expertise in clinical operations and clinical trial conduct. Thank you to Kim Burson, Darren Hassey, Asif Haque, and Matthew Lee for teaching me about the integration between supply chain management and quality in the pharmaceutical space. Thank you to Justin Vitale and Katrien Brak, two of the brightest process chemists and small-molecule CMC team leads I've had the pleasure to work with. Thank you to Tony Delucchi, who has taught me about large-molecule CMC development and given me the space in my career to take on an effort of this magnitude. Thank you to Sarah Miller, who first pushed me to be a better clinical planner and then shared incredible insights on cell therapy supply chain management. Thank you to Jon Miller for sharing his experience building and managing allogenic supply chains. Thank you to Azuka Okeke from the Africa Resource Center for Excellence in Supply Chain Management for spending time with me to discuss the intricacies of clinical trials in Africa. Thank you to Brittney Elko, another supply chain expert, who has often challenged my thinking. Thank you to Lola Fan, one of the smartest clinical trial technologists I've had the pleasure of working with for her contributions to Chapter 6. Thank you to Ben Zuffi, an IT superhero, for providing wisdom on technology strategy. Thank you to my former professor at North Carolina State University, Dr. Rob Handfield, who provided excellent feedback and ideas for consideration.

I'm also a member of the BioSupply Management Alliance (BSMA), an organization that strives to create a stronger network of biopharmaceutical professionals across sponsors, consultants, and suppliers. Thank you to the many people who I've interacted with during my time there through conferences, panels, webinars, and information conversations, especially Devendra Mishra, Bill Coakley, Sebastien Coppe, Luke Moyer, Allan Jacques, Stacey Erickson, Tony Stolis, Brad Graves, Scott Sabalvaro, Laura Shuga, Jan Pieter Kappelle, Antonio Tramontano, Christopher Jacks, Darrell Shiera, Hilary Ross, Rob Cascone, and Irene Xie for leading, participating, and sharing ideas that make our industry better.

Finally, and most importantly, thank you to my family, who have supported me in this journey from inception to publication. To my wife Crisel, I love you to the moon and back, and I'm so appreciative of all the time you've given me to pursue this passion project. And to my kids, Olivia and Alex, I hope this work will provide a source of inspiration that you can accomplish anything and change the world in a positive way when you set your mind to it.

1

Supply Chain Management

1.1 Supply Chain Management

The goal of this chapter is to introduce basic supply chain management principles. Before understanding the nuances of pharmaceutical supply chains, it's helpful to have a general summary of what a supply chain is, the nodes within the supply chain, and the goals of supply chain management as a discipline. The terms and concepts shared in this chapter underpin the remainder of the book, so this is an excellent starting point for those without a theoretical background in supply chain management.

A supply chain is a network of suppliers, manufacturers, and distributors that collaborate to manufacture a good/product and make it available for consumption. Supply chains are responsible for delivering the simplest items, from fresh strawberries at a farmer's market delivered by truck that morning, to the most complex items, like biopharmaceuticals that require dozens of suppliers and manufacturers. Supply chains have existed for hundreds of years, but they have taken on special importance over the last few decades. Technological advances in manufacturing combined with the advent of the internet have enabled supply chains to stretch around the world, and they can react to consumer requirements almost instantaneously. We now take it for granted that we can order an item on Amazon, and it will be delivered within hours or days.

As modern supply chains have stretched themselves further and further across suppliers and continents, they have become more vulnerable to disruption. War, natural disasters, and political–economic turmoil can all derail a supply chain from delivering on its intended promise. These risks became particularly clear during the COVID-19 pandemic and its aftermath when basic goods such as toilet paper and more advanced goods like microprocessors became scarce or weren't available at all. These shortages illustrated how important properly managed supply chains have become to both the global economy and the individual consumer.

The supply chain itself is a sequence of interconnected nodes with a manufacturer standing as the central node in the chain. Upstream of the manufacturer are a set of first-degree suppliers who provide the components needed to make the good. First-degree suppliers may have suppliers of their own – **second-degree suppliers** – who provide them with the raw materials or components they need to make their product. Like a spider web, supply chains can spiral out for several nodes before hitting basic raw materials. Downstream of the manufacturer is the consumer. In most cases, an intermediary such as a retailer, warehouse, or pharmacy will store the good. The connection between these upstream and downstream nodes is why these networks are called supply chains.

Supply Chain Planning for Clinical Trials – A Practical Guide, First Edition. Ryan Mills.

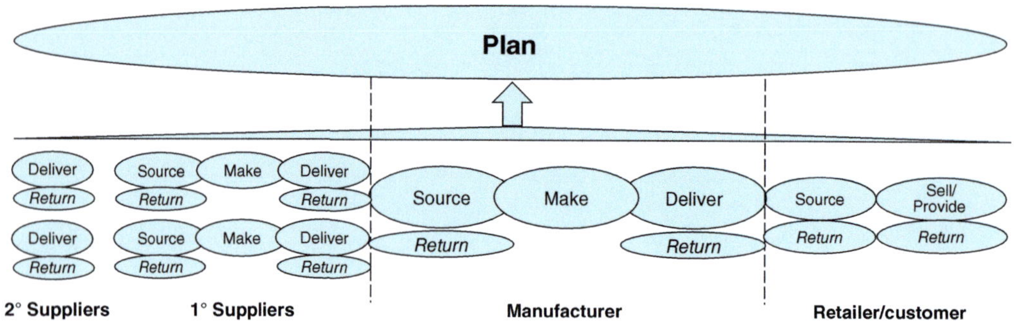

Figure 1.1 Supply Chain Operations Reference Model. *Source:* With permission of APICS[1]

The Supply Chain Operations Reference (SCOR) model that's been developed by the Association for Supply Chain Management (ASCM) provides an excellent diagram of the basic elements of a supply chain.

Figure 1.1 shows the SCOR model, which lays out the four core processes involved in supply chain management:

- *Plan*: The act of developing the supply chain strategy. Planners are responsible for determining how much material will be procured or produced, when sourcing or manufacturing activities will take place, and where goods should be distributed through the supply chain. Given the importance of developing an optimal supply chain strategy, planning is the focus of this book.
- *Source*: The act of purchasing or acquiring a good (e.g. a raw material or component) that is needed to facilitate a downstream manufacturing step. Suppliers provide these goods to downstream parties.
- *Make*: The act of producing a good that will be used in a downstream step of the supply chain or sold as the final product. Manufacturers make the components and products within the supply chain.
- *Deliver/Return*: The act of providing a good to a downstream node in the supply chain such as a manufacturer, retailer, or consumer. Return is the act of returning a good to an upstream node of the supply chain. Transportation service providers (TSPs) and couriers ship raw materials, components, and goods through the supply chain.

Supply chain management is the discipline of planning, coordinating, and executing supply chain activities. It has been recognized as a separate field from manufacturing operations since the 1980s. Following the development of computer hardware and software applications that enabled tracking sales, manufacturing, and logistics activities, it became possible to develop cohesive analyses and strategies incorporating each of the supply chain processes. Supply chain management is now a core function whose goal is to ensure sufficient supply of a product to meet consumer demand at the lowest cost possible.

To achieve this, it's not enough to just plan, source, make, deliver, and then call it a day. The supply chain does not exist in a vacuum, and the external environment around it is constantly changing and evolving. Therefore, another central tenet of effective supply chain management is to appropriately manage supply chain risk. A supply chain risk is anything that can disrupt the supply chain from delivering sufficient supply where and when it's needed. There are many categories of supply chain risk; demand risk, economic risk, environmental risk, sociopolitical risk, and ethical risks are all examples. An effective supply chain management strategy takes these risks into account and attempts to build resilience within the supply chain. Supply chain resilience is the ability of the supply chain to withstand the realization of these risks. If a war closes access to a key market or if a natural disaster shutters a critical supplier or if a distributor is found in breach of contract, does the manufacturer have strategies in place to enable the supply chain to continue functioning?

How much, when, and where are the questions at the heart of supply chain management. To answer these, supply chains need three elements operating in harmony – teams of people to do the work, well-defined business processes that enable efficient execution, and data and technology to facilitate the flow of information: People–Process–Technology. People come first in this equation because supply chain management is the ultimate team sport. Collaboration is the secret ingredient because no actor can succeed without the partnership of their upstream and downstream partners.

A business process is a sequence of tasks designed to accomplish a specific objective and to ensure that information is generated and passed to the stakeholders who need it to act. Business processes provide structure to supply chain activities and ensure consistency across task execution. Meanwhile, technology is the great enabler of supply chain management. The internet, big data, and tailored supply chain systems have combined to put an incredible amount of data at a user's fingertips. The world's best supply chains are armed with well-trained people, well-defined business processes, and well-designed technology solutions that enable processes to run efficiently and people to make good decisions.

We will spend the remainder of this chapter elaborating on the core elements of the supply chain: plan, source, make, and deliver/return. Each section below will describe the purpose of the process; how the process is executed; strategic considerations, including how to manage risk within that process; and the stakeholders responsible for making the process happen. After reviewing each process, we will analyze how data and metrics are used to drive the supply chain and improve supply chain performance. Then, we will conclude the chapter by revisiting supply chain management within the framework of People–Process–Technology before finishing with key takeaways.

1.2 Plan

The definition of planning is "to create a detailed proposal for doing or achieving something.[2]" Planning determines activities within a supply chain that need to be executed, and it takes place at every level. With that said, there is one planning process that governs the entire end-to-end supply chain – demand and supply planning – and it is the primary focus of this book. Demand and supply planning takes a demand forecast for a good and converts that data into a supply plan for production and distribution. This process begins with demand forecast generation, where demand is defined as the quantity of a good required by a consumer of that good. The demand forecast indicates how many units of the good are required within a specified timeframe (e.g. days, weeks, months, or years). The demand forecast for the finished good is also known as independent demand because this demand is independent of any other variable than the consumer's needs. The finished good demand forecast creates a requirement for supply to meet that demand. Supply is the quantity of a finished good that a manufacturer must produce to meet demand, and the supply plan indicates how many units need to be produced or distributed to a location within a specified timeframe.

Once the supply plan has been generated, the supply needs to be allocated to each source of demand. For a commercial good, these sources may be an online store or a brick-and-mortar retail location. In a simplified example, if the total demand for a bicycle over one year is 1000 units and that demand is projected to be spread evenly at 500 bicycles apiece between two locations, the supply plan may call for 1200 bicycles and for 600 to be sent to each location. The additional 200 units of supply is safety stock, and these extra units are intended to protect against demand uncertainty. In other words, what happens if demand does not split evenly according to the forecast, and 600 consumers (each wanting one bike of course!) show up to one store and 400 to another? Or what if demand exceeds expectations and 550 consumers show up to each store?

The finished good supply plan serves another purpose. Where the independent demand forecast creates requirements for supply, the finished good supply plan creates requirements for the components and materials needed to produce those finished goods. To continue the example above, the finished good supply plan calls for 1200 bicycles. Those 1200 bicycles will need 2400 tires, 2400 pedals, and 1200 frames. The need for 1200 frames will create further requirements for metals, and so on. The demand for components created by the finished good supply plan is known as dependent demand because it is dependent on the finished good.

Figure 1.2 shows a demand and supply planning schematic where an independent finished good demand forecast creates a requirement for finished good supply. When the finished good is manufactured, supply from each batch must be allocated to satisfy finished good demand. The finished good supply plan also creates dependent demand for the components that are needed to manufacture the finished good.

The objective of demand and supply planning is to ensure the *right product* is delivered in the *right quantity* to the *desired place* at the *desired time* for the *least amount of money* possible. Meeting these five objectives requires a combination of data, analytics, strategy, and collaboration across the organization. To generate a robust demand forecast and supply plan, it should incorporate the following elements. First, the plan should be end-to-end which means the plan should account for

- A demand forecast window that fits the product's lifecycle.
- All the finished good products that are consumed in the network.
- All critical upstream production or sourcing required to enable the manufacture and distribution of the finished good is accounted for in the plan.

Second, the supply plan should be entirely demand driven. All production and distribution decisions need to be made within the context of demand. If an external event (such as a sale) increases demand, the plan should incorporate this. If the company raises the price of a product and demand decreases, the plan should incorporate this. Similarly, dependent demand creates

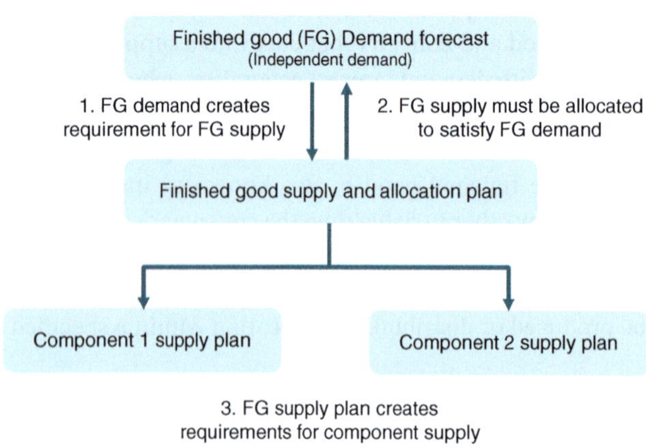

Figure 1.2 Demand and Supply Planning Schematic

requirements for components, and these requirements form the initial basis of the planning process for upstream materials. Requirements for the finished goods and their components will be modified during the planning process, but working without requirements is flying blind and a precursor to waste.

Third, the supply plan should be weeks-(or days ... or months)-of-supply (WoS) based. Weeks-(or days ... or months)-of-supply describes how much on-hand inventory is available within a specific forecast period, and it is the most accurate way to measure and communicate the inventory position of a good. If a manufacturer has 500 units in inventory, that might sound like a lot. But, if demand for that product runs at 200 units/week, that manufacturer will stock out in just three weeks. Conversely, let's assume that the same manufacturer has 50 units of another good in inventory. That doesn't sound like nearly as much as the good with 500 units, but if demand for the second product runs at an average of 1 unit/week, there's enough inventory on hand to last a whole year. Therefore, communicating in WoS – for example, "The product has seven weeks and is starting to run lean" or "The product has 18 weeks and is running long" – tells a more accurate story about the relative strength of the inventory position and what actions may need to be taken within the plan.

Building a plan that's end-to-end, demand driven, and WoS based mitigates the risk inherent to all demand forecasts and supply plans – that the plan will be wrong: more specifically, the risk that the demand forecast will be too low and not enough supply will be produced to meet actual demand, or the risk that the demand forecast will be too high and an excess of supply will be manufactured. When the forecast is too low, this can result in missed sales and lost revenue, but if the forecast is too high, then cash gets tied up in inventory and potentially wasted altogether. Here is a hard truth, the demand forecast will always – I repeat, always – be wrong in one direction or the other. While the theoretical goal is to manage the demand forecast and supply plan to a perfect balance, the reality is that the plan will need to be constantly updated and rebalanced in pursuit of that goal.

When a demand forecast and supply plan incorporates these elements, they give the organization three powerful tools for managing the supply chain – visibility, control, and flexibility. Visibility means an organization has insight to changes in demand signals and supply risks. If the plan is not end-to-end, then a risk to a key component may be missed. If the plan is not demand driven, line of sight to component requirements is lost. And if the plan is not WoS based, changes in demand will not signal a risk of under- or oversupply on the horizon. Visibility is powerful, and without it, the risk that a plan will fail to meet demand or waste money increases exponentially.

Second, if there is no visibility as to where problems may occur, there is no way to control the strategy. There is no worse feeling than the helplessness of finding a problem without an apparent reason for why that problem exists. Without a visible cause, control is lost, so visibility and control go hand in hand. To achieve visibility and control, the plan must be built at the appropriate level of detail. End-to-end planning translates to control over all aspects of the supply chain, and WoS enables better decision-making through the supply chain when events go awry.

Finally, these elements enable flexibility to respond when events don't go as planned. Flexibility has two meanings in the context of demand and supply planning – flexibility in approach and flexibility in strategy. Flexibility in approach means that the plan is tailored to meet the needs of the consumer, the product, and the supply chain. No two supply chains are alike, and the plan must be adaptable to the unique circumstances that it attempts to model. Flexibility in strategy means the plan has a variety of options embedded in it to account for the plan going wrong in either direction. If demand is high, appropriate safety stocks exist to buffer at the finished product level and at the component level. If demand is low, production and ordering can be delayed to a desirable date.

End-to-end plans, demand-driven plans, and WoS-based plans enable this strategic thought process as we will learn in Chapters 7–13.

With all of that in mind, a well-designed plan that lives in isolation is not enough. Business processes must exist to manage the plan, and there are two of these to note. The first is the demand and supply planning process itself. On a regular cadence – weekly or monthly in most cases – the plan needs to incorporate historical demand (e.g. sales) and supply (e.g. production) actuals. Then, the plan needs to incorporate any new demand or supply assumptions before it is updated for the next planning cycle. Second, the entire organization, from sales to operations to finance, needs to align around and endorse the demand and supply plan because it governs both the manufacturing operations of the company and the spend required to support that production. Sales and operations planning (S&OP) is the business process for gaining this alignment and endorsement.

The function responsible for these processes is planning. Within a planning organization, individual planners are typically assigned to a specific product or product line. For highly complex supply chains, these roles may be split into demand planning and supply planning for a single product. The planner serves as a bridge between the sales and operations organizations and must effectively collaborate with both functions to develop an optimal plan. Without this collaboration, the plan will have poor or unrealistic inputs, which then drive poor or unrealistic outputs that the company will march to – garbage in, garbage out as the saying goes. Planners must also think analytically while updating the plan and strategically to place the plan outputs into the overall context of the product. For example, if the plan calls for a surge in supply, but the company is planning to discontinue the product in six months, that plan doesn't make sense.

To summarize, planners should build demand forecasts and supply plans that are end-to-end, demand driven, and WoS based. This structure gives planners the visibility they need to control the plan and execute a flexible strategy that can respond to unforeseen events without jeopardizing product supply. The planner needs to work closely with sales, operations, finance, and product strategy to collect the proper inputs for the plan and ensure the outputs are realistic and in sync with the overall strategy.

1.3 Source

Sourcing is the act of identifying, selecting, qualifying, and engaging with external organizations for the acquisition of goods and/or services. Without raw materials and component suppliers, it's impossible for a supply chain to function. There are many attributes that separate a good supplier from a poor one, and a single supplier may be a strong fit for one manufacturer and a poor fit for another. The goal for all supplier relationships is the same though – to acquire the goods and/or services on time, in full (OTIF), at the expected quality level, and for the negotiated price.

Every supplier relationship follows a basic four-step process – identification, selection, qualification, and management. The sourcing process begins with recognizing the need for a supplier to provide a good or service. In most cases, this decision is obvious. If an outside party can provide a good cheaper than a manufacturer can make it itself, it makes sense to procure that good from that outside party. In some cases, though, it may not be immediately obvious that sourcing a good is cheaper than making it or there may be other factors to consider. In these circumstances, a make-versus-buy analysis may be required to determine whether it makes sense to insource (e.g. make it in-house) or outsource (e.g. procure from a third party) the good in question. Additional factors may include increased control over the process to manufacture it or the opportunity to leverage learnings from in-house production to other areas of the business.

Once the decision to outsource has been made, the supplier identification process will begin. This first step will often include sending out a request for information (RFI) to suppliers who have demonstrated the capability to produce the good or service in question. RFIs are intended to solicit feedback from potential suppliers across a multitude of categories such as location, company size and history, team structure, service levels, quality levels, and cost for the goods rendered. Once the field of candidates has been narrowed to three to five potential candidates, the selection process will begin.

The first step in supplier selection is to develop a request for proposal (RFP) or a request for quote (RFQ). The RFP/RFQ contains a detailed description of the goods or services required by the manufacturer, the timeline by which these goods or services should be delivered, and the expected quality standards. The RFP/RFQs are then sent to each potential supplier for response. After the responses have been collected, the team responsible for making the selection decision will evaluate the suppliers against predefined criteria. While cost is always a part of the analysis, other factors may be more important to consider. For example, if the good or service must be delivered within a narrow timeframe, project management and customer responsiveness may rank highly in the decision.

Once a decision has been made and the selected party has been notified, the contract and qualification processes will begin. If the work will be done on a one-time basis, a simple statement of work (SOW) will suffice. But, if the products will be delivered on an ongoing basis, then a master services agreement (MSA) may also be required. An MSA describes the legal framework within which the two parties – the buyer and the seller – will engage. The MSA is designed to establish ways of working and to protect both parties in case of failure to deliver on the terms of the contract.

In some cases, a supplier will also need to undergo further qualification to ensure they have the capabilities and controls in place to deliver the product or service on a consistent basis. This additional qualification may take place via an audit of the supplier, which is an official inspection of the organization. Audits are especially common in the biopharmaceutical industry because of regulatory requirements.

The final step of the supplier lifecycle is supplier management – monitoring supplier performance over time. If the supplier provides routine goods or services, this may be as simple as tracking budgeted spend. Most suppliers fall into this routine transactional category and don't require much energy or investment. However, there is another category of supplier – those that provide a particularly critical or expensive good to the buyer. For those suppliers, a more robust supplier management strategy is appropriate.

Supplier/strategic relationship management (SRM) is a holistic approach to managing critical suppliers that reframes the buyer–seller relationship from a transaction basis into a partnership basis. Managing these relationships is not just about tracking spend or performance, but also about learning the supplier's values, goals, and ways of working. When both parties understand each other, they can find areas where their objectives may coincide. Using this partnership framework can then increase the value of the relationship for both the buyer and the seller. In the biopharmaceutical industry, SRM fits as a philosophy for outsourced manufacturers who make the drug on the company's behalf. Key tenets of SRM that may be applied to this type of relationship include

- Establishment of MSA and supply agreements.
- Development and implementation of acceptable service-level metrics.
- Provision of production forecasts and negotiation of critical path activity lead times.
- Quarterly business reviews to discuss goals, expectations, and future projects.
- Use of supplier score carding to measure supplier performance.

Alongside individual supplier management, supply chain risk and resiliency must also be considered within the sourcing strategy and selection framework. There are two strategies that firms can use to reduce risk and improve resiliency within their supply chain. The first strategy is to pursue relationships with multiple suppliers who provide the same good. If two such suppliers exist, this is known as dual-sourcing, and if there are more than two suppliers for a specific good, this is known as multi-sourcing. When a firm is dual- or multi-sourced for a particularly critical good, it has an insurance policy against disaster striking one of its suppliers. Dual-sourcing makes the most sense when the suppliers are geographically spread out and belong to separate corporate entities to further reduce risk. There is a downside to pursuing multiple suppliers, though. It is expensive and resource intensive to maintain multiple relationships, particularly if the need to have more than one supplier isn't justified by the demand profile of the final product.

Another strategy firms are pursuing now more than they have in the past is onshoring. Onshoring means to bring an activity that had been outsourced to an international partner back to a domestic partner. Beginning in the 1970s and 1980s, many American and European firms pursued offshoring, particularly to China and Southeast Asia, as a strategy to save money from cheaper labor sources. Offshoring can expose a company to multiple supply chain risks, including geopolitical risks and transportation risks. During the COVID-19 pandemic, these risks were realized for innumerable organizations as COVID restrictions shut down China's ports for months at a time.

Like all supply chain processes, sourcing requires a team of people, but there is one function that owns the sourcing process – procurement. Procurement's responsibilities will vary from company to company, but they always drive the supplier selection process and almost always develop the overall sourcing strategy. Procurement owns contract negotiations, supplier agreements, and budgetary issues and usually manage supplier performance as well. With that said, procurement is not typically responsible for the technical requirements and deliverables needed from the supplier. To return to the outsourced manufacturing example, manufacturing experts need to partner with procurement to provide those technical details.

Once an engagement with a supplier has begun, delivery is the highest priority. Within the supplier, there are three roles worth mentioning. First is the team of people who are directly responsible for delivering the good. Wherever possible, this team should be identified before the work begins to properly evaluate their credentials and working style. Second, and arguably as important as the people delivering the work or good, is the project manager. The project manager can make or break an engagement with a supplier, particularly for highly complex projects such as pharmaceutical development and manufacturing. A strong project manager will set expectations, communicate in a timely fashion, and identify risks before they materialize. A weak project manager will lead to chaos, missed deadlines, and frustration. Finally, the sales/business development lead has an important role to play. They are incentivized to please their customers and can be an advocate within the supplier organization for the buying party.

1.4 Make

Simply put, manufacturing is the production of goods. And while a detailed analysis of specific manufacturing operations is beyond the scope of this book, a general summary of manufacturing resource planning (MRP II) yields lessons that are applicable to the discipline of supply chain management. MRP II encompasses the processes and inputs a plant must consider when planning production, including demand, materials, capacity, labor, and cost. This approach is known as MRP II because it improves upon materials requirements planning (MRP I). MRP I was designed

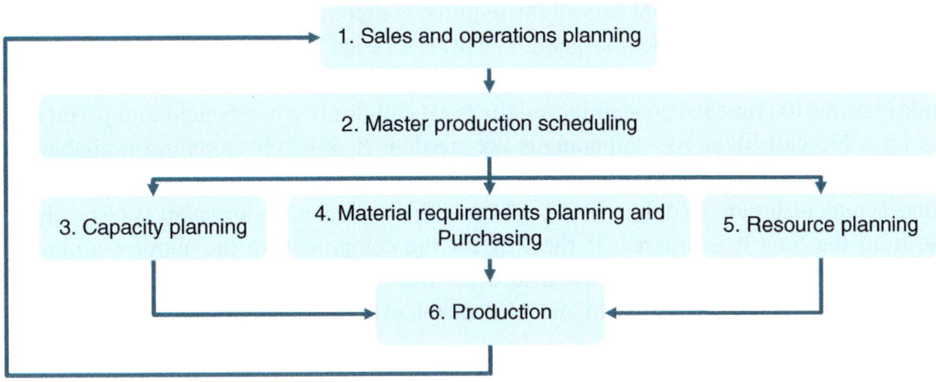

Figure 1.3 MRP II Process Flow

to ensure that sufficient materials were on hand to support scheduled manufacturing while minimizing inventory holding costs. MRP I had serious flaws in that it did not account for capacity or resource requirements. Thus, MRP II was developed to incorporate these inputs and improve plant operations.

A streamlined MRP II process flow is captured in Figure 1.3.

In Figure 1.3, there are six planning, scheduling, and manufacturing processes:

1) *Sales and Operations Planning (S&OP)*: S&OP is a business process where the sales function provides the demand forecast to the planning and operations function. Planning then reviews the forecast and develops the production plan to support that forecast. This process creates alignment across the entire company on what operations can deliver. This alignment is important because it prevents sales from overpromising and gives operations a concrete supply forecast to hit. S&OP is fundamental to successfully managing manufacturing because without it, demand variability can overwhelm operations' ability to respond and provide supply to meet that demand. S&OP principles are applicable to clinical trial supply chain management, and this will be revisited in Chapter 15 – Clinical Supply Chain Planning Processes.
2) *Master Production Scheduling (MPS)*: The master production schedule tells the plant what to produce, how much to produce, and when to produce it. The MPS ties the top-line forecast agreed to during the S&OP process to the detailed operations of the plant. Typically, the MPS is broken down into specific stockkeeping units (SKUs). The MPS takes into consideration plant capacity, component availability, and plant resources to ensure the production schedule is feasible. Due to these requirements, MPS generation is typically managed by a software or cloud-based solution known as enterprise resource planning (ERP), which is the technological backbone of many mid- and large-sized manufacturing companies.
3) *Capacity Planning*: Capacity planning assesses how many units of a good can be produced with the plant's available resources. Two variables used in capacity planning are the number of machines and the maximum utilization of those machines. The specific manufacturing process in question will dictate the analysis, but the key point is that capacity planning places a constraint, or limitation, on the amount of goods that can be produced within a particular machine or equipment train. These constraints must be considered in the development of the MPS because if they are exceeded, the plan is no longer feasible.
4) *Materials Requirements Planning (MRP) and purchasing*: MRP translates the MPS into specific component requirements that support the production schedule. MRP relies on an accurate bill

of materials (BOM) to run. A BOM lists all the required components and the quantities of each component needed to produce the final good. The BOM for each SKU must be accurate because MRP drives the component planning and purchasing process. For example, if the production schedule requires 100 finished good units and the BOM calls for two component unit parts (e.g. wheels on a bicycle), then 200 components are needed. If only 150 units are available in inventory, then a replenishment order for that SKU must be generated. The other variable MRP considers is replenishment lead time (RLT) or how long it will take a component to be received for use from the time it is ordered. If the RLT for the component in the above example is 10 weeks, then MRP will create a request to order that component 10 weeks ahead of the requirement date. Combining the BOM and RLTs enables the creation of a purchasing strategy for requisite components.

5) *Resource Planning*: Like capacity planning, resource planning considers human labor as a potential constraint to the manufacturing process. And, like MRP, resource planning creates a labor requirement for the production schedule. This enables the plant manager to appropriately allocate human resources to the schedule. If those resources are unavailable or not sufficient to meet the schedule, then the plan must be updated to account for this shortage. Planning resources is more difficult than components because human labor cannot be acquired as easily as material goods. Hiring, training, and off-boarding must all be considered in the process.
6) *Production*: Once the equipment, components, and labor are ready, manufacturing can begin. When production ends, the outputs are fed back into the demand and supply plans, and the planning cycle begins anew.

Many different manufacturing strategies exist, but the biopharmaceutical industry largely uses batch production in which a group of identical units of a particular SKU are produced. This is a different approach than job-based production, which produces an individual unit of an SKU, or flow production, which produces identical units on a continuous basis of an SKU. Batch production gives biopharma manufacturers the flexibility to scale production up or down as needed and to meet the strict regulatory requirements of the industry.

Another manufacturing strategy that biopharma companies traditionally pursue is make-to-stock (MTS) manufacturing, as opposed to make-to-order (MTO). In an MTS strategy, stockpiles of inventory are built ahead of demand to ensure sufficient inventory is available. MTS is most appropriate when the supply chain has a long lead time and/or customers require a good as soon as they have demand for it (e.g. buying fresh milk off the shelf). In an MTO strategy, units are produced only after the demand signal is received. MTO works well for supply chains that have fast lead times or for consumers who are willing to wait for a unique product (e.g. a custom laptop). Biopharmaceuticals have long lead times and consumers who need a product immediately and, therefore, MTS has traditionally been favored by the industry. As we'll see in Chapter 18, new drug modalities and approaches to supply chain management are challenging this paradigm.

There are two risks within the production floor that must be balanced. The first risk is a failure to meet the MPS. Any hiccup in operations, from an equipment issue to a component stockout to a labor shortage, may cause a delay in production. The possibility of this risk happening increases as the plant is pushed close to its capacity and room for error is reduced or removed entirely. One approach to mitigate against this failure is to add redundant capacity or additional inventory to ensure production will run smoothly even if a problem occurs. But, this strategy introduces a second risk – wasted spend on unutilized equipment, inventory, or labor. Balancing these risks while maximizing efficiency and minimizing cost is the crux of an effective manufacturing strategy. Three innovations – the theory of constraints, just-in-time, and lean manufacturing – revolutionized production management and helped manufacturing organizations become more efficient, mitigate risk, and reduce costs.

Several functional areas are responsible for production within an organization. For simplicity's sake, we'll refer to this team as manufacturing operations. It's clear from the process outlined above that manufacturing operations partners closely with sales, procurement, and finance to drive the strategy behind the production plan, scheduling, and the manufacturing process itself. Modern manufacturing plants also need technology and software applications to run optimally. ERP is the most important system, but many others may be involved, including manufacturing execution systems (MES) that control the equipment on the floor, procurement systems that handle ordering of goods, and, for those companies without a complete ERP solution, financial systems for managing their accounting processes.

1.5 Deliver/Return

When most people have a vision of supply chain management, they picture the delivery (and return!) of goods. Think of giant cargo ships, airplanes, and trucks laden with goods departing from one port to the next. This is logistics – the discipline of shipping a product from an origin location to a destination. While logistics is the purest form of supply chain execution, it also requires careful planning and coordination between numerous actors to ensure the product is picked up at the desired time and delivered to the destination in a smooth manner.

Logistics shipments can be segmented into domestic and international shipments. Completing a domestic shipment is a relatively straightforward exercise involving three parties: a seller (typically the origin location), a buyer (typically the destination location), and a courier. The seller is the side that agrees to make a sale to the buyer, while the buyer is the side that agrees to acquire ownership of the goods. While the buyer is usually the consignee, which is the party who physically receives the goods, this may not always be the case. This is particularly common for biopharmaceutical companies who act as the buyer but arrange for the shipment of product between one contract organization (e.g. the seller) and another (e.g. the consignee). Meanwhile, the courier physically picks up the shipment and transports it to the destination location. Due to this relative simplicity, domestic shipments can be set up and executed in a manner of days.

International shipments are more complex, time consuming, and costly because when a shipment crosses a border, it must pass through customs and is subject to the duties and taxes of the destination country. To set up an international shipment, the buyer will contract with a TSP who will coordinate the shipment from start to finish. To initiate the shipment, the buyer must prepare a proforma invoice that lists required information about the shipment, including the seller, buyer, consignee (if different than the buyer); a description of the goods; a valuation of the goods; the Harmonized Tariff Schedule (HTS) code for the goods; and the Incoterm of the shipment. The TSP will then forward the invoice to a customs brokerage who will review and "green-light" the shipment for pick up. Only after this green-light approval has been given will the shipment be picked up for delivery. The TSP will arrange for a local courier to collect the shipment and that courier service will then deliver it to an airline or ocean freighter, who will hand it off again to a local courier in the receiving country. All told, 8–10 different parties may be involved in handling an international shipment, and the entire process can take weeks from initiation to delivery. This process is illustrated in Figure 1.4.

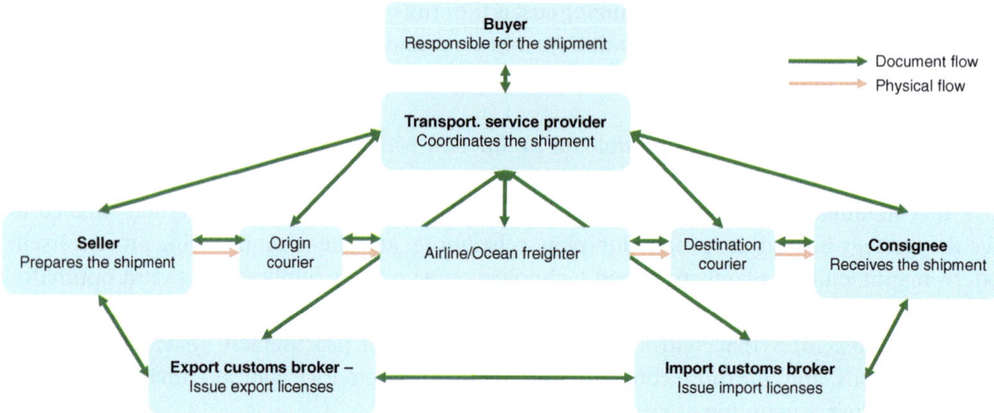

Figure 1.4 International Shipment Framework

Three terms listed above require careful consideration when preparing an international shipment:

- *Valuation*: The monetary value of the goods. This is the basis upon which duties and taxes will be applied.
- *HTS Code*: HTS codes are a classification system of goods that determine which duties will be applied to a shipment.
- *Incoterm*: The Incoterm dictates when the transfer of goods ownership takes place between seller and buyer and defines the respective obligations of each party with respect to cost, insurance, and risk.

Temperature-controlled products, such as biopharmaceutical and agricultural products, add yet another layer of complexity to the shipping process. Storing products at a prespecified temperature range requires the use of a validated shipping container that has demonstrated its ability to maintain the desired range. A temperature monitor (TempTale) must also be included in the shipment to record the temperature for the duration of the shipment. If the product falls out of range and becomes too hot or too cold, it may no longer be suitable for use and must be thrown away.

Within the shipping process, there are numerous risks to manage. Human error ranks high when so many different parties are involved. A courier can forget to include a temperature monitor, or a brokerage house can issue an incorrect license, or a freighter can lose a shipment. It takes a dedicated team of experts, including logisticians and trade compliance experts to reduce these risks to the greatest extent possible. One key to success is the relationship with the TSP responsible for the shipment. Things go wrong in logistics. When the TSP communicates well and works collaboratively to solve problems, problems become much more manageable to work through. Clinical stage biopharmaceutical products offer myriad challenges with respect to international shipping. They are high-value products, require temperature controls, and are often shipped across international borders. This topic is revisited in detail in Chapter 21 – International Logistics for Clinical Biopharmaceutical Products.

1.6 Data, Analytics, and Metrics

Supply chain management lives on data, analytics, and metrics. Decision-making within each of the business processes is data driven, whether it's updating a demand forecast, ordering a raw material, or hiring more labor to run a manufacturing plant. There are a handful of data elements that deserve further attention and explanation because they underpin the demand and supply planning process:

- *Demand*: Demand is a numerical estimate of the projected consumption of a good and is measured in units. Demand is usually calculated by analyzing historical consumption data for a product and then applying a mathematical methodology (e.g. predictive or conjoint analysis) to predict future consumption. Forecasting clinical trial demand does not require these methodologies. Rather, the clinical trial design provides all the information needed to develop the demand forecast.
- *Inventory Targets*: Inventory targets (also known as min–max levels) set the safety stock level (minimum) and target resupply level (maximum) of a product. Inventory targets can be measured in time (e.g. 18 weeks of safety stock) or in quantity (e.g. 500 units of safety stock). Time-based inventory targets are superior because they account for demand variability. Inventory targets are used across industry and within clinical trial supply chain management.
- *Order Quantity*: An order quantity is the desired amount of a good that will be delivered as part of that order and it is also measured in units. There are multiple types of order quantities, including finished good production orders, finished good distribution orders, and component purchasing orders. Order quantities are generated by the supply plan, and all these order quantity types are used in demand and supply planning for biopharmaceutical products.
- *BOM Quantity*: The BOM quantity is the number of units of an upstream component required to enable production of a downstream good. BOM quantities are universal to supply chain management and are applied throughout the clinical trial supply chain as well.
- *Production Yields*: A yield is the percentage of component productivity expressed as the number of units produced/number of units entered in the manufacturing process. Another way to state this is the yield reflects the amount of the component that's wasted during the manufacturing process. There are two types of yields – the theoretical yield and the actual yield. For example, if production begins with 900 tablets (component) and these are being packaged into 30-count bottles (finished good), then the theoretical yield is 30 bottles ($30 \times 30 = 900$) or 100% (900 tablets out/900 tablets in). If production finishes with only 28 bottles, the actual yield is 93.3% (840 tablets out/900 tablets in).
- *Lead Times*: The lead time is the time between the initiation and completion of a supply chain process. Replenishment (sourcing) lead times measure the amount of time it takes a component to be received for use from the time it is ordered. There are also production (make) lead times and distribution (deliver) lead times. Lead times are measured in days, weeks, months, or years and are ubiquitous in supply chain management.

Notice that all the data elements listed above are measured in units (e.g. how much) and time (e.g. when). These answer two of the fundamental questions identified in the demand and supply planning process.

While data forms the backbone of supply chain execution, it's also used to analyze supply chain performance. Metrics, also known as key performance indicators (KPIs), use data to measure how a process is functioning. Supply chain metrics specifically measure how the supply chain is performing. There are hundreds of potential supply chain metrics. Many of these are designed for traditional commercial supply chains and, therefore, have limited utility for clinical trial supply chains. However, nine metrics stand out as particularly applicable, and they can be grouped into four categories – forecast accuracy, order fulfillment, inventory management, and supplier performance. These metrics can also be categorized as diagnostic, meaning they provide insight into how the supply chain is currently performing, or as prescriptive, meaning they provide guidance on action to remedy or prevent an issue. Each metric is described in greater detail below.

1) *Forecast Accuracy*: Forecast accuracy is an important metric in consumer supply chain management because it indicates how reliable the demand forecast is. When the forecast is accurate, this drives the appropriate amount of inventory to satisfy demand. Conversely, when the forecast is inaccurate, this leads to missed sales due to insufficient inventory or excessive inventory build-up that ties up capital from supporting other activities.

   Order fulfillment metrics measure how well a supply chain delivers its purchase, production, and distribution orders:

2) *On Time, In Full (OTIF)*: OTIF measures a percentage of orders that are delivered on time (when) and in full (how much). OTIF is the single most important diagnostic metric because it measures the most basic function of the supply chain – delivery of goods. OTIF is calculated by

$$\text{On Time In Full} = \frac{\text{Number of Orders On Time In Full}}{\text{Total Number of Orders}}$$

   OTIF can be applied to every level of the supply chain from purchase orders to production orders to distribution orders. The "perfect order" metric builds on OTIF and incorporates damaged goods as a variable alongside timing and quantity.

3) *Order Lead Time (OLT)*: OLT measures how long it takes to complete a production or distribution order from the moment the order is received to the moment it's delivered. OLT is both a diagnostic and a prescriptive metric because it measures current lead time performance and gives insight to potential lead time reduction efforts. Lead times are a core input to the planning process, and they are never static. Like OTIF, OLT can be applied to the entire supply chain. OLT is calculated by

$$\text{Order Lead Time} = \text{Average Lead Time of a Production or Distribution Activity}$$

   OLT is sometimes one of the more challenging metrics to maintain because there can be a high degree of variability from order to order. This usually smooths out over time, but this requires attention to capture.

   Inventory management metrics measure how well inventory is being managed through the supply chain:

4) *Weeks-of-Supply*: As discussed in the planning section, WoS is a prescriptive output of the planning process and tells the most comprehensive story about how much inventory is available at a given point in time. WoS is calculated by

$$\text{Weeks of Supply} = \frac{\text{On Hand Inventory}}{\text{Average Weekly Demand}}$$

   Calculating WoS for commercial goods can be tricky because average weekly demand may vary from week to week. Determining the WoS for clinical trial material is more straightforward because the demand curve of a clinical trial product is bounded by the clinical protocol where the demand curve of a consumer good is only constrained by demand itself. See Chapter 9 – Finished Product Supply Planning: The Production Plan and Chapter 11 – Finished Product Supply Planning: The Distribution Plan for more on WoS in clinical trial supply planning.

5) *Stockout Rate*: Stockout rate is a diagnostic measure of the percentage of items or orders not available when needed for consumption. The stockout rate is calculated by

$$\text{Stockout Rate} = \frac{\text{Number of Unavailable Orders}}{\text{Total Number of Orders}}$$

Traditionally, stockout rate has been considered the most important clinical trial supply chain performance metric. This is because a patient missing a dose can compromise their eligibility in the trial and the use of that patient's data in the overall trial analysis. It's a big deal when a stockout happens, so the goal of many clinical trial planners is to avoid them at all costs. This can lead to overproduction and is an excellent segue to the final inventory metric.

6) *Waste*: Waste is the amount of unconsumed good on hand when a supply chain is discontinued. Waste is calculated by

$$\text{Waste} = \frac{\text{Number of Unused Finished Good Units}}{\text{Number of Finished Good Units Produced}}$$

This is another diagnostic metric that can be used to identify improvement opportunities for future supply chains. Waste is inevitable in every supply chain because it's impossible to produce exactly what is needed, where it's needed, and when it's needed when demand is variable. With that said, waste should obviously be minimized to the greatest extent possible, and this can be accomplished through strong supply planning and inventory management. Waste reduction is an important strategic goal in biopharmaceutical supply chain management because of the high cost of clinical trial goods.

7) *Working Capital in Inventory*: This is the total amount of cash tied up in inventory, and it's calculated by summing the total cost of manufacturing with the inventory holding cost. The inventory holding cost includes the total amount of storage fees associated with holding inventory. Measuring the amount of working capital in inventory is important because these are dollars that cannot be used in other areas of the business such as product innovation, sales, or marketing.

Supplier metrics measure supplier performance:

8) *Defect Rates*: A defect rate measures how often an order is defective and can serve as a diagnosis of supplier quality issues. Defect rate is calculated by:

$$\text{Defect Rate} = \frac{\text{Number of Defective Orders}}{\text{Total Number of Orders}}$$

When a buyer purchases a good, the seller agrees to provide it under certain predefined criteria. These criteria are known as product specification, and if the product does not meet this specification, then it is considered defective. For example, if a farm delivers an order of milk but the milk arrives spoiled, this would be a defective order. Defect rates are critical for biopharmaceutical companies, especially in clinical development where each batch of product requires a significant investment and is needed for downstream activities.

9) *Customer Service Level*: Customer service is the lone subjective metric on this list, but it's worthy as a prescriptive metric for improving supplier performance. Extracting the most value out of a supplier means measuring not just the quantitative outputs but also the qualitative experience. This is because poor service leads to a greater resource burden for the buyer to manage that supplier and because poor service can signal a potential decline in future service levels. Project management, overall responsiveness, and flexibility to solve problems are all categories that can feed into a broader customer service-level rating.

Chapter 14 returns to this topic and discusses which of these metrics matter the most in clinical supply chain management and why. The information covered between then and now provides relevant context for that topic.

1.7 The People–Process–Technology Framework

A supply chain is a network of suppliers, manufacturers, and distributors who come together to deliver goods to a consumer base. Effectively managing a supply chain requires a group of people and organizations executing business processes using data inputs and outputs that are housed within technology solutions. A simpler version of this statement is that supply chain management operates within the intersection of People, Process, and Technology. All three are necessary components of supply chain management, but their role and relative importance also matter.

People come first because people run business processes; people design the technological solutions used to enable these processes; and people must come together across organizations, cultures, and continents to get the work done. When designing processes or technological solutions, the people who interact with them must be kept front and center. That means hiring well, training well, and seeking feedback constantly from the people in every supply chain function about what's working and what's not.

Process comes next because business processes provide the structure within which people do their jobs. If that structure is well built, then sound decision-making and positive results are the outcome. Well-designed processes ensure that everyone has access to the information they need when they need it, and the outputs of those processes enable the supply chain to operate smoothly and efficiently. On the other hand, if the process structure is weak, then errors, delays, and dysfunction will reign. Poorly designed processes limit information availability, have poorly defined outputs, or have redundant non-value-added steps. All these characteristics increase confusion and inefficiency leading to poor outcomes.

Technology's role is to facilitate the execution of business processes by providing data and information to people when and how they need it. Notice that technology's role is not to replace a person or perform the process on their behalf. While artificial intelligence and big data are exciting concepts, they are not yet able to replace the power of human relationships and human decision-making. Technological solutions never exist in a vacuum. They are always used by people in support of a specific goal or outcome. If the technology has that goal or outcome in mind when it's designed and implemented, the result will be positive. Implementing a system that's not rooted in business process enablement, on the other hand, only creates an impediment to process execution. A wise person said it well: "Technology only makes a good process better and a bad process worse."

People–Process–Technology, in that order. For supply chain planning, that means centering people first. The top priority of supply chain leadership is hiring, training, managing, and leading people. Next is designing and implementing well-thought-out business processes that enable people to act efficiently and thoroughly across the supply chain. Only then should technology solutions be layered in to facilitate access to information and data. Respecting this prioritization results in optimal supply chain performance.

1.8 Key Takeaways

1) A supply chain is an interconnected network of suppliers, manufacturers, and distributors that collaborate to manufacture a product and make it available for consumption. There are four supply chain processes: plan, source, make, and deliver/return. Planning accounts for activities that take place within each of the other three processes.

2) Demand and supply planning is the principal planning process in supply chain management. It results in a supply plan that attempts to satisfy the demand forecast. Demand and supply plans should be end-to-end, demand driven, and WoS based to yield the best results.
3) Sourcing is the act of procuring goods and services provided by suppliers. Every supplier relationship follows a three-step process – identification, selection and qualification, and management. There are many sourcing strategies to consider, including make versus buy, single versus dual, and onshoring versus offshoring.
4) MRP II provides important concepts that can be applied to demand and supply planning. These include BOMs, yields, lead times, and constraints.
5) Logistics involves transporting material between a seller and a buyer. International logistics has high stakeholder complexity with up to 8–10 different parties involved in shipping a good across borders.
6) Data, analytics, and metrics underpin all aspects of supply chain management. Demand, inventory targets, order quantities, BOM quantities, yields, and lead times are data inputs of particular focus for demand and supply planning.
7) Supply chain metrics measure performance and can be thought of as either diagnostic or prescriptive. There are seven metrics worth mentioning, and they include OTIF, order lead time, WoS, stockout rate, waste, defect rate, and customer service level.
8) People–Process–Technology is another paradigm of supply chain management. People are the top priority, then process, and technology should be thought of as an enabler to the first two.

Notes

1 http://www.apics.org/docs/default-source/scor-training/scor-v12-0-framework-introduction.pdf?sfvrsn=2.

2 https://languages.oup.com/google-dictionary-en/.

2

Introduction to Clinical Trials

2.1 Introduction to Clinical Trials

The goal of every pharmaceutical, biotechnology, and cell therapy company is to produce medicines that are both safe and efficacious at treating disease. To accomplish this, a series of experiments are required to demonstrate the drug in question meets these goals. A clinical trial is a research study in which one or more human subjects are prospectively assigned to one or more interventions (which may include a placebo or another control) to evaluate the effects of those interventions on health-related biomedical or behavioral outcomes.[1] This chapter introduces clinical trials and their design elements to the reader. Clinical trial design is a highly complex space because trials are made up of a dizzying array of interconnected elements. While the theory behind how trial designs come together is beyond the scope of this book, understanding the separate elements is foundational to becoming an effective clinical planner. This is because the trial design, protocol, and environment in which they take place determine the demand forecast for the study.

We will open this chapter with a conversation introducing the clinical trial elements that impact the supply chain. Then, we'll turn to a deeper review of the specific intrinsic and extrinsic elements of study design that inform the demand forecast and supply requirements needed to support a study. To learn more about how trials are designed, we'll briefly review the preclinical and clinical development processes. Next, the chapter moves to a deeper look at different trial types and how the intrinsic and extrinsic design elements typically play out within them. We conclude the chapter with a review of trends in clinical development and how these are adding more complexity to the clinical trial space.

2.2 A Conversation About Clinical Trials

Let's begin with an introduction to the elements and their vocabulary. The below section follows a conversational question and answer format between a new clinical planner and a more experienced counterpart. Every topic introduced in this conversation is relevant to clinical trial supply chain management, and it's a lot to digest! But, rest assured, these topics are revisited again and contextualized as they're applied in later chapters.

- *Question*: "What is a clinical trial?" – A clinical trial is "a research study in which one or more human subjects are prospectively assigned to one or more interventions (which may include placebo or other control) to evaluate the effects of those interventions on health-related

Supply Chain Planning for Clinical Trials – A Practical Guide, First Edition. Ryan Mills.

biomedical or behavioral outcomes."[1] Clinical trials are either observational, meaning subjects are measured over time without the use of a drug, or interventional, meaning that a drug is administered to the subject during the trial. This book will focus exclusively on interventional clinical trials where investigational treatments must be provided.

- *Question*: "How are 'drugs' and 'diseases' referred to in clinical development?" – A drug is often referred to as a treatment, but the most formal term is investigational medicinal product (IMP). Drugs may also be referred to as therapeutics, investigational agents, or active/placebo treatment. Another term of importance is the treatment indication which consists of the disease area, the patient population, the treatment, and the desired outcome of the study.
- *Question*: "What is a clinical protocol?" – The protocol is a detailed plan of a scientific or medical experiment, treatment, or procedure. The protocol explains the desired outcomes of the trial and how those outcomes will be achieved, and it describes all the treatments that will be administered and the procedures that will be performed. Most of the information needed to generate a demand forecast can be found within the protocol.
- *Question*: "When does a clinical trial officially begin?" – There are three "firsts" a clinical planner needs to follow. The first site activation (FSA) date is the date the first site in the study is ready to begin screening a subject for entry into the trial. The FSA is usually when IMP is expected to be available to ship to a clinical site. The first subject screened (FSS) date is when subjects begin evaluation for entry into the trial. And, the first patient in (FPI) date is when the first participant is randomized into the trial and/or when the first patient dose is administered to a subject.
- *Question*: "How are subjects chosen to participate in a trial?" – All subjects must go through the screening process which is based on a set of inclusion criteria – a set of qualifying factors that a potential subject must meet to be included. There are also exclusion criteria and, if a subject meets one of these, they are disqualified from participating in the trial. When this happens, it's referred to as a screen failure. Some inclusion/exclusion criteria are inherent characteristics, such as age. Others require assessments to be performed prior to the subject's entry in the study. These assessments may take days or weeks to analyze. The period over which these analyses take place is referred to as the screening window. Just prior to study entry, subjects may go through one final round of baseline assessments before treatment begins.
- *Question*: "That seems like a lot of work to qualify a subject?" – Indeed. Subject recruitment and enrollment are some of the most unpredictable aspects of clinical trial operations and execution. Subject recruitment is the process of attracting potential participants to enroll in a specific trial, and enrollment is the process through which participants formally enter the trial. Recruitment and enrollment are forecasted and measured through the enrollment rate, expressed as the number of subjects per site per month (patients/sites/month or p/s/m) or subjects per month (p/m) will enter the study. The enrollment rate is another critical input for the clinical supply planner.
- *Question*: "What are the predefined measures mentioned earlier?" – These are known as endpoints. There are three types of endpoints – primary, secondary, and exploratory. The primary endpoint(s) is/are the basis for concluding that the study met its objectives.[2] The secondary endpoint(s) is/are an additional endpoint that may extend the understanding of an effect related to the primary endpoint or provide evidence of a distinct clinical benefit and should be included in the prospective statistical analysis plan if they provide evidence of additional effects of the drug. An exploratory endpoint seeks to collect additional data on a clinical metric.
- *Question*: "So how do we measure safety?" – Through the frequency and severity of adverse events that are observed in participants of the trial. Adverse events are defined as "An unexpected medical problem that happens during treatment with a drug or other therapy"[3] and may be mild, moderate, or severe, as examples.

- *Question*: "And, how do we measure efficacy?" – Efficacy is usually measured by comparison to a control arm. An arm refers to a group of subjects who are receiving the same treatment under the same regimen. Another commonly used term for arm is cohort, although these two terms have different meanings. A cohort is a group of subjects who share a particular characteristic such as age, gender, and pretreatment status entering the trial.[4] The control arm can sometimes be a placebo to the active product or sometimes an available active treatment. When analyzing the study objectives, subjects who were given the test drug are compared to those who received the control to conclude if the endpoint(s) have been met and the result(s) are statistically significant.
- *Question*: "Is it ever unethical to give subjects placebo?" – In some cases, yes. If the disease is severe and if there is already an approved drug for that disease, known as a "standard of care" available, then the clinical trial may be designed to compare the test IMP against that standard of care or in addition to the standard of care. In this case, the standard of care is known as an "active comparator" treatment.
- *Question*: "If a placebo or comparator product is used, will the subject know which drug they are taking?" – In almost every case, the answer is no. To minimize bias in the statistical analysis, clinical trials are designed to "blind" the subjects and the investigators from knowing which arm of the study, active or placebo or comparator, they've been assigned. Every blinded clinical trial has a detailed strategy describing who is blinded and how the blind will be maintained through the duration of the study. Enabling the blind is a key component of the supply chain strategy for the study.
- *Question*: "When a blinded trial completes, can subjects choose to continue on with treatment?" – If the clinical trial offers what is called an "extension," subjects (if eligible) may continue active treatment for as long as the extension period is open. Extensions serve two purposes. First, they are useful for gathering longer-term safety data. Second, extensions are a mechanism to provide access to active treatment to all participants after the blinded/controlled period of a study has ended. This is attractive to potential participants and provides the sponsor with additional data to analyze whether or not the study met its intended objectives.
- *Question*: "Are there aspects of endpoint selection and the statistical analysis that impact supply chain management for the trial?" – Endpoint selection and biostatistics are beyond the scope of this text, but the number of subjects – referred to as the sample size or "n" – required to sufficiently conduct the statistical analysis is another variable the clinical supply planner needs to know.
- *Question*: "How is the data collected to make safety and efficacy measurements?" – Over the course of the clinical trial, subjects will undergo a series of assessments. Assessments are procedures or tests that measure a subject in some way. Examples of assessments include blood draws, urine draws, or metrics of disease progression. Assessments will take place before the subject is dosed, at regular intervals through the duration of the trial, and during the final visit of the trial.
- *Question*: "If a study has more than one arm, how are subjects assigned?" – The process of cohort assignment is referred to as randomization. Before the study begins, the ratio of subjects assigned to each arm is predetermined. A common randomization scheme for a two-arm study is 1:1, meaning that 50% of subjects are assigned to the active cohort and 50% to the control cohort. Another example is a 2:2:1 scheme for a 3-arm study. This would translate to 40% of subjects assigned to the first arm, 40% to the second arm, and the remaining 20% to the third arm.
- *Question*: "How often are subjects dosed?" – The protocol specifies how often a subject is dosed. Common dosing schemas include once a day (QD), twice a day (BID), three times a day (TID), weekly (QW), biweekly (Q2W or QOW), every three or four weeks (Q3W or Q4W), bimonthly

(Q8W), quarterly (Q12W or Q13W), annually, or one time only. Notice that each of the abbreviations start with "Q" which is Latin for "quaque" meaning every. Thus, QD = every day, QW = every week, and so on. Administration frequency is derived from the safety, pharmacokinetic (PK), and pharmacodynamic (PD) profiles of the drug. Pharmacokinetics is defined as the movement of the drug through the body.[5] Pharmacodynamics is defined as the body's biological response to the drug.[6]

- *Question*: "How is the amount of drug administered and administration frequency determined?" – Determining dose and regimen is a complex analysis to minimize safety/side effects while maximizing the potential benefit of the therapeutic.
- *Question*: "Can the amount of dose administered or the dosing frequency change during the trial?" – Yes, particularly for Phase I clinical trials where the purpose of understanding the safety, PK, and PD profiles is a priority. The dose can also change within cohorts through two mechanisms. First, dose titration is a planned increase in dose at scheduled intervals. Second, down-dosing, which may be planned or unplanned, is a decrease in dose.
- *Question*: "Do all trials end after a specified period of treatment?" – In many cases, yes, but not always. Clinical endpoints can measure different outcomes such as overall survival, progression-free survival, change from baseline assessment, or delay of onset. Due to this variety, some trials require subjects to stay on treatment for a fixed duration, while others will keep all subjects on treatment until a total number of "events" have been reached. Events can be anything from tumor expansion in oncology studies to delay of symptom onset in immunology or neurology trials.
- *Question*: "What is a visit?" – Visits are predetermined timepoints where a subject has assessments performed on them, receives IMP for administration, or both. The calendar of visits is known as the visit schedule.
- *Question*: "What happens if a subject can't make the exact visit date?" – To account for this, the visit schedule will allow subjects to make the visit within an allotted range of days, known as a visit window. For example, if a visit is scheduled for a Wednesday, the protocol may allow the visit to happen anytime from Monday to Friday. This would be a $\pm$ 2d visit window.
- *Question*: "Where do clinical trials physically take place?" – Clinical trials traditionally take place at clinical trial sites. Sites can range from traditional medical centers such as hospitals or clinics to academic centers specializing in research to nonprofits. If a location houses the requisite resources and expertise to support the study, it can be qualified. More recently, clinical trials have become decentralized, meaning that some or all aspects of the clinical trial, including administration of the drug to the subject, are brought directly to the subject's home or to a nearby facility.
- *Question*: "So where do clinical trials geographically take place?" – All over the world! Some studies can be managed out of a single site and country. Others require setting up dozens or even hundreds of clinical sites on every continent but Antarctica. As of 2022, clinicaltrials.gov[7] listed more than 64,000 actively recruiting studies in more than 200 countries. Of those, 33% took place in the United States only while 61% took place outside the United States. Six percent or more than 3800 trials were taking place on a global scale. The country list names all the participating countries in a clinical trial. This country-level breakdown is important because each country's regulatory body must approve the study before it can begin.
- *Question*: "How does clinical trial drug get to the clinical site or to the subject?" – Shipping product from a distribution depot to the clinical site or to the subject's home may be performed through a manual request but is usually controlled through a technology system. This solution is called the integrated response technology (IRT) or randomization and trial supply management

(RTSM) platform. IRT/RTSM controls site activation, subject randomization, and drug supply from depot to site for a clinical trial and is integral to successful supply chain management for the clinical trial.

- *Question*: "Do clinical trials require regulatory approval?" – Yes. Putting untested drugs into human bodies has led to an extensive regulatory framework. To gain approval to conduct a trial, a sponsor company must submit a battery of data and documents to the national Competent Health Authority (CHA) that may include nonclinical data, manufacturing information, and clinical data, including the clinical protocol itself. This submission package is called an investigational new drug (IND) application in the United States and a clinical trial application (CTA) in many countries outside the United States. In addition to the country-level package, clinical trials also need to be reviewed by local human review boards. Their job is to ensure that patient safety is at the forefront of the study and that safeguards are in place to ensure ethical conduct of the trial. Approval from both the national CHA and the local ethics committee (EC)/institutional review board (IRB) is required to initiate the trial.
- *Question*: "Aside from dose frequency and administration, is it possible to change other aspects of the trial?" – Yes, nearly every design element listed above can be changed, depending on the circumstances. Inclusion/exclusion criteria can be modified to protect participant safety or spur enrollment, or cohorts can be added or removed, or the n can be increased, or the dose scheme can be updated. Even, the presentation of the drug itself can change. To make these changes, an amendment is often submitted to the regulatory body responsible for approving a trial. The amendment will describe the proposed changes, justification for the change, and implementation strategy. Some amendments can be processed immediately whereas others can take months to approve.
- *Question*: "Can a subject leave the trial early?" – Yes. Sometimes a subject can decline or withdraw participation at any point for any time. And, other times, this happens when a subject's disease progresses to the point they can no longer continue participating in the study. When a subject drops, this is referred to as attrition. The rate at which subjects drop out over an entire trial is known as the attrition rate.
- *Question*: "When does a clinical trial end?" – The date the last subject enrolls is referred to as last patient in (LPI), and when that patient receives their final dose, it's referred to as last patient out (LPO). Once LPO is reached, all the data generated through the course of the trial is collected and organized in the trial database for analysis. The process of finalizing this data is called database lock (DBL). Once that's complete, the sponsor will review the data and determine if the trial has met the endpoints outlined in the clinical protocol and if the drug should progress to the next stage of development.

This conversation gives a thumbnail sketch of nearly all the factors that determine how much supply is required to support patient needs over the course of the study. Given the large number of variables at play, some of which are beyond the sponsor's control, every clinical trial is unique in both design and conduct. The clinical planner must learn the particulars of their specific trials.

2.3 Intrinsic Elements of a Clinical Trial

Clinical trials are highly regulated and structured events. Every clinical trial proposal must be submitted to a regulatory body and ethical committee in the country in which the trial will be conducted. Within the submission package, the sponsor must submit the clinical protocol which describes the purpose of the study and how it will be conducted. Most of the design elements that impact supply chain management are contained within the protocol, and it will nearly always contain the below sections:

- Protocol summary.
- Disease of interest and study rationale.
- Summary of available nonclinical data.
- Summary of available clinical data.
- Risk–benefit assessment.
- Study objectives and outcome measures.
- Study design including study schema.
- Study population including inclusion/exclusion criteria.
- Treatment summary including treatments administered, blinding strategy (if applicable), dose preparation and administration, and dose modification (if applicable).
- Assessments and procedures performed.
- Statistical considerations.
- Schedule of assessments (SoA).
- Additional examinations (if applicable).

While not all the content listed impacts demand or supply directly, many do. At the minimum, the clinical planner should read and familiarize themselves with the study design, study schema, study treatment(s), and the SoA.

The Study Design: The study design is a comprehensive overview of how the study will be conducted. It describes, at minimum, the phase of the clinical trial, whether it will take place at a single location or multiple locations, the intended endpoints, the number of subjects who will participate, the number of arms or cohorts involved in the study, the drugs that will be administered to the participating subjects, and the duration of treatment. The study design may also include other trial characteristics that will influence the eventual demand forecast for the trial. If the study is blinded, meaning that the subjects and/or the investigator and/or the sponsor do not know which arm or cohort the subjects are assigned to, then a blinding or unblinding plan may be included. Criteria for changing a subject's treatment arm, changing the dose, or stopping subject dosing altogether are described here. If a concomitant medication is to be administered, that would also be found in the study design. To summarize, this should be the first stop in the protocol to understand the treatment administration strategy over the course of the study.

The Study Schema: Where the study design outlines the protocol in words, the study schema is a graphical representation of the trial. Study schemas are particularly useful when trial complexity is high. If the visit schedule is nonlinear, if there are multiple arms, or if there are blind and open label segments of the study, the schema is a helpful tool to understand how the trial will unfold. Two example study schemas are presented in Figures 2.1 and 2.2.

The study schema in Figure 2.1 shows an example Phase I clinical trial with an ascending dose structure and three cohorts testing different treatment administration patterns. Cohort A follows a single ascending dose (SAD) structure. Within each sub-cohort, eight subjects will receive a single dose of the drug. Only after each sub-cohort is completed can the next sub-cohort open and enroll new subjects. Once Cohort A3 has completed, then Cohort B1 will open. Cohort B follows a multiple ascending dose (MAD) structure. Twelve subjects within each of the three sub-cohorts will receive treatment every day (QD) for seven days. Like Cohort A, each sub-cohort must complete treatment before the next can enroll. Finally, Cohort C will open following completion of Cohort B1. Cohort C also follows a MAD structure but will administer drug twice a day (BID) instead of QD. Once all sub-cohorts have enrolled, the study is complete.

Figure 2.1 Example Phase I Study Schema

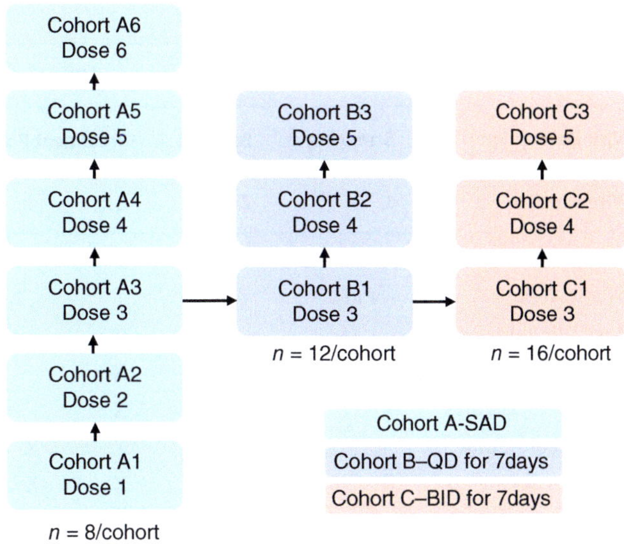

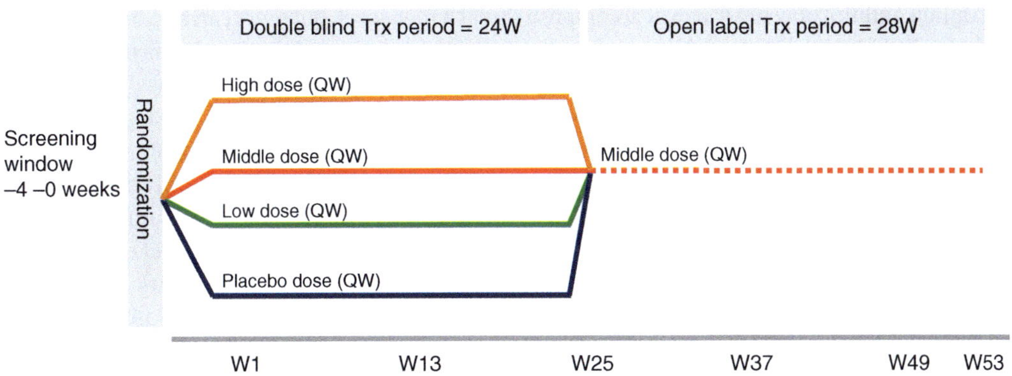

Figure 2.2 Example Phase II Study Schema

Figure 2.2 diagrams an example Phase II clinical trial studying three active doses followed by an open label extension (OLE) period. There is a four-week screening period followed by randomization into one of the four arms. All subjects will dose once a week for 24 weeks during the double-blind period at which point all subjects will roll over into a planned OLE at the middle dose.

Study Treatment(s): This section describes the IMP(s) that will be supplied during the trial, and its content extends beyond a simple description of the treatments. All active, placebo, or comparator treatments that will be administered are included. Details of each treatment are also provided, including how the material will be packaged, how it should be administered to a subject, and how potential dose modifications should be handled. This section is usually authored by or results from information supplied by the chemistry, manufacturing, and control (CMC) team. The supply chain function may also be responsible for elements of this section. See Chapter 3 – Introduction to the Clinical Trial Supply Chain for more information on the CMC team and supply chain's role.

Blinding Strategy: To maintain the blind, the appearance of the active IMP and placebo or comparator IMP should be identical to the subjects, investigators, and other clinical trial staff. This means that the drug itself, the packaging of the drug, the labels on the packaging, and the technological systems used

Table 2.1 Example Schedule of Assessments

	Treatment Period Description (e.g. Blind or Open Label)							
Visit Description	**Screening**	**Baseline**	**Treatment Phase**					**Follow-Up**
Visit Number	**1**	**2**	**3**	**4**	**5**	**6**	**7**	**8**
Study Day	−28 to −2	−1	1	7	14	21	28	35
Study Week	−4 to 0	0	1	1	2	3	4	5
Visit Window		0	0	±2d				
Randomization			X					
General Procedures	X	X	X		X		X	X
Treatment Admin			X	X	X	X	X	
List of Assessments	X	X	X	X	X	X	X	X

to support trial conduct must not reveal the treatment administered. Within the protocol, the blinding description outlines how the trial will accomplish this at a high level, but there can also be a separate, more detailed blinding strategy document for the trial. To preserve study integrity, it is important to document those who are blinded, how the blind will be maintained, and what happens if unblinding is required. The blinding strategy is an important component of the supply chain because it dictates how the final product should be provided to the clinical sites and subjects for treatment.

The SoA: Along with the study schema, the SoA table within the clinical protocol contains vital information for the clinical supply planner. The SoA is exactly as it sounds, a diagram of the visit schedule, associated assessments, and IMP administration dates. The SoA can be laid out with the visits arranged in columns and the assessments arranged in rows with an X to mark the date an assessment will be performed. Treatment administration is a single row within the SoA and identifies all the visits where IMP will be given to a subject. Visit windows are also included in the SoA with each visit date.

In the study laid out in Table 2.1, subjects screen for up to 28 days before Visit 3, their randomization visit. At the randomization visit, subjects will be administered IMP for the first time. Subjects will then return to the clinic on a weekly basis for the next four visits (Visits 4–7). Subjects may come in up to two days before or wait up to two days after their scheduled visit before they are considered out of window, which would potentially jeopardize the use of that subject's data in the analysis.

Between the study design, study schema, study treatments, blinding strategy, and SoA, the clinical planner has access to all the information needed to create a subject-specific demand forecast. We will work through this process in more detail in Chapter 8 – Finished Product Demand Planning: Basic Concepts.

2.4 Extrinsic Elements of a Clinical Trial

Beyond the design elements discussed above, there are a few factors that are sometimes, but not always, stated in the protocol that directly impact the supply chain. We'll refer to these as extrinsic design elements, and there are four to focus on here – the enrollment rate, the number of clinical sites, the site activation schedule, and the country list of those sites.

Enrollment Rate: The largest determinant of site selection is enrollment. Sponsors can control most, if not all, of the intrinsic design elements listed above, but they can't control this most important attribute – finding subjects who fit the inclusion/exclusion criteria of the trial and are interested in participation. It is the great unknown of clinical trial execution and the single biggest driver of variability in the supply chain. If subjects enter the study as predicted, the plan is smooth because demand and supply stay in harmony, costs stay in budget, and the trial is completed on time. However, this almost never happens! Enrollment usually strays and changes happen, and this has downstream impacts on the other elements we'll describe as well as on the plan itself.

Number of Clinical Sites: The number of clinical sites selected is driven largely by the enrollment needs of the study. As the number of subjects required to support the trial increases, the number of clinical sites will often also increase. One exception to this is Phase I healthy volunteer studies with a SAD structure, which may only require a single site that specializes in these types of studies. In most other cases, if the study plans to enroll dozens or hundreds of subjects over time, dozens or hundreds of clinical sites will be required. Other considerations include the prevalence of the disease and the number of competing trials. As the disease becomes rarer and/or the number of trials being conducted in that disease increases, the number of sites required to enroll the necessary subjects also increases.

Site Activation Schedule: Clinical site selection and activation is an enormous undertaking. First, sites must go through a feasibility process in which they demonstrate they have the capabilities required to support the study. This can include having subjects pre-identified, a qualified investigator and trained staff on hand, and relevant equipment and facilities to support study-specific assessments. Once a site passes feasibility, there are several more steps including

- Negotiating and signing contracts.
- Filing all required regulatory submissions including country-level agencies and local ethics review boards.
- Setting several vendor relationships from sample collection to data management.
- Training the site on all the processes and procedures associated with trial conduct.

Given all the required activities, each site will be ready to activate and begin screening and enrollment at different points in time. Understanding the overall number of sites and the site activation schedule is imperative to ensure that IMP is available to support each site when they activate. The site activation schedule also influences the enrollment forecast for the study.

Country List: The country list is a complete list of all the countries with active clinical sites participating in the trial. The country list can be dictated not only by a desire for strong enrollment, but also by broader regulatory and commercial considerations. If a clinical trial is conducted in a specific country, it may strengthen the case for or accelerate the timing of a marketing approval in that country. The country list is of particular importance to the clinical supply planner for several reasons. The country list determines the distribution depot strategy, the label that will be applied to the final product, and the regulatory application strategy for the study. At a high level, the broader the country list, the more complex the clinical supply chain becomes.

2.5 Preclinical Drug Development

Prior to beginning clinical trials, years of research are required to understand the disease, the drug, and how the drug will potentially behave in the human body. At the highest level, preclinical drug development breaks down into the following phases:

- *Target Identification*: The first step of drug development involves identifying a biological target that can be modulated. This modulation is hypothesized to be associated with the behavior or progression of a disease.
- *Molecule Identification*: Once a target is identified, the process begins to find molecular structures that can bind to/modulate the target. The nature of the target (e.g. receptor, gene, and protein) can influence the modality of the drug (e.g. chemical, protein, and antibody).
- *Lead Molecule Selection*: During the lead molecule selection process, anywhere from dozens to thousands of potential structures are analyzed as potential candidates for further development. These suitability characteristics can include molecule performance, stability, reproducibility, and manufacturability among others.
- *Preclinical Studies and IND/CTA Preparation*: Preclinical studies begin once a lead molecule or molecules has/have been selected. The purpose of preclinical development is early understanding of the potential safety risks and PK and PD profiles of the drug. As data is gathered from these studies, the initial application to test the drug in humans is written and prepared for submission to a regulatory body where the Phase I first-in-human clinical trial may be conducted.

After completion of preclinical studies, the IND or CTA is ready for application submission. As part of this regulatory package, the clinical protocol must also be provided for review. Once that review has been completed and approved, the clinical trial sites participating in the study have been trained and onboarded, and the drug is available for dispensation, the first clinical study is ready to begin.

2.6 The Clinical Development Plan

While preclinical studies are taking place, a strategy is put together outlining how a drug will be taken through the clinical development process from Phase I to Phase III. This strategy is known as the clinical development plan (CDP). The CDP describes the clinical trials required to ultimately support registration of the drug. Details include which trials will be conducted, how long these trials will take to conduct, and the primary endpoint of each trial. The CDP is the anchor to which every functional group, including supply chain, attaches themselves to perform their work.

From a supply chain management perspective, monitoring and understanding the CDP is necessary. The CDP enables the clinical planner to not only forecast short-term demand but also prepare end-to-end lifetime product forecasts. It also informs the process development and manufacturing strategies that will influence how the physical supply chain is selected and managed. As the CDP evolves, the supply chain strategy needs to evolve with it, or the supply needs of the overall program won't be met.

In a traditional CDP, at least three distinct clinical trials are planned[8]:

- *Phase I*: Purpose is to test the safety and potential dosage of the drug. Also known as "early-stage" trials.
- *Phase II*: Purpose is to initially test efficacy, find side effects, and find the optimal test dosage of the drug. Also known as "mid-stage" trials.
- *Phase III*: Purpose is to confirm efficacy and monitor long-term safety of the drug. Two Phase III trials are typically required to prove efficacy. Also known as "late-stage" trials.

Figure 2.3 shows a basic CDP for a drug. A Phase I study lasting approximately a year is followed by a Phase II study lasting approximately two years, and then clinical development concludes with

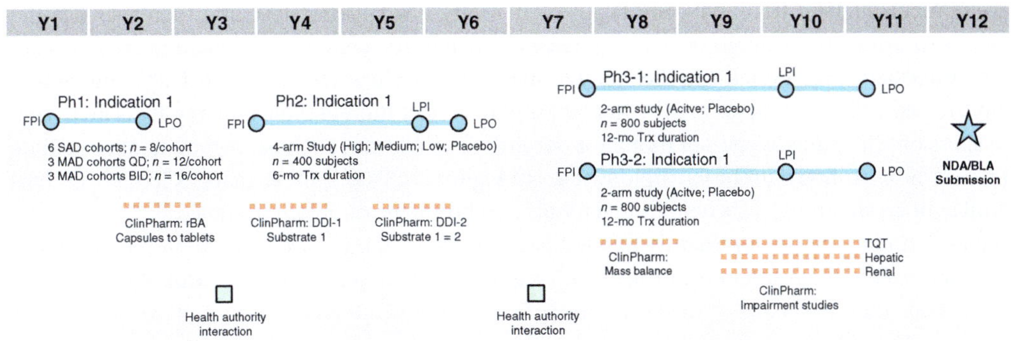

Figure 2.3 Example Clinical Development Plan

twin Phase III studies taking place over a 3+ year period. As the program progresses, the trials increase in both size and duration. Also notice the gaps between studies. This time is spent analyzing data generated from the trial and then using that data to develop and write the submission package for the next study.

Below the main studies are dashed orange bars outlining a variety of clinical pharmacology (ClinPharm) studies. First, a relative bioavailability study will be conducted before the Phase II to introduce a tablet formulation. This is followed by a pair of drug–drug interaction (DDI) studies during Phase II development and then the ClinPharm segment ends with a mass balance study to fully understand the absorption, distribution, metabolism, and excretion (ADME) profile of the drug and further safety studies on potential heart, liver, and kidney impairment.

Finally, there are interactions with health authorities (HA) to review the proposed clinical trial design prior to study initiation. Feedback from HAs will inform the sponsor if the study is acceptable or if changes need to be made before the trial application will be accepted. These preparation periods can take months to years to complete. From start to finish, the CDP for this program lasts more than a decade!

Like an individual clinical trial, the CDP is not static. While the CDP won't change on a week-by-week basis the way a clinical trial can, it will evolve over time as the sponsor seeks to optimize the path to commercialization. Internal factors such as emerging data, resource constraints, and pipeline prioritization can impact how studies are designed and which studies should be prioritized. External factors such as market conditions and the competitive landscape can also force changes to the CDP.

Following the Phase I–II–III path offers the likeliest path to success, but speed and cost considerations have pushed the industry to find ways to streamline the clinical development process. There are infinite CDP possibilities for a clinical program, given the numerous variables at play within a specific clinical trial, let alone across multiple trials. For example, adaptive trial designs attempt to combine all three phases into a single study design. With that said, every clinical trial within the CDP contains common intrinsic and extrinsic design elements. Intrinsic elements are defined by the clinical protocol, while extrinsic elements are defined by the external environment within which the trial is conducted.

2.7 Phase I Clinical Trials

The clinical trial process begins with Phase I trials. From an intrinsic design perspective, Phase I trials have the fewest number of variables and are therefore the most straightforward to forecast. Primary endpoints testing solely for drug safety are most common at this early stage of development. Phase I trials typically begin with a very low starting dose and then increase the amount of drug administered or the frequency of dosing. As a result of frequent dose increases, there are usually amendments to the study design as more is learned about the drug's interaction with the body. Phase I studies are the first attempt to collect human data, and as data is analyzed, the sponsor will want to test additional or varying dose regimens. These regimens may include MAD cohorts, as shown in the example schema in Figure 2.1. Due to these changes, Phase I studies enroll cohorts spanning anywhere from a handful to hundreds of potential subjects.

Phase I studies are predominantly blinded because the endpoints are focused on measurement of adverse events as a result of the drug, which requires comparison to a control arm. When cohorts are opened, they will often start with two to three sentinel subjects who are dosed before the broader cohort. These sentinels offer an extra layer of risk protection to ensure that any severe adverse events are caught before too many subjects receive the drug. Phase I studies also commonly use healthy volunteers as participants rather than patients as subjects. Healthy volunteers are more readily available, and they provide a lower-risk opportunity to evaluate safety concerns than going straight into patients.

From an extrinsic perspective, Phase I studies usually enroll quickly because there is a much wider population of healthy volunteers available to choose from than studies requiring actual patients. However, Phase I trials can also go on for months or even years as different doses, regimens, or formulations are introduced for testing. Over the past few decades, some clinical trial sites have evolved to specialize in Phase I studies. They have recruitment efforts in place to ensure healthy volunteers are available and specialized capabilities to rapidly start up and execute these trails. As a result, most Phase I studies are run at single sites in a single country, further simplifying the supply chain.

Phase Ib trials are a subcategory of Phase I trials. The potential endpoints of a Phase Ib study expand from safety to characterization of the PK and PD profiles of the drug. A Phase Ib trial is also likely to test multiple dose regimens (either dose quantity, frequency of administration, or both) over much longer periods of time than in a Phase I design. These studies also include blinded control arms to obtain a baseline comparison of how the drug behaves in the body. The result of this increased scope is a more complex design. Phase Ib trials also more commonly take place across multiple sites and countries, further increasing supply chain complexity.

2.8 Clinical Pharmacology Trials

Clinical pharmacology studies are a specialized branch of Phase I study that seek to define the PK and PD profiles of a drug. ClinPharm studies, as they're known colloquially in the industry, fall into a handful of subcategories:

- *Relative Bioavailability (rBA) Study*: Designed to compare the PK profiles of different drug formulations. As clinical development progresses, the formulation (e.g. capsule, tablet, or oral solution) may also change. When that happens, the sponsor must work to understand how similar the PK profiles are to ensure the optimal dose is selected when introducing the new formulation into mid- or late-stage studies.
- *DDI Study*: Designed to understand how an IMP will interact with another drug that may increase or dampen the effects on each other.
- *Radio-labeled Mass Balance Study*: These studies attach an isotope-tracer to the drug which allows it to be tracked through the entire ADME process.

- *Thorough QT, Hepatic, and Renal-impaired Studies*: Designed to understand how a drug may impair cardiovascular function and how liver or kidney impairment impacts PK and/or PD of a drug. These are the most common organ-systems affected by drug processing and may be required to be run prior to filing a marketing application.
- *Bioequivalence (BE)*: A BE study is a more rigorous version of an rBA study that is required for certain formulation or manufacturing changes of a drug the drug development or post-approval stages.

Except for rBE studies, ClinPharm studies are usually small ($n < 50$) and run at single-site centers. While less complex from a planning perspective, they still need to be accounted for from a demand perspective, and they can offer unique challenges with respect to execution. DDI studies often require sourcing non-IMP, a topic that is covered extensively in Chapter 17 – Comparator and Ancillary Planning. Radio-labeled studies are made more difficult if the drug substance is not stable or needs to be temperature controlled during transport to the study site.

Bioequivalency studies serve a variety of purposes in pharmaceutical development. They can be used to introduce a new formulation of a drug or to demonstrate that other manufacturing process changes yield a product with similar PK and PD properties. Generic drug approvals also typically rely on bioequivalency studies to show that the generic performs similarly to the branded product it seeks to replace. In certain situations, a single bioequivalency study is sufficient to gain an abbreviated new drug application (ANDA) approval.

2.9 Phase II and III Clinical Trials

If a drug is proven well tolerated and shows PD activity in a Phase I trial, the next step is to begin testing for efficacy in a broader population of patients. In standard clinical development, a Phase II trial is that next step; however, in many indications such as oncology, immunology, and neurology, it is now common to skip a Phase II trial and proceed directly to a Phase III trial. From a supply chain management perspective, there is little distinction between the two phases because studies in both phases feature many of the same design elements that impact demand and supply planning.

First, the clinical protocol will set primary endpoints focused on efficacy and will also include secondary endpoints for long-term safety and supportive efficacy. Many more subjects, ranging from several dozen for rare diseases to hundreds or thousands for common maladies, are included in the design, and the duration of treatment will often stretch for weeks or months. Studies in Phases II or III can last for years. Placebo or comparator control arms are also common in later stage trials, and as the number of arms increases, so too does the planning complexity.

Two other common elements of Phase II and III trials are the potential use of crossover designs and OLEs. In a crossover design, subjects are provided two or more treatments at different times in a prespecified sequence. Assignment to a particular sequence is randomized though. Crossovers are particularly challenging from a demand and supply planning perspective because of the need to always have all treatments available to all subjects through the duration of the trial.

OLEs are used when there is no standard of care available for the disease or if the drug is perceived to have an efficacious effect. Once subjects complete the blinded arm of the study, they are offered an opportunity to continue or transition to active treatment. In Phase II trials, OLE periods may have a limited duration, but in Phase III trials, these can last until commercial launch of the drug. OLEs can greatly increase total demand for treatment during the clinical development process because all the active and placebo subjects from the trial are eligible for treatment.

One area of difference between a Phase II and a Phase III study is the number of doses that will be tested. Phase II studies are often used for dose-ranging requiring multiple active arms, like the study schema in Figure 2.2. Meanwhile, by the time a Phase III study begins, the potential commercial dose is usually locked in and can be tested in isolation against the control. When the number of doses administered increases, this can further increase the number of IMPs required.

Phase II and Phase III trials are also quite similar as it relates to extrinsic factors. The main difference here is that the largest Phase III trials can enroll anywhere from several hundred to thousands of subjects. The Phase III COVID trials run by Pfizer and Moderna enrolled tens of thousands of subjects. To enroll so many subjects, these trials are conducted on a global scale involving hundreds of sites in countries around the globe. It's not uncommon for a Phase II or III trial to have sites located in North America, Europe, the United Kingdom, Asia-Pacific, and Latin/South America. When combined with the increased number of treatments that need to be supplied, this geographic complexity presents a huge challenge for the clinical planner to manage.

2.10 Adaptive Trials

Mid- and late-stage clinical trials usually follow prescriptive designs where the intrinsic design elements of the study are adhered to through its duration. Adaptive trial designs (ATDs) offer an alternative approach where a trial design incorporates preplanned changes. These changes can include refining the sample size (n), adding arms, changing the dose and dose frequency, or stopping arms for futility[9]. ATDs have the potential to make the clinical trial process much more efficient, but they are challenging to design and even harder to predict. This is because many of the preplanned changes revolve around enrolling enough subjects to perform analyses that subsequently un-gate the planned changes.

From a supply chain perspective, this introduces another complicating factor. Increased demand uncertainty means potentially increasing supply to ensure IMP availability to support changes at the earliest possible moment. However, enrollment rarely goes as planned, and if these changes are delayed, then supply will likely be wasted and eventually replaced to support ongoing dosing.

2.11 Master Protocol Designs – Platform, Basket, and Umbrella Trials

Platform, umbrella, and basket trials operate under a different design paradigm than standard clinical trials. Instead of testing a single active treatment against a control arm, these studies use a master protocol to test subcategories of treatments or diseases[10]:

- *Platform Trials*: Test multiple active treatments in an ongoing manner.
- *Umbrella Trials*: Test multiple therapeutic agents against a single disease.
- *Basket Trials*: Test a single therapeutic against multiple disease types.

Platform trial protocols allow for introduction of new treatments and futility analyses to eliminate ineffective treatments. Platform trials are often sponsored and conducted by independent research institutes that work alongside the industry to bring various treatments into the study over time. Due to this structure, platform trials offer a different set of challenges than standard or adaptive trials. In addition to traditional demand and supply planning, the rules of engagement with the sponsoring party need to be negotiated. For example, determining which party will conduct demand and supply planning, who will be responsible for packaging and labeling the drug, and who will manage the IRT/RTSM all need to be decided.

Umbrella and basket trials are most common in the oncology space. Umbrella trials need extra attention from the clinical planner because of the high number of combination treatments that may be required. Adding multiple therapeutics also increases the total number of drugs that need to be managed from a demand–supply perspective, increasing the resource intensiveness of managing these studies. Often, these additional treatments are comparators which must be sourced from a third party, and these can be very expensive. Therefore, the clinical planner needs to pay extra attention to both planning and execution of these studies to ensure supply availability and minimize cost. Meanwhile, basket trials may provide some reprieve in the total number of comparator treatments, but each indication studied should be considered as an additional arm with an unpredictable enrollment rate. Fortunately, supply can be pooled for each arm, which makes the job easier from a supply standpoint. We'll return to this topic in greater detail in Chapter 18 – Make-to-Order and Just-in-Time Supply Chains in Clinical Trials.

2.12 Decentralized Clinical Trials

With the outbreak of SARS-CoV-2 (COVID-19) and its impact on the ability to conduct clinical trials in face-to-face environments, decentralized trials (DCTs) grew significantly in popularity to accommodate the pandemic. This trend is here to stay though because patients are often deterred from taking part in a clinical trial by the need for frequent travel to and from a clinical site. This is especially true for trials that may only have a few sites scattered across a country or even the globe. It's not uncommon for patients to relocate entirely to participate in some trials. DCTs will continue to push sponsors toward innovative solutions that reduce patient burden and improve retention, protocol compliance, and patient engagement.

From an operational perspective, incorporating remote study visits and drug administration is not a one-size-fits-all approach. Every clinical trial and patient population have unique needs as the objectives and safety considerations for the population may differ. Therefore, the operational format of clinical trials must be customized to accommodate the needs of the trial and patient population to ensure success.

From a supply chain perspective, DCTs are known by another term, direct-to-patient (D2P), implying that the drug is shipped directly to the patient for administration. There are three approaches to D2P:

- *Site-to-Patient*: The product ships from a clinical site to the subject's home.
- *Pharmacy-to-Patient*: The product ships from a depot to a licensed pharmacy and then to the subject's home. In the United States, shipments to a subject's home must come from a licensed pharmacy.
- *Sponsor/Depot-to-Patient*: The product ships directly from sponsor control (i.e. from a distribution depot) to the subject's home. This model is more common outside the United States.

Selecting the most appropriate model is driven by geography, regulatory, and legal considerations. In the United States, the pharmacy-to-patient model is popular because federal regulation requires investigational products to be prepared and shipped via a licensed pharmacy. The pharmacy must also be licensed to distribute in the state where the subject lives. If a subject lives in a different state from the site where they're enrolled in the study, then a prescription must be sent to a pharmacy that is licensed in that state and the IMP must be distributed from that pharmacy. Several pharmacies in the United States are licensed in all 50 states, and shipping IMP from a single location can simplify overall complexity. However, this approach does add another node to the supply chain.

In the EU and other parts of the world, the only models that may be available are site-to-patient or depot-to-patient depending on where the subject lives and where the clinical site is located. The regulatory framework continues to evolve for D2P trials as well. Direct-to-patient supply chain models are covered extensively in Chapter 19 – Direct-to-Patient Supply Chain Management.

2.13 Post-marketing Clinical Trials

After the Phase III trial(s) have been completed, the sponsor is required to submit a marketing application for approval of the drug. Once this has been accepted, reviewed, and approved, the drug is ready to launch in that commercial market, but the clinical development process does not necessarily stop here. The sponsor may choose or be required to run a confirmatory post-marketing clinical trial where longer-term safety and efficacy effects are studied. These studies are referred to as Phase IV studies or post-marketing surveillance trials. Phase IV studies vary in their size and scope. In some instances, they are larger and longer than Phase III trials in terms of subjects, countries, and duration of treatment required. With respect to supply, a decision looms whether to use clinical or commercial supply. This strategy will be further explored in Chapter 23 – Integrating the Clinical Supply Chain into a Commercial Launch.

Investigator-sponsored trials (ISTs) are another type of post-marketing study, but instead of the sponsor planning and conducting the trial, an outside investigator takes on that responsibility. Other descriptions include investigator-initiated trial (IIT), physician-sponsored trial, or nonindustry-sponsored trial. Investigators may be research institutes, an individual doctor, or an external party with the resources to conduct a small-scale clinical trial. If a third-party investigator wants to run a trial, they can request a sponsor to provide IMP. These requests represent another demand stream on the drug. While the demand for an individual IST is usually small, treatments for indications with large patient populations can draw significant interest from within the medical community and increase demand for the drug. Establishing a process to collect and aggregate these requests is needed to ensure supply availability.

2.14 Clinical Trial Stakeholder Summary

Drug development is a team sport. Dozens of experts, ranging from scientists to doctors to financiers, are required to execute clinical development. The list of stakeholders below is by no means exhaustive. Rather, these are the people and organizations the supply chain function will most commonly interact with and work alongside. These stakeholders are best categorized into the internal functional areas responsible for specific aspects of trial design or conduct and external groups or organizations that help sponsor companies execute their trials. Pre-commercial or small biopharma companies tend to rely more heavily on third-party organizations for support because they may not have the internal infrastructure or expertise in house yet. That said, almost every biopharma will have an internal function to oversee operationalization and conduct, and this team is called clinical operations.

Clinical operations, also known as ClinOps or CLO, has a large job to do. The ClinOps team works closely with all the other stakeholders to ensure an understanding of the protocol and that party's role in executing it. For example, ClinOps partners with clinical sites from feasibility to training, activation, enrollment, and study closure. Due to their responsibilities, ClinOps has knowledge of and access to all the intrinsic and extrinsic design elements that impact the demand forecast. ClinOps is specifically tasked with developing or providing the enrollment projections for the study, the list of potential sites, and the countries in which the trial should operate. There is no more important functional area for the clinical planner to develop a relationship with from the outset of the trial planning through study completion.

Other functional areas that support a clinical trial include:

- *Clinical Science (ClinSci)*: ClinSci oversees development of the clinical trial design and is responsible for providing medical monitoring and patient evaluation services during the study. ClinSci may provide trial design assumptions alongside ClinOps.
- *Clinical Regulatory (ClinReg)*: ClinReg develops the regulatory strategy and manages the regulatory submission process. ClinReg will provide the submission timelines that inform when the study will start and consequently when supply will need to be available.
- *Biostatistics (BioStats)*: BioStats develops the statistical analysis plan for the trial. BioStats is responsible for the randomization scheme of the study, a useful data point when building the demand forecast.
- *Data Management (DM)*: DM captures and organizes the clinical data generated by the trial. Supply chain will often work with DM on the IRT/RTSM design.
- *Clinical Pharmacology (ClinPharm)*: ClinPharm influences the clinical trial design and analyzes the data generated by the trial. ClinPharm may provide design assumptions for the ClinPharm studies outlined in the previous section.

Collectively, the internal functional areas that operationalize and enable conduct of the trial are known as the study management or study execution team (SMT/SET). Supply chain is usually included as a core team member on SMT/SETs and is expected to provide updates on supply availability and related issues. The SMT/SET is also an excellent source of information for the clinical planner through the duration of the study.

There are multiple external groups involved in making trials happen:

- *Clinical Subjects*: Participants in the trial.
- *Clinical Sites*: Location where trials are conducted.
- *Health Authorities (HA)*: Regulatory body that reviews and approves CTAs.
- *Ethics Committees (ECs) and institutional review boards (IRBs)*: A second group of regulatory bodies that also must review and approve the clinical trial before it can begin.
- *Third-party Suppliers*: Organizations that provide specialized services to enable a clinical trial. Examples include clinical labs, data warehouses, and clinical assessment specialists to support assessment execution.

There are two third-party suppliers who are worth special consideration. First is the clinical research organization (CRO). The CRO is ClinOps' partner in organizing and executing the clinical trial, and they may take on a wide range of activities across the trial. These activities include site feasibility, site start-up, regulatory submission preparation and execution, and site monitoring. In many cases, CROs are also tasked with developing projected enrollment rates for the study and providing updates to the sponsor on how enrollment is faring and what can be done to accelerate or improve enrollment. As such, they are another partner for supply chain to liaise with alongside ClinOps to ensure alignment across the assumptions that will inform the demand forecast.

Second is the IRT/RTSM supplier. The IRT/RTSM is the technological platform that connects data related to the site, subject, and supply for the study. From a supply chain perspective, IRT/RTSM is central to planning and execution. See Chapter 6 – RTSM and Its Role in the Clinical Supply Chain for a complete analysis.

2.15 Trends in Clinical Trial Design and Conduct

Clinical development – spanning drug discovery, preclinical research, interventional clinical trials, market application, and commercial launch – is a lengthy and costly endeavor. The entire process can take upward of 12–15 years and cost well over a billion dollars by the time a marketing application is filed. Over the last decade, sponsors – referred to as such because they take responsibility for the initiation, management, or financing of the clinical trial – have come under increasing pressure to deliver approved medicines faster than ever before. The reasons for this pressure are threefold. The first reason is the most obvious – the faster a company can push a drug to market, the faster they can recoup the enormous investment cost.

The second reason is that advances in biology, genetics, and manufacturing have enabled drug companies to develop and produce increasingly targeted medicines for highly specific patient populations. Biomarker-driven development, establishing a clear link between biological activity and disease response, has become a more popular approach as scientific advances have enabled it. This specificity comes at a hidden cost though. From the 1980s through the 2000s, the biopharma industry relied on generalized medicines targeting broad patient population. When these drugs reached the commercial market, they reaped billions of dollars a year in profits that fueled the sponsor's growth. Drug development has now become more targeted, and the market sizes for these smaller patient populations are less profitable than the blockbuster model big pharma had previously relied on. The result is less profit for each approval, further motivating sponsors to move faster and reduce cost wherever possible.

Third, the competitive environment has grown tremendously. There are now thousands of biopharmaceutical companies, employing more than 5.5M people worldwide. By 2022, more than 9000 compounds were in clinical development[11]! This explosive growth has fueled competition across the entire industry, forcing companies to achieve "first-in-class" status, "best-in-class" status, or ideally both. Differentiation not only in safety and efficacy, but also in convenience, price, availability, and a host of other factors matter to a drug's ultimate success once it reaches the commercial marketplace.

All these trends have converged to make clinical development programs and their associated trials more complex and therefore more difficult to effectively manage from a supply chain perspective. To achieve greater speed, sponsors are trying to do more with less. Instead of only studying the short-term safety of a product in a Phase I trial, that trial may be designed to also capture biomarker data and PD data or generate a longer-term safety profile. Or a company may opt to combine a Phase II and a Phase III trial together, resulting in multiple doses being tested across large patient populations. When a sponsor tries to squeeze more data out of fewer trials, the complexity of the trial increases and it becomes more challenging to execute.[12]

Two additional trends have revealed themselves since the COVID pandemic. First is the concept of patient-centricity and the decentralized clinical trial design. The decentralized model inverts the paradigm from bringing the patient to the site to bringing the site to the patient's home or local treatment center. This approach means supplying the drug directly to a home-setting, even in cases where the drug may require a complex preparation and administration process.

DCTs have added another variable for the clinical planner to untangle in both planning and execution.

The second trend is human resource scarcity driven by the "Great Resignation" from 2020 to 2022. Clinical sites have fewer trained staff than ever to support more clinical trials than ever. Consequently, sites around the world have struggled to enable this massive increase in the overall number of clinical trials. This has impacted site start-up timelines (which are longer), enrollment rates (which are slower), and study execution (resulting from more inexperienced staff). Planning and executing trials have always been difficult tasks. Now it's akin to conducting an orchestra with earplugs while the different instrument sections are playing from different sheets of music. Something or many things are bound to go wrong and deviate from the plan.

If it weren't enough to manage all the intrinsic and extrinsic design elements while dealing with these complicating trends, clinical trials offer a final challenge – their near-constant evolution. As the trial unfolds and new data emerges, or competition increases, or enrollment does not go as planned, or the unforeseen comes to pass, the trial design will be amended. Sometimes, the protocol itself can change, and other times, the geographic location of the study can change, and yet other times, the drug itself can change. And, sometimes, these changes can all cascade and happen together. In this way, clinical trials are akin to a living organism, constantly reacting to its internal and external environment. Keeping up with those changes and successfully adapting to them are the hallmarks of excellent clinical trial supply chain management.

2.16 Key Takeaways

1) Clinical trials have dozens of design elements that inform the demand forecast for how much drug is needed over the life of the study. Understanding clinical trial design and how to translate those elements into a forecast is fundamental to clinical trial supply chain management.
2) The clinical protocol contains the intrinsic design elements that will inform the demand forecast. Key sections for the clinical planner include the study design, study schema, study treatment, blinding strategy, and SoA.
3) There are extrinsic design elements that are not identified in the clinical protocol that will also impact the demand forecast. These include enrollment rate, number of clinical sites, site activation schedule, and country list.
4) There is tremendous variety in clinical trial designs. From early-stage Phase I trials to late-stage Phase III trials, each stage of clinical development offers unique challenges. The clinical planner needs to learn and understand the design elements of each trial to be successful.
5) Specialty trials, such as adaptive designs, master protocol designs, and DCTs, require special consideration from a supply chain management perspective.
6) The CDP outlines the clinical trials required to move a drug from early-stage clinical trial to a potential marketing application. The CDP documents which trials will need to be forecasted.
7) ClinOps is the most important stakeholder that supply chain partners with over the course of the trial. ClinOps or CLO is responsible for providing the study assumptions, both intrinsic and extrinsic, that impact the demand forecast.
8) Clinical trials are becoming more complex due to the increasing cost of drug development, the advent of precision medicine, and competitive pressure. COVID-19 and the rise of decentralized, patient-centric designs have also contributed to increasing trial complexity.

Notes

1 https://grants.nih.gov/policy/clinical-trials/definition.htm [Date Accessed March 14, 2024].
2 www.fda.gov/media/162427/download.
3 https://www.cancer.gov/publications/dictionaries/cancer-terms/def/adverse-event.
4 https://www.cancer.gov/publications/dictionaries/cancer-terms/search/cohort/?searchMode=Begins.
5 https://www.oxfordreference.com/display/10.1093/acref/9780198529170.001.0001/acref-9780198529170-e-15178.
6 https://www.oxfordreference.com/search?q=pharmacodynamics&searchBtn=Search&isQuickSearch=true.
7 https://www.clinicaltrials.gov/ct2/resources/trends#LocationsOfRecruitingStudies.
8 https://www.fda.gov/patients/drug-development-process/step-3-clinical-research.
9 https://bmcmedicine.biomedcentral.com/articles/10.1186/s12916-018-1017-7.
10 https://www.fda.gov/media/174976/download.
11 https://www.ifpma.org/publications/facts-and-figures-2022-the-pharmaceutical-industry-and-global-health/.
12 https://www.nature.com/articles/nrd.2017.65.

3

Introduction to the Clinical Trial Supply Chain

3.1 Introduction to the Clinical Trial Supply Chain

This chapter introduces the clinical trial supply chain, clinical manufacturing steps, and drug modality types. To state the obvious, the biopharmaceutical supply chain is responsible for the supply half of the demand–supply equation. Clinical drug manufacturing and distribution relies on one of the most technically complex and expensive supply chains in the world. It's not unusual for the end-to-end process to span 12–18 months of time and for the drug itself to move across multiple continents before its ready for introduction to the clinic. There are two primary drivers for this.

First, biopharmaceutical manufacturing is highly specialized. It requires dedicated equipment trains, numerous starting materials and components, and highly trained personnel. Building a plant dedicated to drug substance (DS) or drug product manufacturing can easily cost \$100–200M+. Due to this specialization and to the high costs, manufacturing sites tend to perform only one or two steps of the process for a single-drug modality type, such as chemical or biologic manufacturing. With limited options, it's not unusual for a product's supply chain to end up spread around the world.

Second, biopharmaceutical manufacturing is a heavily regulated space. There are strict requirements enforced by national health authorities to document every step of the manufacturing process and analyze the drug for product quality before it can be released for human consumption. Each manufacturing step yields a unique batch of material. Every batch of DS and drug product intended for human use must be analyzed and then released for use. The rigor of this batch release process is unique to biopharmaceuticals. While necessary for patient safety, these regulations add cost and time to the overall process. When the analytical and quality requirements are considered, each step takes weeks or even months to complete from the beginning of manufacture to the final release of the product.

Production of a clinical drug requires the following steps:

- Sourcing of raw materials required in the manufacture of the DS.
- DS or active pharmaceutical ingredient (API) manufacturing where the underlying active compound is made.
- Drug product (DP) manufacturing where the drug is formulated into its final state for human consumption.
- Primary packaging (PP) where the drug is placed into a container closure system such as a bottle, blister pack, or glass vial.

Supply Chain Planning for Clinical Trials – A Practical Guide, First Edition. Ryan Mills.

- Secondary packaging (SP) where the PP is placed into an outer layer of packaging or bundled with other products that enable its consumption.
- Labeling where a clinical label is applied describing study and treatment-related information. This step results in the finished product (FP).
- Distribution through a network of depots prior to delivery to the clinical site that will administer the final form of the drug to the subject.

Within almost every clinical stage biopharmaceutical company, there are a team of people responsible for delivery of drug to the clinic. Chemistry, manufacturing, and controls (CMC) is the most common abbreviation for this team and is taken from the days when small-molecule production dominated drug pipelines. Today, CMC is still a popular moniker, but the CMC team may also be referred to as technical operations, manufacturing operations, or some other variation.

We'll spend this chapter exploring the clinical supply chain (CSC). First, we'll begin with a summary of the most common drug modalities, small molecules, biologics, and cell and gene therapies. Then, we'll turn to the supply chain itself and examine the manufacturing steps, analytical operations (AO), process development, and distribution activities. We'll also delve into CMC strategies and how they provide the background within which the supply chain will take shape and operate. An analysis of CSC trends is provided, and the chapter concludes with a breakdown of the internal and external stakeholders that make up the CSC.

3.2 Drug Modalities

There are three primary types of drug modalities – chemical compounds known as small molecules; compounds synthesized or manufactured from biological sources, known as biologics or large molecules; and treatments where whole cells are infused into a subject, known as cell therapies. Historically, small molecules have dominated as the modality of choice for drug development. Over the last 40 years, since the invention and approval of insulin in 1982, biologics have emerged as a popular alternative because they open a wider range of therapeutic targets, especially in the areas of oncology, immunology, dermatology, and ophthalmology.[1] Cell therapy products, a subtype of biologics, have also rapidly gained momentum over the last 10 years.[2] Scientific advances have enabled drug companies to genetically modify cells to target specific biological pathways within the body. The first cell therapy product, Kymriah (tisagenlecleucel), was approved in 2017, and there are now dozens of cell therapy candidates in various stages of clinical development.

Of the three modalities, small molecules are the most straightforward to produce and characterize. The production process begins with API manufacturing, in which chemical raw materials are synthesized into the API. Next, the API is formulated into a DP. In some cases, multiple small-molecule APIs can be combined into a single DP formulation. Due to their molecular size and ability to be absorbed through the stomach and gastrointestinal-tract, small molecules can take a variety of forms for human consumption. Capsules, immediate release tablets, sustained release tablets, and solutions are the most common formulation types developed for oral use.

Figure 3.1 shows an example supply chain for a small-molecule (chemical) product. The end-to-end process takes almost a year to complete and begins with sourcing three raw materials to support API manufacturing. The product is then shipped to another site for DP manufacturing and completes with a packaging operation and a labeling step. At this point, the product is ready for distribution to clinical sites.

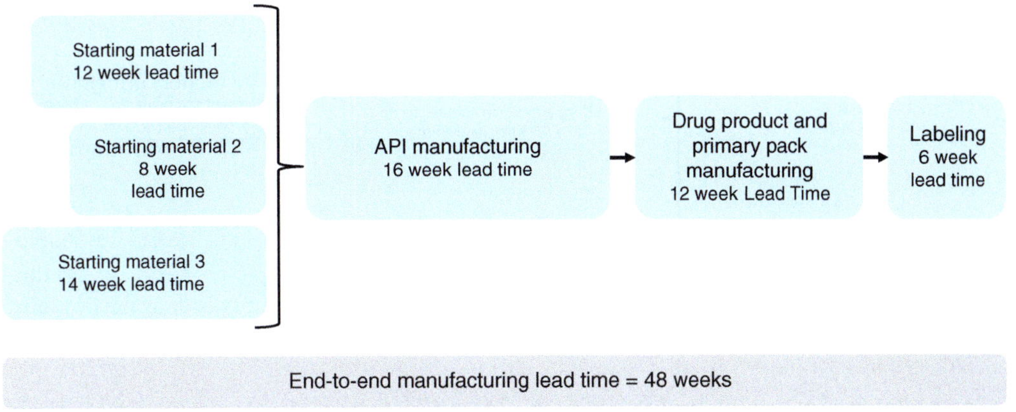

Figure 3.1 Example Small-Molecule Supply Chain

Once DP manufacturing has completed, small molecules are packaged into their primary container closure system. Small molecules are also varied in the types of PP available. Bottles, blister packs, stick packs, ampoules, and syringes are all capable of delivering small molecules to the body. Some packaging types, particularly those made of glass like vials, syringes, and ampoules, require a secondary outer container before giving the product to a clinical subject. SP is required when there is physical risk to the primary container (e.g. ampoules and syringes) or when the dose may not be easily accessible from the PP (e.g. blister packs).

A subcategory of small molecules, oligonucleotides (oligos), has also been gaining momentum in drug development. Oligonucleotides are fragments of nucleic acids, such as DNA or RNA, that have been genetically modified to correct DNA or RNA alterations that result in mistranslated proteins which lead to disease onset and progression. Unlike pure chemical compounds, oligos are larger in size and harder to deliver to the target of interest. Therefore, they are often conjugated, meaning added to, a biological material such as a protein or lipid.[3] As a result of this combination, oligos are typically formulated into a liquid and delivered through injection or infusion. Oligos are an example of how scientific discovery is blurring the lines between traditional small-molecule and large-molecule manufacturing processes.

Where small molecules are typically made through chemical syntheses, large molecules are produced from living cells. This approach enables a plethora of treatments including vaccines, proteins, enzymes, and antibodies. Like small molecules, the manufacturing process for large molecules begins with DS and then moves onto DP. However, unlike small molecules, when biologics are formulated, they are immediately put into their primary container closure system. Packaging the product right away is necessary because biologic DPs are liquid based at the time of manufacture as opposed to most chemicals which exist in a solid state. Large-molecule container closure systems can be vials, ampoules, or prefilled syringes. For every large-molecule product, SP is required to protect the PP, which is typically made of glass to prevent any interaction with the underlying DS.

One subclass of large molecules that have been particularly effective at fighting cancer is antibody drug-conjugates (ADCs). ADCs are made when an antibody is joined by a linker to a cytotoxic drug that is specifically targeted to kill cancer cells. The manufacturing process for ADCs is particularly complex and involves making the antibody, the linker, and the cytotoxin, followed by a step which connects them all together. ADCs are another example of manufacturing processes becoming more specialized, complex, time consuming, and costly.

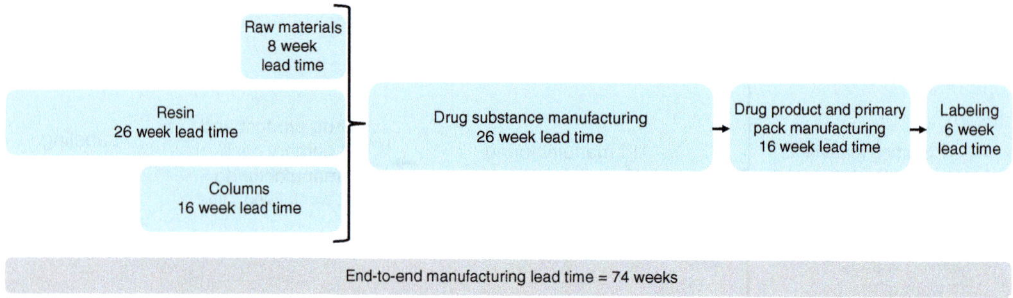

Figure 3.2 Example Large-Molecule Supply Chain

Figure 3.2 shows an example end-to-end supply chain for a large-molecule compound. Note that the raw materials for DS manufacturing typically have longer lead times than their small-molecule counterparts. The DS manufacturing process itself is also typically lengthier for biologics while the DP and FP lead times are similar to small molecules. Note that these are all general statements though. There are small-molecule supply chains with lengthy raw material lead times and biologic products which are made using resins and columns that have shorter lead times. Lead times also vary on a batch-to-batch basis for both raw materials and drug substances.

Like oligos, cell therapies have emerged as a viable modality within the past two decades. The premise of cell therapy treatments is to use genetically modified human cells to repair diseased cells, tissues, and/or organs. There are two subtypes of cell therapies that have advanced into human treatments. Autologous therapies take cells directly from an individual for modification before reintroducing them back into the same person. This is the definition of precision medicine, a true one-for-one treatment.

Autologous supply chains operate under a different paradigm than standard small or large-molecule manufacturing. The latter usually follows a sequential, make-to-stock production strategy where a batch of DS is processed into a batch of DP that can be delivered to many subjects. In autologous supply chains, the individual patient's blood is drawn, shipped to a processing site, cells are genetically modified and made into the final DP, and this is shipped back to the subject for administration. In most cases, these activities must take place in a matter of days or weeks, a stark contrast to the months or years of lead time required for small and large molecules. Due to this incredibly rapid turnaround time, autologous cell therapy supply chains are challenging to execute and relatively expensive on a per-unit basis when compared to standard small and large-molecule manufacturing. Autologous products are the only cell therapy treatments to have achieved a commercial approval. But, a result of their complex manufacturing process and supply chain is that the market price for autologous therapies soars into millions of dollars for a single treatment.

Alternatively, allogenic therapies take cells from a different person than the eventual recipient of the treatment. Allogenic therapies offer hope for cell therapy supply chain simplification and cost reduction. This is because allogenic therapies follow a more traditional large-molecule production strategy where products are considered "off the shelf" because they can treat many subjects instead of the individual. Allogenic products offer potentially superior economies of scale to their autologous counterparts and could theoretically be offered at a much cheaper price, but no allogenic product has moved past clinical development due to ongoing safety issues. We'll take a deeper look at allogenic and autologous cell therapy supply chains in Chapter 18 – Make-to-Order and Just-in-Time Supply Chains in Clinical Trials.

Gene therapy is a separate branch of cell therapy treatments where the goal is to permanently correct a genetic defect. This is accomplished by sending viral vectors into the body to perform gene insertions or deliver premodified genetic material in vivo (e.g. in a human being). Gene therapies are particularly challenging from a manufacturing perspective because they are usually made at −80°C and must be stored throughout the product's lifecycle in this temperature range. This extremely cold storage condition then limits product labeling to the manufacturing site and makes expiry extensions nearly impossible. While gene therapies have incredible promise as individually tailored treatments, they also experience challenges with raw material sourcing, manufacturing issues, and product toxicity profiles. This combination has limited their commercial viability to date to a handful of approved products.

Over the next several sections, we break down each phase of the manufacturing process and examine modality-specific considerations including raw materials, lead times, and cost.

3.3 Starting Materials and Drug Substance

DS/API manufacturing has the same primary goals across modality – to maximize purity and production yields at a minimal cost. In Phase I, the goal is to produce material in time to support the initial clinical trials. As the program evolves and demand increases for API supply, process improvements are implemented to achieve these goals. With those commonalities noted, there are vast differences in how small molecules, biologics, and cell therapies are made.

Small-molecule DS production uses starting materials as chemical building blocks to synthesize the final API. In a very simple API synthesis, there may be two or three manufacturing steps, while complex syntheses may take dozens of reactions to reach its final form. The first step begins with mixing starting materials to yield a desired intermediate compound. Subsequent steps using additional starting materials, intermediates, or reagents are performed until the desired compound is reached. The raw materials themselves may also require an entire manufacturing process to make. Therefore, planning API production often requires months of lead time to ensure the starting materials are available in sufficient quantities and there is enough time to manufacture the API to enable DP manufacturing.

As small-molecule production moves from lab-scale batch sizes to clinical batch sizes, manufacturing and analytical process development activities lead to improved process reliability and consistency. Meanwhile, economies of scale drive the cost of goods down. This is because the quantity of inputs and outputs increases, but the number of steps performed remains the same. Manufacturing DS is relatively expensive in early-stage development, but as the process is optimized and scaled up, costs fall. By the time a small molecule makes it to late-stage production, the cost may range from \$2–50,000/kg or \$500k–2M for a 100 kg batch with exceptions on the low or high side if a process is particularly simple or complex.

Large-molecule DS production, also known as bioprocessing, requires growing cells that have been engineered to express the desired biological compound. Like small molecules, DS manufacturing takes months to complete and must be planned for years in advance to ensure all the raw materials are available. Bioprocessing is divided into two steps:

- *Upstream Processing (Cell Culture)*: The process of identifying, isolating, and growing cells that have been genetically modified to express the desired biological compound. Prior to manufacturing the first batch, the host cell must be selected and isolated into a master cell bank (MCB). When production begins, cells from the MCB are placed into a medium of nutrients,

hormones, gases, and growth factors that support cell growth and protein production. As the cells multiply, they begin to express the desired DS molecule. Once the cell culture process finishes, the media is harvested and then staged for purification.
- *Downstream Processing (Purification)*: Purification is a multistep process designed to remove impurities such as host cell proteins, viruses, fragments, and aggregates. The media is sent through a variety of filtration and chromatographic steps to filter out everything but the DS. Once purification is complete, the DS is then frozen for use downstream or may be immediately filled into DP.

Bioprocessing depends on a wide variety of raw materials and components. Two of these components stand out for their relative importance and cost to source, resins, and media. Both are required for cell culture and can take up to a year or more to source and cost over a million dollars each.

Large molecules typically have wider yield variances from batch to batch than small molecules, especially early in the clinical development process. This is because biologics DS manufacturing has more process variables to control than in small-molecule manufacturing. On the cost front, biologics manufacturing is also considerably more expensive than their small-molecule counterparts. One study estimated biologic manufacturing at $95–225k/kg with higher costs associated when using a contract organization for support.[4] In early-stage development, a single batch of DS may cost upward of $3–4M.

Autologous cell therapy and gene therapy DS manufacturing is formidable on multiple fronts. First, autologous cell therapy manufacturing capacity is limited because every patient requires a unique batch of drug and the timing remains uncertain. This means that a sponsor must be ready to execute the entire manufacturing process within days of patient identification. As a result, many companies in this space are choosing to invest in internal capacity to manage these tight timelines and to ensure the requisite capacity is available. Second, product stability is a huge concern. Cell therapy products have expiry periods that range in days or weeks compared to standard biologics that are measured in months or years. Short stability periods can strain the ability to test and release the product before it expires. Finally, for gene therapy products in particular, sterility is a challenge. Gene therapies often rely on viral vectors for delivery of the modified gene, and it's necessary to check that contaminants have not soiled the viruses.

3.4 Drug Product

DP manufacturing is the next step in the process where API is transformed into a formulation suitable for human use. Formulating a drug involves combining the API with excipients, inactive materials used as a medium for the DS, to enable drug delivery to the body. Most small molecules are formulated for oral consumption, while biologics need to enter the bloodstream or treatment area directly and thus must be administered via intravenous, subcutaneous, or other non-oral methods. The primary goals of DP manufacturing are the same across modalities – to develop a formulation that is safe to consume and hits the desired target to enable an efficacious response. As the program evolves from Phase I to Phase III, finding a formulation that is patient centric (e.g. easy to administer and reduces burden) and a manufacturing process that is repeatable and cost effective are important secondary priorities.

Like API manufacturing, small-molecule DP manufacturing is usually a multistep process. The exception to this sometimes occurs during early development, when a simple API powder-in-capsule formulation with no excipients may be most appropriate. As the program moves into later phases, the target product profile (TPP) emerges which states the desired characteristics of the product such as the ideal route of administration. For example, if the drug is processed quickly by the body, it may be desirable to develop a sustained release formulation to help the drug stay in a person's system longer for increased effect. More complex formulations, such as tablet manufacturing or oral solutions, require a multistep manufacturing process to combine the excipients with the DS into the final desired state. The entire small-molecule DP manufacturing process ranges from weeks to months depending on the number of steps required. Fortunately, small-molecule DP manufacturing is cheap relative to API production. Capsule or tablet costs can be as low as $0.20–0.50/unit, and the total cost of a manufacturing campaign can range from $10 to 200k.

Large-molecule DP manufacturing is colloquially known as the fill–finish operation because the DS is literally filled into the PP container and then finished by closing off that container. DP manufacturing requires thawing the frozen DS into liquid form, blending the DS with excipients designed to support stability and preservation, and then filling the DP into a vial, ampoule, or other glass container.[5] There are two common DP formulation types, liquids and lyophilisates (Lyo). Lyo products are made through the process of lyophilization where the drug is freeze-dried into a powder form after it's been placed into its container closure system. The lyophilized drug can then be reconstituted into a liquid form prior to administration.

Biologics formulation development follows a similar path to small molecules in that sponsor companies want to make the final commercial formulation as patient centric as possible. In a hypothetical scenario, a large molecule may start a Phase I trial formulated as a lyophilisate for reconstitution and administration intravenously. As the program progresses, the goal may be to develop a more concentrated formulation that would be suitable for subcutaneous injection, a more patient-friendly delivery mechanism. For both small and large molecules, formulation changes drive supply chain complexity, especially if multiple formulations are being supplied to ongoing trials concurrently.

Relative to small-molecule DP manufacturing, biologics use fewer excipients, and these are generally easy to source. The lead times for large-molecule DP production are like small molecules, measuring anywhere from 8 to 16 weeks before the product is released and ready for the next manufacturing step. Like DS production, biologics and small molecules again have significant cost differences. Biologics are more expensive than their chemical counterparts, ranging from $300 to 600k for a batch of DP, with exceptions to the low and high side.

3.5 Analytical Development and Characterization

As discussed at the top of the chapter, a primary goal of the DS and DP manufacturing process is to produce the drug in its intended state. Analytical development is the process of defining the intended state of the drug and is a complementary piece to manufacturing the drug. The goals of analytical development are to:

- Define a specification for the DS and DP.
- Develop the test methods to verify that the quality attributes in the DS and DP specifications are met.
- Establish a stability profile for the DS and DP.

A specification defines the attributes and the predefined ranges, or acceptance limits, of values to which those attributes must conform. For small molecules, those attributes include appearance, dissolution, purity, and stability. For large molecules, product activity and potency are also usually included. Within the specification, the attributes need to be identified and acceptance ranges developed for each of them. A batch of reference standard, which is a highly characterized sample of the DS or DP, is also needed as a point of comparison for each subsequent batch.

Once the specification has been set, the next step is to define the test methods that will be used to measure if the product meets each criterion. Many test methods are standard, but if an unusual criterion is identified, developing a new method takes weeks or months of work to complete. In pharmaceutical development, test methods must also be validated. Validation means that the test method gives the appropriate results consistently and accurately. If a test method isn't fully validated, there is risk that the product may not be accepted by a competent health authority and therefore cannot be used in the clinical study. Chapter 5 – Regulatory Considerations covers the information that must be submitted to and reviewed by national regulatory bodies.

Tables 3.1 and 3.2 show example specification tests, acceptance limits, and methods. Specifications often start with relatively broad criteria during early-stage development, but as process knowledge is gained through manufacturing campaigns, the specifications are usually tightened. International guidelines[6] have been established for setting new specifications for small-molecule and large-molecule DS and DP products.

Finally, a stability profile must be established for the drug. There are two noteworthy aspects of the stability profile. First is the establishment of the retest date or expiry period of the product. These dates represent the point at which a DS or DP may no longer be used. The expiry date assigned to the FP the patient receives is based on the date of manufacture (DoM) of the underlying DP batch. For example, if the DP was manufactured in June, 2025, and it has a 24-month expiry period, the expiry date of the FP will be listed as June 30, 2027, on the clinical label. The terms expiry date and shelf life are often used synonymously in CSC management. Note too that the shelf life printed on a label is typically rounded to the last day of the month.

Table 3.1 Partial Small-Molecule Specification Table

Test Description	Acceptance Limits	Method Description
Appearance	White to light yellow	Visual inspection
Identity	Confirmation by method	High-performance liquid chromatography (HPLC)
Assay (Purity)	95.0–105.0%	HPLC
Total Impurities	Not more than (NMT) 2%	HPLC

Table 3.2 Partial Large-Molecule Specification Table

Test Description	Acceptance Limits	Method Description
Characteristic – Opalescence/Clarity	<Reference standard	Compendial method
Characteristic – Protein Concentration	Confirmation by method	Compendial method
Identity	Confirmation by method	Non-compendial method
Activity/Potency	75–125% relative activity	Non-compendial method
Purity – HPLC	Main peak: >97% Aggregates: <3% Fragments: report	Non-compendial method

Stability studies are the mechanism through which retest and expiry period dates are established. A stability study takes samples of the DS, DP, or packaged DP and then tests them under different environmental profiles at predefined time points for degradation. If the product has not degraded, the retest or expiry date is extended based on predetermined criteria within the stability study protocol. If the product has degraded, then the retest date or expiry period is limited to the time point before degradation occurred. Like specifications, international guidelines have been established outlining the appropriate structure of stability studies and the parameters by which the retest or expiry dates may be extended.

In the stability study shown in Table 3.3, the product will be run at two standard conditions, 25°C/60% relative humidity (RH) and 30°C/75% RH, and one stressed condition 40°C/75% RH. Per International Council for Harmonization (ICH) guidelines, the stressed condition can be used to extend retest and expiry periods beyond the time point of the sample pull. For example, a T = 3mo sample can be used to set a 12mo expiry period for small molecules and a 6mo expiry period for large molecules. Table 3.4 lays out an example expiry period assignment schematic.

There are three events that happen with each stability time point:

1) *Time Point Date*: Day samples are pulled from the stability sample pool for analysis.
2) *Analysis Complete Date*: Day the analysis is passed to the quality team for assessment. This date can take anywhere from a few weeks to several months to complete from when the time point date is reached.
3) *Extension Date*: Day the retest date or expiry period is extended and updated shelf life is justified. This date can take anywhere from a few days to a few weeks from the analysis complete date.

Table 3.3 Example Small-Molecule Stability Study Table

Time Point	*T* = 1 mo	*T* = 2 mo	*T* = 3 mo	*T* = 6 mo	*T* = 9 mo	*T* = 12 mo	*T* = 24 mo	*T* = 36 mo
40°C/75% RH	X	X	X	X				
30°C/75% RH	X	X	X	X	X	X	X	X
25°C/60% RH	X	X	X	X	X	X	X	X

Table 3.4 Expiry Period Assignment Table

Modality	Time Point	Expiry Period
Small Molecule	3 mo	12 mo
	6 mo	18 mo
	12 mo	24 mo
	24 mo	36 mo
Large Molecule	3 mo	6 mo
	6 mo	12 mo
	9 mo	18 mo
	12 mo	24 mo
	24 mo	36 mo

Table 3.5 Example Extension Date Table

Time Point	Time Point Date	Analysis Complete Date	Extension Date
3 mo	15-Jan-2023	1-Mar-2023	15-Mar-2023
6 mo	15-Apr-2023	1-Jun-2023	15-Jun-2023
9 mo	15-Jul-2023	1-Sep-2023	15-Sep-2023
12 mo	15-Jan-2024	1-Mar-2024	15-Mar-2024
24 mo	15-Jan-2025	1-Mar-2025	15-Mar-2025

Table 3.5 lays out the key extension dates to monitor. Each of these is critical to understand because expiry assignment relies on the extension date, not the time point date, to assign the shelf life on the clinical label. If it takes months to get from time point date to extension date and that lead time is not well understood, a poor packaging and labeling strategy will result. Expiring product may not be extended leading to a stockout, or unnecessary resupply batches will be produced that could have been avoided by an expiry extension activity.

The second aspect of the stability profile is the storage condition. Storage conditions are the temperature range at which the drug must be kept. Many small molecules are stored at 15–25°C, otherwise known as controlled room temperature (CRT). However, some small molecules and most biologics must be stored at 2–8°C, otherwise known as a refrigerated condition. Other large molecules require even colder conditions, all the way down to −80°C. Any product with a storage condition at 2–8°C or below is considered a cold chain product. As the temperature goes down, supply chain complexity goes up. This is because cold chain products must be stored in refrigerated or frozen conditions through the entirety of the supply chain. If the product experiences a temperature excursion (e.g. exposed to a temperature outside of the prescribed range), it may no longer be suitable for use.

Both expiry and storage condition have implications on the CSC. Expiry considerations influence the supply plan, as a strategy must be developed to deal with the expiring material. Can the material be extended or will it need to be replaced? Chapter 10 – Finished Product Expiry Planning provides answers to these questions. Meanwhile, the product's storage condition impacts the distribution plan. Since cold chain products are at higher risk of experiencing temperature excursions, the distribution plan may be more conservative to prevent the loss of material in the event of an excursion.

All small molecules, biologics, and cell therapies undergo analytical characterization and require stability testing. Small molecules are highly characterized, have the tightest specifications, and, if they are relatively stable, can achieve expiry periods of 3–5 years. Biologics tend to be relatively well characterized, but their specifications have wider acceptance criteria than small molecules. Biologics also tend to be less stable, particularly in liquid formulations, and expiry periods typically range from 2 to 3 years. Cell therapies are the least well characterized because of how quickly the drug must be processed into its final form, and their expiry periods are measured in weeks.

3.6 Manufacturing Process Development

All DS and DP manufacturing processes change through clinical development to enable larger yields and to ensure the processes themselves are robust and reproducible. These changes are known as process development, and they take place between the manufacturing campaigns that

Table 3.6 Summary of Manufacturing Process Development Changes

Drug Substance Changes	Drug Product Changes
Solvents or reagents for small molecules	New formulation
Media, resins, or columns for large molecules	Dose strength
Starting materials	Excipients
Container closure system	
Equipment train	
Specifications or test methods	
Manufacturing or testing sites	

produce what's needed for the clinical trials in which they'll be tested. The largest changes take place between Phase I and Phase II as human data comes in and improvement opportunities are identified from one campaign to the next.

Process development changes touch the manufacturing process itself, analytical specifications and methods, or the physical supply chain where the product is made and tested. Table 3.6 documents several example changes.

Process development is managed by the technical expert responsible for DS or DP manufacturing. But, from a supply chain perspective, it's important to understand what's happening in this area for a few reasons. First, process changes often impact what needs to be planned. If a new manufacturing site is brought online to support DS manufacturing, there are now two DS sources to plan against. Or, if a new DP formulation is developed, that DP needs to be incorporated into the supply plan.

Second, process changes impact the yield of future batches, and these batch size changes impact the plan for the downstream supply chain. If the most recent batch of DP yielded 1000 vials, it may be tempting to assume the next batch will also yield 1000 vials, but if the yield increases to 1500 vials, that change impacts the packaging strategy. Future packaging runs may be delayed or made smaller as a result. If the yield decreases to 750 vials, then future packaging runs may need to be pulled in or made larger to make up the deficit.

Third, lead times may be impacted in two ways by process development. Process changes can shorten or lengthen the amount of time it takes to make or test the product before it can be released. And the act of conducting process development may push out the expected start date of the next campaign. The supply plan needs to account for both potential outcomes.

Finally, there are regulatory implications with process development. If the process changes are significant enough, the competent health authority where the trial takes place must be notified and approval to implement the change must be granted before the material manufactured under the new process can be introduced to the study. This topic is explored more fully in Chapter 5 – Regulatory Considerations.

3.7 Packaging

Packaging is a critical aspect of the CSC for four reasons. First, packaging protects the drug from damage or contamination. Second, for biologics and cell therapy products, the product's packaging enables drug administration. Third, the optimal packaging type enables patient compliance with respect to consumption of the appropriate dose, every time. And, finally, a well-thought-out packaging design can reduce waste. There are two types of packaging, primary and secondary. These packaging types are named because the PP encloses the product and is the final container closure system of the drug, and the SP supplements the PP.

With respect to timing, small molecules may be packaged as soon as DP manufacturing finishes or held as bulk with PP occurring at a future date. Biologics are always placed into their PP configuration (e.g. vial, ampoule, and syringe) during the fill–finish process. SP and labeling nearly always happen as a joint operation for efficiency and cost purposes.

Key drivers for PP selection include the drug modality, route of administration, and patient compliance. The route of administration is the mechanism by which the drug is delivered to the subject's body (e.g. orally, intravenous, subcutaneous, and intrathecal). Biologics are commonly delivered through intravenous injection to avoid the hyper-acidic environment of the stomach. A popular alternative route of administration is subcutaneous injection because it can be taken at home versus IV injections which require a qualified site and trained professionals to help administer.

Solid-state small molecules taken orally are most likely to be packaged in bottles or blisters. If the clinical trial is blinded and mixing two formulations is required to achieve the desired dose, blisters are the best choice to ensure patient compliance. This is because the packaging and labeling on the secondary carton does not indicate which strength is which, and the chances of a subject mixing up their dose by taking the incorrect number of capsules or tablets from a bottle increase. In other cases such as open label trials or trials where active and placebo are not mixed in a single dose, bottles are less costly and easier to produce.

Another strategic consideration for solid-state small molecules is the number of capsules or tablets to package within a single packaging container. This number dictates how many days or weeks-of-supply are provided to a subject when the drug is dispensed. If this quantity doesn't match the prescribed dose, visit schedule, and visit windows, then drug wastage will increase. Let's use an example with the following assumptions to illustrate this:

- *Packaging Configuration*: 30-count bottle configuration with tablets.
- *Dose*: two tablets are taken once daily (QD).
- *Total Supply Provided Per Bottle*: 15 days (e.g. 30 tablets/2 tablets per day).
- *Visit Schedule*: 7 days (weekly visits).
- *Visit Window*: ±3 days.
- Total potential time between visits = Up to 13 days.

At first glance, one bottle should be enough to cover two weekly visits because there are enough tablets per bottle to cover 15 days of dosing. However, the visit window complicates this analysis. What happens if a subject comes back to the clinic 2 days later than projected per visit window allowance? Now the subject has taken 9 days of Investigational Medicinal Product (IMP) and only has enough for 6 more days, not enough to cover the total potential time between visits. Therefore, one bottle must be dispensed at each dispensing visit to cover the total potential time between visits, but this approach also guarantees anywhere from 4 to 16 out of the 30 tablets are thrown out with every bottle dispensation. This interplay between packaging configuration, dose, visit schedule, and visit window is a key opportunity to reduce waste in a clinical trial. The CMC team and clinical operations need to work together on the PP configuration so the study design elements and the packaging design are as closely aligned as possible.

The final packaging step is SP, which encloses the primary packaged drug in its final state for delivery to a subject. A single SP unit is known as a kit. Nearly, every type of PP requires SP to protect the underlying drug, group the desired treatments to meet the study design, and provide enough space for clinical label application. The one notable exception is bottle packaging for small molecules. Bottles don't require SP because they are made of plastic, and they have enough surface area to support clinical label application. Examples of SP include:

- Blister packs (PP) → wallets (SP) containing blister strips.
- Stick packs or sachets (PP) → cartons (SP) with dividers and/or inserts.
- Vials, ampoules, or prefilled syringes (PP) → cartons (SP) that contain vial(s).

Like small molecule packaging configuration design, SP configurations should minimize drug wastage while still meeting the clinical design requirements. For example, biologics should always be packed as single units. This is because the dose for biologics is often weight based, expressed as milligram/kilogram. In other words, a subject who weighs 60 kg will receive a different amount of drug than a subject who weighs 80 kg and therefore may need another vial of drug to meet the desired dose. If a kit contains multiple vials, there is a risk that lighter subjects will not consume all the vials which creates unnecessary waste.

3.8 Labeling

The final manufacturing step in the CSC is the application of a label that describes the study and drug information to both the investigator and subject. When this step is completed, it results in the FP. There are six steps involved in the labeling process as shown in Figure 3.3.

The first step is to identify the label text that is required for inclusion on the label. The scope and content of clinical label text requirements are dictated by the national competent health authority of each participating country. The European Union has the clearest set of requirements, as outlined in Regulation No. 536/2014[7]:

1) Name, address, and telephone number of sponsor.
2) Pharmaceutical dosage form, route of administration, quantity of dosage units, and product identifier if the clinical trial uses an open label design.
3) Batch and/or code number.
4) Trial reference code (e.g. protocol number).
5) Trial subject identification number/treatment number and visit number.
6) Investigator name.
7) Directions for use (reference to a pharmacy manual or other document is acceptable).
8) The caution statement: "For clinical trial use only."
9) The caution statement: "Keep out of the reach of children" if the product will be taken home for use.
10) Storage condition of the drug.
11) Period of use or expiry date of the drug.

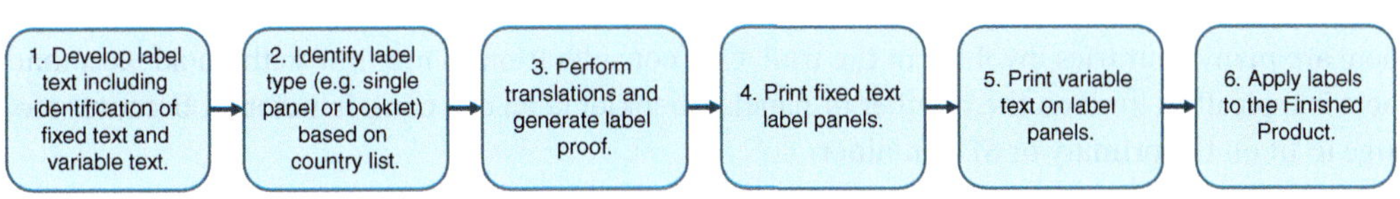

Figure 3.3 Label Development, Printing, and Application Process

When secondary packaging is used, the outer container must have all 11 points listed while the PP must contain at least points 1, 2, 3, 4, and 5. If the expiry date is printed on the secondary label, this means the label(s) must be updated to reflect the new shelf life of the product. This additional labeling step is known as an expiry extension or shelf-life extension. There is a trade-off that needs to be evaluated for all shelf-life extensions. Does it make more sense to invest in performing the expiry extension or to replace the expiring supply? Expiry extensions are costly and complex to execute, but if the underlying DP is scarce, expensive, or both, then performing the expiry extension may be necessary or worth the investment. On the other hand, if the underlying DP is plentiful and relatively cheap, allowing the labeled product to expire and planning replacement supply is likely the better strategy. This decision-making process is fully explored in Chapter 10 – Finished Product Expiry Planning.

Within the United States, clinical labeling requirements are largely the same with two exceptions. First, the caution statement in the United States must read: "Caution: New Drug – Limited by Federal (or United States) law to investigational use." Second, the expiry date is not required to be printed on the label of either the PP or the SP. Instead, the United States allows expiration dating to be controlled through other mechanisms such as documents (e.g. memos to the clinical site) or technology (e.g. the IRT/randomization and trial supply management (RTSM) system). Leaving the expiry date off the label is advantageous because the product's shelf life can be extended without having to perform an expiry extension labeling step, thus saving time and money.

The final piece of the first process step is to identify the fixed text and variable text components of the label. Fixed text is consistent across every batch of material produced to support that trial. Examples of fixed text include sponsor information, dosage form, route of administration, and the protocol number. Meanwhile, the variable text information changes from one batch of FP to the next and includes the batch number, subject identification number, visit number, and expiry date of the drug. One strategy that adds flexibility to small-molecule label text is to make the dose strength variable text. Using this approach makes the bulk fixed text labels applicable to multiple DPs because the strength can be added as variable text for a specific batch. For example, if the plan is to supply a 50 mg dose strength and a 100 mg dose strength, this would typically require two different sets of bulk labels, one for each strength. Making the dose strength variable text means only one bulk label can cover both formulations because the strength can be added to the label during the variable text printing process.

The second process step is to identify the label type as either a single-panel label or a booklet label. A single-panel label is a country-specific label and so named because only one page or panel is required. Booklet labels contain multiple pages, with each page containing core label text that has been translated into the language of each country in which the trial will be conducted. Single-panel labels are appropriate for single-country clinical trials or if it makes sense to segregate one country's supply from the other countries. A good example of when it makes sense to create this separation is for a global clinical trial that is starting in the United States and has supply that will undergo expiry period extensions. When these conditions are met, creating a US-only single-panel label reduces the lead time to have FP supply ready because of the reduced label printing lead time. It also reduces cost because the US-only supply will not need an expiry extension operation performed to continue extending the material. On the other hand, booklets are most efficient when there are many countries involved in the trial. One consideration to monitor with booklets is that there is typically a limit on the number of panels a booklet label can contain before it becomes too large to fit on the primary or SP container.

Developing translations of the fixed text is the third step in the process. Translations are based on the country list of the study, and they are typically performed by a third party specializing in this space. During the translation process, additional country-specific text requirements may emerge that also need to be added to the label. The translated pages also typically need to be included in the country-specific clinical trial application to initiate the study. Once the translations are complete, the final label proof is generated which contains all of the country-specific label panels, the locations of the variable text, and the dimensions of the label.

The fourth and fifth steps involve the two printing operations. The first operation produces the single-panel or booklet labels with the fixed text, and the second operation places the variable text onto the label in the designated locations. There are a few elements to consider with the first operation. It's usually more economical to print large batches of labels in a single event, and these labels can then service multiple FP packaging campaigns. From a lead time perspective, single-panel labels can be translated and printed in a few days or weeks, booklets can take much longer, especially if the number of countries is greater than 10. Translating, approving, and printing a booklet for a study with 10+ countries can take months. In the second operation, the variable text is added during a second operation that depends on the batch-specific information for that set of labels. This typically takes place as part of the SP and labeling production event because the batch number and expiry date aren't available until this point, and it's always advantageous to print the best possible shelf life on the label.

The sixth and final step is the label application process. It typically accompanies the SP operation, but it can also be executed as its own independent step. The total lead time may range from 2 to 8 weeks on average to complete. Packaging and labeling campaigns can cost anywhere from $10 to 100k per batch depending on the complexity of the packaging design and number of countries included on the label. There is one unique attribute to labeling though, the retention (e.g. retain) sample. Every batch of labeled product must set one kit aside that cannot be used in the clinical trial. The retention sample acts as an example of the FP batch and can be used as a test article in the event of a product quality issue.

3.9 Distribution

After the FP has been labeled and released for use, distribution is the final step in the supply chain before the product reaches the clinical site. The simplest version of distribution happens when the FP manufacturing site is in the same country as the participating clinical sites. A domestic shipment of IMP avoids the challenge of international logistics, so these shipments only take a few days to process and receive. Wherever possible, this approach should be taken to minimize distribution complexity.

Complexity, lead time, and cost all increase when the supply chain goes international. This is because shipments of IMP are subject to both good distribution practices (GDPs) and an array of international customs requirements that vary from country to country. While GDPs are legally enforced for commercial products, they also apply to IMPs, and the sponsor is responsible for assuring "that IMPs are stored, transported, and handled in a suitable manner.[8]" This responsibility entails keeping the product in the appropriate storage condition for the duration of time it is in transit. Therefore, all IMPs must be shipped in qualified shipping containers that are designed to hold the prespecified temperature range, and all shipments must include a temperature monitor that can provide a temperature readout of that shipment. If the product goes out of the prescribed range at any point during the shipment, this is known as a temperature excursion. When an excursion happens, the product must be assessed against the stability data on hand and a determination made if the product is still good for use. Refrigerated and frozen products are more prone to temperature excursions, and if the product's stability profile is not strong, the risk of losing product increases. This risk is heightened further if the product must be shipped to hot climates in Latin America, Asia, or Africa.

While the details of how to distribute a drug into a specific country for a particular study is situation dependent, there are general documentation and lead time considerations to keep in mind for each region:

- *United States*: Importing product into the United States requires a few additional documents, most notably the investigational new drug (IND) application. Without an approved IND, the product will be held at customs unless it is headed to a bonded warehouse for further processing. US customs also require an end-use letter stating the intended purpose (e.g. for research and development, animal testing, or human use). Lead times to import to the United States are generally quick, taking a week or less from shipment pickup to delivery at the intended destination.
- *European Union and the United Kingdom*: Within the EU and the United Kingdom, a QP certification of the product must occur after the IMP has physically arrived. While the QP certification process is explained in Chapter 4 – Quality Considerations, it's worth noting here that the process adds at least 2–3 weeks of lead time before the product can be shipped to a clinical site.
- *Asia-Pacific*: Within the Asia-Pacific region, document and lead time requirements range widely. Some countries (e.g. Singapore and Taiwan) have minimal additional document requirements and are relatively quick to import. Other countries (e.g. China) place restrictions on the drug itself in terms of having specific stability requirements met. And, yet others (e.g. Thailand and Vietnam) require complex import licenses that can add weeks of time to the import process.
- *Latin America*: Like Thailand and Vietnam, the major Latin American countries including Brazil, Colombia, Mexico, Argentina, Peru, and Chile require an import license application to be completed prior to shipment delivery. These are some of the most difficult countries to get product into, so having local regulatory and logistics partners in the country who are familiar with the country-specific details is recommended.
- *Africa*: This is the most difficult region to assess from a documentation and lead time perspective. The regulatory framework and customs institutions vary dramatically from country to country. To best understand requirements and lead times, it's recommended to work closely with an in-country manufacturer or other partner to facilitate the distribution process.

Due to regulatory and legal requirements, the distribution node of the supply chain is composed of local depots located within each region or country where the product must be received before it can be sent on to clinical sites. For example, a study taking place in the United States, the EU, the United Kingdom, Brazil, and Russia has up to five clinical depots, one for each listed country/region. Distribution planning is covered in Chapter 11 – Finished Product Supply Planning: The Distribution Plan, and the mechanics of international logistics are detailed in Chapter 21 – International Logistics for Clinical Biopharmaceutical Products.

3.10 Clinical Trial Supply Chain Summary

Every investigational drug has a unique supply chain in terms of geography, lead time, and cost, but the steps to get the FP to a clinical site are similar. Tables 3.7–3.9 summarize this chapter's discussion so far.

Table 3.7 Small-Molecule Lead Times and Average Cost by Manufacturing Step

Small Molecule			
	Manufacturing Lead Time	**Analytical and Batch Review Lead Time**	**Average Cost**
Drug Substance	4–8 weeks	6–8 weeks	\$2–50k/kg
Drug Product	1–2 weeks	6–8 weeks	\$50–500k/batch
Primary Packaging	1–2 days	3–6 weeks	\$5–50k/batch
Finished Product (Secondary Pack + Label)	1–4 weeks	1–4 weeks	\$25–50k/batch
Distribution	1–12 weeks	N/A	\$1–3k/shipment

Table 3.8 Large-Molecule Lead Times and Average Cost by Manufacturing Step

Biologics incl. Allogenic Cell Therapy and Gene Therapy			
	Manufacturing Lead Time	**Analytical and Batch Review Lead Time**	**Average Cost**
Drug Substance	6–8 weeks	12–24 weeks	\$2–50k/kg
Drug Product and Primary packaging	1–2 weeks	8–16 weeks	\$250–500k/batch
Finished Product (Secondary Pack + Label)	1–4 weeks	1–4 weeks	\$25–50k/batch
Distribution	1–12 weeks	N/A	\$1–5k/shipment

Table 3.9 Autologous Lead Times and Average Cost by Manufacturing Step

Autologous Cell Therapy			
	Manufacturing Lead Time	**Analytical and Batch Review Lead Time**	**Average Cost**
Drug Substance to Finished Product (One Continuous Step at One Manufacturing Site)	1–2 weeks	1–4 weeks	\$100–300k/batch
Distribution	1–3 days	N/A	\$1–5k/shipment

3.11 CMC Development and Manufacturing Strategy

At each phase of development, the CMC team creates a strategy linking each manufacturing step – DS, DP, analytical, process development, packaging, labeling, and distribution – to ensure the drug is available in the desired form, packaging, and quantity to support all ongoing and planned clinical trials. These strategies are always closely linked to the clinical development plan and typically span a three- to five-year time horizon given the long lead times involved. To develop the CMC strategy, the CMC works together with the project team to understand the strategic context of the molecule. In clinical development, the project team oversees all aspects of the program and sets that context.

For example, in a Phase I study, the strategic driver might be speed to data, in which case the simplest formulation and packaging configuration is likely best. In a Phase II or III program where the dose and PK profile are known, developing a specific strength and delivery system may take priority. Competitive advantage is another driver in late-stage development. If the drug can be dosed in a more patient-friendly manner, this could strengthen its position on the commercial market. The final output of these strategic discussions is typically a quality target product profile (QTPP) capturing the DP characteristics. The clinical development plan (CDP) and QTPP provide necessary structure for the CMC team to use as it develops the CMC strategy for the program.

Figure 3.4 illustrates a CMC strategy based on the example clinical development plan presented in Figure 2.3.

Years before the drug enters the clinic, process development activities begin on the DS. For small molecules, this includes identification of the initial synthesis, starting material manufacture, and generation of an initial demonstration batch. For large molecules, this includes cell line development, establishment of an MCB, and developing the upstream and downstream manufacturing processes. DP process development is also necessary at this stage for large molecules but can be optional for small molecules if the Phase I will be run with a pure API powder-in-capsule formulation.

Process development culminates with production of the first batches of DS and DP which are not initially made under Good Manufacturing Practice (GMP). This is because the goal is to demonstrate the processes at small scale and identify the characteristics that will make up the specification for the first GMP campaign. These non-GMP batches, also known as Demo, Tox, or Good Laboratory Practice (GLP) batches, may be used for animal-based toxicology studies or other nonclinical work. In the meantime, analytical development is in full swing to characterize both the substance and the product. Critical process parameters will be identified, specifications developed, and test method selection and validation are all happening during this time. The initial non-GMP batches are put on stability, and these batches form the basis for expiry period assignment for the first FP campaign.

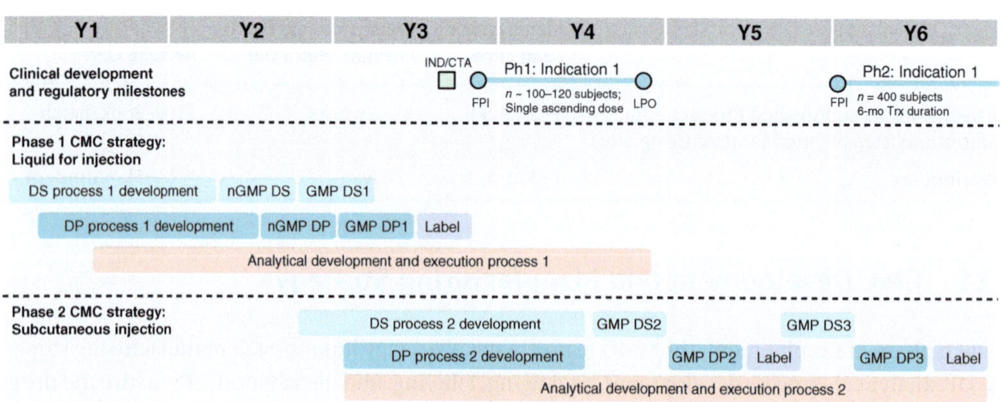

Figure 3.4 Preclinical to Phase II Biologic CMC Strategy

As soon as the initial GMP batches are completed, work begins on DS process improvements. The Phase II study requires a 25-fold increase in the amount of drug required, which may cause the team to examine scaling up the DS equipment train to produce more material or find other ways to improve the overall yield. Specifications, methods, and stability work may also be needed to support the new process. From a DP perspective, the Phase I strategy is to manufacture a liquid formulation for intravenous injection. As the program moves from Phase I to Phase II, the formulation strategy changes from IV to subcutaneous injection, a more patient-friendly approach. To accomplish this, the team needs to make changes to the DP and PP processes to enable this strategy. When the formulation and/or packaging strategy change, an updated analytical strategy is also needed. New specifications and a new stability study top the list, but new methods may also be necessary. Multiple manufacturing campaigns are also planned to support total demand from the Phase II study.

The CMC strategy is important to CSC management for three reasons. First, the strategy creates visibility into the key manufacturing activities and timelines that drive DS and DP supply availability. Second, the CMC strategy illustrates the analytical strategy and shelf life the material will receive over the duration of the program. Third, the strategy creates visibility to nonclinical demand for DS and DP that will need to be included in the overall demand and supply forecast for the program.

The CMC strategy is also influenced by the CSC function, particularly as the program moves from early-stage to late-stage development. This is because, at minimum, CSC owns the demand forecast and FP supply plan for the program. These plans drive the requirements for DP and DS. If supply chain does not communicate or is not asked to provide this input, there is risk the CMC strategy will overproduce material or, worse, underproduce what the program needs to be successful.

3.12 CMC Stakeholder Summary

Before covering the various roles that make up a CMC team, it makes sense to begin with the CSC function. There are typically three broad groups within CSC:

- *Planning*: CSC planning owns the end-to-end demand forecast and supply plan for all levels of the supply chain from FP to DS. Raw materials planning may be included as well.
- *Execution*: CSC execution usually owns SP, labeling, and FP distribution. Supply chain will take on different responsibilities depending on whether these steps will be performed internally or will be outsourced to a contract development and manufacturing organization (CDMO). If performed internally, then CSC typically owns the scheduling and prioritization of manufacturing jobs and distribution orders. If outsourced, CSC owns the relationship with the CDMO from supplier selection to project oversight.
- *Technology*: CSC technology provides input on the RTSM platform which controls subject randomization and dispensation of IMP to sites and subjects. While clinical operations typically owns the RTSM supplier relationship, CSC is heavily involved with the development and implementation of the system. See Chapter 6 – RTSM and Its Role in the Clinical Supply Chain for more. All aspects of the RTSM that impact drug supply typically fall under CSC's purview. This function may also cover planning solution enterprise resource planning (ERP) oversight responsibilities where applicable. See Chapter 22 – Clinical Supply Chain Planning Tools and Technology for more.

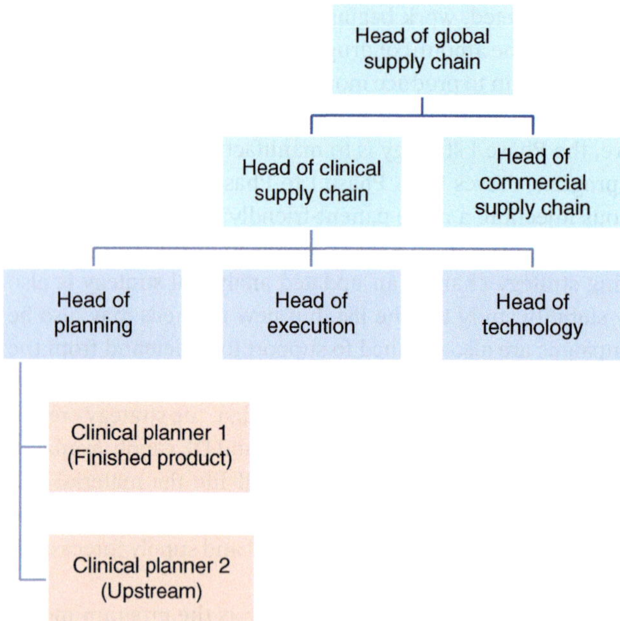

Figure 3.5 Example Clinical Supply Chain Organization Chart

This is not a comprehensive list of all the activities the CSC function may own, but it does cover the primary areas where CSC interacts with the other functions listed in this section. Depending on the size and maturity level of the sponsor, the CSC function may be a single person or it may be a sizeable organization made up of several specialized sub-functions. See Figure 3.5 for an example organizational chart. Notice how there are distinct roles for the head of CSC, head of clinical planning, and individual clinical planners. We'll touch on these roles again in Chapter 15 – Clinical Supply Chain Planning Processes.

Within the CMC team, supply chain partners with every other function to collect the inputs it needs to operate and provide the outputs required for the team to operate. Below is a breakdown of the internal and external stakeholders with a description of their role and interaction with the clinical planner for the program:

- *CMC Project Lead*: The CMC project lead drives the CMC strategy and serves as the primary liaison to the overall project team. It's their job to ensure all the other functional activities are tethered to the CMC strategy and to identify potential risks and mitigation strategies for the program. The project lead drives alignment between the overall project strategy and the CMC team. In smaller organizations, the CMC project lead may fulfill any of the other functional roles outlined below from project management to CDMO management, to regulatory strategy, as needed. The clinical planner will work with the project lead to understand the CMC strategy and to provide the DS and DP supply requirements for the clinical trials in the program. The planner may also work closely with the project lead on defining the supply chain strategy (e.g. single- versus dual-sourcing) and developing a supply chain risk assessment.
- *CMC Project Manager (PM)*: The CMC PM drives execution of the CMC strategy. They develop the detailed timelines for the project and ensure tasks among the team are completed on time. The CMC PM may also liaise with other teams as needed when cross-functional alignment is required. The PM keeps the planner informed of DS and DP availability dates, while the planner provides the PM with FP supply availability targets.

- *Small-molecule DS Lead(s) – Process Chemist*: The process chemist develops the chemical synthesis of the API and oversees its manufacturing from raw material sourcing through the final step. Their goal is to scale the manufacturing process from the lab to market readiness, an effort that typically translates to 100–1000× increases in batch size over the life of the clinical program. The planner works with the process chemist to understand manufacturing and release lead times, estimated yields, process development requirements, and any raw material challenges that may be impacting the supply chain.
- *Large-molecule DS Lead(s) – Upstream and Downstream Bioprocessing*: On large-molecule teams, the DS lead roles are often broken up between upstream (e.g. cell line development, cell culture, and harvest) and downstream (e.g. purification). Like their small-molecule counterparts, these functions work to optimize the manufacturing process from campaign to campaign. Due to the longer lead times and greater variability in large-molecule process performance, it's especially important for the clinical planner to stay in touch with these colleagues on lead times, yields, and other manufacturing-related issues that may impact the plan.
- *Drug Product Lead(s) – Formulation Development*: The DP lead is responsible for all aspects of DP formulation development. This includes designing the manufacturing process, selecting excipients, and engaging with CDMOs if needed. The DP lead leads selection of the dose strengths that will be supplied for the trials. If there is not a packaging engineer on the team, the DP lead will often take responsibility for PP selection. As with the DS leads, the planner needs to understand lead times, yields, and manufacturing risks.
- *Packaging Engineering (PE)*: The packaging engineer owns all aspects of the PP design, manufacturing process, and component selection. This role is especially valuable on large-molecule programs that have complex delivery systems such as syringes for subcutaneous injections. The packaging engineer and clinical planner align on packaging operation lead times and potential component supply risks.
- *Analytical Operations (AO) and Quality Control (QC)*: AO builds the entire analytical strategy for the DS and DP including specifications and stability protocols. In some situations, there may be dedicated analytical leads for DS and DP due to the wide scope of the role. QC is a branch of AO that reviews batch release and stability data for potential issues within the production process or to establish the expiry period of the product. The planner works with analytical to understand and document the expiry strategy, expiry period schedule, storage condition, and specification changes that impact the supply strategy.
- *External Manufacturing Lead(s)*: External manufacturing manages CDMO relationships within biopharmaceutical companies that use contract services. While the DS and/or DP leads handle the technical aspects of the relationship, the external manufacturing group manages the business aspects of the relationship. These activities include supplier selection, contract negotiation, and business performance management. In smaller organizations, CSC may act as the external manufacturing lead for SP, labeling, and distribution CDMOs. If there is a dedicated external manufacturing role, the planner can learn about CDMO network reliability with respect to lead times, quality, and project delivery.
- *Quality Lead*: Quality is the single most important relationship for the clinical planner to maintain within the CMC team. The quality lead is responsible for the overall quality strategy of the program and will review and approve all GMP documentation related to the program. Their role is so integral to the function of the clinical trial supply chain that the entirety of Chapter 4 – Quality Considerations is dedicated to their scope of work. Supply chain works closely with quality on a wide variety of topics ranging from batch release lead times to supplier qualification status to temperature excursion management.

- *Regulatory Affairs (RA)-CMC*: RA-CMC is responsible for the regulatory submission strategy and timelines. This strategy is vital for the clinical planner to plug into because without regulatory approval, supply cannot be released for use to clinical sites. RA-CMC's relevance and importance to CSC management is broad enough to also have a dedicated analysis, which is captured in Chapter 5 – Regulatory Considerations.
- *Finance*: Finance is usually not a core member of the CMC team, but they are responsible for tracking the CMC budget. While CMC costs aren't the largest spend category in drug development, they can still run into the tens or hundreds of millions of dollars. The clinical planner can potentially act as the primary liaison with finance from CMC to provide projected spend for the program. This topic is explored fully in Chapter 14 – Metrics, Summary Reporting, and Budget Forecasting.

Figure 3.6 puts the broad internal stakeholder picture together. The clinical planner sits as a core team member on the study execution team to collect demand information and on the CMC team to collect supply information. Program strategy is relayed to each of these teams from the project team. Supply chain may also drop into other teams (e.g. project team) as needed to address supply chain strategy and risks as needed.

CSC also needs to interact with external stakeholders where they are used to enable the supply chain. Commonly used external service providers are listed below:

- *DS and DP Manufacturing CDMOs*: These organizations support manufacturing process development and production. A majority of the CMC team will work with the CDMO including the relevant DS or DP technical lead, AO, QC, external manufacturing, and quality. See Chapter 20 – CDMO and Supplier Management for more on the role these organizations play and how to best manage them.
- *Packaging, Labeling, and Distribution CDMOs*: These organizations support packaging, labeling, and distribution of FP through the supply chain. CSC and/or external manufacturing take the lead in managing these relationships.
- *Contract Testing Organizations (CTOs)*: CTOs provide dedicated analytical release and stability testing services. They are an important player from a supply chain management perspective because they conduct GMP operations and must be qualified and documented as part of the regulatory submission and approval process. AO typically owns the relationships with the CTOs when used, but quality also engages with them.

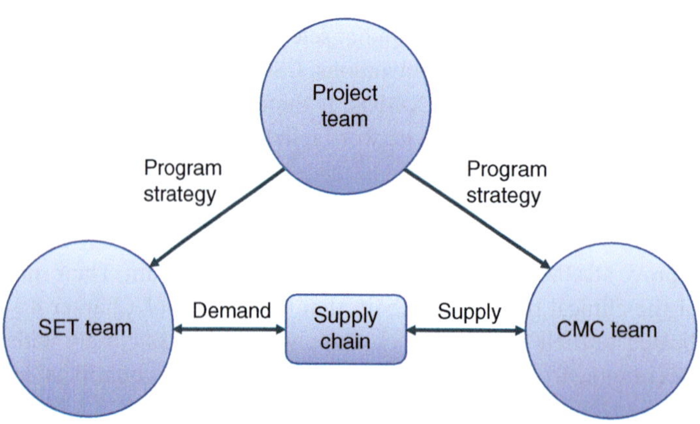

Figure 3.6 Clinical Supply Chain Functional Touchpoints

- *Transportation Service Providers (TSPs)*: TSPs offer logistics services to move drug from one node to the next in the supply chain. This is a particularly complex activity when drug needs to cross international borders which necessitates involvement of a customs authority. Therefore, many TSPs also offer customs brokerage services to ensure smooth shipment and release of materials. See Chapter 21 – International Logistics for Clinical Biopharmaceutical Products for more on TSPs and the international logistics process.

3.13 Trends in Clinical Manufacturing and Supply Chain Management

Biopharmaceutical supply chains have experienced many of the same trends and pressures as the clinical trials they are meant to support – scientific advances leading to specialization, an increased need for speed, and greater competition. These pressures have manifested themselves in different ways though.

First, scientific advances have driven modality diversification. Alongside traditional small molecules and biologics, ADCs, oligonucleotides, and cell and gene therapies have emerged. These new modalities require highly specialized materials and manufacturing capabilities to produce and characterize them. As a result, many sponsors now rely on external CDMOs to execute some or all of the supply chain on their behalf. CDMOs are vital partners because building in-house capacity is so expensive and time consuming. If that capability already exists somewhere else, it usually makes sense to take advantage of it. Some companies may only rely on one or two suppliers while others may outsource the entirety of the GMP and GDP supply chain. At the same time, CDMOs have grown exponentially in terms of their sheer number and their geographic spread.

Given their importance, the CDMO selection and management processes often pave the way to success for a program. In turn, poor selection or management decisions lead to hindrances and delays. CDMOs also hold leverage over sponsors because once a selection decision has been made, it is very difficult and costly to transfer a program between CDMOs. In most cases, the sponsor is effectively trapped into using the CDMO they've selected once the selection has been made and the contract is signed. This trend toward outsourcing differentiates biopharmaceutical manufacturing and supply chain management from many other industries. There is as much emphasis on collaboration, relationship-building, and communication as there is on the delivery of materials that normally drive supply chain planning.

As demand for specialized CDMO capabilities has increased, manufacturing capacity has become an issue. This is particularly true for cell and gene therapy products, but also for other types of biologics. Some biopharma companies have chosen to build their own in-house manufacturing capabilities as an alternative to putting themselves at the mercy of a CDMO. The result of this trend is that many sponsors need to manage both internal and external supply chain networks.

Meanwhile, COVID-19 has also had a large impact. Shortages of raw materials, excipients, and packaging materials manifested during the pandemic and still need to be monitored. If a key component or material isn't available, that's just as damaging to the supply chain as a manufacturing plant shutting down, particularly if production is happening at a CDMO. Capacity may be limited and getting a production campaign back into the schedule can take weeks or months. International logistics also struggled mightily due to lockdowns in China and other parts of the world. TSPs saw long delays at ports and, like many other sectors, experienced higher turnover rates and staff shortages. Lead times to ship pharmaceutical products have always been measured in weeks, but the variation in customer service from TSPs has generated countless headaches for clinical planners throughout the industry.

To mitigate these material and logistics risks, many industries have begun to reconsider offshoring as the increased cost of manufacturing domestically can offset the risk of having supply chain nodes spread around the world. Onshoring has grown in popularity through the pandemic and will continue through the next decade. Interestingly, biopharma has largely resisted these efforts so far.[9] This is because of the extraordinarily high costs of building capacity and executing a product transfer from one site to another. Biopharmaceutical manufacturing plants and their associated supply chains can take years to build. While there won't be a rush to onshore or co-locate these supply chains in the next few years, this will be a trend worth consideration over the next few decades.

Finally, geopolitical risk has been on the rise and will continue to increase over the next several years. As of the writing of this book, war continues to rage between Russia and Ukraine, Chinese tensions with Taiwan continue unabated, and climate change has wreaked havoc on global economies. Any one of these issues taken alone would cause concern but, taken together, they paint a stark picture of future events that could upheave supply chains at any time. These conditions emphasize how important supply chain planning is to a biopharmaceutical organization. There are so many opportunities for disruption that having a well-thought-out approach to planning, sourcing, manufacturing, logistics, and risk management gives the sponsor organization an opportunity to successfully work through disaster.

3.14 Key Takeaways

Biopharmaceuticals are manufactured and distributed through one of the lengthiest and most technically complex supply chains in the world. It's composed of at least five to six distinct manufacturing steps – raw materials, DS, DP, packaging, and labeling – before distribution to clinical sites and subjects occurs. The total lead time to produce an FP ranges from 6 to 18+ months.

1) There are a wide variety of drug modalities in the clinic ranging from traditional small molecules and biologics to oligonucleotides and cell therapies. The newer modalities require even more specialized manufacturing processes and capabilities to enable their production and release.
2) The DS and DP manufacturing steps are the costliest and lengthiest segments of the supply chain. CDMOs specializing in DS and DP production have proliferated over the last two decades to support smaller and mid-sized companies in this space.
3) Analytical development produces the specification, the stability strategy, and the storage condition. All three have an important role to play in effective supply chain management. If a batch of material does not meet the specification, it cannot be released for shipment. The stability strategy dictates the expiry period of the product, and the storage condition is the temperature range at which the product must be maintained through manufacturing, distribution, and storage.
4) The packaging and labeling steps are the final manufacturing operations and are typically overseen by the CSC function. These are the least technically complex manufacturing operations, but still offer a variety of strategic considerations to manage from kit design to "at-risk" packaging to label type.
5) The CMC strategy is a three- to five-year plan showing the process development and manufacturing operations required to support the overall clinical development plan. CSC provides inputs to this strategy and must incorporate the activities and timelines into its own plans.

6) The CMC team is composed of a project lead, PM, and several functional groups, including CSC. The project lead is responsible for the CMC strategy, but every function contributes to its development and execution over the life of the program. CSC works directly with every function in one aspect or another to obtain the information it needs to operate.
7) Specialization, competition, and COVID-19 have all placed greater pressure on the clinical trial supply chain. It's not uncommon for a product to travel across continents and through several individual companies before delivery to a patient. Geopolitical risk has left biopharmaceutical supply chains more exposed than ever, necessitating robust risk assessments and mitigation strategies to avoid disaster.

Notes

1 https://medicalaffairsspecialist.org/blog/therapeutic-areas-transformed-by-biologics.
2 https://asgct.org/global/documents/asgct-pharma-intelligence-quarterly-report-july-20.aspx.
3 https://www.nature.com/articles/s41573-020-0075-7.
4 https://www.forbes.com/sites/theapothecary/2019/03/08/biologic-medicines-the-biggest-driver-of-rising-drug-prices/?sh=56df0f6518b0.
5 https://www.ncbi.nlm.nih.gov/pmc/articles/PMC7316246/.
6 https://www.ich.org/page/quality-guidelines.
7 https://health.ec.europa.eu/system/files/2016-11/reg_2014_536_en_0.pdf.
8 https://www.gmp-compliance.org/gmp-news/imps-how-do-gdp-guidelines-apply.
9 https://www.fiercepharma.com/manufacturing/pharma-pushes-back-u-s-legislation-to-bring-drug-manufacturing-stateside.

4

Quality Considerations

4.1 What Is Quality?

"Product quality is the foundation that allows patients and consumers to have confidence in the safety and effectiveness of their medications."[1] In the United States, the Food and Drug Administration (FDA) regulates the biopharmaceutical industry, and it has made quality a cornerstone for drug manufacturers so that people know the treatments they receive are what has been described to them. The clinical planner benefits from a thorough understanding of quality because it is important at a strategic, tactical, and procedural level for clinical supply chain planning and execution. The rest of this chapter focuses on quality and gives the reader an overview of both the what and the why.

The quality and clinical supply chain functions are deeply intertwined within the biopharmaceutical supply chain. This is because all stages of pharmaceutical manufacturing are governed by good manufacturing practice (GMP) regulations. GMPs are a set of legal regulations designed to ensure the quality of a product and protect clinical trial subjects from harm. Before GMPs were established, it was common for drugs to be manufactured with impurities or other defects. The first American regulatory action to correct this was the passage of the Food and Drug Cosmetic Act (F&DC Act) in 1938. The F&DC Act was a first attempt to prevent the adulteration and false marketing of products made for human consumption, and while the act was an important step forward, it did not place overt controls on the manufacturing process. This was troublesome because the manufacturing process is where impurities are most likely to form and cause harm if they go undetected.

By the mid-1970s, legislation was taken up to further regulate the pharmaceutical space, and in 1978, the US Congress passed the 21 Code of Federal Regulations (CFR) Parts 210 and 211. Part 210 establishes current GMP (cGMP) adherence as a legal responsibility of drug manufacturers and Part 211 outlines all the components of cGMPs that must be considered. CFR Part 210 outlines the definitions of key terms and the application scope of GMP. The regulation currently includes small molecules, biologics, and cell-derived products. Part 211 documents the specific areas that encompass GMP[2]:

- Organization and establishment of a quality unit within a manufacturer and the sponsor company.
- Buildings and facilities where drug manufacturing takes place.
- Equipment used in the manufacture or testing of a drug.
- Components including container closure systems.

Supply Chain Planning for Clinical Trials – A Practical Guide, First Edition. Ryan Mills.

- Production and process controls including deviations and sampling.
- Packaging and label control.
- Warehouse and distribution procedures.
- Lab controls including testing and release of the product and stability requirements.
- Records and reports including batch production records and lab records.

The quality and quality control (QC) functions are the quality unit outlined in the top bullet above, and they are responsible for developing, implementing, and managing the quality management system (QMS) within a sponsor company. Figure 4.1 illustrates the scope of the QMS, which is a systematic approach and a proactive infrastructure aimed at ensuring the quality of pharmaceutical products throughout their lifecycle, from early clinical development through post-commercial availability. Elements of the QMS include the organizational structure of the quality unit and a collection of standard operation procedures (SOPs) that govern the work performed by the quality unit. By this definition, the quality unit's scope is extensive and touches every aspect of the supply chain from sourcing to manufacturing to delivery of the product. For the clinical planner, understanding the QMS is important because it directly impacts the demand and supply planning process as well as the production and distribution plan outputs.

Through the rest of this chapter, we explore cGMPs, the QMS, and its influence on clinical supply chain management. The chapter begins with an overview of GMPs and how that translates to the quality unit's responsibilities as a function. Then, we'll turn to the sourcing and manufacturing processes where the quality unit is responsible for several activities from supplier qualification to batch release. Next is a discussion on how the QMS fits into product distribution, and we'll also discuss quality's impact on the planning process. Finally, we'll introduce the qualified person (QP) and their role in supply chain qualification and batch certification within the European Union and the United Kingdom.

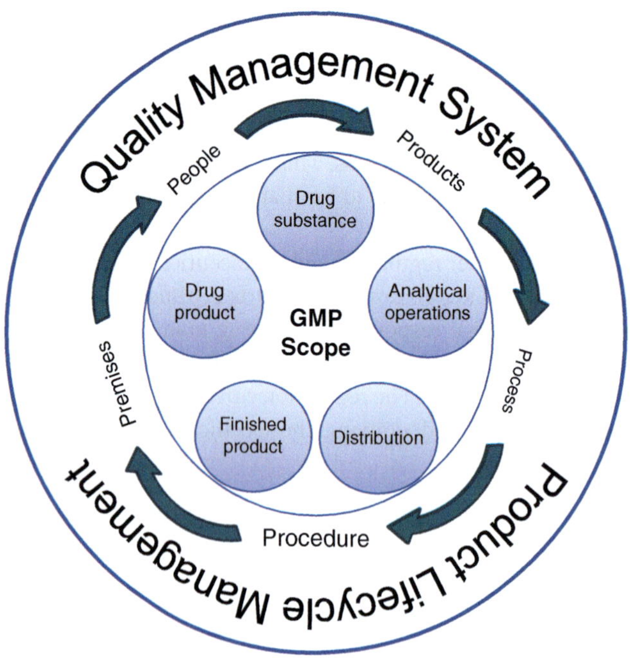

Figure 4.1 Scope of GMP and the Quality Management System

4.2 Good Manufacturing Practice: An Overview

GMPs cover every aspect of the clinical trial supply chain. GMP adherence is mandated for drug substance (DS), drug product (DP), packaging, labeling, and all the analytical work required to characterize, quantify, and release a batch. Even distribution of the product requires a subset of GMP known as good distribution practice (GDP). Complying with cGMPs requires taking a holistic approach with respect to each step of the supply chain. One useful paradigm segments cGMP compliance into the "five Ps":

1) *People*: Everyone who performs a job that touches manufacturing, analytics, or batch release activities is expected to understand and follow cGMP. Training must take place on all processes and procedures to ensure employees know what to do and how to do it. The leadership teams within sponsor companies are also responsible for installing a quality unit (e.g. quality) and reinforcing the importance of quality practices throughout the organization.
2) *Products*: Every DS and DP batch manufactured within the supply chain must be rigorously analyzed against its specification. Packaging and labeling activities are also thoroughly documented. While raw material and component manufacturing do not fall under the scope of GMP, they are still expected to be tested and analyzed against their own specifications before use in downstream DS or DP manufacturing.
3) *Process*: All manufacturing and analytical process steps must be thoroughly documented and made available to the team of people responsible for executing them. Process adherence should be monitored by the quality unit on a regular basis for gaps and improvement opportunities.
4) *Procedures*: Standard operating procedures (SOPs) document processes contained within the QMS such as supplier qualification, specifications, expiry period assignment and management, and temperature excursions. Procedures contain the scope of the process, the responsibilities of each function within the process, and step-by-step instructions for how to execute the process. Procedures establish control over the activities and ensure consistency in process performance.
5) *Premises*: Every location that performs a GMP activity must be qualified for use. The qualification process checks the building (e.g. cleanliness), equipment (e.g. calibration and validation), materials (e.g. warehouse storage locations), and staff (e.g. training) for compliance with cGMP standards.

Adherence to GMPs is a requirement for biopharmaceutical companies not only in the United States, but also around the world. In Europe, cGMPs are documented in Eudralex Volume 4 – Good Manufacturing Practices.[3] Most developed nations outside North America and the EU also have codified GMP requirements including Japan[4] and China.[5] There are variations between nations' GMP regulations, though. To assure alignment and boost adherence across countries, the International Conference of Harmonisation (ICH) was founded in 1990 with a mission to draft and publish a common set of guidelines around quality and GMP for all biopharmaceutical companies to follow. This harmonization effort is especially valuable given the global nature of clinical trials and the multinational structures of the companies that run them. Since its inception, ICH has published 14 quality guidelines.[6] Of those, five are most pertinent to clinical supply chain management:

ICH Q1A-Q1F – Stability Testing and Data Packages[7]: ICH Q1 is divided into seven documents that describe stability testing guidelines for DSs and DPs and how to evaluate the stability data generated within those studies. The ICH Q1 guidelines are instrumental in determining how expiry periods are set for DS and DP, and this directly impacts the shelf life of the finished product (FP), which the clinical planner needs to manage.

ICH Q6A and Q6B – Specifications for Chemical Substances and Biological Products[8]: ICH Q6A and Q6B lay out recommendations for how to establish specifications for small and large molecules, respectively. They include guidance on how to set and justify acceptance criteria and how to select test procedures for new DSs and DPs. These guidelines inform what the product's specification needs to contain.

ICH Q7 – Good Manufacturing Practice for Active Pharmaceutical Ingredients[9]: ICH Q7 documents how GMPs should be applied in the context of active pharmaceutical ingredient (API) manufacturing for small molecule and biologic products to ensure product quality. The scope of Q7 is broad and covers personnel, facilities, documentation, materials, in-process controls, storage, and validation among others. Within Q7, there is a subsection dedicated to material intended for use in a clinical trial, which is worth quoting below:

- *Section 19.11*: "The controls used in the manufacture of APIs for use in clinical trials should be consistent with the stage of development of the Drug Product incorporating the API."
- *Section 19.20*: "Appropriate GMP concepts should be applied in the production of APIs for use in clinical trials with a suitable mechanism of approval of each batch."
- *Section 19.70*: "Changes are expected during development, as knowledge is gained, and the production is scaled up. Every change in the production, specifications, or test procedures should be adequately recorded."

Section 19.11 means the QMS and its controls should be aligned with the stage of development of the product. This translates to flexibility in terms of how rigorous the overall QMS needs to be with respect to API and downstream nodes of the supply chain by association. Section 19.20 establishes that GMP controls need to be applied to API production and batch release. Finally, Section 19.70 states that change is a fundamental aspect of clinical development and implies that changes made in clinical development do not need to be controlled to the same extent as post-marketing changes.

ICH Q8 – Pharmaceutical Development[10]: ICH Q8 lays out suggested content for the DP pharmaceutical development section within the regulatory application for the drug. The guideline covers components of the DP, including the DS and excipients; the manufacturing process of the DP; and the container closure system. While clinical supply chain is not usually responsible for this content, it does impact product availability. This topic is fully explored in Chapter 5 – Regulatory Considerations. ICH Q8 also discusses the quality target product profile (QTPP) and critical quality attributes (CQA) of the product. Per ICH Q8, CQAs are "... a physical, chemical, biological, or microbiological property or characteristic that should be within an appropriate limit, range, or distribution to ensure the desired product quality."

ICH Q10 – Pharmaceutical Quality System[11]: ICH Q10 recommends a comprehensive, phase-appropriate approach to QMS development and implementation. ICH Q10 deserves special attention because its recommendations are designed to facilitate regulatory approval of the product, establish a state of control over the product, and facilitate continual improvement of the product. ICH Q10 attempts to bring together the concepts in ICH Q7, ICH Q8, and ICH Q9 – Quality Risk Management and describes how to apply them across the four stages of a drug's lifecycle: pharmaceutical (clinical) development, technology transfer, commercial manufacturing, and product discontinuation. There are a few recommendations that are applicable across all stages:

- *Management Responsibilities*: Discusses the importance of senior leadership establishing a quality policy and ensuring the quality strategy is well communicated throughout the company.
- *Management of Outsourced Activities and Purchased Materials*: Confirms that quality oversight extends to contract partners who provide manufacturing services or materials on behalf of the sponsor.

- *Pharmaceutical Quality System Elements*: Lists four elements that enable a state of control and continual improvement, process performance and product quality monitoring system, corrective action and preventive action (CAPA) system, change management system, and management review of process performance and product quality.

So far, we've looked at the scope of cGMP (CFR Parts 210 and 211), how to approach cGMP management (the "five Ps"), and recommendations for how to interpret and apply GMPs in the context of DS, DP, and the pharmaceutical quality system (ICH guidelines), but we have not discussed how GMP requirements translate into the specific procedures that make up the QMS. To do this, it makes most sense to look at the QMS within the context of the supply chain processes the quality function supports. There are 14 quality processes spanning the four dimensions of the supply chain introduced in Chapter 1 – Supply Chain Management:

- *Sourcing*: Quality must ensure that every third-party supplier of materials, manufacturing services, analytical services, or distribution services has been identified (1. Supply chain Identification) and qualified for use (2. Supplier qualification and 3. Quality agreements) before services are rendered.
- *Make*: Quality must ensure the manufacturing process has been documented prior to production start (4. Master batch records (MBRs)) and after every manufacturing event (5. Batch release and 6. Quality review board). Quality must also review and approve the product-specific specifications (7. Specifications) that enable batch release and any issues that occur during the manufacturing or distribution process (8. Deviations and CAPAs). Within packaging and labeling specifically, quality oversees clinical label generation (9. Clinical labels) and shelf-life assignment to FP batches (10. Stability and expiry period assignment).
- *Deliver and Return*: Quality must review and approve shipments (11. GMP shipments) between manufacturing sites if the batch has not been fully released. Quality must review all temperature excursions (12. Temperature excursion management) and disposition batches impacted by them. In the event of a product quality issue, quality manages the recall (13. Recalls) process.
- *Plan*: Quality owns the change management (14. Change management) process that must be executed each time a change is made to the manufacturing process, analytical strategy, or supply chain supporting the process. Quality also provides some of the most important inputs to the demand and supply planning process.

Finally, the quality unit must implement an annual review of quality processes and procedures. This process is called the quality measurement review (QMR), and it's an opportunity for the quality unit to share how well the QMS is performing with respect to compliance and resource management. QMR topics include audits and audit findings, batch releases, CAPAs, change controls, deviations, nonconformance events, and training compliance. Within the Quality System Measurement Review (QSMR) process, clinical supply chain may be responsible for providing data and trends on GMP shipments, temperature excursions, and recalls, if applicable.

To summarize, there are numerous processes and procedures where the clinical supply chain and quality functions must work together. As clinical development progresses from early stage to late stage, the QMS grows and evolves, too. For a company in early-stage development, a handful of people in quality may oversee the entire QMS. By the time a company has commercialized its first product, quality usually divides into several subfunctions with more specific responsibilities. This means a clinical planner may not be directly involved in all, or even a few, of these processes, but they need to be aware of the impact each of the processes has on the supply chain.

4.3 Source: Supplier Identification and Qualification

Recall that biopharmaceutical companies tend to rely heavily on contract organizations (contract development and manufacturing organizations (CDMOs)) to perform manufacturing and testing for DS, DP, packaging, labeling, and distribution. Each of these activities fall under the scope of GMP, and therefore every supplier who is brought on to perform work on behalf of the sponsor must be qualified by the sponsor's quality department. Qualifying a supplier for GMP work means assessing the capabilities of that supplier within the context of GMP requirements. Per ICH Q9, this is a risk-based assessment that sometimes results in an on-site audit in which the sponsor company sends quality representatives to inspect the supplier in question over the course of a two- to three-day period.

Conducting this risk-based assessment gives the sponsor a chance to examine every aspect of the supplier's services and QMS. The scope of an audit for a CDMO will touch on personnel, training, facilities, equipment, validation, and all the procedures that are used to control the work that takes place at that site. A tour of the facility usually takes place as well as an opportunity to witness the activities that are under the scope of the audit. At the audit's conclusion, the sponsor notes any observations or issues it sees with adherence to procedures or compliance with GMP requirements and files these in a formal audit report that is provided to the company under audit. Once the audit report is completed and any responses and corrections are appropriate and implemented, the company is considered qualified for use.

In addition to the audit, a quality agreement (QAG) needs to be established and agreed to between the sponsor and CDMO. The QAG is a legally binding document that spells out the roles and responsibilities that each party must uphold with respect to GMP compliance. The QAG includes commitments from both parties on activities that are important to managing the supply chain. Examples of this include lead time commitments for provision of stability data and documenting responsibilities for inbound and outbound shipments from the supplier.

Over time, sponsors that rely on third parties build a pool of CDMOs that are qualified to perform DS, DP, analytical, packaging, and labeling steps. When new products move forward into clinical trials, the chemistry, manufacturing, and controls (CMC) team is responsible for identifying who will perform manufacturing and testing activities for the product. Options may include in-house capabilities (e.g. internal manufacturing sites), third-party suppliers who have already been qualified, or new suppliers altogether. Given the complexity of biopharmaceutical supply chains, quality needs clarity from the CMC team on exactly who will do what, so they have an opportunity to review the qualification status of the selected sites. Without that review, it's possible for a supplier to slip through the qualification process. This is problematic because if a supplier has major observations during the audit, it may not be possible to qualify them which puts the entire supply chain for the product at risk.

To mitigate this, the clinical supply chain function needs to identify and document the entire GMP supply chain for the quality unit. Names for this document include supply chain map, supply chain diagram, or supply chain flowchart. No matter the name, the business process should produce a document that lists

- Clinical trial protocol number.
- All interventional products that will be used in the study.
- Every GMP manufacturing, testing, and storage site that will be used to produce and release the interventional products.
- An assessment by quality of the qualification status of each supplier listed.

The supply chain map should be given to the quality unit so they have visibility to a comprehensive summary of all the GMP sites that will be used to support the clinical study. The timing of this is important because it can take months to schedule and execute an audit for a new supplier. Besides giving quality notice, the supply chain map has the added benefit of providing the necessary content for two sections of the regulatory submission for the product: S.2.1 – DS Manufacturing and Testing Sites and P.3.1 – DP Manufacturing and Testing Sites. When changes are made to the GMP supply chain, such as adding a new manufacturing site or a new formulation to the study, the supply chain map should be updated to reflect those changes.

4.4 Make: Batch Records, Sampling, Batch Release, Deviations, and Disposition

GMP regulations mandate that all biopharmaceutical manufacturing processes be documented and controlled to help ensure product quality. To meet cGMP requirements, several activities are needed prior to production start, during the production event itself, and especially after production concludes. Figure 4.2 provides a summary view of these processes and their relationship to each other.

Before production begins, MBRs and specifications must be created and test methods must be validated. The MBR describes the production process and specific steps and actions taken within that process as well as the in-process controls and sampling requirements of the process. During and after production, samples are taken for analysis to check the final batch for adherence to the specification. If the product does not have an out-of-specification (OOS) result and no out-of-trend (OOT) results are observed, a certificate of analysis (CoA) is issued stating that the batch has met the thresholds contained within the specification.

Meanwhile, the technical lead responsible for the batch in question and the quality representative conduct a thorough review of the executed batch record (EBR) to check for any issues or problems that arose during manufacturing. EBRs for DS and DP manufacturing can range from hundreds to thousands of pages so the review process is time consuming. This additional lead time makes biopharmaceutical products uniquely challenging from a supply predictability perspective, and this problem is further exacerbated when deviations happen.

During the EBR review, quality checks the batch record for completeness and accuracy, that it contains all applicable in-process and final test results, and that it has been signed by the supplier.

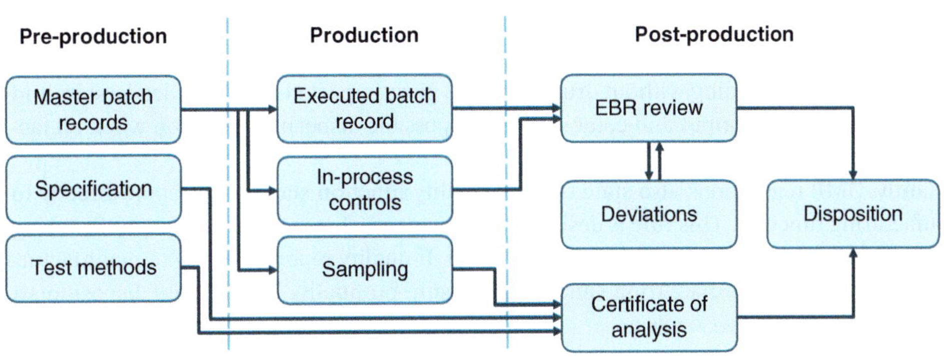

Figure 4.2 Quality Processes in Biopharmaceutical Manufacturing

For labeled products, the shelf-life assignment must also be verified against the stability profile of the underlying DP. When an outsourced supplier performs the manufacturing activity, the supplier's quality unit must perform their own EBR review. Under certain conditions, a sponsor company can accept the supplier's review and disposition in lieu of performing a complete EBR review of its own.

If no problems are found during the EBR review and the CoA has been issued, the quality unit dispositions the batch for further use. Batch disposition is the final step of the manufacturing process from a quality perspective, and there are three categories: "released," "conditionally released," and "rejected." To release a batch means to authorize the use of that batch in the next step of the supply chain, so a DS batch can be used in DP manufacturing, a DP batch in packaging, and so on. When the FP is released, that means the batch has been approved for consumption in a clinical trial. Conditional release means allowing the DS, DP, or FP to move forward into the next level of the supply chain while a final test result or other item required to disposition the batch is pending. Conditional release is rarely used and only if the condition is considered to have minimal risk to product quality or patient safety. Once the condition is met, the batch is moved to "released" status. A "rejected" batch status means that the batch cannot be used in a clinical trial.

If a deviation is discovered that impacts product quality or if the batch fails to meet the specification, the batch is rejected and cannot be used. Deviations typically fall into three different categories that are distinguished by their potential impact to product quality and patient safety. Minor deviations are unlikely to impact a quality attribute. Major deviations are likely to impact a quality attribute. Critical deviations will impact a quality attribute. All deviations require investigation to assess their impact before a batch that's experienced a deviation can be released for use. When deviations happen, quality and clinical supply chain need to communicate with each other on the potential impact the deviation will have on resolution timing and potential availability of the batch.

Deviations potentially trigger two separate actions. First, a CAPA may be initiated. CAPAs live up to their namesake in that they describe actions required to correct the immediate deviation and recommendations for preventing future instances of the deviation. Like deviations, clinical supply chain and quality partner where actions are required from the clinical supply function to support CAPA implementation and closure. Second, for deviations that are major or critical and that possibly impact product quality, a quality review board (QRB) meeting may be called. The QRB is a key piece of the QMS because it creates a cross-functional forum to review and assess deviations and product nonconformances. Quality has the right to request a QRB at any point during the batch release process, and they are responsible for preparing materials related to the QRB with input from technical leads and/or clinical supply chain as needed. Senior management is also asked to attend and weigh in on the risk–benefit profile of rejecting or releasing the batch. The quality unit – typically the head of the quality function – is always the final decision-maker within a QRB. It's not hard to imagine the pressure a quality organization may feel to release a batch if rejection leads to a patient going without drug. The QRB gives impacted functions, leadership, and quality a chance to provide input and come to the best possible disposition decision when all factors are considered.

Importantly, GMP regulations also state that the quality function should not report directly to the manufacturing function. This rule is designed to prevent conflicts of interest if manufacturing and quality disagree on the disposition status of a batch. If quality reports directly to manufacturing, there is a risk that manufacturing can apply pressure on quality to favorably disposition a batch that they would otherwise reject.

Unlike commoditized consumer goods which are typically manufactured and then shipped immediately to the next node in the supply chain, DS and DP lots must undergo a rigorous analysis before disposition. There is a strategy sponsor companies may employ to reduce total lead time, which is manufacturing or packaging "at risk" or "under quarantine." These terms mean DS may proceed to DP manufacturing or DP may proceed to packaging and labeling production before a full review of analytical results and batch records have been completed. The FP batch must then be held in a quarantine status, meaning it is not dispositioned for release to a clinical site, until the upstream DS and DP batches have been reviewed and released. At-risk manufacturing reduces the total lead time of the supply chain by enabling concurrent downstream production and upstream analytical/quality review. But, if the product fails its specification or there are issues with documentation, the money spent on packaging and/or labeling is wasted. The best strategy is to avoid using an "at-risk" or "under-quarantine" process unless absolutely necessary to get the product to the clinic for subject dosing.

There is a second "at-risk" manufacturing activity – labeling product with a shelf-life date that is not yet supported by the underlying stability data and then gating the final release of the FP batch to the availability of the supportive stability data. This strategy is useful when on-hand inventory is tight and there may not be time to wait for the stability data to come in before the final packaging and labeling step. Like manufacturing DS or DP at risk, this strategy saves weeks of time because labeling takes place ahead of the expiry extension. And, the primary risk is the same: if the stability data does not support the assigned shelf life, the batch cannot be dispositioned for use and must be relabeled with the appropriate shelf-life assignment.

Within the packaging and labeling manufacturing processes, which clinical supply chain typically owns, quality plays an important role in preproduction readiness and postproduction release. Prior to manufacturing, quality not only reviews the MBR but also checks the final clinical label proof against the original master text for accuracy and is responsible for verifying the shelf life assigned to specific batches. After production is over, quality limits release of the product to countries that have already approved the sponsor's regulatory application. If a country has not approved the application, the batch cannot be shipped to that country.

4.5 Deliver: Good Distribution Practice, Temperature Excursions, Returns, and Recalls

The World Health Organization published a detailed guideline on GDPs in 2021.[12] Remember that while GDP was initially developed for commercial products, its applicability extends to clinical products. GDP encompasses a broad scope of activity that sounds familiar in the context of the five Ps discussed earlier in this chapter. The guideline covers personnel and training requirements, the quality system governing distribution, warehouses and storage unit expectations, vehicles and equipment, complaints, recalls, and returns. There are several notable takeaways from this document:

- All parties involved in the distribution of pharmaceuticals including sponsors, manufacturers, and logistics providers fall within the scope of GDP. All parties need to have the appropriate controls in place to ensure that product quality is maintained by avoiding product mix-ups or contamination.
- A quality system should be implemented within each location where the product is stored or transported.
- At sites that store quarantined material, quarantine areas must be clearly marked, and access restricted to appropriate personnel.

- Systems need to be in place to ensure that a first expired, first out (FEFO) distribution methodology is followed to minimize drug waste. However, pharmaceutical products should not be distributed so close to their expiry date that the product expires by the time it reaches a patient.
- Temperature monitoring and control is always required.
- Repackaging and relabeling are allowed at distribution or storage sites only if that site has been qualified to perform that GMP operation.

GDP requires temperature monitoring of GMP products through the supply chain. A product must be stored at its required condition, and this introduces the possibility of an excursion when a shipment's temperature range exceeds or falls below the product's required storage condition. Excursions can impact a batch at any time, either during storage of the product or during a shipment. When an excursion happens, the clinical supply chain function is typically notified of the occurrence. The temperature data from the shipment must then be provided to analytical and quality to determine the impact of the excursion to make a disposition decision. Stability data collected per ICH Q1 guidelines forms the basis for making these disposition decisions. If the excursion is extreme or no data exists, the entire shipment may be lost at a cost of thousands or even millions of dollars. This risk is a major reason why stability studies are run on DS and DP at accelerated and/or stressed (e.g. high-temperature and high-humidity) conditions. These studies provide the data required to inform a disposition decision on these batches.

FP temperature excursions are found when a distribution depot or a clinical site receives the product. Whereas DS and DP excursions are worrisome because they impact high-value batches, FP excursions are concerning because they are time sensitive. If a site receives a batch with an excursion and the batch is rejected for use, that site may not have enough drug to support upcoming patient needs. This situation results in a scramble to get more FP to the impacted site as soon as possible, a position no clinical planner wants to find themselves in.

Another aspect of GDP where clinical supply chain is heavily involved is FP returns and product recalls. Here, the guidelines refer to having "control" over the product and its supply chain. One way to demonstrate this control is through Investigational Medicinal Product (IMP) accountability, which takes place at the end of a clinical trial. Returns are very common in clinical trial management because not all clinical sites have IMP destruction capabilities, and it's important that a noncommercial drug not be left out for accidental consumption. Therefore, IMP needs to be returned to the clinical depot from whence it shipped at the conclusion of the trial. The clinical supply chain function often partners with clinical operations to execute this process and document every kit of IMP that's been returned (or destroyed).

Product recalls are a less common but much more serious event. Recalls happen if a product quality issue is discovered with a product or batch after it has been released for use. Product quality issues may stem from emerging safety data or through a defective batch. Moderna suffered this fate in 2022, and they were forced to recall multiple batches of COVID-19 vaccine after foreign substances were discovered in the vials of the product.[13] When a recall is needed for a clinical product or batch, clinical supply chain is at the center of the action because they own the supply plan and can identify where the impacted product is stored within the supply chain. Supply chain needs to provide a complete list of batches and locations to clinical operations and quality so the product can immediately be quarantined, and the clinical planner needs to quickly pivot to a new resupply strategy for the study. Given the need for speedy and decisive action across multiple functions, sponsors are required to have a mock recall procedure in place to ensure organizational readiness for such an event.

4.6 Plan: Planning Inputs and Change Management

There are two areas where the QMS impacts the demand and supply planning process. First, quality is responsible for some of the assumptions and inputs that are used in the supply plan. These include batch release lead times, batch disposition status, batch-specific sampling requirements, and stability study execution. The most important quality input to the planning process is the batch release lead time assumptions. These lead times are particularly important because they dictate when manufacturing of an upstream step needs to begin to have that material ready for the next step in the supply chain. Batch release lead times are also multidimensional and challenging to predict on a batch-by-batch basis. There are a handful of key dates to keep in mind when determining batch release lead times:

- *Date of Manufacture*: The production start date.
- *Date EBR Available for Review*: The date the EBR is made available for review. If an internal manufacturing site is used, this may be available in a few days, but if an external manufacturing site is used, these may take weeks to produce.
- *Date CoA Available for Review*: The date the final CoA is issued which is a requirement to disposition the batch for use.
- *Further Processing Date*: If the product can be sent under quarantine for further processing (e.g. "at risk"), this is the date when that next production event can begin or when the product can be shipped to the next downstream manufacturer in the supply chain.
- *Batch Release Date*: The date the quality organization formally releases the batch for use.

With those five dates in mind, there are two potential batch release lead times that impact the supply plan. The first is the further processing lead time which is the lead time between the date of manufacture and the further processing date. This lead time is equal to the total time to manufacture the batch plus the total time for EBR availability and initial review for any major or critical deviations. The second is the batch release lead time which is the lead time between the date of manufacture and the formal batch release date. This lead time is equal to the total lead time to manufacture the batch plus the total lead time to perform all batch release activities.

While the manufacturing lead time is relatively stable, there will always be minor variations in the further processing lead time and greater variation in the batch release lead time. This is especially true for DS whose manufacturing process is the most complex within the supply chain. Theoretically, DP has more consistency with respect to lead times, but variations occur at this stage too, especially for biologics. OOS and OOT results are often culprits here. Due to these lead time inconsistencies, it's especially important for clinical supply chain and quality to communicate on a regular basis throughout the EBR review process to ensure alignment on when a specific batch will be dispositioned to keep the supply chain moving.

Closely related to the batch release lead time is the batch disposition status. Quality is responsible for performing the disposition decision and just as importantly, they need to communicate that decision broadly within the organization. This seems like a simple step, and it usually is for DS and DP. The story gets more complicated at the FP level, where the disposition status needs to be documented not only on a batch-specific basis, but on a country-specific basis as well. More to come on this topic in Chapter 5 – Regulatory Considerations.

A third key input into the planning process is the batch sampling requirements. During and after each manufacturing event, samples are taken for a variety of reasons including in-process controls, analytical testing, and stability studies. Sampling requirements reduce the amount of a batch available for processing in the next step of the supply chain. If these requirements aren't accounted for, the clinical planner ends up with a lower-than-expected delivery against their supply plan which may jeopardize downstream inventory positions. The clinical planner needs to work closely with the technical lead responsible for production, QC, and quality to document the sampling requirements and incorporate them into their plan.

The final direct inputs from quality are the assigned retest periods for DS and DP and the shelf-life assignment for FP. This is a critical space for clinical supply chain, QC, and quality to work together because expiry extensions are almost always performed while the trial is taking place. The timing of expiry extension issuance, the newly assigned expiry period, and any potential risks to expiry extension issuance all need to be proactively managed between these functional areas. This collaboration enables the clinical planner to develop an optimal packaging and labeling strategy. These expiry management strategies are expanded upon in Chapter 10 – Finished Product Expiry Planning.

The second area where the QMS impacts demand and supply planning, albeit indirectly, is change management. In this context, changes are updates to the manufacturing process or the supply chain of a product. For commercial products, formal change control is a key component of GMP and is a regulatory requirement to ensure changes to biopharmaceutical manufacturing processes are implemented in a controlled manner that reduces risk to product quality and patient safety. Clinical products require a modified version of the change control process due to the nature of manufacturing process development for clinical stage drugs. As we'll cover in Chapter 5 – Regulatory Considerations, both clinical and commercial products are bound to follow the manufacturing processes covered within their respective regulatory applications.

Figure 4.3 shows a sequential commercial change control process where a potential change to a product's GMP supply chain, manufacturing process, or analytical strategy is identified and recommended for assessment. The proposed change is then assessed for its potential impact on product, processes, procedures, personnel, premises, and the regulatory applications that are filed for that product. Once the impact is assessed and documented, an implementation plan is made covering all the activities, resources, and costs associated with the proposed change. Only then is the change formally evaluated and a decision is made on whether to proceed with implementation. Assuming the recommended change is approved, implementation begins, and all documented actions in the implementation plan are monitored through completion. The sequential nature of the process enables a thorough understanding of the change's risks, costs, and benefits, and that the entire process is controlled from start to finish.

This strict, sequential approach does not work well for managing CMC changes in clinical development because change happens so broadly and frequently. Trying to control every proposed change in this manner would quickly overwhelm a change control system. Fortunately, ICH Q10 recognizes "change as an inherent part of the development process" and recommends that the change management process be "consistent with the stage of pharmaceutical development." This statement leaves room for interpretation, so sponsors are allowed to implement a risk-based, phase-appropriate change management process.

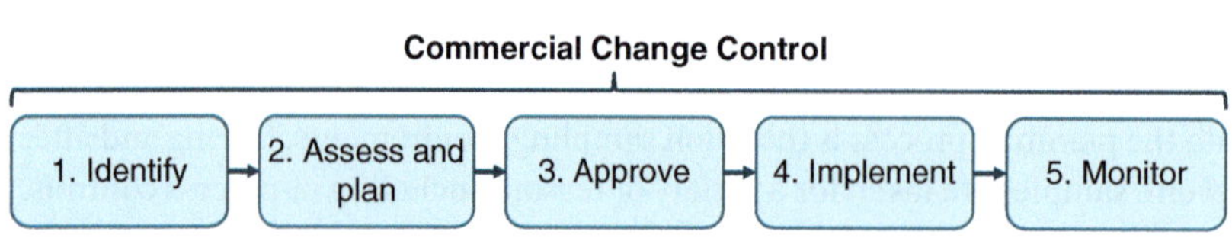

Figure 4.3 Commercial Change Control Process

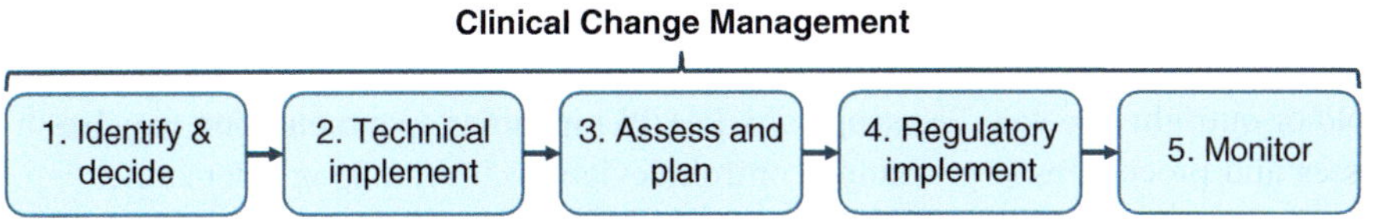

Figure 4.4 Clinical Change Management Process

In early-stage development (e.g. Phase I clinical trials), simple documentation of the change is usually sufficient.

But when a product enters late-stage development, a more robust change management process is appropriate. This is because a litany of CMC information needs to be submitted within the regulatory applications for every country that participates in a clinical trial. When CMC content in the regulatory application changes, the sponsor is obligated to notify each regulatory body so they can review the changes. For programs in early-stage development, there may only be one regulatory application, and the first batch of DS or DP may be enough to supply the entire trial, so CMC changes don't impact that application. Recall that late-stage trials often take place in many regions and countries so when CMC changes are planned and implemented, they also need to be assessed for potential impact to each country's regulatory application. Keeping this process well organized and clearly documented is why a strong change management process matters.

Figure 4.4 shows the clinical change management process where there are two distinct implementation steps – a technical implementation followed by a regulatory implementation. This is in contrast to the commercial change control where regulatory implementation typically precedes technical implementation. A well-designed clinical change management process allows for technical implementation prior to regulatory implementation to enable speed in process development. That said, the clinical change management process also requires assessment and monitoring of regulatory implementation to ensure compliance with regulatory applications.

Change management impacts the clinical supply chain because regulatory approval gates the introduction of supply that was manufactured using the processes contained within a clinical trial application. If proposed changes have not been approved by a particular country's competent health authority, then supply made using those changes is not eligible to be shipped into that country. The regulatory implementation strategy needs to be incorporated into the supply plan to prevent noncompliant material from being shipped into a country and to prevent the risk of a stockout if compliant material is unavailable for a particular country.

4.7 Preapproval Inspection

Another aspect of GMP adherence and QMS development warrants discussion, the preapproval inspection (PAI) or prelicensing inspection (PLI). In the United States, GMP regulations vest the FDA with the right to inspect and audit companies that submit a marketing application for a drug. PAI/PLIs may be conducted at the site of DS manufacturing, site of DP manufacturing, analytical release testing sites, primary packaging sites, and/or at the applicant themselves. The FDA's goal during the PAI/PLI is threefold. First, they seek to ensure the applicant and the GMP supply chain for the product are ready to meet GMP commercial requirements. Second, the FDA examines the applicant's conformance to their marketing application. Third, the FDA seeks to confirm the data submitted within the application is complete, accurate, and of high integrity.[14]

During a PAI, the sponsor needs to demonstrate they have the appropriate procedures and controls in place for all activities within the scope of GMP. If they fail, their commercial application can be placed on hold or outright rejected. The scope of a PAI/PLI includes documentation (e.g. batch records), processes and procedures (e.g. change control, deviations, CAPAs), and personnel (e.g. qualifications and training compliance). The FDA provides guidance on how to prepare for a successful PAI/PLI, and this contains useful advice for clinical supply chain and quality. From the FDA, the guidance is "Quality and Operations work together to investigate deviations/issues ... Quality and Operations work together to best present significant issues during inspections."

To expand on this, quality and operations (specifically clinical supply chain) need to work together to develop and implement the QMS. This work begins in early-stage development and continues through commercialization. While quality owns the QMS, clinical supply chain plays a fundamental role in QMS implementation and execution. Creating a successful partnership between the two functions is critical for inspection readiness and to effective supply chain management.

4.8 The Qualified Person

There is another quality stakeholder within the biopharmaceutical universe who deserves special consideration given their unique impact on the clinical supply chain, the qualified person. Colloquially known as the QP, this role only exists in the European Union, the United Kingdom, and Israel. The QP is responsible for two activities within the supply chain: declaring the GMP supply chain meets EU regulatory requirements and certifying each batch of product for release to a specific trial and country. Essentially, the QP acts as a quality unit on behalf of an entire country or region. QPs are required to go through extensive training and qualification activities, and they must be licensed to perform their duties.

Within the EU and the United Kingdom, the GMP supply chain inclusive of DS, DP, primary and secondary packaging, labeling, and analytical testing sites must meet EU GMP compliance requirements. Sites located in the EU and the United Kingdom are issued a GMP certificate by their local regulatory agency following inspection and demonstration of compliance with GMP regulations. If all the GMP sites within a supply chain are in the EU or the United Kingdom, provision of GMP certificates to the QP is sufficient. A QP declaration is required if one or more GMP sites are located outside the EU or the United Kingdom, and it is a necessary supplement to the regulatory application for the clinical trial.

The QP declaration states that all sites listed in the regulatory filing meet EU GMP standards. To issue a QP declaration, the QP typically requires evidence to support this claim and that evidence comes from audits performed at the sites in question by qualified auditors. Without an audit, the QP will reject the claim and refuse to issue the declaration. If quality has not audited a site that needs to be listed in the QP declaration, this can delay the regulatory application and the start date of the entire trial. This point emphasizes the importance of the supply chain map process outlined in the source section of this chapter. The supply chain map provides quality with a complete description of the proposed GMP supply chain and gives quality time to assess if a supplier needs to be audited to enable QP declaration issuance in advance of the trial. Additionally, the sites listed in the QP declaration application must be the physical location where the manufacturing or testing activity takes place, and these addresses must match those listed in the regulatory application. The supply chain map also plays a key role in providing a source of truth for the regulatory application and the QP declaration.

Table 4.1 QP Certification Documentation Requirements

Batch Documentation	Regulatory Documentation
Executed batch records for DS and DP	Clinical Trial Annex 1 Form (trial regulatory application) and cover letter
Deviations	Current IMPD (quality regulatory application)
Certificates of analysis for DS and DP	Country-specific approvals
Certificates of analysis for excipients and packaging materials	QP declaration
Approved label text	Current audit reports (if applicable)

The second responsibility of the QP is to certify each batch of product intended for use in a clinical trial. This certification states that the batch was manufactured and released in accordance with the approved regulatory application for that trial and meets EU GMP requirements. EU regulations stipulate that the QP takes legal responsibility for the release of each batch. This means that if a product quality issue is discovered after the batch has been released, the QP may be held legally liable for the consequences of that issue. Due to this potential legal exposure, QPs are rigorous in their approach to batch certification and release.

To complete a QP certification, the QP reviews the entire GMP supply chain including DS manufacturing and release, DP manufacturing and release, primary packaging, secondary packaging, and labeling production. In addition to the batch documents, the QP also needs the applicable regulatory documents which are documented in Table 4.1.

The QP analyzes all the listed documents above and has the right to ask questions and request supportive documentation for missing or incomplete information. QPs often focus heavily on deviations, OOS results, and audit report findings that are deemed major or critical. Any issue that could impact product quality or risk patient safety requires explanation or the batch is rejected by the QP for use. QP certifications are also issued on a batch- and country-specific basis. For each new batch that is supplied into the EU or the United Kingdom, the QP must review the documents specific to that batch. And, as regulatory approvals occur, the approval documentation needs to be provided to the QP so they can release the batch (or batches) to those countries. Clinical sites who receive IMP in the EU need to have the relevant QP certificates on hand to dispense IMP to subjects participating in that trial.

In certain situations, a QP may not be qualified or able to certify the entire supply chain. For example, if a clinical site in the EU or the United Kingdom performs the final packaging and labeling step, they may not be able to certify the upstream DS and DP manufacturing steps. In these situations, a QP–QP agreement needs to be established between the sponsor who is responsible for supplying the product, the upstream QP responsible for DS and/or DP certification, and the downstream QP responsible for FP certification. The QP–QP agreement assigns clear roles and responsibilities for each step of the supply chain and needs to be signed by all parties to go into effect.

Given the large number of documents to review and the potential for QP–QP arrangements, the QP certification process usually takes weeks (or months) to complete. The clinical planner needs to allocate lead time in their planning assumptions for the QP certification process. This is important for two reasons. First, it's common for the clinical planner to reside in a geographically distant location from continental Europe or the United Kingdom. This limits the amount of time to address queries which can extend the lead time further. Second, in smaller biopharmaceutical organizations, QPs are often contracted from third-party organizations to provide their services. This means they are prioritizing activities across multiple clients and may not be able to respond to queries or issues right away. For this reason, commercial-stage biopharmaceuticals often hire their own QPs to ensure batch certifications are prioritized appropriately.

4.9 Key Takeaways

1) GMPs are a legal framework designed to ensure the quality of a product. The quality function is responsible for monitoring adherence to cGMP regulations, and they do so through employment of a QMS. The QMS is composed of a broad set of SOPs that touch every aspect of the GMP supply chain.
2) Compliance with cGMP requirements demands a holistic approach to managing the clinical supply chain. One paradigm through which to segment cGMP compliance is the five Ps: people, products, process, procedures, and premises. A set of international guidelines published by the International Council for Harmonisation (ICH) provides additional input on how to meet cGMP. ICH Q7 (Good Manufacturing Practice for Active Pharmaceutical Ingredients), ICH Q8 (Pharmaceutical Development), and ICH Q10 (Pharmaceutical Quality System) are the most pertinent to the supply chain.
3) Within the sourcing step of the supply chain, supplier identification, qualification, and audits are a core element of the QMS. To ensure quality has visibility to the supply chain and adequate time to perform supplier qualification activities, the clinical supply chain function should provide trial-specific supply chain maps that outline all the GMP manufacturing and testing sites used in support of that trial.
4) Numerous QMS procedures dictate oversight of all cGMP manufacturing and analytical testing activities. The processes include MBR generation, specification development, test method validation, EBR review, deviations, and CAPAs. The manufacturing process culminates in quality issuing a disposition decision to approve or reject a batch of material.
5) Packaging at risk is a potential strategy used in clinical development and takes two forms. The first form is to package DP that has not been fully released and quarantine the FP until all upstream production has completed. The second form is to label product with a shelf-life that is not yet supported by the underlying stability data.
6) Distribution activities are governed by GDP regulations, and these behave similarly to cGMPs. FP must be held at its prescribed storage condition through the entire supply chain. If there is a product quality issue, then a recall may be issued. In these situations, clinical supply chain and quality must work closely together to identify where impacted batches are in the supply chain and block their distribution.
7) The QMS impacts the demand and supply plan in two ways. First, quality owns or provides multiple inputs to the plan itself, including further processing lead time and batch release lead time. Second, quality runs the change management process. When changes are substantial, they require regulatory amendments, and supply is gated to the approval of these changes.
8) The QP is a central figure to cGMP adherence in the European Union, the United Kingdom, and Israel. The QP is responsible for issuing a QP declaration which states that all manufacturing and testing sites used outside the EU meet EU GMP requirements. The QP also issues batch certifications that state a batch meets EU GMP standards and is aligned to the regulatory application under which the product was approved.

Notes

1 https://www.fda.gov/drugs/development-approval-process-drugs/pharmaceutical-quality-resources.
2 https://www.accessdata.fda.gov/scripts/cdrh/cfdocs/cfcfr/CFRSearch.cfm?CFRPart=211.
3 https://health.ec.europa.eu/medicinal-products/eudralex/eudralex-volume-4_en.
4 https://www.pmda.go.jp/english/review-services/gmp-qms-gctp/0001.html.
5 http://english.nmpa.gov.cn/2019-07/25/c_390613.htm.
6 https://www.ich.org/page/quality-guidelines.
7 https://database.ich.org/sites/default/files/Q1A%28R2%29%20Guideline.pdf.
8 https://database.ich.org/sites/default/files/Q6A%20Guideline.pdf.
9 https://database.ich.org/sites/default/files/Q7%20Guideline.pdf.
10 https://database.ich.org/sites/default/files/Q8%28R2%29%20Guideline.pdf.
11 https://database.ich.org/sites/default/files/Q10%20Guideline.pdf.
12 https://www.fda.gov.ph/wp-content/uploads/2021/03/World-Health-Organization-Good-Distribution-Practices.pdf.
13 https://www.fiercepharma.com/pharma/moderna-recalls-covid-19-vaccine-batch-over-foreign-body-found-cdmo-made-vial-again.
14 https://www.fda.gov/files/drugs/published/FDA%E2%80%99s-Pre-Approval-Inspection-(PAI)-Program-and-How-to-prepare-for-a-successful-outcome.pdf.

5

Regulatory Considerations

5.1 Introduction to Regulatory Applications

Regulatory affairs is another cornerstone of clinical trial supply chain management. In basic terms, a clinical trial application (CTA) describing the study, the drug, and supporting data needs to be approved before the trial can begin and drug can be imported to the clinical trial site. Oftentimes, clinical planners' interactions with their regulatory counterparts are limited to a notification of approval to proceed. However, managing regulatory activity is a deceptively complicated endeavor. Not only does the regulatory application need to be approved, but the batch documentation for each lot of material supplied for that country needs to match the application as well. This chapter unpacks this reality and shows how clinical supply chain (CSC), quality, and regulatory form a unified trinity to ensure product movement through the supply chain efficiently and in compliance with country-specific regulations.

Submitting the first regulatory application for a new investigational drug takes years of work and represents the culmination of the preclinical development process. To achieve this milestone, the sponsor authors a CTA. In the United States, this application is called an investigational new drug (IND) application, but in all other countries it's known as a CTA. The IND/CTA contains a wealth of data and plans about the clinical trial(s), the drug, and the justification for dosing in human subjects. The completed application is submitted to the national competent health authority (CHA) and to a human subjects review committee (e.g. ethics committee (EC) or investigational review board (IRB)) within the country or countries where the clinical trial will take place. The submission is only the first step toward obtaining approval though. During the application review, the HA and/or review committees have an opportunity to ask clarifying questions and provide feedback. The sponsor is then obligated to respond or make updates to the application, and only after their queries have been answered satisfactorily is approval granted so the trial may proceed.

But, the work in this space doesn't stop with the initial submission and approval. Every time there is a change to an existing protocol or to the chemistry, manufacturing, and controls (CMC) information contained within the application, the CTA needs to be updated via an amendment for review by each participating country's regulatory body. Some amendments are simple administrative changes that can be implemented with a simple notification to the relevant regulatory bodies, but others are far more complex and require prior approval before implementation may proceed. Like the initial application, feedback may be provided during the amendment process, and the sponsor must respond adequately or make changes to the amendment to gain approval for implementation. If a trial has several countries participating, managing the amendment process becomes a large undertaking to

Supply Chain Planning for Clinical Trials – A Practical Guide, First Edition. Ryan Mills.

manage because amendments can overlap and because months may pass between the time when an amendment is prepared to when that amendment is reviewed and approved in all countries.

With this potential complexity in mind, every clinical program needs a regulatory strategy that is guided by the clinical development plan and the CMC strategy. There are key inflection points within the Clinical Development Plan (CDP), such as emerging data from an existing study or a new study altogether, that dictate when a new CTA needs to be prepared or an amendment needs to be introduced. Similarly, the CMC strategy lays out planned changes to the product and its supply chain including transitions to new manufacturing site, introduction of new formulations, or specification updates that necessitate amendments. Even with an optimal plan and regulatory strategy, unforeseen events may occur that lead to unplanned amendments.

Understanding submission content, the application process, amendments, and the regulatory strategy for the program are all important to the clinical planner. First, the CSC function is responsible for some of the content contained within the application itself. Second, the initial application dictates when the clinical trial can start, and these timelines are an important input to the supply plan. Third, regulatory amendment approval is required to introduce supply made from changes contained within the amendment. Fourth, regulatory feedback during the review process can limit which batches of drug can be supplied to a specific country, and the supply plan needs to incorporate these restrictions to ensure compliance.

Through the rest of this chapter, we'll dive into the regulatory application, the difference between INDs and CTAs, and the quality module components that cover the supply chain, manufacturing, and analytical activities. Then, we'll turn to the application and amendment processes as well as potential CHA interactions that can influence the overall program strategy. Next is a discussion of regulatory commitments, common restrictions, and strategies for managing each of these. We'll finish the chapter by introducing the batch and investigational medicinal product (IMP) lifecycles and outlining how process version management (PVM) enables and ensures regulatory control over the CSC.

5.2 Regulatory Application Content and the CTD

The purpose of the regulatory application is to provide sufficient information for regulators to evaluate patient safety and, in later-stage studies, to determine if the clinical trials appropriately measure efficacy. In the United States, regulation 21CFR312 describes each of the sections required in an IND application.[1] Like the laws governing GMP requirements, every country uses slightly different verbiage in the regulations for clinical trial initiation. Fortunately, the International Council for Harmonisation (ICH) addressed these discrepancies when they published Guideline M4: The Common Technical Document (CTD). Figure 5.1 is taken from ICH and shows the CTD framework.

The main purpose of the CTD is to provide a common template for the biopharmaceutical industry to use, no matter which countries will participate in the trial. The CTD was originally published in 2000 and has now been implemented in the United States, European Union, Canada, Japan, China, South Korea, and Brazil among others. The CTD divides the regulatory application for a new drug into five modules:

- *Module 1 [Not Part of the CTD]*: Regional or administrative information that is not part of the CTD.
- *Module 2*: Summary information including quality overall, nonclinical overview, and clinical overview.

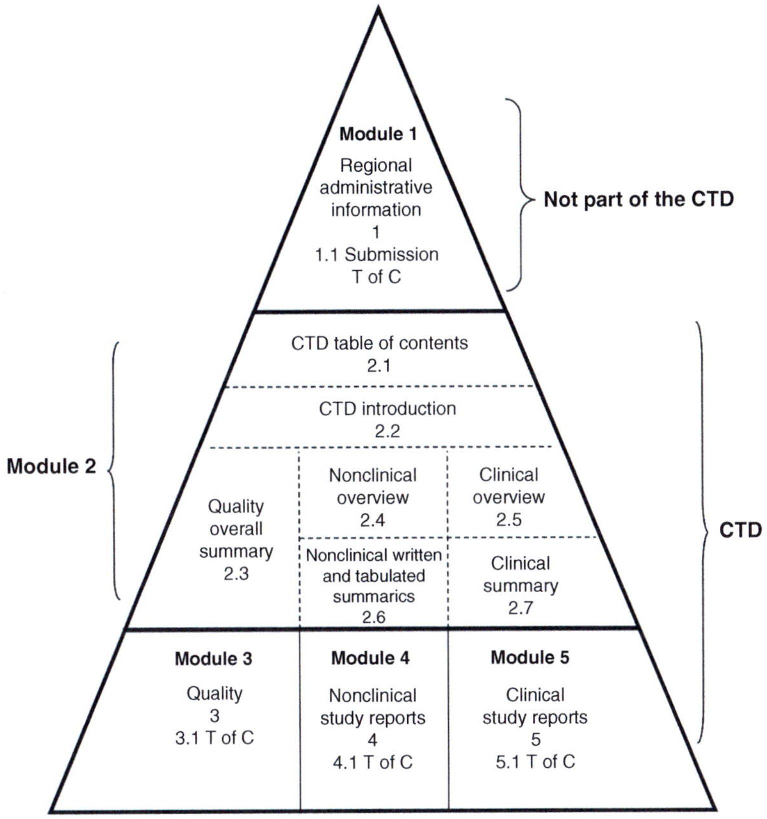

Figure 5.1 The Common Technical Document Framework. *Source:* Reprinted with Permission from the International Council for Harmonisation

- *Module 3*: Quality module describing CMC content.
- *Module 4*: Nonclinical study reports describing animal studies and toxicology profile of the drug.
- *Module 5*: Clinical study reports describing data gleaned from human trials to date.

Module 3 captures CMC information for the drug substance (DS), drug product (DP), and placebos that are in the scope of the application. Within the European Union, the quality module is called the **Investigational Medicinal Product Dossier (IMPD)**. The EU has published extensive guidance on the IMPD and notes that it aligns with the CTD Module 3 structure.[2] Meanwhile, ICH guideline M4Q(R1) describes the specific sections within the module, which are summarized in Table 5.1.[3]

Listing the subsections, 21 in total for DS and 27 for DP, shows the depth and breadth of Module 3 and how many considerations there are to ensuring product quality. The supply chain, manufacturing process, materials, specifications, test methods, batch history, container closure, and stability profile are captured for both DS and DP. Within the DP section, additional detail is required for excipients and for the container closure system of the bulk DP and packaged DP. Individual Module 3 documents are required for each product that will be supplied by the sponsor in the trial, so placebos and comparator products need their own respective sections.

In general, the level of detail required within each section increases as the product moves through the phases of development. One strategy many sponsors pursue is to provide minimal

Table 5.1 CTD Module 3 Sections

Section 3.2.S.X.X – Drug Substance	Section 3.2.P.X.X – Drug Product
• S.1.1 – Nomenclature • S.1.2 – Structure • S.1.3 – General properties • S.2.1 – Manufacturer(s) • S.2.2 – Manufacturing process and process controls • S.2.3 – Control of materials • S.2.4 – Control of critical steps and intermediates • S.2.5 – Process validation and/or evaluation (Biologics) • S.2.6 – Manufacturing process development • S.3.1 – Elucidation of structure and other characteristics • S.3.2 – Impurities • S.4.1 – Specification • S.4.2 – Analytical procedures • S.4.3 – Validation of procedures • S.4.4 – Batch analyses • S.4.5 – Justification of specification • S.5 – Reference standards or materials • S.6 – Container closure system • S.7.1 – Stability summary and conclusions • S.7.2 – Post-approval stability protocol and commitments • S.7.3 – Stability data	• P.1 – Description and composition • P.2.1 – Components of the drug product • P.2.2 – Drug product • P.2.3 – Manufacturing process development • P.2.4 – Container closure system • P.2.5 – Microbiological attributes • P.2.6 – Compatibility • P.3.1 – Manufacturer(s) • P.3.2 – Batch formula • P.3.3 – Manufacturing process and process controls • P.3.4 – Control of critical steps and intermediates • P.3.5 – Process validation and/or evaluation • P.4.1 – Excipients: specifications • P.4.2 – Excipients: analytical procedures • P.4.3 – Excipients: Validation of procedures • P.4.4 – Excipients: Justification of specifications • P.5.1 – Drug product: Specifications • P.5.2 – Drug product: Analytical procedures • P.5.3 – Drug product: Validation of procedures • P.5.4 – Drug product: Batch analyses • P.5.5 – Characterization of impurities • P.5.6 – Justification of specifications • P.6 – Reference standards or materials • P.7 – Container closure system • P.8.1 – Stability data • P.8.2 – Post-approval stability protocol and commitments • P.8.3 – Stability data

process information and wide specification ranges in early-phase development. This is because regulatory bodies recognize that little is known about the product before Phase I and that process improvements will be made. Providing broad information also helps to limit the number of amendments as changes are inevitably made to the manufacturing process and specifications. From an agency perspective, the Food and Drug Administration (FDA) has provided written guidance to Module 3 while the European Medicines Agency (EMA) references ICH M4Q(R1).[4]

While the United States adheres to the CTD structure, there are notable differences between the IND application and the CTA application structure used in the rest of the world.[5] The initial IND application for a program, US CFR312 requires the inclusion of supplemental documents such as the **Investigators Brochure (IB)**, which contains a summary of all clinical and nonclinical data gathered on the product at the time of filing. CTAs do require country-specific documents, but these aren't usually as extensive as an IND. Another consideration that separates the IND is that it is a living application. After the initial IND application is cleared by the FDA, additional clinical trial protocols can be added to that IND without filing a separate application. All other countries require individual applications for each clinical trial that will be conducted in that country.

5.3 Health Authority Interactions

Prior to filing an IND or CTA, it can be helpful for a sponsor to communicate with the national CHA about ambiguous or potentially controversial aspects of the submission package. Possible topics include the nonclinical or toxicology profile of the drug, elements of the trial design, or clarification on aspects of the quality module. Requesting input prior to application enables the sponsor to incorporate feedback ahead of the review process and improve the chances of application approval. In the United States, there are six types of meetings that sponsors may request with the FDA.[6]

1) *Type A Meeting*: Discussion needed to move a stalled product (e.g. on clinical hold) forward.
2) *Type B Meeting*: Discussion prior to IND, New Drug Application (NDA), or Biologics License Application (BLA).
3) *Type B End Of Phase (EOP) Meeting*: End of Phase I, end of Phase II, or pre-Phase III.
4) *Type C Meeting*: Any other meeting type to review aspects of the clinical development program.
5) *Type D Meeting*: Discussion of a narrow set of issues (one to two focused topics) at key decision points to move a program forward.
6) *INTERACT Meeting*: Intended for novel products and development programs that present unique challenges in early development.

While the EMA does not hold meetings with sponsors, the EMA Scientific Advice Working Party does offer scientific advice and protocol assistance. The sponsor may submit questions about any specific component of the CTA for feedback. This approach is recommended by the EMA for new medicines and in situations when there is high unmet need for the product or international guidance is insufficient to address a specific question.

Interestingly, different regulatory bodies often provide a variety of responses to the same question. This is especially true when the sponsor is proposing something that does not have precedent and/or has an ambiguous risk–benefit trade-off. For example, in diseases with high unmet need (e.g. few or no approved medications on the market), one regulatory body can find a potential toxicological finding to be an acceptable risk while another regulatory body will not find that acceptable. Therefore, in early-stage development, feedback from the FDA, EMA, or an individual country can heavily influence which region the program moves forward. In late-stage development, where it's often necessary to conduct the study in many countries, interacting with each of the major CHA bodies is necessary to ensure the application is harmonized and broadly acceptable. If feedback from an important country is missed and that country requires a significant change to the trial application, that means not moving forward in that country or creating a separate version of the application. One scenario adds time to the development program and the other adds complexity and cost, and neither is a desirable outcome.

5.4 The Clinical Trial Regulation in the European Union

The most significant shift in the global clinical trial regulatory landscape over the past decade is the European Union's transition to the Clinical Trial Regulation (CTR).[7] In 2014, the European Union adopted a new regulatory framework across all EU member states which seeks to streamline and harmonize the regulatory application and approval process. Beginning in 2024, all new clinical trials must follow CTR, and by 2025, all ongoing clinical trials must complete their transition to CTR. The most significant change contained within the CTR is the elimination of country-specific regulatory applications in favor of a single application that's submitted to a centralized EU portal for consideration. There are both pros and cons to this approach as it relates to CSC management.

On the positive side, CTR eliminates a tremendous amount of country-specific complexity. There is only one regulatory application to monitor for approval instead of several countries' worth. This has two significant positive implications. First, the number of potential country-specific restrictions and commitments decreases. Second, the number of qualified person (QP) certifications required to distribute product within the EU also decreases. Both of these are major wins for clinical planners. However, this streamlined process does come at a cost which is an increased amount of time to gain an approval from the EU to start a clinical trial.

While the CTR represents a seismic shift of how CTAs are submitted and processed, the requirements around packaging, labeling, and QP certification were left largely unchanged from the CTD framework. Expiry dating is still required on the outer packaging of all clinical trial material; clinical labels must be applied to primary packaging (PP) and secondary packaging when the primary container can be removed from the outer packaging; and QP declarations and certifications are still necessary. Also keep in mind that as of 2024, the United Kingdom is still not a member of the European Union and has its own regulatory framework to navigate.

5.5 Application Authoring, Submission, and Approval

Figure 5.2 diagrams the IND/CTA application process flow.

The application process begins with authoring, an intensive effort for numerous reasons not least of which is the sheer amount of content that needs to be generated. An early-phase IMPD for a single DS–DP combination can reach a few hundred pages of information while late-phase filings can stretch to thousands of pages once the process development, batch histories, and stability data profiles are all included. If DS and DP manufacturing and testing are insourced, the DS, DP, and analytical leads will author their respective sections. But, if those activities are outsourced, the Contract Development and Manufacturing Organization (CDMO) may offer regulatory authoring services for those respective sections.

There are a few sections within Module 3 for which the CSC function may be responsible for providing content. Typically, CSC is responsible for providing the information listed in S.2.1 and P.3.1. All manufacturing, release testing, and analytical testing sites for DS, DP, packaging,

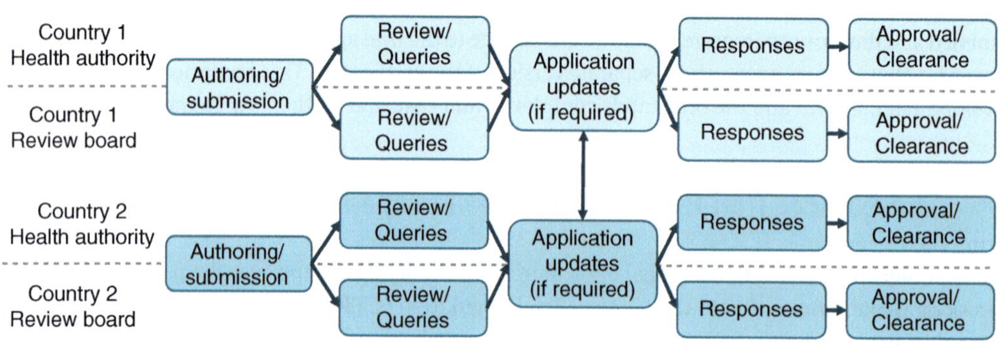

Figure 5.2 Regulatory Application Process

labeling, and, if applicable, QP certification activities need to be listed. The supply chain map document, discussed in Chapter 4 – Quality Considerations, is an excellent source of this information. If the PP step is outsourced, CSC may also review the content in Section P.7.

One common strategy used for Section P.3.1 is to list all qualified secondary packaging and labeling sites as opposed to only listing the primary sites where those activities will take place. This strategy enables switching back and forth between sites in the event of capacity issues or a natural disaster without having to file an amendment to the regulatory application. The only drawback to this approach is the need to keep all sites qualified, but if they belong to the same organization, this should not be overly difficult.

After the first draft of the application is complete, nearly the entire CMC team will review their respective sections. Importantly, all data provided in the application must be verified against source documentation by quality, and this process alone can take multiple rounds of review and several weeks to complete. Once Module 3 is finalized, the entire document is signed off and combined with the other CTD sections for submission to the national CHA and human subjects review board for their respective reviews. When the application is received, the national CHA will assign a team of experts to examine each of the modules provided by the sponsor. The reviewers will then develop a set of questions, considerations, and required updates to the application for the sponsor to respond to.

In the United States, this process takes 30 days, while in other countries, the timelines can vary widely. Historically, countries within the EU accepted applications on an individual basis and typically provided within 45–60 days. In 2014, the region passed a new law, Clinical Trial Regulation 2014/536, that consolidates the review process for all countries. The total lead time from application submission to approval now ranges from 60 to 120 days. The submission–approval lead time varies broadly in other regions. Latin America is notorious for lengthy review times of up to 6 months, but the region has been working hard to reduce these to be more competitive with the United States and EU.

Human subject review boards are also required by law to review the clinical protocol with a focus on protecting the rights and welfare of participants. In the United States, 45 CFR 46 was passed in 1974 for this purpose, and the regulation created the IRBs as a separate entity to review clinical trials. The European Union has taken a similar approach, but names these bodies ethics committees (ECs), and other countries such as China, Japan, and Brazil have taken similar approaches. The IRB process mimics CHA review in that there is a defined window for the review to take place and an opportunity for the IRB to provide queries and request updates to the clinical protocol. Notably, requested changes from the IRB can impact the schedule of assessments which in turn impacts how the drug may need to be supplied for the study.

Once feedback has been received from HAs and review boards, the sponsor needs to decide how to incorporate requested updates. This can be especially challenging if one country requests significant changes to the trial design such as endpoints or assessments because this feedback may create the need for a unique protocol design within a particular country. While manageable, it's not desirable to use this approach, and sponsors much prefer to consolidate feedback to the greatest extent possible to avoid added complexity and cost.

For late-stage studies, it's not uncommon for different regions to provide conflicting responses during the review process given the amount of data presented in a CTA and the vast number of permutations clinical designs can take. To gain insight into the potential feedback and to reduce the risk of application rejection, sponsors have the option to request pre-application meetings with national health authorities to ask specific questions about the trial design or other components of the submission package that may be controversial. These HA interactions are particularly helpful

Table 5.2 Example Submission and Approval Targets

Country	Submission Date	Approval Date
United States	October, 2024	November, 2024
European Union	November, 2024	March, 2025
Japan	December, 2024	February, 2025
Brazil	March, 2025	September, 2025

for programs transitioning into late stage where the bar is higher for demonstrating patient safety and potential efficacy of the drug.

There are three potential outcomes to an IND application: hold, partial hold, and clearance.[8] If the FDA finds a serious-enough risk to patient safety, they will place a hold on the application, meaning further development under the IND cannot proceed. In the event of a hold, the FDA asks for additional information and/or data they feel are needed to resolve the identified risks. Only after acceptable responses have been given and the FDA removes the hold may the clinical study begin. If the FDA issues a partial hold, this means the trial may proceed, but with exceptions, such as testing a high dose that may not be fully supported by the toxicology profile of the drug. Finally, if the IND achieves clearance status, clinical trials may begin immediately without restrictions.

Other regions take a similar approach. For example, in the United Kingdom, the MHRA designates three identical outcomes: acceptance (e.g. clearance), acceptance subject to conditions (e.g. partial hold), and grounds for nonacceptance (e.g. hold).[9] And, in the European Union, the CTR application process results in acceptable (e.g. clearance), acceptable subject to conditions (e.g. partial hold), or not acceptable (e.g. hold).[10]

One final consideration with the regulatory application process is the timing of submissions when multiple countries are involved in a study. Submissions almost never happen on the same day, but rather on a rolling basis over a matter of weeks or months for a few potential reasons. First, each country or region requires unique information within their specific application, and it may not be feasible to prepare all that content simultaneously. Each country's application must also be translated into the local language, and this effort adds weeks of time to the process. Second, sponsors who have taken advantage of pre-application meetings with the FDA, EMA, or other CHA bodies may receive feedback at different points in time and with conflicting messages, and this feedback can also complicate submission timelines. Finally, the country list itself can change over time with new countries added to the mix at later dates, which also leads to extended submission dates. The result of these factors is a submission/approval timetable that might resemble Table 5.2.

5.6 Amendments

As previously defined, an amendment is a proposed change to a previously approved regulatory application.

For Module 3 content, there are two general types of amendments, substantial and non-substantial. The EMA defines[11] a substantial amendment as any modification that has a significant impact on

- The security or physical or mental integrity of the subject.
- The scientific value of the trial.

- The organization and execution manner of the trial.
- The quality and security of each IMP used in the trial.

Logically, a non-substantial amendment is defined as any modification that does not have a significant impact on the bullets listed above. Notice the first three bullets consider patient safety, the endpoints of the study (e.g. safety and efficacy measurements), and the intrinsic design elements of the trial, and the last bullet specifically addresses Module 3. The EMA recently issued updated guidelines related to quality module content for small molecules[12] and biologic[13] products. Within these respective guidelines are helpful tables outlining which changes qualify as substantial versus those that are non-substantial. Examples of substantial changes include:

- *Manufacturing Supply Chain*: Addition of manufacturing and/or testing sites for DS, DP, and packaging.
- *Distribution and Release Supply Chain*: Addition or change of importing site; addition or change to batch release (QP) certification sites.
- *Small-molecule DS Manufacturing Process*: Changes to the synthesis or physicochemical properties of the DS.
- *Biologics DS Manufacturing Process*: Changes to the cell line, cell bank, production scale, cell culture, purification process, or viral clearance steps.
- *Formulation*: Introduction of a new formulation or change to the formulation composition including changes in the active substance concentration and/or excipients.
- *Small-molecule DP Manufacturing Process*: Significant changes to the manufacturing process and critical process controls; scale-up for nonstandard processes (e.g. lyophilization; aseptic processing).
- *Biologics DP Manufacturing Process*: Significant changes to the manufacturing process.
- *Specifications*: Extension of acceptance criteria, deletion of tests, or addition of new tests for a quality/safety reason.
- *Analytical Methods*: Addition of new test methods or new test conditions.
- *Analytical Profile*: Changes to storage condition or extension of shelf life outside the agreed to stability protocol or without prior commitment or reduction in shelf life due to stability concerns.
- *Small-molecule Container Closure System*: New container closure system is introduced with less protection than previous system.
- *Large-molecule Container Closure System*: New container closure system is introduced.

Amendment categorization between non-substantial and substantial matters because this determines the review process and implementation timing of the changes. Colloquially, non-substantial amendments are known as "do and tell" amendments because they can be implemented before notification to a CHA. Substantial amendments are referred to as "tell and do" or "prior approval" amendments because they require review and approval by the applicable CHA before products made using those changes can be released for shipment into the country.

Amendment strategy and timing have important implications for the clinical planner because this information is necessary to accurately plan when new material may be introduced to a study. Like the initial application, amendments typically occur on a rolling basis for many of the same reasons (e.g. amendments need to be translated to local languages). Additionally, in some countries, if an amendment is proposed while a CHA is reviewing the trial application or another amendment, the first review must complete before the next amendment can be processed. In other countries, amendments may be reviewed in parallel.

Another factor influencing amendment timing is how expensive and time consuming they are for the sponsor to execute. To minimize costs and resource burden, there is nearly always a desire to bundle multiple proposed amendments into a single submission for the study. This may not always be possible though as amendments related to trial conduct can take precedence over proposed CMC changes to the supply chain or manufacturing process. One additional consideration for CMC amendments (e.g. changes to Module 3) is that they can impact multiple ongoing trials at the same time, further complicating the submission strategy. The team then must decide which study should take priority over the other(s).

All this to say that developing an optimal regulatory amendment strategy is a difficult task! It's a cross-functional affair that depends on input from the project team, the study management team, and the CMC team. The regulatory function synthesizes all potential amendments against the regulatory landscape of the program to develop a recommended approach. While this process takes place, the clinical planner needs to stay in close communication with the regulatory–CMC representative to understand the strategy and ensure it is incorporated into the supply plan for the program.

5.7 Commitments and Restrictions

CHAs have an opportunity to review and provide feedback on both the initial application and on amendments. Following their review, the CHA usually sends specific questions and recommendations back to the sponsor for response before approval takes place. There are two categories of CHA feedback to which sponsors must pay special attention. The first is a request for additional data or a recommended change to the application at a future date. If the sponsor responds to the CHA agreeing to provide the data or make the change, this is known as a regulatory commitment. Commitments are necessary to track because if they are not met, it increases the risk that a future amendment to an existing study or a new CTA for the next phase of development will be rejected by the CHA.

The other type of feedback is a regulatory restriction on a specific segment of the application. Restrictions occur when a CHA limits or restricts a component of the regulatory application. This is known as a restriction because it restricts importation of a batch that fails to comply with the feedback outlined in the CHA response. Within Module 3, there are a few areas where restrictions are usually placed. The first category is a restriction on the DS or DP specification where a CHA may require tighter acceptance limit ranges or inclusion of acceptance tests that weren't originally included before the product can be released for shipment. The sponsor has three options to choose from to manage a requested specification change:

1) Update the global specification and file a non-substantial amendment in the other countries participating in the trial. This is usually the most straightforward because there is only one specification to manage on a go-forward basis. This approach is most appropriate when there is confidence that the manufacturing process can deliver product that meets the new specification requirements.
2) Create a country-specific specification with the tighter limits and/or expanded tests. This creates additional complexity at the point of DS or DP batch release because each batch must be released against multiple sets of active specifications. This approach is appropriate when there is less confidence that the manufacturing process can deliver product that meets the new specification requirements, but the country is important to the clinical trial.
3) Decline to tighten the specification. This approach is appropriate if the country is not important to the conduct of the clinical trial and there is low confidence that the manufacturing process can meet the requested update.

If the sponsor chooses to create country-specific specifications, documenting which specifications the applicable DS and DP batches are released against becomes a critical activity. If a specific batch does not meet a country-specific specification, that batch must be blocked from entering that country, and the clinical planner must be notified of this restriction.

Expiry periods and shelf-life assignment are another area where CHAs can request commitments or place outright restrictions on the sponsor. These restrictions tend to happen when new formulations are introduced going into late-phase development. This is because sponsors often attempt to take stability data from a previous formulation and apply it to a new formulation by arguing the two formulations are suitably comparable. If the CHA disagrees with the sponsor's comparability assessment, they can request additional stability data from the sponsor before agreeing to the expiry period claimed in the application. See Chapter 10 – Finished Product Expiry Planning for a detailed discussion on how to manage expiry period restrictions.

In all scenarios, it's critical to document which countries have approved which expiry periods. This enables the clinical planner to develop an appropriate labeling strategy and quality to restrict release of finished product (FP) batches with lengthier expiry periods to those countries that have approved that expiry period.

One of the most challenging restriction types is one placed on a manufacturing or testing site. This restriction prevents import of material that was manufactured or tested at the site in question, and they arise when a site has not been appropriately qualified or has had recent quality issues. The reason these restrictions are so challenging is because of the amount of time it takes to transfer manufacturing activities from one site to another. If there is no material available made outside that supply chain, this can lead to months-long delays to the program as a new site is qualified and brought online to support the program. To avoid this situation, it's critical to get quality involved early via the supply chain map process outlined in Chapter 4 – Quality Considerations.

A less common but equally tricky type of "restriction" is one placed on the label text. These types of restrictions are better thought of as required changes to the label text which then block the use of the submitted text. Examples of label text restrictions include needing to update the storage condition, changing the route of administration description (e.g. from injection to infusion), or adding a missing caution statement. Any feedback that forces a change to the label text means the current version of the label cannot be imported to that country. The clinical planner needs to account for this restriction by not allocating product with the existing label to that country and then creating a new label to meet the request of that country. The effect of this is added complexity to manage multiple label types in the FP supply chain.

5.8 Regulatory Status Tracking

Applications, feedback, approvals, amendments, commitments, restrictions – it's enough to make a clinical planner's head spin, especially because these activities often overlap with each other. In early-phase development, this usually isn't as much of an issue because Phase I studies are smaller, are faster, and take place in fewer countries. Once a program hits late-phase clinical trials though, it's imperative for the regulatory function to track and communicate the regulatory status of each country participating in the clinical trial. If this information is not kept current and broadcast throughout the CMC team, managing supply becomes a more difficult task, and the risk of accidentally shipping noncompliant product increases.

A basic regulatory application tracking document includes the following information:

- Protocol identifier
- Country
- Application title, version, and description
- Application type (e.g. initial application or amendment)
- Application status (e.g. pending submission, under review, approved (cleared), and on hold)
- Approval date
- Commitments
- Restrictions

Each of these fields contains important information, but the application version and description fields deserve extra attention. These fields capture the sequence of amendments through the title and a summary of changes through the description that have been made to the application over time. This is how the CSC and quality functions can begin to associate DS and DP batches to the regulatory application status of their respective manufacturing and analytical release processes. As we'll see in the remainder of this chapter, if those associations don't exist, FP batch release has the potential to become an inefficient mess at best and risks noncompliant shipments at worst.

5.9 Batch and IMP Lifecycle Management

So far, we've explored the steps within the CSC, the quality management system that controls the supply chain, and the CTA which documents the supply chain. The batch lifecycle is the process by which a specific batch of IMP is manufactured, released, filed, and approved in a country to enable that batch to ship to a clinical site. Figure 5.3 diagrams the upstream activities that need to take place to get the product to a clinical site and how the manufacturing, quality, regulatory, and CSC functions collaborate to make this happen.

Before manufacturing begins, the DS and DP leads work together with the designated internal/external manufacturing sites to define the materials, process, and equipment that will be used to make and release batches of the product. The final output of this is the master batch record (MBR). In parallel, the analytical team works to define the specifications for the batch including the test methods that will be used to analyze the batch at the completion of manufacture. Only after both the MBR and the specification are approved can the manufacturing process begin.

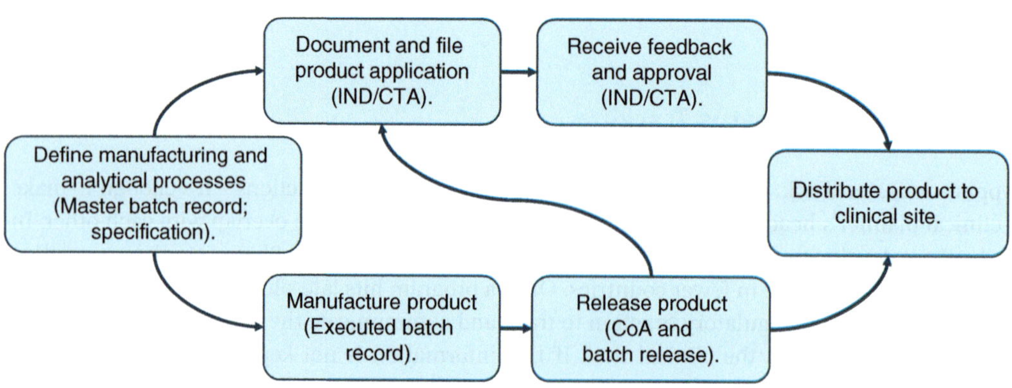

Figure 5.3 The Batch Lifecycle

The manufacturing process generates a batch or batches of material which are then sampled and analyzed against the specification. Quality then reviews the executed batch record (EBR) and the test results, and if there are no major deviations or out-of-specification (OOS) results, they release the batch for further processing. Keep in mind that the DS and DP batches that went into manufacture of the FP all need to be released before the final batch can be shipped.

In parallel to the manufacturing workflow, the CMC team generates and aggregates information for inclusion in the IND and/or CTA. The supply chain, manufacturing process, specifications, stability profile, and batch results are written into the application, and this is eventually submitted for approval to the CHA. Only after quality has released the DS, DP, and FP batch and the IND/CTA has been approved by the CHA and human subjects review committee (e.g. IRB or EC) is CSC able to initiate shipment of the first FP batch to the clinical site.

But, the batch lifecycle is only beginning! As we've discussed in Chapter 3 –Introduction to the Clinical Trial Supply Chain, manufacturing and analytical process changes are inherent to product development. This means the CMC team and the clinical planner need to manage not only the batch lifecycle identified above, but the entire IMP lifecycle. The IMP lifecycle is the process by which all batches of IMP are manufactured, released, filed, approved, and distributed while accounting for supply chain, manufacturing, and analytical changes to the product. When the IMP lifecycle is well managed, it answers two fundamental questions for the clinical planner:

1) At the point of FP release, which countries can the batch be shipped to and which not?
2) At the point of CTA or amendment approval, which batches can be shipped and which not?

The IMP lifecycle outlined in Figure 5.4 builds on the batch lifecycle by incorporating changes, which come from four primary sources:

- *Process Development*: As the program scales from early phase to late phase, the sponsor will proactively make changes to the manufacturing process by tweaking process steps, materials, or equipment. Specifications may be tightened, test methods changed, or stability protocols added as process understanding improves and regulatory requirements increase. It's rare during clinical development for there to be no changes from one DS campaign to the next, and these changes need to be documented for potential regulatory impact.

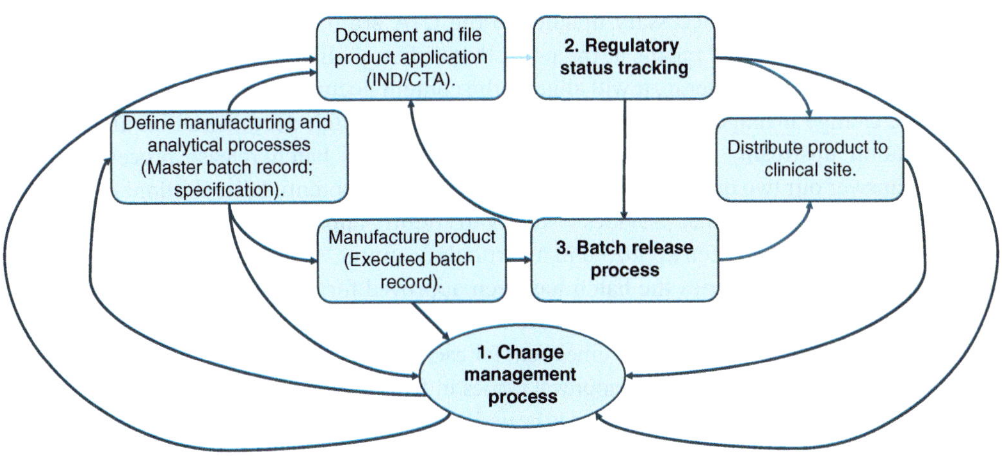

Figure 5.4 The IMP Lifecycle

- *Manufacturing*: Sometimes, there are unexpected changes that occur during the manufacturing process itself. If something unexpected happens during production, this can lead to a process change that needs to be filed with a regulatory body before the batch can be used.
- *Supply Chain*: It's common for manufacturing, testing, and distribution sites to be added through clinical development. This can happen in between clinical trials but also during the trials themselves. Each time this happens, a substantial amendment is required.
- *Regulatory Action*: CHAs can also be a source of change. Regulatory commitments and restrictions need to be accounted for in the manufacturing process, analytical strategy, and/or future regulatory applications.

Each of these changes could necessitate a substantial amendment to the CTA, and if there are multiple trials and several countries in flight, the amount of batch release and regulatory activity increases exponentially. The IMP lifecycle diagram illustrates how complex managing this information can become and highlights the importance of three critical processes during late-phase development – change management, regulatory status tracking, and batch release. Multiple changes may occur from one batch to the next and these three processes combine to ensure the supply chain can function effectively through change. At the point of FP batch release or CHA approval, they ensure the clinical planner has the answers to the two questions highlighted above.

Of the three, the most important process is change management because without robust and efficient change management the two questions can't be answered accurately, and this can have serious consequences. In clinical development, changes will happen with or without a change management process in place. Poor or absent change management only means those changes will not be appropriately assessed and the regulatory impact may not be caught until the FP batch is at the point of release and distribution.

These points are especially true in the EU where the QP sits. Recall from Chapter 4 – Quality Considerations that the QP evaluates the batch against EU good manufacturing practice (GMP) requirements and the approved regulatory application. If the sponsor has made changes to content within the CTA such as revising a specification or updating a label, and the CTA does not reflect those changes, the QP will not certify the batch until an amendment has been filed. The situation becomes worse if the missed change is substantial and requires prior approval, further delaying supply availability and increasing the risk of a stockout situation where no supply is available for sites or subjects.

Conversely, effective change management ensures that a complete and timely regulatory assessment occurs and that necessary updates to the CTA are made before batches arrive for release and distribution. This means at the point that either quality or a QP reviews the batch-specific documentation for release, it will align to the content documented in the relevant IND or CTA. Where change management drives alignment between batch documentation and regulatory application upstream, the regulatory status tracking and batch release processes work together to answer our two questions at the point of release and potential distribution. A comprehensive regulatory status tracker provides visibility to country approval status and to the version of the IND or CTA that has been approved in a particular country. At batch release, quality needs to document in which countries the batch has been approved for use so CSC knows where the product can be shipped to. The regulatory status tracker gives quality this information.

Importantly, an FP release is not a one-time event for each batch. Recall that CTA submissions happen on a rolling basis, so every time an approval comes in from a new country, the regulatory status tracker and batch release documentation need to be updated to reflect this. The regulatory status tracker notifies quality and CSC as approvals happen, and quality can then update the batch release documentation to indicate the approval status of that country. When these processes work in harmony, they answer the second question – at the point of an IND or CTA approval for an initial application or an amendment, quality and CSC will know which batches can be shipped to that country.

5.10 Process Version Management

Clearly, quality, regulatory, and CSC must work closely together to enable IMP lifecycle management and the end-to-end clinical trial supply chain. Change management, regulatory status tracking, and batch release are the three fundamental processes needed to enable IMP lifecycle management and to ensure batch distribution is both controlled and timely. But, there is a fourth process that can make the entire effort more efficient, PVM.

Think of how FP batch release has been described to this point. On the batch release documentation, quality needs to list which countries have approved the underlying DS and DP manufacturing process and analytical control strategy that went into that FP batch. The regulatory status tracker tells quality which countries have approved the clinical trial and which version of the IND/CTA is approved, but the tracker does not contain any batch-specific information. At the point of FP release, quality also needs to check underlying the DS and DP batch documentation to ensure it is aligned to the regulatory application documentation. Figure 5.5 captures this resource-intensive process that can slow batch release and down distribution significantly. PVM solves this issue.

Implementation of the PVM is a four-step process. First, every product needs a batch genealogy table (BGT). The BGT connects each FP batch to the upstream batches of PP, DP, and DS that went into it. Building a genealogy for every clinical product is a best practice from both a supply chain management and a quality control perspective. At minimum, the batch genealogy should capture the product description (e.g. Product 1 drug substance), batch number, date of manufacture, quantity manufactured, and unit of measure.

The second step is to add the regulatory-relevant attributes of each batch as in Table 5.3.

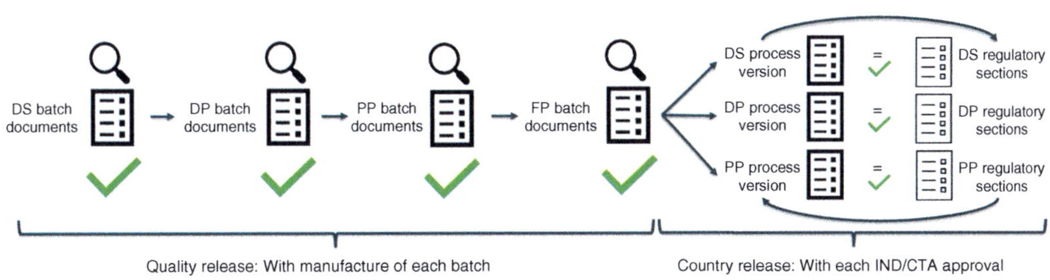

Figure 5.5 Batch Release, Quality Release, and Country Release

Table 5.3 Regulatory-Relevant Attributes of Drug Substance, Drug Product, and Primary Packaging

Drug Substance	Drug Product	Primary Packaging
• Production Site • Master batch record version # • Specification version # • Methods version # • Container closure • Scale (biologics only)	• Production site • Formulation • Master batch record version # • Specification version # • Methods version # • Container closure	• Production site • Master batch record version # • Container closure

Table 5.4 Process Version Number Assignment

Drug Substance Batch #1 = Process Version DS1.0	Drug Product Batch #1 = Process Version DP1.0
• *Production Site*: ABC DS manufacturing • MBR V1; Specification V1; Methods V1 • *Container Closure*: Glass bottle	• *Production Site*: ABC DP manufacturing • MBR V1; Specification V1; Methods V1 • *Container Closure*: Prefilled syringe

After the attributes have been added to each batch within the table, the third step is to create the process version codes for each DS, DP, and PP as in Table 5.4. The process version code summarizes the combination of regulatory-relevant attributes of each DS and DP batch into a single data element. Nomenclature for the process version code captures the supply chain step (e.g. DS, DP, or PP) and the process version itself (e.g. 1.0, 1.1, and 2.0). In Table 5.4, let's assume the first DS and DP batches for a biological product have the attributes as given in the table.

When changes are made to any of the regulatory-relevant attributes, the change management process triggers and a regulatory impact assessment is conducted. If the change(s) to the attribute(s) is non-substantial but requires up-versioning the MBR, specification, or methods, then the process version moves from 1.0 to 1.1 or 1.1 to 1.2 (or 2.0 to 2.1) and so on. If the change to the attribute(s) is substantial (e.g. addition of a manufacturing site), then the process version moves from 1.0 to 2.0 or 2.1 to 3.0 and so on. When changes are made, a best practice is also to capture the change record number in the BGT for reference.

Let's continue the example from above and assume a substantial change has been made to the DS (e.g. change of manufacturing site) and a non-substantial change has been made to the DP (e.g. non-substantial change to MBR and specification). In Table 5.5, the DS process version revises from DS1.0 to DS2.0 and the DP process version revises from DP1.0 to DP1.1.

The fourth and final step is to associate the DS and DP process version that was filed to each CTA within the regulatory status tracker. The initial CTA for a Phase I trial may have DS1.0 and DP1.0, but as the program moves deeper into development, the process versions initially filed for Phase II and III studies will mature with the program. Late-stage development often features DS and DP changes that are implemented months or years before the FP batches are completed that are made from those upstream changes. Building process versions creates a clear link between the process version of the DS and DP batches that have been made (e.g. in the BGT) and the process version that's been filed

Table 5.5 Process Version Number Revision

Drug Substance Batch #1 = Process Version DS1.0	Drug Product Batch #1 = Process Version DP1.0
• *Production Site*: ABC DS manufacturing • MBR V1; Specification V1; Methods V1 • *Container Closure*: Glass bottle	• *Production Site*: ABC DP manufacturing • MBR V1; Specification V1; Methods V1 • *Container Closure*: Prefilled syringe
Drug Substance Batch #2 = Process Version DS2.0	**Drug Product Batch #1 = Process Version DP1.1**
• *Production Site*: DS manufacturers R US • MBR V1; Specification V1; Methods V1 • *Container Closure*: Glass bottle	• *Production Site*: ABC DP manufacturing • MBR V2; Specification V2; Methods V1 • *Container Closure*: Prefilled syringe

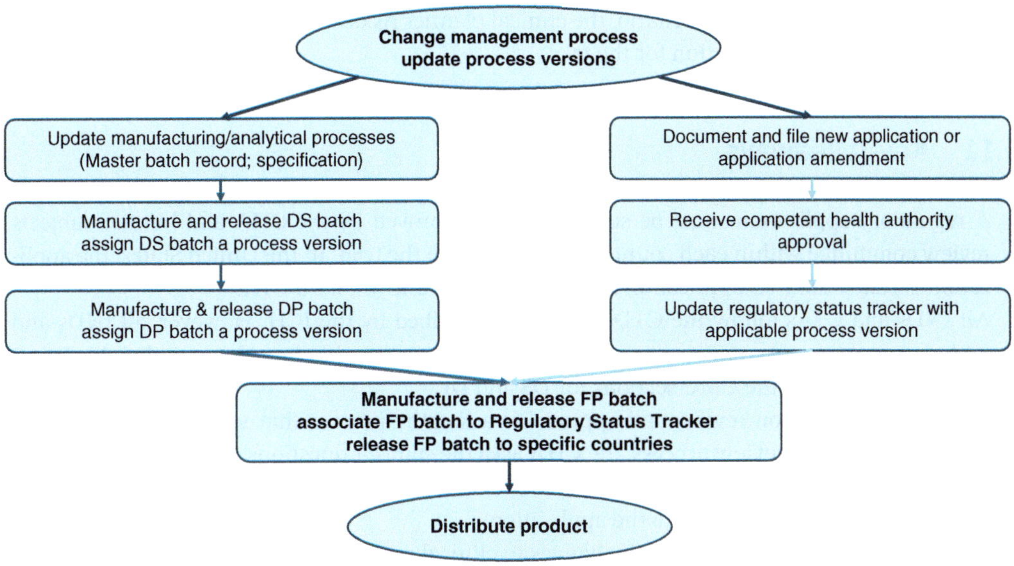

Figure 5.6 Managing the IMP Lifecycle with Process Versions

and approved in each country (e.g. in the regulatory status tracker). This association is incredibly helpful for quality team members at the point of FP batch release when trying to figure out which version of the DS and DP processes is filed and approved in each country.

Notice that change management serves as the entry point for both process version and IND/CTA updates. After the process version has been updated, each subsequent batch made from the updated process can be assigned its respective process version within the BGT at the point of that batch's release. And, when the revised IND/CTA is approved, the regulatory status tracker is updated to capture that approval and the respective process versions contained within that application. Then, when quality is ready to release FP batches made off that DS and DP, the association is made. When all the processes are integrated together as in Figure 5.6, it creates a structure out of the chaos that is IMP lifecycle management.

Developing PVM, the surrounding quality management system processes, and a regulatory documentation process infrastructure takes time, energy, and resources. For clinical-stage biopharma companies that have relatively small pipelines with only a handful of active trials, the level of process control prescribed in this section may not be needed. However, as the organization's pipeline grows, so grows the number of changes to manage and the concurrent amount of work it takes to align FP batches to regulatory application content. At a certain point, the upfront investment of creating process versions, BGTs, and robust change management systems begins to pay significant dividends downstream in the FP batch release process.

PVM offers a final benefit, stimulation of conversation and alignment between the CSC, quality, and regulatory functions on the regulatory strategy for each clinical protocol. When the clinical planner and the quality lead have visibility to when regulatory applications will be submitted for approval, which process versions are captured in those applications, and if any restrictions need to be accounted for across the country list, they can plan supply appropriately and execute each batch release efficiently. Even if process versions aren't formally implemented, gaining alignment within the CMC team on the regulatory strategy for initial applications, amendments, commitments, and restrictions is critical. Without this alignment, the batch release process becomes much more inefficient, and in a worst-case scenario, the clinical planner risks allocating a noncompliant batch that results in a stockout situation for the trial.

5.11 Key Takeaways

1) A regulatory application must be submitted and approved by the CHA and human subjects review committee within each country participating in the trial. In the United States, the application is called the IND application and in all other countries, it's a CTA.
2) All INDs and CTAs follow the CTD structure prescribed by the ICH. However, all INDs and CTAs have additional country- or region-specific information that must be provided. Module 3 of the CTD contains the CMC sections for DS and DP.
3) Regulatory application review and approval is a lengthy process that varies from country to country. During the review process, the CHA asks the sponsor questions and provides feedback on specific segments of the application. The sponsor then has an opportunity to respond and/or incorporate that feedback into the application.
4) Over the course of the clinical trial, changes to clinical protocol design or to the CMC content must be submitted to the CHA for approval. These changes are captured in amendments, which can be broadly categorized into non-substantial amendments (do and tell) and substantial amendments (prior approval).
5) Sometimes, a CHA will require the sponsor to provide further information at a future time (e.g. make a commitment) or restrict a specific aspect of the application. Within Module 3, these restrictions are most often placed on specifications, expiry periods, manufacturing/testing sites, and label content.
6) To best manage regulatory activity, a regulatory status tracker should be used. The regulatory status tracker captures country, application title and version, application type, application status, approval date, and any commitments or restrictions resulting from that application. The regulatory status tracker should be updated with each regulatory action and communicated to CSC and quality.
7) Each batch of material manufactured in support of a study must be released by the quality function in alignment with an approved IND/CTA. The batch lifecycle is the process by which a specific batch of IMP is manufactured, released, filed, and approved in a country to enable that batch to ship to a clinical site.
8) The IMP lifecycle incorporates change management into the batch lifecycle and enables the clinical planner to understand which countries a newly released batch can be shipped to and which batches can be shipped following a new CTA or amendment approval. Change management, regulatory status tracking, and batch release are the three business processes required to enable IMP lifecycle management.
9) Process versions summarize the regulatory-relevant attributes (e.g. sites, manufacturing process, specifications, methods, and container closure) of DS and DP batches. PVM simplifies the FP batch release process by creating a line of sight from the process version of each manufactured batch to the process version that's been filed and approved in the IND/CTAs of each country.

Notes

1 https://www.accessdata.fda.gov/scripts/cdrh/cfdocs/cfcfr/CFRSearch.cfm?CFRPart=312&showFR=1&subpartNode=21:5.0.1.1.3.2.
2 https://www.imp-dossier.eu/imdp_guidance/.
3 https://database.ich.org/sites/default/files/M4Q_R1_Guideline.pdf.
4 https://www.fda.gov/media/113760/download.
5 https://ascpt.onlinelibrary.wiley.com/doi/full/10.1111/cts.12635.
6 https://www.fda.gov/media/172311/download.
7 https://www.ema.europa.eu/en/news/clinical-trials-transition-new-eu-system-one-year-left.
8 https://www.fda.gov/drugs/investigational-new-drug-ind-application/ind-application-procedures-clinical-hold.
9 https://www.gov.uk/guidance/clinical-trials-for-medicines-apply-for-authorisation-in-the-uk.
10 https://www.ema.europa.eu/en/human-regulatory-overview/research-and-development/clinical-trials-human-medicines.
11 https://health.ec.europa.eu/system/files/2016-11/list_sa_be_eng_en_0.pdf.
12 https://www.ema.europa.eu/en/documents/scientific-guideline/draft-guideline-requirements-chemical-pharmaceutical-quality-documentation-concerning_.pdf.
13 https://www.ema.europa.eu/en/documents/scientific-guideline/draft-guideline-requirements-quality-documentation-concerning-biological-investigational-medicinal_en-2.pdf.

6

RTSM and Its Role in the Clinical Supply Chain

6.1 Randomization and Trial Supply Management

When purchasing a consumer good, a customer typically selects an item randomly from an existing pool of inventory, pays for the good, and receives their new purchase. Clinical supply chains deliver investigational medicinal product (IMP) to trial participants in an entirely different way. Instead of subjects choosing what they want, IMP is strictly assigned on a unit-by-unit (e.g. kit-by-kit) basis. Trials are set up this way for a few reasons. First, from Chapter 2 – Introduction to Clinical Trials, each subject is allocated and/or randomized into one arm of the clinical trial, so they can only receive the treatment that supports that arm. This is especially important for blinded trials where active and placebo kits are labeled identically and neither subject nor investigator knows which treatments are administered to a specific subject. Additionally, some trial designs allow for subjects to titrate up in dose, reduce the dose, or even switch arms from one treatment to another. It's important that subjects only receive the treatment they should and no more or less than they should at any given visit.

Second, the sponsor and clinical sites have an obligation to document IMP location and status during the trial. Throughout the duration of the study, each site keeps a dispensation and IMP accountability log that lists which drugs were administered to each subject at each visit. This information demonstrates that every subject within an arm received the IMP they were supposed to receive. At the end of the study, every unit of IMP supplied to a site is reconciled and accounted for as dispensed, damaged, or destroyed. If a product quality event occurs and a specific batch must be recalled, IMP accountability enables the sponsor to identify where all the impacted product is through the supply chain so it can be quarantined and blocked from further use.

For these reasons, inventory management in clinical trials requires control, visibility, and traceability down to the finished product kit level. In a small Phase I study involving one site and one type of treatment, this can be achieved via manual processes. The sponsor supplies a single batch or a few batches of drug, the site maintains the dispensation and accountability log, and all units of the drug can be accounted for and reconciled at the end of the study. If there's a product quality issue, there's only one site to contact and a relatively small amount of inventory to trace. For larger studies though that involve multiple treatment types and sites spread around the globe, manual processes are not feasible to manage subject randomization, IMP dispensation, accountability, and reconciliation. Enter the randomization and trial supply management (RTSM) database technology platform.

Today's RTSM systems are the latest iteration of an automated solution dating back to the 1990s. Back then, these were phone based and known as interactive voice response (IVR) systems. In the

Supply Chain Planning for Clinical Trials – A Practical Guide, First Edition. Ryan Mills.

2000s, they moved onto the internet and became known as interactive web response (IWR). Over time, functionality was added, and by the 2010s, they became known as integrated response technology (IRT) systems. IRTs not only randomized subjects and allocated supply but they also integrated with other platforms such as packaging depots and electronic data capture (EDC) systems which are the central source for capturing participant data during the trial. Now, there are dozens of RTSM solutions and suppliers on the market offering ever-expanding functionality and more user-friendly platforms than ever. Here, "RTSM" encompasses all contemporary IRT and RTSM solutions.

To summarize, RTSMs facilitate subject, site, and IMP management within the clinical trial and control:

- Activation of clinical sites.
- Documentation of site and subject status through the duration of the study.
- Randomization of subjects into specific treatment arms.
- Management of changes to subject dose level.
- Assignment and administration of IMP to each subject.
- Initial supply and resupply shipments of IMP from a depot to clinical sites.
- Distribution of IMP from one depot to another depot.
- Prevention of expired IMP to any subject.
- Accountability and reconciliation of IMP at study closure.

RTSMs are a virtual representation of the physical supply chain and indicate where finished product (FP) inventory exists throughout the depot and clinical site network. Therefore, it's a critical tool to incorporate into clinical supply planning and execution. On the planning side, some settings within the RTSM need to be incorporated as inputs into the supply plan, and the RTSM generates detailed reports that feed into the demand and supply planning process. On the execution side, there are settings within the IRT that directly impact movement of product through the supply chain, product availability, and on-hand inventory. The RTSM is central to clinical trial supply chain management and understanding how these systems work is required for effective demand and supply planning.

We'll begin this chapter with a broad overview of RTSM scope and functionality and then take a deeper look at kit identification numbers and how they support RTSM functionality. Then, we'll take a deep dive into IMP management within the RTSM, including a review of initial supply values, resupply strategies, and how the RTSM accounts for IMP expiry. Next is an overview of RTSM transactions, alerts, and reports, with a focus on how reports provide data that informs both the demand forecast and supply plan for a study. A few additional IMP management considerations are offered before the chapter finishes with a discussion of RTSM supplier selection and exciting future trends in the RTSM space.

6.2 RTSM Scope and Functionality

RTSM solutions connect the nodes that make up the "last mile" of the clinical supply chain: the supplying central or regional depot, the clinical site, and ultimately the subject. The scope of RTSMs reflects these connections and can be thought of in the following buckets: supply chain network design, site management, subject management, and IMP management. In terms of structure, RTSMs are transaction based with respect to their functionality and role based with respect to user access. Figure 6.1 illustrates both the RTSM scope and structure.

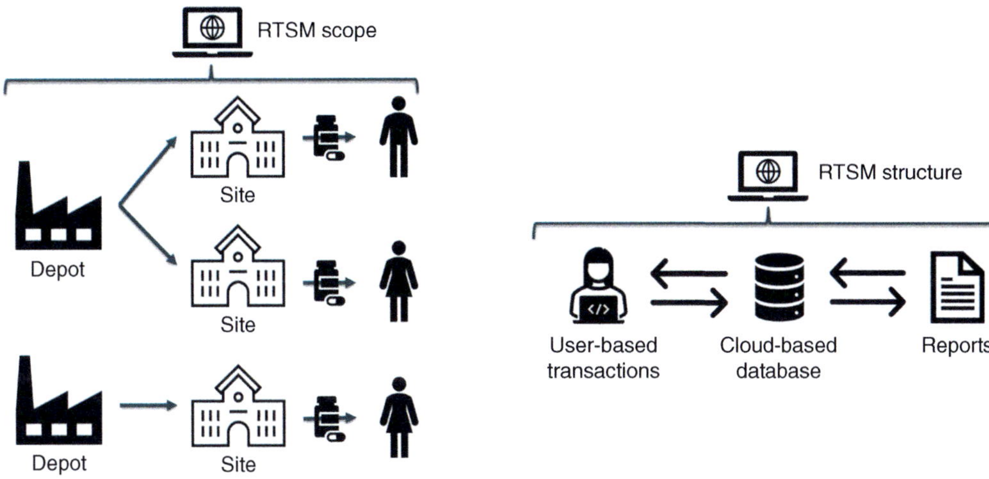

Figure 6.1 RTSM Scope and Structure

RTSMs are not out-of-the-box solutions, but instead require extensive customization to meet the needs of the specific trial design. A common approach to RTSM development and rollout is to

1) Perform requirements gathering from the study sponsor which usually takes place over a series of meetings to discuss the clinical protocol, schedule of assessments (SoA), and the supply chain network.
2) Develop a specification for the system design that incorporates all the intrinsic elements of the trial that touch on site, subject, and IMP management.
3) Program the system to meet the specification.
4) Perform user acceptance testing (UAT) on the transaction functionality, alerts, and reports.
5) Execute the "go-live" where the RTSM database is made available to users.

The entire design–build–implement process takes anywhere from 8 to 20 weeks depending on the supplier and the complexity of the system.

To properly design an RTSM, the study elements that impact how subjects will be managed and how IMP will be administered are needed. These elements include the number of study participants, a list of countries, the supplying depots for each country, and the number and type of treatments that will be administered. Of paramount importance is the IMP dispensing sequence, which is tied to the SoA from the clinical protocol. See Chapter 2 – Introduction to Clinical Trials for a refresher on the SoA. Another consideration is if the study is blinded and/or has an extension phase continuing treatment. If an extension phase is part of the design, then the rollover strategy from double-blind to extension phase is also needed. While the elements listed above aren't exhaustive, they do illustrate how any aspect of the clinical protocol design that touches IMP administration and storage must be incorporated into the RTSM specification.

Detailed information on subject and site management requirements is also central to the design process. Subject and site management encompass any update to a subject or site's information or status during the study. Subject management includes entering a new subject into the system, screening, randomization, and any changes to subject status (e.g. screen failure, withdrawal, or completion). If the study is blinded, the system can also control planned unblinding situations. Site management includes entering a new site or updating a site's status (e.g. activated, inactivated, or suspended). Within a site, different RTSM users (e.g. clinical site staff who will interact with the system) are also defined.

Finally, the IMP management requirements are gathered. These are typically quite extensive and cover

- Initial site supply strategy development.
- Site resupply strategy development.
- IMP expiry management.
- IMP status updates (e.g. available, expired, or destroyed) at depot and site.
- Material release (e.g. quality release and country-specific release).
- Material relabeling (e.g. for expiry extensions).
- Drug accountability, returns, and destruction.

Next, the necessary transactions such as site activation, subject randomization, and manual resupply shipments are defined. RTSMs are transaction-based systems because every action taken within the system needs to be recorded. Subject randomization and IMP administration are central to determining if the trial meets its endpoints. This means there is a transaction associated to every activity within the scope of subject, site, and IMP management. Core IMP management transactions are covered later in this chapter.

RTSMs are also role based, meaning that access to specific information or the ability to execute specific transactions is limited by the role assigned to a user. Roles are needed to shield blinded team members from unblinding information (e.g. subject cohort assignment). Common roles within an RTSM include site user (blinded), clinical supply (unblinded), and clinical depot (unblinded). Roles and their associated accesses are customizable and can be designed to the needs of the study and sponsor.

Finally, since RTSMs are databases they can generate alerts about upcoming events and detailed reports on subject, site, and IMP status. Alerts are particularly useful with respect to inventory levels and upcoming expiry events. Meanwhile, RTSM reports are an incredibly valuable tool to the clinical planner because of the wealth of information they provide that informs both the demand forecast and the supply plan for the study. Both alerts and reports are also customizable during the design phase, and given their importance, these segments of the RTSM are covered in depth later in this chapter.

6.3 The Kit Identifier

Before jumping further into IMP management, it's helpful to have a better understanding of the kit identification (Kit ID) number and how they're used in clinical trial supply management. A Kit ID is a unique code assigned to a single unit of IMP. Kit IDs are typically numeric, range from four to seven digits in length, and are created via a master kit identification list (MKIL). The MKIL contains all the Kit IDs for the study and organizes them into treatment groups. Treatment group is a term for a specific arm within the study. For example, if a study has three arms with three different doses that will be supported by three unique kits, then three treatment groups with three associated Kit ID ranges are created:

- *Treatment Group 1 (500 mg Dose)*: Kit IDs = 00001–10000
- *Treatment Group 2 (250 mg Dose)*: Kit IDs = 10001–20000
- *Treatment Group 3 (Placebo Dose)*: Kit IDs = 20001–30000

In open label designs, Kit IDs may be assigned and printed sequentially on the label, and this approach is often preferred for simplicity's sake. But in blinded trials, the Kit IDs are scrambled (e.g. mixed randomly) to avoid giving information about which arm of a trial the subject is randomized into to clinical site staff. To scramble Kit IDs without losing track of which Kit IDs have been assigned to the packaged kit, the MKIL contains sequence numbers that are paired with each Kit ID. The sequence number is literally a predefined sequential range (e.g. 00001–10000) while the Kit IDs are randomized and linked to each sequence number. Let's revisit the previous example for additional clarity:

- *Treatment Group 1*: Sequence range = 00001–10000, Kit IDs 00001–30000 (scrambled)
  - Sequence number 00001 = Kit ID 15478
  - Sequence number 00002 = Kit ID 00549
  - Sequence number 00003 = Kit ID 22890
- *Treatment Group 2*: Sequence range = 10001–20000, Kit IDS 00001–30000 (scrambled)
- *Treatment Group 3*: Sequence range = 20001–30000, Kit IDs 00001–30000 (scrambled)

Every kit receives a unique Kit ID during the final packaging/labeling step. Kit IDs are assigned to each label in sequence number order during the label computerization process described in Chapter 3 – Introduction to the Clinical Trial Supply Chain. So, if 1000 kits of each treatment group are packaged, then the Kit IDs associated to sequence ranges 00001–01000, 10001–11000, and 20001–21000 are printed on the labels. In parallel, the MKIL is uploaded into the RTSM prior to system go-live. When the RTSM is ready to support the trial, it associates inventory at the kit level to the appropriate treatment group. As subjects randomize into a specific treatment group, the RTSM assigns the designated kits, and the clinical site staff picks that kit off the shelf and provides it to the subject for take-home use or prepare the kit for administration on site. With this approach, the Kit ID enables the RTSM to function and protects the blind through the study.

At the conclusion of the trial, the Kit ID is also central to the drug accountability and reconciliation process. The RTSM system has a record of every uploaded Kit ID and can provide a report of which kits were administered to subjects, which kits were unallocated, which kits expired on the shelf, and which kits were returned and destroyed during trial conduct. This system eliminates the need to comb through thousands of pages of manual records and improves drug accountability completeness and accuracy.

6.4 IMP Supply Strategies

As discussed in the previous section, there are several IMP management design elements within the RTSM that warrant attention and analysis. Determining how IMP will be administered to each subject and how sites will be supplied and resupplied is interrelated and requires careful thought to develop an optimal set of supply strategies for how clinical sites will be initially supplied and resupplied through the duration of the trial. When the RTSM is programmed in a way that fits the study design and supply plan, the "last mile" runs smoothly and inventory is not wasted. But, RTSMs are often poorly understood, and this can lead to understocking some sites while overstocking others. Almost every experienced clinical planner has an RTSM horror story to share where the system did not behave as they thought it would and either a stockout or a significant waste of drug occurred.

Creating a smart RTSM IMP management design begins with documenting how IMP will be administered at the subject level in an IMP dispensing table. Subject-level IMP administration is

Table 6.1 Example 1 RTSM IMP Dispensing Table

Visit Number	Visit Name	Dose	Treatment Group	Material Description	Quantity Kits Dispensed
1	W1 – scheduled	100 mg	1	Drug 100 mg	1
		Placebo	2	Placebo drug	1
2	W2 – scheduled	100 mg	1	Drug 100 mg	1
		Placebo	2	Placebo drug	1

conceptually equivalent to the subject-level demand forecast, and this topic will be covered extensively in Chapter 8 – Finished Product Demand Planning: Basic Concepts. To briefly summarize that discussion, IMP administration quantities depend on the desired dose, the visit schedule and visit windows within the SoA the frequency of dosing, the packaging design of the FP (e.g. kit design), and whether the product will be taken home or administered in the clinic.

IMP dispensing tables are visit based. Table 6.1 captures a basic two-arm study (active and placebo) where each subject receives one kit of drug on a weekly basis.

Note that this example IMP dispensing table captures not only the visit number, visit name, and dose, but also the treatment group of each material. The treatment groups should align with those created in the MKIL discussed earlier. Dispensing tables may also capture unscheduled visits, which are visits that happen outside (e.g. before or after) the planned visit schedule and visit windows. Visits can also be categorized as static or dynamic. If a visit is static, that means the forecasted visit date is assigned based on the randomization date. If a visit is dynamic, that means the forecasted visit date is based on when the previous visit date occurred. Static and dynamic visit designations can influence the site resupply strategies and are another system configuration element to heed during the RTSM design phase.

After the IMP dispensing table is built, the next step is to determine the site supply strategies which dictate how sites will be initially supplied and resupplied over the course of the study. Three goals should be kept in mind when developing a site supply strategy. First, ensure the site always has enough IMP on hand to service subject visits. Second, minimize the amount of drug wastage due to IMP expiry or lack of demand. Note that these two goals must be carefully balanced against each other because as more IMP is sent to a site to service future visits, the amount of potential wastage also increases. Conversely, if less IMP is sent, then potential waste decreases but stockout risk increases. Third, if the study is blinded, reduce unblinding risk to the investigator or site staff based on the resupply shipments the site receives.

When developing site supply strategies, remember that clinical sites behave differently in terms of recruitment and enrollment. Some sites may only screen and enroll one or two subjects (or no subjects if that subject screen fails!) while other sites may successfully recruit dozens of subjects (or more depending on the size of the trial!). There can be a huge range of enrollment outcomes in terms of both timing and the number of patients. As such, there is almost never a one-size-fits-all approach for multisite studies. RTSMs account for this by allowing the sponsor to design as many site strategies as they see fit. The most common approach is to build three strategies – low, medium, and high – but more can be added as needed.

The first value to determine in a site supply strategy is the initial supply value. Three variables should be considered when determining the initial supply value:

1) *Site Replenishment Lead Time (SRLT):* The lead time to replenish the clinical site from the supplying depot (e.g. 3 days and 1 week).

2) *Subject Demand Over SLRT*: The number of kits dispensed to an individual subject over the SRLT.
3) Number of subjects who will potentially screen and enroll over the SRLT.

The trick with determining initial supply values is that variables 1 and 2 in the list above are known, while variable 3 is a guess based on feedback from clinical sites during the start-up process. Let's build on the example from the IMP dispensing table above and assume the lead time to replenish a site is less than one calendar week (a reasonable assumption for most clinical trial sites). If variable 3 is small (e.g. 1–2 subjects screening at a time over the SRLT), then setting an initial supply value of 2 kits (2 = 1 kit per week * 2 subjects) is reasonable. As the number of subjects who will potentially screen and enroll over the 1-week lead time increases, so too does the initial supply value. If clinical operations projects that some sites could screen and enroll 1–2 subjects, some sites 3–7 subjects, and some sites 8–15 subjects, then it might be fair to set the low initial supply value at 2 kits, the medium value at 7 kits, and the high value at 15 kits, respectively. Obviously, there is a fair amount of subjective judgment in these estimates, and once the study starts enrolling, the estimates may change. For this reason, the clinical planner and clinical operations need to stay in close communication to ensure these values are updated as needed to ensure sufficient supply for high-demand sites and prevent excess waste at low-demand sites.

The largest influence on subject recruitment and enrollment projections is the target indication or disease area. For broad indications in virology and immunology (e.g. COVID-19, HIV, hepatitis C), there may be hundreds or thousands of available subjects ready to screen and enroll on the first day of a study at a given site. This situation can translate to establishing a wide range of initial supply values and more overall site supply strategies such as a maximum category that accounts for a super-high enrolling site. Meanwhile, targeted and/or rare indications with smaller patient populations feature fewer patients screening and enrolling at the same time. Under these conditions, creating much narrower initial supply value ranges (e.g. low = two kits; medium = four kits; high = six kits in the baseline example) are appropriate and/or having only one or two total site supply strategies (e.g. low and high only).

There is a common misconception about the initial supply values that's important to dispel. They do not reflect screening and enrollment projections over the duration of the study, but rather the maximum number of subjects who are expected to screen and enroll over the SRLT at site activation. Under the IMP dispensing table assumptions documented above, if a site only has the capacity to screen two subjects at any point in time and the screening window is greater than the lead time to replenish the site, the site should be set to "low" even if the site plans to enroll 10 subjects. This is because the site will never have more than two subjects screening when it opens for enrollment. It's a common mistake to set a site to a "high" setting if they project 10 total subjects for the study and watch inventory fly out of the depot and then sit on the shelf when it could be used elsewhere in the study. Initial supply values are often set in the RTSM by colleagues (e.g. clinical operations or the Contract Research Organization) who do not understand this subtlety. With that said, it's the clinical planner's responsibility to not only calculate the initial supply values but to communicate how they work to the team members working within the RTSM, particularly if the planner does not control initial supply setting assignment.

With the IMP dispensing table and initial supply values set, the next step is to select the type of resupply strategy for the study. There are three general approaches to resupply strategies:

1) *Manual Resupply Model*: A user tracks inventory levels at the site and then manually triggers a resupply transaction to a clinical site in the RTSM. This scenario only makes sense when there is one clinical site and only a small number of resupply shipments are needed for the duration

of the study. Otherwise, manually tracking site inventory becomes an extremely resource-intensive task and introduces a needless risk of someone forgetting to resupply a site leading to a stockout situation. Phase I studies may sometimes fall into this category.

2) *Min–Max Resupply Model*: A simple automatic resupply model that sets a minimum inventory level for a particular treatment group and then resupplies the site up to a predetermined maximum inventory level. The challenge with this model is that it can easily lead to excessive waste if a patient drops out unexpectedly, and it's not robust enough to handle more complex study designs such as planned dose titrations or reductions. Eventually, RTSM service providers developed a more intelligent approach, the predictive threshold resupply model (PTRM), which is now used almost universally for studies that do not use a manual resupply model.
3) *PTRM*: This approach combines min–max-level functionality with the concept of looking ahead in time at future enrolled subject visits to dynamically calculate how many kits a site should be resupplied with over a predetermined forecast window.

Most clinical studies that utilize RTSM solutions now use the PTRM. The methodology offers the best of both worlds – ensuring supply and minimizing waste – by automatically restocking sites to account for future visits while also not overstocking a site when a subject's time on the trial ends. The PTRM determines when resupply shipments are needed and calculates how much inventory to include in each shipment based on a handful of variables that interact with each other over time. The PTRM algorithm typically runs daily and begins with the following inputs:

- *Quantity of Kits Dispensed per Subject Visit (QKD = # kits)*: Taken from the IMP dispensing table.
- *Number of Enrolled Subject Visits (NESV = # subjects)*: The number of subject visits for enrolled subjects over a predetermined period of time.
- *Site on-Hand Inventory (SOHI = # kits)*: Amount of on-hand available inventory at the site when the PTRM algorithm runs.

There are four additional inputs that compose an individual PTRM strategy:

- *Floor Quantity (F Expressed in # kits)*: The safety stock buffer that needs to be maintained.
- *Ceiling Quantity (C Expressed # kits)*: The maximum supply quantity to be shipped.
- *Short Window (SW Expressed in # days)*: How far ahead in time the RTSM checks for IMP requirements.
- *Long Window (LW Expressed in # days)*: How far ahead in time the RTSM will resupply IMP to cover.

The PTRM then makes two demand calculations based on the NESV within the two windows, short and long:

1) Short window demand (SWD expressed in # days):

$$\text{SWD}\left(\text{units}\right) = \text{QKD} \times \text{NESV}_{\text{SW}}$$

2) Long window demand (LWD expressed in # days):

$$\text{LWD}\left(\text{units}\right) = \text{QKD} \times \text{NESV}_{\text{LW}}$$

Once the demand calculations are made, the PTRM makes a resupply order decision (ROD) by checking if the site on-hand inventory is greater or less than short window demand + the floor quantity:

3) ROD = Resupply order not required if

$$\text{SOHI}\left(\text{units}\right) > \text{SWD} + \text{F}$$

4) ROD = Resupply order required if

$$\text{SOHI}(\text{units}) < \text{SWD} + \text{F}$$

Finally, if a resupply order is required, the PTRM calculates the resupply order quantity by adding the long window demand to the ceiling value and then subtracting the site on-hand inventory:

5) ROQ = Resupply order quantity (expressed in # kits)

$$\text{ROQ}(\text{units}) = \text{LWD} + \text{C} - \text{SOHI}$$

Table 6.2 is an illustrative example showing a PTRM supply strategy in action, assuming the dispensation strategy in the IMP dispensing table above – 1 kit per visit per subject; visits happen weekly – and the values as in the table.

In Figure 6.2,

- SWD = 2 = 1 (QKD) * 2 (NESV_{SW})
- LWD = 4 = 1 (QKD) * 4 (NESV_{SW})
- ROD = "Not required" because 5 > 2 (SWD) + 2 (F)

What if the Site On-Hand Inventory quantity is only three kits instead of five kits?

- ROD = "Required" because 3 < 2 (SWD) + 2 (F)
- ROQ = 4 (LWD) + 6 (C) – 3 (SAI) = 7 kits

How does the example change if a new subject enrolls in W1 as in Figure 6.3?

Table 6.2 PTRM Example Values

Site On-Hand Inventory	5 kits
Floor	2 kits
Ceiling	6 kits
Short Window	14 days
Long Window	28 days

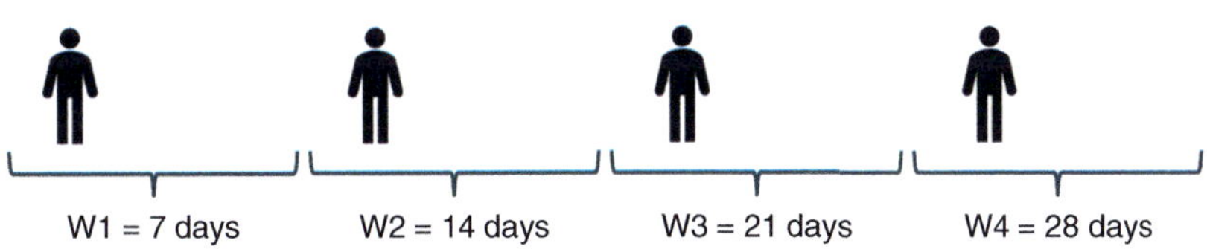

Figure 6.2 PTRM Example Week 1

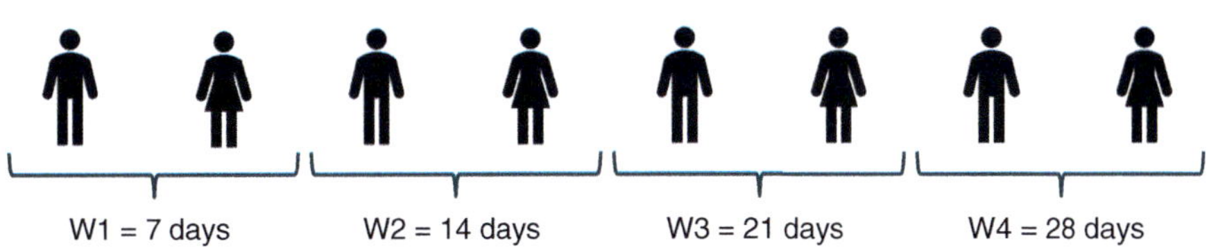

Figure 6.3 PTRM Example Week 2

Assuming the subject randomizes and is dosed on the same day, they will consume one kit out of the site on-hand inventory, taking that quantity from five kits down to four kits. The remaining calculations play out like this:

- SWD = 4 = 1 (QKD) * 4 ($NESV_{SW}$)
- LWD = 8 = 1 (QKD) * 8 ($NESV_{SW}$)
- ROD = "Required" because 4 < 4 (SWD) + 2 (F)
- ROQ = 8 (LWD) + 6 (C) – 4 (SAI) = 10 kits

Each input plays a role in maintaining the balance between maintaining enough supply without wasting inventory or overwhelming the site with resupply shipments:

- *Floor:* The floor setting acts as a buffer stock for unexpected events. The most common such event is when a quiet site suddenly has a burst of screening and enrollment activity. Note that the PTRM only accounts for enrolled subjects in its calculations and can't "see" subjects that are in screening. Thus, the PTRM does not plan resupply shipments to support those subjects' first dose. Consider Example 2 above but with an SAI of three kits (the floor value) and three new subjects randomizing to make four subjects total in Week 1. Demand within Week 1 is now 4 kits which is >3 kits of SAI. This is a stockout situation. For this reason, it's never advisable to set a floor quantity of only one kit. If one unexpected event occurs, such as a small burst in enrollment or if a kit gets damaged or lost, the risk of a stockout increases sharply.
- *Ceiling:* The ceiling value drives both the reorder quantity and the frequency of resupply shipments. The lower the ceiling quantity value, the lower the reorder quantity and the more frequent resupply shipments will occur. As the ceiling value increases, so too does the reorder quantity and the duration of time between shipments leading to less frequent shipments overall. Beware though because a high ceiling value will also increase the number of kits left unused at the end of the study.
- *Short Window:* The short window triggers the ROD; 14 days is a common setting for the short window because this creates a resupply order while there is sufficient site on-hand inventory to satisfy short window demand and to cover unexpected events before a resupply shipment arrives. Most clinical sites in the United States, the United Kingdom, EU, Canada, Japan, China, and Latin America can be resupplied by a depot within one week (or 2 weeks at absolute most). The 14-day short window ensures the resupply order will arrive from the depot before site on-hand inventory is consumed by patient demand over that duration. If the short window is set as less than 14 days, it increases the risk that if a resupply shipment is lost or damaged, the site could stockout before a replacement resupply shipment could arrive. If the short window is set greater than 14 days, this increases the minimum amount of inventory the site will carry through the study.
- *Long Window:* The long window behaves exactly like the ceiling value in driving reorder quantities and frequency of resupply shipments. The common settings are 28 days and 56 days for the long window because each limits the amount of inventory sent with any specific shipment to the next month or 2 months of demand.

Setting floor and ceiling values is largely driven by two factors – the IMP dispensing table (e.g. QKD at each visit) which is known and the level of recruitment and enrollment activity at the clinical site, which is more unpredictable. Due to enrollment variation, there is no magic combination of floor/ceiling values that work "best" for all sites in a study. Instead, different strategic combinations are needed to account for the range of enrollment outcomes different sites will experience.

To manage this range of outcomes, multiple PTRM strategies (e.g. low, medium, and high) can be created within the RTSM, and individual PTRM strategies can be updated. The best PTRM systems enable a user to adjust PTRM strategy inputs in real time. Chapter 13 – Advanced Concepts in Demand and Supply Planning returns to the subject of setting optimized floor/ceiling values in the context of the demand profile for the study and the inventory strategy through the upstream supply chain.

Recall that for initial supply values, the key consideration is assessing how many subjects could be expected to screen and enroll at site activation over the SRLT (e.g. within a 7-day window). When setting a floor value for PTRM, the consideration shifts to considering how many subjects could screen and enroll at any given point in time over the SRLT. For highly active sites with several potential participants, the floor should be higher than for a site with only one or two participants. Meanwhile, ceiling values should also be given careful consideration to appropriately manage how often IMP shipments will be sent. In most cases, sites prefer to receive shipments every —four to 6 weeks.

Table 6.3 captures a more complex example that ties together the IMP dispensing table, site enrollment activity, and PTRM strategies for a study.

In Table 6.4, the initial supply values are taken by multiplying the high estimate for max site enrollment projection at activation by the QKD. The floor values take the same approach with the

Table 6.3 Example 2 RTSM IMP Dispensing Table

Visit Number	Visit Name	Dose	Treatment Group	Material Description	Quantity Kits Dispensed (QKD)
1	W1 – scheduled	100 mg	1	Drug 100 mg	3
		Placebo	2	Placebo drug	3
2	W2 – scheduled	100 mg	1	Drug 100 mg	3
		Placebo	2	Placebo drug	3
3	W3 – scheduled	100 mg	1	Drug 100 mg	3
		Placebo	2	Placebo drug	3
4	W4 – scheduled	100 mg	1	Drug 100 mg	3
		Placebo	2	Placebo drug	3

Table 6.4 Example PTRM Strategy Development for Active or Placebo

	Low	Medium	High
Max Site Enrollment Projection at Activation (# Subjects)	1–2	3–5	6–8
Initial Supply Value (# Kits)	6	15	24
Max Site Enrollment Projection at Any Time (# Subjects)	1–2	3–4	5–6
Floor (# Kits)	6	12	18
Ceiling (# Kits)	12	24	36
Short Window (# Days)	14		
Long Window (# Days)	28		

maximum site enrollment projection at any time field, and the ceiling value simply doubles the floor value. It's clear from this example how important the site enrollment projections are to developing the PTRM strategies, so where do those numbers come from?

Clinical operations is responsible for recruitment and enrollment forecasts. When developing the PTRM strategies, it's worth stating again how important it is that the clinical planner work closely with clinical operations to get the best estimate of how enrollment might play out before the study starts. It's also worth reemphasizing that recruitment and enrollment are fluid over time and that means sites can move from low to medium or high and vice versa over time. Ownership of the PTRM strategies within the RTSM can belong to either clinical operations or clinical supply chain. If it belongs to clinical operations, clinical supply chain needs to be informed when site supply strategies are changed. If it belongs to clinical supply chain, the clinical planner needs to collect site activity information on a regular basis from clinical operations. Either way, collaboration between the functions is critical. Unfortunately, it's a common situation that PTRM parameters are not well understood by one group or the other and this leads to inventory mismanagement and stockout situations.

6.5 IMP Expiry Management Settings

Beyond the PTRM strategies, there are three additional IMP-related settings related to expiry management that impact how the RTSM controls inventory transfers from depot-to-site and allocation of drug from site-to-patient. These settings are known as the do not dispense (DND), do not ship (DNS), and do not count/include (DNC/DNI) parameters. The DND, DNC, and DNI all subtract a prespecified number of days from the assigned expiry date to ensure that IMP at or near expiration is not assigned to a subject or unnecessarily included in a shipment to a clinical site.

The DND parameter covers the time between subject visits to prevent a subject from receiving IMP that will expire before their next visit. For studies where IMP is administered on site (e.g. most biologic drugs), the DND parameter should be set to the number of days it takes to prepare and administer the product. In most on-site administration scenarios, this is —one to two days. For take-home products (e.g. most small-molecule drugs) though, there can be much longer durations of time between visits. In this case, the DND parameter needs to be set to the longest duration of time between visits + the visit window. For example, if the longest period between visits is 4 weeks and the visit window is ±2 days, then the DND parameter should be set to 30 days (e.g. 28-day time between visits + 2-day visit window).

The DNS parameter covers the time it takes to ship a product from a supplying depot to a clinical site. Where the DND parameter is static across all subjects, the DNS parameter can and should be customized for each supplying depot. This is because different depots may have different lead times to supply sites within their region. For example, if the lead time for a particular depot to resupply a site is 2 weeks, then the DNS parameter should be set to 15 days. This way if product expires within that 2-week window, the DNS prevents the site from receiving the product. The DND and DNS parameters are cumulative values such that if the DND is 30 days and the DNS is 15 days, the effective expiry date of a batch is reduced by a total of 45 days.

Finally, the DNC (aka include) parameter can supplement either the DND or DNS parameters or both. The DNC dates are useful for further refinements to the expiry management strategy. Adding DNC days to the DND parameters translates to triggering resupply shipments to clinical sites earlier, which is useful if sites take several days to receive and process those shipments. Adding DNC days to the DNS parameter means that depot-to-site shipments trigger earlier which is useful if depots take several days to prepare and ship resupplies to the sites.

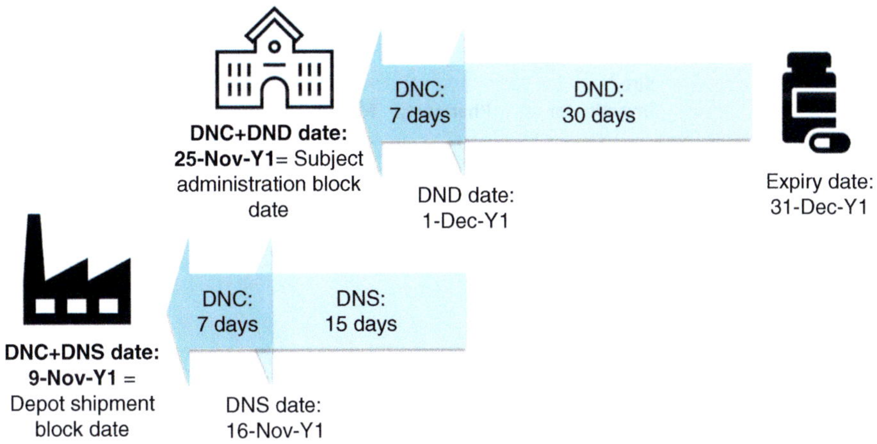

Figure 6.4 RTSM Expiry Parameters

Figure 6.4 shows how the DND, DNS, and DNC values work together to shorten the expiry date of IMP. In this example, kits of IMP will expire on December 31. The 30-day DND and 7-day DNC parameters are applied and this translates to the IMP effectively expiring on November 25. IMP at the clinical site will not be allocated to a subject after that date. Meanwhile, the RTSM "sees" that the material will no longer be available after November 25 and should trigger a resupply of material with a longer shelf life before that date. For inventory at the depot, there is a 15-day DNS parameter and another 7-day DNC parameter added to it. The DNS + DNC is cumulative to the DND and pushes the effective expiry date of this batch all the way back to November 9. Thus, IMP from this batch will not be available to resupply a clinical site after November 9.

RTSM suppliers often use different terminology when describing predictive threshold resupply methodology inputs and DND, DNS, and DNC/DNI parameters. While this chapter aims to provide a general overview of these methodologies and nomenclature, every RTSM solution, clinical trial, and supply chain will have unique nuances. Therefore, when setting up the RTSM for a study, there are a few best practices for the clinical planner to consider:

- Review both the terminology used and how the RTSM formulates the strategies and parameters described above. Variation is the norm!
- Build an offline model that reflects the PTRM supply strategies and IMP management parameters and test the inputs under a variety of conditions before the RTSM goes live to understand how the system will behave.
- Conduct those tests with the team of people (e.g. clinical operations and/or the CRO) who are responsible for controlling the inputs in the RTSM.
- Once the study begins, continuously monitor how the PTRM and IMP management parameters behave and adjust the strategies and parameters when needed.

6.6 Transactions, Alerts, and Drug Accountability

RTSMs are both transaction-based and role-based systems where the user role dictates which transactions are available to execute. Common roles include a site user who is responsible for subject-related transactions and IMP accountability, a site pharmacist for IMP dispensation

Table 6.5 RTSM Roles and Transaction Responsibilities

Transaction	Site/ Investigator	Pharmacist	Monitor	Clinical Supply	Clinical Depot
User Access			X		
Site Status			X		
Subject Screening and Randomization	X				
Subject Unblinding	X				
IMP Management Transactions		X		X	
Accountability and Reconciliation	X		X		
Returns and Destruction	X				X

transactions, a site monitor to grant user and site access, and a clinical depot role for returns and destructions. There is also a dedicated clinical supply role, and within that role, there are several transactions the clinical planner (or other clinical supply chain function) is responsible for executing. Table 6.5 lays out role-based access to several potential transactions.

The most common IMP management transactions are documented below in sequential order of potential execution from the point a batch of IMP is released and ready to be entered into the RTSM:

- *Material Release*: Material release is the most important transaction from a quality perspective because material release in the RTSM must be linked to the batch release process discussed in Chapter 4 – Quality Considerations. Only after quality has released the FP batch should the clinical planner release the batch in the RTSM. The batch should also be released at the depot location where the batch sits, and any retention samples (e.g. samples from the batch that must not be used in the study) should not be included in the release process. The most straightforward way to prevent a retain sample from shipment is to simply not enter its kit ID/sequence number in the RTSM at all.
- *Country-specific Release*: The country-specific release is an extension of the material release transaction and acts as the final gate to IMP availability to a particular country. Only after the FP batch has been released by quality and the competent health authority approval has been received can the clinical planner complete the country-specific release. The RTSM transaction represents the final step to the batch and country release-related processes introduced in Chapter 5 – Regulatory Considerations. Due to the compliance-related nature of material and country-specific releases, these transactions are typically set up in an operator–verifier structure where two independent approvals are needed to complete the transaction.
- *Depot-to-Depot Transfer*: The depot-to-depot transfer is a simple transaction to move a batch in whole or in part from one depot to another. The sequence ranges from the kit Identification list of a particular batch are needed to specify exactly which kits should be moved. For this reason, the clinical planner should always request a list of the sequence ranges assigned during the packaging process, and the sequence number should be a data field within the RTSM that can be added to inventory reports.
- *Depot Shipment Confirmation and/or Cancellation*: This transaction confirms that a depot-to-depot transfer has been successfully completed. During the transit process, kits can break, so this transaction allows the clinical planner to damage out kits that may not be useable. The other

common occurrence during a depot shipment is for a temperature excursion to take place. When this happens, the shipment is quarantined via this transaction so an assessment can take place of the excursion and the batch's suitability for use.

- *Depot Quarantine Release*: When a temperature excursion happens and the batch is dispositioned as OK for use, this transaction removes the quarantine status and makes the batch available again.
- *Manual Site Resupply*: In most cases, the initial supply values and the PTRM strategies control site shipments, but this is still an important transaction. Sometimes, a site will not be available to receive shipments over a prespecified duration of time (e.g. the holiday season) and will request additional IMP to be sent in advance. Other times, the PTRM may not catch a surprise surge in enrollment, and the clinical planner needs to act outside the automated functionality to resupply the site.
- *Site Material Status Update and Site Quarantine Release*: These transactions behave similarly to the depot shipment confirmation and depot quarantine release transaction, except for site shipments. One difference is that these transactions are usually executed by the site or pharmacist user role where the depot transactions are usually executed by the clinical supply or depot user roles.
- *Material Status Update*: Another important transaction that allows the clinical planner to update the status of the kit. There are situations where a kit is lost, damaged, or needs to be used for a purpose outside of treatment (e.g. to prime a syringe pump), and the impacted kit(s) need to have their stats reflected appropriately in the RTSM, so they aren't accidentally assigned to another subject.
- *Material Relabeling*: If an expiry extension event is anticipated during the study, the RTSM needs to have a transaction available to enable this activity. Two points are of note here. First, when the material is extended, keep in mind that the underlying batch number of the batch does not change, only the expiry date of the existing batch number. Second, the material and country-specific releases need to be executed again for the extended batch. If any regulatory restrictions on shelf-life extension have been made, those need to be reflected in the RTSM to prevent distribution of the extended batch.

While the accountability, reconciliation, returns, and destruction transactions are not usually performed by the clinical planner, it's still important to understand the RTSM workflows for each of these. This is because each of these transactions can impact inventory levels at either the site or depot, and these inventory levels are direct inputs to the clinical supply plan. It's also important when the study ends because a complete accounting of inventory status and location may be requested as part of a preapproval inspection, which was covered in Chapter 4 – Quality Considerations. The RTSM is the primary source of this information, and the results of these transactions form the basis of that analysis.

RTSMs offer a final useful bit of functionality, the ability to generate alerts that notify the clinical planner when key events happen. These events fall into a few categories. First are notifications when transactions have been successfully (or unsuccessfully) completed, such as site activation, subject randomization, subject withdrawal, or subject dispensing. More important though are alerts that notify the clinical planner about failed transactions. Specifically, the planner must be notified when a subject dispensing fails or when a site shipment fails. The first indicates an inventory issue at the site level and the second an inventory issue at the depot level. If these alerts are received, the clinical planner must take immediate action to investigate and correct the failed dispensing or shipment.

The second category of alerts is about upcoming events. Three alerts in this bucket should be made available to the clinical planner – when material is expiring at a depot or site, when inventory levels reach a predefined minimum threshold at a depot or site, and when subject randomization caps have been hit. While none of these alerts dictate immediate action, they serve as a reminder to the clinical planner that a future strategy or action is needed, whether that's replacing expiring stock, planning a resupply shipment, or revisiting the demand forecast for the study, respectively.

6.7 Reports and Their Link to Demand and Supply Planning

Another key feature of the RTSM is its ability to produce detailed reports on site status, subject enrollment, and inventory levels. All of these are useful inputs to the clinical planner in one way or another. Below are common reports, a brief description of the information contained within them, and how they can influence the demand and supply plan:

- *Subject Summary Report*: The subject summary report provides subject-specific information including the site at which they enrolled, the arm to which the subject has been randomized, the treatment type they receive, and their most recent visit number and visit date. The subject summary also provides information on subject status which is important to document when a subject changes treatment arms or withdraws from the study. This report primarily informs the demand forecast for the study but can also inform whether a resupply to a particular site is needed based on expected/unexpected changes in treatment type.
- *Subject Enrollment Report*: The subject enrollment report is a more specific iteration of the subject summary report that captures the enrollment date of each subject, the site at which they enrolled, and the country in which those sites reside. This information can then be rolled up to calculate the total number of subjects that have enrolled within a given country over a select period (e.g. weekly or monthly). The country enrollment data can then be tied back to the supplying depot. This extrapolation creates demand against the appropriate distribution node of the supply chain. Both the subject summary and subject enrollment reports can also be used to calculate dropout/withdrawal rates for the study and the average dose of patients by country and/or depot. Dropout rates are important inputs for trials in severe disease indications, while the average dose is useful in studies with multiple arms or dose titration schemes.
- *Subject Dispensing Report*: The subject dispensing report details the IMP dispensation record for each subject. This report includes the subject, their respective site, the treatment group and batch number of the kits dispensed, the quantity of kits dispensed, and the associated kit ID numbers for each kit dispensed. The subject dispensing report provides a complete history of subject demand, which is usually the largest source of demand for a clinical trial. Like the subject enrollment report, subject dispensing demand can be rolled up at the country level and then tied back to the supplying depot.
- *Site and Depot Inventory Reports*: Site and depot inventory reports show inventory levels and status at each site and depot location in the study. Inventory status types include "available/on-hand," "dispensed," "expired," "in transit," "destroyed," or "other," Where the subject enrollment and subject dispensing reports provide information to the demand forecast, the site and depot inventory reports are inputs for the supply side of the plan. The clinical planner can use both reports to determine if study demand and inventory levels are behaving as expected based on the PTRM parameters that have been set up for the sites. The site inventory report also provides information on expiring inventory that the clinical planner can use to accurately forecast expiry events.

Table 6.6 Suggested RTSM Report Data Fields

Report	Field 1	Field 2	Field 3	Field 4	Field 5	Field 6	Field 7	Field 8
Subject Summary	Subject ID number	Site ID number	Site country	Country supplying depot	Subject status	Cohort assignment	Last visit # and date	Next visit # and date
Subject Enrollment					Enrollment date	—	—	—
Subject Dispensing					Treatment group	Kit ID number	Batch number	—
Site Inventory	—					Quantity kits	Inventory status	—
Depot Inventory	Supplying depot	Treatment group	Quantity kits	Inventory status	—	—	—	—
Shipment Summary		Site ID number	Site country	Shipment status	Shipment date	Receipt date	Batch number	Quantity kits

- *Site Shipment Summary Report*: The site shipment summary report communicates depot-to-site transit details including the site, country, batch number and quantity of kits shipped, shipment date from the depot, receipt date at the site, and the shipment status. This report provides another trove of information. It enables calculation of depot-to-site lead times, which can vary from country to country. Recall that the lead time to replenish a clinical site is a key input for initial supply values and the short window of the PTRM. The site shipment summary also confirms which PTRM strategies have been selected if the clinical planner doesn't own that responsibility.

During the RTSM design process, clinical supply chain has an opportunity to provide input on report structure and the specific data fields they want captured. RTSM suppliers usually have pre-built reports and fields so if the reports don't have the required information, customization should be requested. Table 6.6 summarizes the data elements by report.

Notice that in every report where site ID number and country are listed, the supplying depot for that country is also listed. This is not a typical RTSM field, but it's one of the most important fields to ask for during the programming process because it will be referenced again and again during the planning process. The other fields to keep an eye on are dates and date formats. Most planning cycles take place weekly or monthly, so requesting a week number field or segregated month and year fields also comes in handy when analyzing RTSM reports during the planning process.

6.8 Additional IMP Management Considerations

There are a few other RTSM considerations the clinical planner should take care to understand. First is the trigger for initial site shipments. A common approach is to have the initial IMP shipment sent at site activation. This ensures the site is supplied with the applicable treatments prior to the first subject screening and enrolling. However, this also creates an opportunity for waste if a site never has a subject screen and enroll. Another option is to have shipments trigger at first patient screening. This ensures supply isn't wasted on a non-enrolling site. Triggering shipments at site activation makes the most sense in large trials where most sites will be active or studies that have very short screening windows (e.g. 1–2 days), where there wouldn't be time to supply a site with IMP at the first screening signal.

Another input to the RTSM is the subject stratification and randomization strategy of the trial. Stratification is the grouping of subjects with similar characteristics to a specific type of treatment.[1] There are instances where a clinical trial may be stratified by region or country, meaning that the RTSM will randomize equal numbers of subjects to the different arms of the study (e.g. active and placebo or high dose and low dose) within a specific geography. This scenario is helpful in supply chain management because it reduces demand variability. For example, if there is regional stratification in a two-arm study and 10 subjects enroll in a particular region, the RTSM will randomize half the subjects into the active arm and the other half into the placebo arm. With global (or no) stratification, those same 10 subjects could be randomized 6:4, 7:3, 8:2, 9:1, or even 10:0 based on enrollment in the other countries and regions. The increased predictability of regional stratification is always preferred!

RTSMs also play an integral role in maintaining the blind. This is for reasons both obvious, such as randomization of subjects to active and placebo arms and control of treatment assignment, and subtle. One specific unblinding concern is the potential for partial unblinding of an investigator or site staff due to the PTRM resupply strategy. Partial unblinding occurs when an investigator determines that multiple subjects are on the same arm or on different arms of the study with certainty.[2] This can happen if the investigator has access to the PTRM resupply inputs (e.g. floor and ceiling values) and then tracks when shipments are triggered and how many kits are contained within the resupply. For this reason, Yu and Coleman recommend a blinded group ordering (BGO) strategy to combat this risk where additional "random" kits are included in the initial and resupply shipments to prevent partial unblinding. Their analysis is worth consideration in trials that anticipate slow enrollment and low PTRM inputs (e.g. a floor setting of two kits).

Finally, and most importantly, the clinical planner must stay engaged with the RTSM and its behavior as the study enrolls and subjects receive treatment. This is especially true for studies that have complex dose dispensation schemas where dose titration, dose reduction, and treatment switching are enabled by the protocol. When this is the case, the RTSM settings need to be checked regularly to ensure IMP is distributed on-time, in efficient quantities, and that waste is minimized. The RTSM reports provide valuable information about dosing trends and yield insights the clinical planner can take advantage of to further optimize their supply plans.

6.9 RTSM Supplier Selection

This chapter may give the impression that RTSM suppliers and solutions are largely similar. While that's a mostly fair assumption as it relates to IMP management, the truth is that there are a wide range of RTSM suppliers and solutions to choose from, and while RTSMs share common functionality and transactions, the underlying technology and the user experience with the solution will vary dramatically. Given the range of internal (e.g. clinical supply chain and clinical operations) and external (e.g. clinical site and depot) stakeholders who interact with the RTSM and the criticality of the data it warehouses, selecting a strong RTSM supplier and an optimal technology platform for the trial is critical.

When selecting an RTSM, there are a few factors that deserve special consideration, and these can be divided into technology- and supplier-related buckets. With respect to the system, the first check should be on basic functionality. Most RTSMs have all the IMP management capabilities and transactions listed above, but special attention should be paid to alerts and reports. Not every RTSM can issue alerts without special programming (and associated cost). Similarly, most RTSMs do not come with every data field outlined in the table above. Instead, a set of standard reports are common, and some RTSM providers also offer an ad hoc report-building tool. The key with reports though is not the presentation of the report or the possibility of adding predefined fields to customized reports; it's the ability to create customized data fields (e.g. country supplying depot) that may not be part of the standard offering. The ad hoc reporting tools are only as valuable as the individual data fields that can be added to them.

The second consideration with the RTSM platform is how flexible the system is when making updates to data captured by the RTSM. Over the course of a clinical study, it's inevitable that the RTSM will not keep up with events taking place during the trial so updates to subject, site, or inventory data will need to be made to keep the RTSM aligned with reality. The data update process should be relatively quick (e.g. less than a day from request to update) while also capturing an explanation (e.g. in a comment field) for why it was made. This creates an audit trail of data changes and gives people who are transitioning onto a study an opportunity to understand a history of what happened. The level of control the clinical supply user has when making updates to the IMP management parameters (e.g. PTRM strategy inputs and DND and DNS dates) is also an important consideration. With some RTSMs, these updates can only be made by programmers on the back end which adds time and cost. More modern RTSM suppliers allow these parameters to be updated directly by the user on the front end of the solution, and this is a more desirable situation for the planner as changes can be implemented quickly.

The RTSM supplier also deserves scrutiny during the selection process. No matter how strong the technology is, unexpected events happen, and the team behind the technology plays a critical role in keeping the RTSM running smoothly. The ability to quickly handle data change requests and programming updates should be a top-ranked category so the team structure, staff turnover rates, and geographic location of the help desk and programming teams should also factor into the decision.

6.10 RTSM Trends and Future Capabilities

A few exciting trends have emerged with respect to RTSM capabilities. First is the integration of RTSM solutions with demand and supply-planning technology platforms, a trend that has emerged over the last decade. These integrations hold promise because they automatically transmit subject enrollment and IMP dispensation data into the planning system. The integration eliminates the need for the clinical planner to pull reports from the RTSM, analyze the data, and then manually transmit that information back into the planning tool of choice (e.g. spreadsheets or a dedicated planning software). While there is definite upside to this approach with respect to saving time, the clinical planner is still responsible for understanding the "why" behind the values they're seeing in their planning solution. Site and depot inventory levels and resupply shipment quantities still require careful monitoring and, depending on how the system integration is set up, this information may not be easily accessible without accessing the RTSM reports directly.

The second RTSM trend is the ability to perform more advanced demand and enrollment forecasting (e.g. stochastic modeling of enrollment outcomes) based on historical and real-time data. This functionality is exciting for trials that have constrained supply or expensive comparators where hard decisions need to be made about how much inventory should be purchased and stocked on a shelf. Cutting-edge RTSMs can perform Monte Carlo-like analyses that show the likelihood of when and where subjects will enroll to help inform depot and site stocking choices as well as optimal PTRM inputs.

6.11 Key Takeaways

1) RTSM systems link the "final mile" nodes of the clinical supply chain, the supplying depot, the clinical site, and the subject. RTSMs facilitate subject, site, and IMP management within the clinical trial and control several activities including clinical site activation, subject randomization, IMP administration, and IMP accountability and reconciliation at the conclusion of the study.
2) Modern RTSMs are cloud-based data warehouses, and they are transaction based and role based to provide an audit trail and ensure data protection across RTSM users. The system development process requires the clinical protocol, SoA, and collaboration with clinical operations, clinical supply chain, the CRO, and the RTSM provider.
3) The kit identification number is a unique code assigned to a single unit of IMP and is the basis for IMP management within the RTSM. Kit IDs are created through the MKIL and are associated to specific treatment groups within the MKIL. Kit IDs are randomized through an association to a sequence number that's also generated within the MKIL.
4) IMP supply strategies determine how clinical site will be supplied initially and resupplied through the trial. The basis for IMP supply strategy development is the IMP dispensing table and projected screening and enrollment activity within the clinical site network. If a range of enrollment activity is expected across sites (e.g. some sites will enroll many subjects while others will enroll only a few), then multiple supply strategies may be developed.
5) The PTRM is the most common resupply methodology used in RTSMs today. A PTRM strategy is composed of five variables: the initial supply value, floor value, ceiling value, short window value, and long window value. These variables work together to determine when a resupply order is needed and the quantity of kits that should be included in that resupply order.
6) The RTSM also accounts for IMP expiry through three settings, the DND, DNS, and DNC parameters. Each of these subtracts days from the IMP shelf life to guarantee that IMP is not administered to a subject or shipped to a site when that IMP is on the verge of expiring.
7) There are many transactions within the RTSM that the clinical supply role needs to perform. These include material releases, depot transfers, manual site resupplies, and material status updates. The material and country-specific release transactions are linked to the FP batch release and competent health authority approvals, respectively.
8) The RTSM also generates alerts and reports. Alerts inform the clinical planner to a failure that requires immediate attention or to a future event that requires an action to prevent a failure. Reports are a fundamental input to the demand and supply planning process. During RTSM development, a field that links site country to supplying depot should be created.
9) Additional IMP management-related considerations include the trigger for initial site shipments, the subject stratification and randomization strategy of the study, and maintenance of the blind.
10) Not all RTSM suppliers and solutions are the same. The flexibility of the solution to make data changes and the user-friendliness of the interface matters and should be evaluated during the RTSM selection process. Characteristics of the supplier such as team structure and location should also be considered.

11) The RTSM is not a set it and forget it system. It requires continual monitoring to ensure the system functions optimally and to prevent stockouts. Collaboration with clinical operations and the RTSM supplier is central to making this happen.

Notes

1 http://www.bccancer.bc.ca/our-research/participate/clinical-trials/questions/randomization-stratification-blinding.

2 https://www.ncbi.nlm.nih.gov/pmc/articles/PMC5935824/.

7

Supply Chain Planning for Clinical Trials

7.1 Framing the Road Ahead

As we transition from the basics of clinical trial supply chain management to the various aspects of planning, let's take a moment to summarize the journey so far. The "Foundations" section covered the following topics:

- *Chapter 1* – Supply Chain Management introduced supply chain management principles based on the supply chain operations reference (SCOR) model of plan, source, make, and deliver and the use of data and metrics to measure and optimize supply chain performance.
- *Chapter 2* – Introduction to Clinical Trials introduced clinical trials generally, their intrinsic and extrinsic design elements, trial types, and key stakeholders.
- *Chapter 3* – Introduction to the Clinical Trial Supply Chain introduced the clinical supply chain (CSC) including drug modalities, the stages of manufacturing, and chemistry, manufacturing, and controls (CMCs) strategy and stakeholders.
- *Chapter 4* – Quality Considerations introduced good manufacturing practice (GMP) regulations and framework and how it impacts the CSC at each stage.
- *Chapter 5* – Regulatory Considerations introduced the regulatory application process, amendments, commitments, restrictions, and the batch and investigational medicinal product (IMP) lifecycles.
- *Chapter 6* – RTSM and Its Role in the Clinical Supply Chain introduced the randomization and trial supply management system and how it connects and controls the "last mile" of the CSC.

All of these concepts are fundamental to supply chain planning, and through the remainder of this chapter, we'll begin tying them together. Before we jump straight into the mechanics of supply chain planning, we'll frame our approach through the answers to four questions:

1) What are the objectives of the clinical planner and/or the clinical planning function?
2) What activities does the clinical planner undertake to meet these objectives?
3) What are the individual elements that compose the clinical trial demand forecast and supply plan?
4) What value proposition does the clinical planner offer to the organization?

Answering these questions provides important context for learning about demand forecasting and supply planning, why each of the foundational components matters in planning for clinical trials, how the foundational components connect each other, and how the Planner uses the demand and supply planning process to deliver value to the clinical trial subject and to the sponsor.

Supply Chain Planning for Clinical Trials – A Practical Guide, First Edition. Ryan Mills.

We'll begin this chapter with a discussion of planning objectives and how these objectives set the stage for what the clinical planner needs to do to manage the clinical trial supply chain. These actions include building the plan, refreshing the plan, and communicating the plan so the supply chain can be effectively managed through continuous change. This framework tees up a discussion on the different mathematical approaches to modeling demand and supply for clinical trials and a summary of the elements of demand and supply planning. The elements of demand and supply planning then flow into the structure of the next several chapters that compose the "Planning" section of this book. Following this, we'll introduce the business processes required to keep the plan current and aligned to the product and corporate strategy. We'll conclude this chapter with a discussion of the clinical planning function and how it drives value for the sponsor.

7.2 Clinical Planning Objectives

So what are the objectives of the clinical planning function and role? The first objective is to guarantee uninterrupted supply for each clinical trial within the program. From Chapter 1, the purpose of demand and supply planning is to ensure the *right product* is delivered in the *right quantity* to the *desired place* at the *desired time* for the *least amount of money* possible. To rephrase that in terms of clinical trial demand and supply planning, the clinical planner should ensure that *IMPs* are delivered to *all clinical trial sites and depots* in *sufficient quantities* to support *all IMP dispensing visits* when those visits happen and *accomplish this for the least amount of monetary investment possible*. Put simply, don't stockout, but don't waste money either.

To meet this objective, the clinical planner incorporates basic supply chain management principles with the intricacies and nuances of the biopharmaceutical industry. But what does that mean, exactly? It means analyzing the clinical protocol and its intrinsic and extrinsic design elements to create a demand forecast for the study. It means taking the drug modality, the physical supply chain, and the broader CMC strategy into account when building the supply plan. It means layering in quality considerations such as batch release lead times and expiry period extensions into the supply plan. It means understanding the product's regulatory strategy and outcomes to ensure IMP is shipped to all sites in a compliant manner throughout the IMP lifecycle. And, it means developing a thoughtful RTSM design that is connected to the demand and supply plan.

Supporting individual trial supply is not the only goal of the clinical planner, though. From Chapter 2, the clinical planner needs to balance the supply requirements of multiple ongoing clinical trials within a program simultaneously. And, from Chapter 3, the clinical planner also needs to plan sufficient drug substance (DS) and drug product (DP) supply to support manufacturing process development and analytical operations (e.g. method development and stability studies). Thus, the second objective of the clinical planner is to ensure all program material demands are met. To accomplish this, the planner takes many of the concepts listed above – supply chain management, demand and supply planning, quality considerations, and the product's regulatory strategy – and places them in the context of the product's clinical development plan and CMC strategy. The clinical development plan (CDP) and the CMC strategy identify where program-level demands come from with respect to the clinical trial landscape and manufacturing process development activities.

The third and final objective of the clinical planner is to minimize the amount of spend required to meet the first two objectives. Traditional approaches to clinical trial supply chain management place more emphasis on ensuring uninterrupted supply and meeting program material requirements. Buffers get added to the demand forecast, and then more buffers are added to the supply plan, and the result is excess production, unused inventory, and a missed opportunity to reinvest those dollars into other value-added activities within the company. The goal should be to avoid this outcome wherever possible. To summarize, the three primary objectives of the clinical planner are to:

1) Ensure uninterrupted supply of all ongoing trials within the program.
2) Ensure all program material requirements are met.
3) Minimize the amount of investment required to meet the first two objectives.

7.3 Clinical Planning Activities

If these are the objectives of the clinical planner, then what actions should the clinical planner take to meet them? To answer this, we should consider what makes clinical trial supply chain management a unique endeavor when compared to consumer supply chains or even to commercial biopharmaceutical supply chain management. These differentiating factors are best thought of in terms of how they impact the demand forecast and the supply plan for the program.

On the demand side, the good news is that the total amount of individual subject demand over the duration of the trial is known. There is a clear formula to calculate how much drug a subject consumes during the trial from data contained in the clinical protocol, and there is a clear sequence in which IMP is administered contained in the schedule of assessments. This is in contrast to a commercial drug supply chain where total patient demand is unknown and must be developed based on historical data combined with advanced statistical techniques such as Monte Carlo simulations or regression-based analyses. Unlike most commercial goods though, there is almost no historical information to rely upon when creating a clinical trial demand forecast. While a huge number of clinical trials have been conducted over the past 30–40 years, each trial is completely unique when considered through the lenses of protocol design, indication, target patient population, and competitive environment. Additionally, enrollment rate data is not publicly accessible or easily available for past trials run by outside sponsors.

The simplicity of deriving demand from the protocol and schedule of assessments is deceptive though in terms of understanding the total demand profile of the study. That's because demand variability does exist. This variability is driven by uncertainty around where subjects will randomize, when they will randomize, and which arm of the study they will randomize into. On top of these unknowns, the demand profile of a trial may change at any time due to intrinsic factors changing (e.g. addition of subjects or changes to the dosing structure) or extrinsic factors changing (e.g. changes to enrollment rate or addition of countries). Put another way, the demand forecast for a clinical trial is inaccurate from the day it's made and will continue to be inaccurate until enrollment for the study is complete and the entire demand profile of the trial is certain. Figuring out how to manage this wrongness is central to effective supply chain planning.

On the supply side, there are also multiple planning inputs that change over time during clinical development. First, the manufacturing yield from batch to batch often varies. This is driven by the relative newness of the manufacturing process because without historical production information, it's challenging to correctly predict the final yields. Only with manufacturing process experience do yield projections become relatively accurate, but even then the process is constantly getting tweaked to improve performance and to scale up to meet increased demand. Second, batch release lead times also differ from one batch to the next. The quality team does not have as much familiarity with the product or processes early in development which adds time to the batch release process, and early manufacturing runs may also experience more unplanned deviations which require additional time to close. In clinical development, it is difficult to estimate how much supply will be available from a given batch and when that batch will be available for further processing.

Supply-side variances in quantity and timing are often exacerbated when manufacturing activities are outsourced to external partners. Contract manufacturing organizations (CMOs) balance the needs of multiple clients, and this has implications for both process yields and batch release lead times. CMOs have a finite amount of available capacity to spread across multiple clients, and this balancing act often leads to unplanned schedule changes. Visibility to the manufacturing process itself is typically more limited when production is outsourced, and unforeseen issues may also arise that lead to lower-than-expected yields. Accessing documentation from a CMO can also be challenging which further delays the ability to review and release batches.

A final supply-impacting variable is that IMPs have a dynamic expiry period while the product is on the shelf at a clinical site or depot. This is not the case for commercial medications or any other degradable product. It's impossible to imagine walking into a grocery store and asking if the milk's expiration date will be extended later that day! But in clinical trial supply chain management, that possibility is a core tenet of the supply planning process.

Beyond demand and supply, the regulatory landscape is also dynamic. From Chapter 5, we know that clinical trial applications occur on a rolling basis and that supply to a particular country must be gated to the competent health authority (CHA) approval. With that said, a CHA approval sometimes leads to regulatory commitments and/or restrictions on a particular type of supply, and this has implications for the entire supply strategy of the trial. Managing regulatory requirements and developing a supply plan that's integrated with the overall regulatory strategy of the program creates another layer of complexity to manage and incorporate into the plan through the duration of the trial.

There is a common theme across demand, supply, and the regulatory landscape: change. Almost every variable of the trial can change at any point in time either before the study officially starts or while it's happening. Change also means that when a demand forecast and supply plan are built, they are already inaccurate. There are too many variables at work for the plan to reflect them all accurately all the time. This constant change means the clinical planner needs to take three actions to meet the objectives outlined above.

The first action is to build a holistic planning model that incorporates all the inputs, variables, and assumptions of the supply chain that impact demand and supply. From Chapter 1, the demand and supply plan should be end-to-end, demand-driven, and weeks-of-supply based. In clinical trial supply chain management, end-to-end means the plan includes finished product (FP), primary packaging, drug product (DP), and drug substance (DS) as well as any other GMP manufacturing or distribution steps. Demand driven means the plan needs to start with the demand forecast (e.g. the clinical protocol and trial design elements) to determine the supply requirements. And, weeks-of-supply based means that resupplies are driven by preestablished safety stock and reorder targets.

When the planning process is complete, it generates two primary outputs. The first output is the production plan which is a series of production orders for DS, DP, primary packaging, and FP that dictate how much product to produce, where to produce it, and when to produce it. The second output is the distribution plan which is a series of distribution orders for FP that dictate where FP needs to be shipped, when material needs to arrive, and in what quantities. Taken together, the production and distribution plans represent the overall supply plan for the trial, and these are the activities the CMC team is responsible for executing to ensure uninterrupted supply and to meet program demands.

Building an initial demand and supply planning model is only the first step though. The second step is to refresh the plan on a regular basis to incorporate updated inputs as subject enrollment, product yields, batch release lead times, and country-specific approvals change. If this doesn't happen, the plan quickly becomes outdated and unable to meet the program's evolving needs.

The most well-designed plans are useless though if internal and external stakeholders are not aligned with and/or acting in accordance with the plan. Therefore, the third action for the clinical planner to take is to communicate the plan and gain buy-in from the people responsible for its execution. This means the clinical planner must build strong relationships with key internal (e.g. clinical operations and CMC team members) and external (e.g. CMOs and the RTSM provider) stakeholders. Implementing a business process that facilitates plan updates on a regular cadence and supports communication of the freshly updated plan to impacted internal and external stakeholders enables this and further strengthens the ability to meet the objectives of the function/role. This is because the structure a business process provides ensures the appropriate parties are communicated to and that there is alignment across the plan and the activities required to implement it.

These three actions – build, refresh, and communicate – combine to help the clinical planner navigate the supply chain through continuous change. Building the plan, refreshing the plan, and communicating the plan give the clinical planner visibility, control, and flexibility (VCF) over the supply chain. A plan that provides end-to-end visibility of all the activities taking place within the supply chain enables control of those events as assumptions change. The clinical planner can then act prospectively if a protocol adds subjects or if there is a supply shock that constrains supply availability. A flexible approach means the clinical planner is able to adapt the structure of the plan and the planning assumptions to fit the needs of the trial designs and supply chain while also designing a supply plan that gives the planner options when the demand forecast is wrong (which it will be) or when manufacturing campaigns don't go as expected. Flexibility in the plan lowers the risk of one of these unplanned events leading to a stockout situation.

A well-designed plan accounts for all the variables impacting demand and supply, and it gives the planner VCF over the supply plan and its outputs – the production plan and the distribution plan. This translates to an optimal supply plan that ensures uninterrupted supply while also meeting other program material requirements, even in the face of constant change. And, when the plan is well communicated, there is an opportunity to evaluate the supply plan and identify further optimization opportunities to avoid spend where it doesn't make sense. Figure 7.1 illustrates the result of this framework which leads to an optimal supply plan that meets all three objectives – ensuring uninterrupted supply and all program demands are met while also minimizing unnecessary investment.

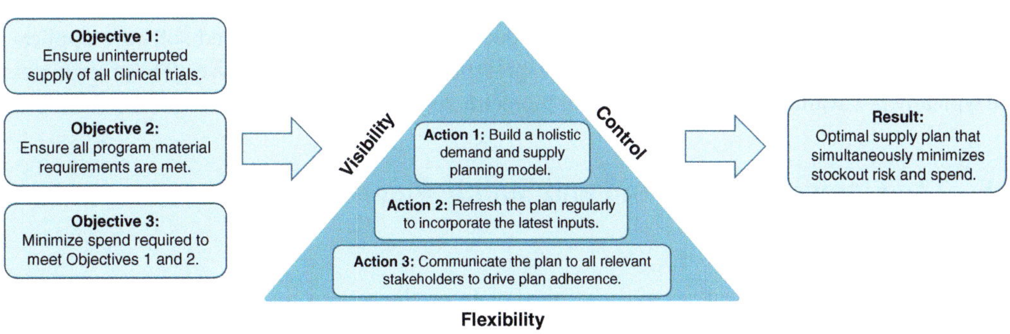

Figure 7.1 Clinical Planning Objectives and Actions Framework

7.4 Mathematical Approaches to Clinical Planning

There are three mathematical approaches used in the demand and supply planning process – simplistic, deterministic, or stochastic. The simplistic approach is exactly as it sounds, using basic addition and multiplication techniques to determine the demand forecast and supply plan for the trial. This entails calculating the total demand for the study in a single step, assuming a straight-line enrollment curve, and/or planning a single DS, DP, and FP production campaign to satisfy demand for the study. The simplistic approach is most appropriate when the intrinsic and extrinsic elements of the trial are also simple. If the trial is relatively short (e.g. <12 months in total duration), is open label (e.g. only requires a single FP configuration), and is conducted within a single country at a single clinical site, then the simplistic approach works well. Many Phase I studies (but not all!) fit this profile well and are usually appropriate for simplistic planning.

As studies get larger and more complex, the deterministic approach becomes more appropriate. Deterministic planning involves taking a mathematically based, step-by-step approach to forecasting demand and planning supply that eventually arrives at a set of clearly defined outputs. Using the deterministic approach is especially useful for planning clinical trials with complexity. That complexity may express itself on the demand side (e.g. multiple cohorts, dose titration/reduction, and geographic complexity) or on the supply side (e.g. multiple formulations and multiple production sites). The deterministic approach gives the clinical planner the tools they need to analyze this complexity and make good decisions with respect to planning supply.

The phrase deterministic is taken from the mathematical concept of eliminating randomness within the context of a closed system. Within that context, all demand forecasts and supply plans are deterministic because they are attempting to simplify highly complex systems with random elements into a set of basic arithmetic equations. The clinical planner needs to recognize that even though they are planning with a deterministic model, randomness still exists, and the plan will always be imperfect. Understanding this fact prevents the clinical planner from becoming overconfident in their plan and neglecting to keep it current as the trial moves through time and assumptions turn into reality.

Finally, the stochastic approach involves applying complex statistical methods to model the randomness that's inherent in large, complex clinical trials. Stochastic modeling generates a distribution of potential outcomes with respect to subject enrollment in terms of location and timing. These probabilities enable the clinical planner to make more educated decisions about how much inventory they should produce, which depots they should commit inventory to, and when inventory should move through the supply chain. The mechanics of stochastic modeling are beyond the scope of this book, but they are most useful in trials that are highly complex and/or have drugs that are especially expensive or difficult to provide. In these situations, trade-offs may need to be made between supplying sufficient drug and risking missing patient doses. Stochastic modeling is more resource intensive than deterministic planning because it requires specialized software applications and statistical experts who are able to interpret the outcomes of the analyses. These resources are typically not available in pre-commercial biotechnology companies.

7.5 The Elements of Demand and Supply Planning for Clinical Trials

We now understand the objectives of the clinical planner, the actions needed to meet these objectives, and the benefits of the deterministic approach to planning that we'll carry forward through the rest of this book. The actions and approach are centered around building an effective demand

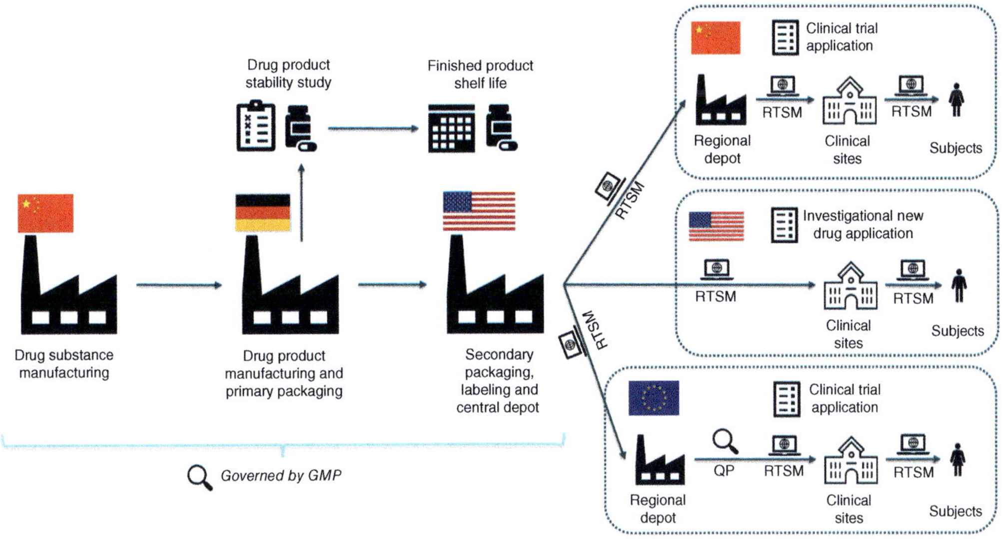

Figure 7.2 Example Small-Molecule Clinical Trial Supply Chain

forecast and supply plan, maintaining that plan through change, and communicating the plan through the organization to ensure it is implemented. We'll now turn to the individual elements of the demand forecast and the supply plan. Let's begin with Figure 7.2 which diagrams an example small-molecule clinical trial supply chain and the key activities happening within it.

In this supply chain, let's assume that all activities happen in a standard, sequential flow, meaning that nothing is done "at risk" before the upstream release has taken place. With that assumption in mind, DS manufacturing begins in China. Once the DS manufacturing campaign completes and the batch is released, the DS is shipped to a DP manufacturing site in Germany. At the German site, the DP is manufactured and packaged into its primary container. The packaged product is then put up onto stability studies to establish the shelf life of the FP. After the DP and primary packaging batches are released, the batch is shipped to the secondary packaging and labeling site in the United States to produce the FP that will ultimately be supplied to the clinical sites and participating subjects. The FP batch is labeled with an expiry date based on the date of manufacture of the bulk DP lot and the expiry period assigned via the stability study. Following the release of the FP batch, the material is now ready for distribution through the supply chain.

At this point, the FP batch is loaded into the RTSM system for distribution through the local depot network. In this supply chain, there are three groups of clinical sites – those in China, those in the United States, and those in the European Union (EU). There is a local Chinese depot to support distribution to sites in that country; the secondary packaging site also acts as the local distribution depot for sites in the United States; and, finally, there is a local depot in the EU that services clinical sites spread throughout that region. Notice that in the EU, the FP batch must be certified by the qualified person (QP) before it can be released in the RTSM for distribution to individual countries. Note too that the respective regulatory applications must also be filed and approved before shipments from the depots to the clinical sites can take place. After the product reaches the local depot and is ready for shipment, the RTSM supply strategy parameters dictate how much product is shipped to each site. Finally, subjects come to the clinical sites, go through the screening process, randomize into a specific arm of the study, and are ready to receive the appropriate kits of IMP.

Note how each of the component parts from Chapters 2 to 6 – clinical trial design, manufacturing and analytical strategy, quality requirements, regulatory applications, and the RTSM platform – is layered together to enable the end-to-end supply chain. Manufacturing can't happen without quality releasing the batch. The FP batch can't be labeled without the stability study and resulting shelf life. Distribution from depot to site and from site to subject is controlled by the RTSM. RTSM inputs are determined by the clinical protocol design. IMP delivery to sites and subject screening and enrollment can't begin without the appropriate regulatory applications in place. These are the most obvious examples, but they drive home the point that analyzing the supply chain and creating an effective plan require thinking through how a change to one planning element impacts the others.

Now let's review Figure 7.3 which diagrams that supply chain through the lens of the elements that make up clinical trial demand and supply planning.

Figure 7.3 illustrates the five elements of demand and supply planning: demand planning, distribution planning, FP supply planning, expiry planning, and upstream supply planning. These elements are described in the reverse order of how the supply chain functions because while the physical supply chain flows from DS to DP to FP to distribution to administration, the demand and supply planning process begins with demand planning and then flows back through the physical supply chain in reverse order. From Chapter 1, the demand forecast is the basis of the supply plan, and, in clinical trials, demand comes from three primary sources: subject demand, site demand, and distribution demand. Once a demand forecast is generated, the next step is to develop a distribution plan for how the product will be supplied from the secondary packaging and/or labeling site to the local depots responsible for supplying in-region sites. It's clear from this figure that the RTSM supply strategies discussed in Chapter 6 factor into both the demand forecast and the distribution plan.

After the demand forecast and distribution plan are in place, developing the FP production plan is the next step. The FP production plan documents the production orders intended to service

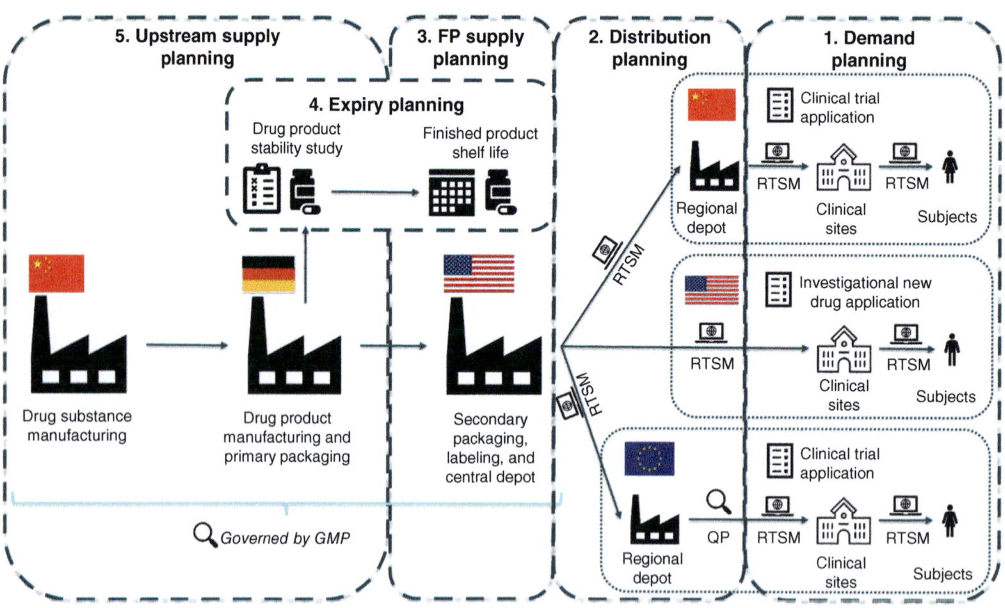

Figure 7.3 Example Small-Molecule Clinical Trial Supply Chain with Planning Elements

distribution, site, and subject demand requirements, and it forms the basis of the upstream supply plan for DP and DS. Before tackling that step though, it's important to layer the expiry management strategy into the FP supply plan. Remember that FP expiry periods may change (or not) and that the FP supply plan needs to incorporate the expiry management strategy chosen by the clinical planner. Only after expiry planning has been incorporated into the FP plan should DP and DS planning begin.

Going forward, we will also reference planning levels. This terminology refers to the segment of the CSC under discussion, so there will be a subject planning level, clinical site level, distribution depot level, FP level, primary packaging level, DP level, and DS level. Using the term level helps distinguish which aspect of the supply chain a concept, principle, or action applies to as we move into the mechanics of planning.

Following the diagram outlined in Figure 7.3, we'll introduce the various elements that make up the demand and supply plan in the order listed below:

- *Chapter 8* – Demand Planning: Basic Concepts
- *Chapter 9* – Finished Product Supply Planning: The Production Plan
- *Chapter 10* – Finished Product Expiry Planning
- *Chapter 11* – Finished Product Supply Planning: The Distribution Plan
- *Chapter 12* – Upstream Supply Planning: Primary Packaging, Drug Product, and Drug Substance
- *Chapter 13* – Advanced Concepts in Demand and Supply Planning

The content in these chapters positions us to address the first action outlined in Section 7.3 – building a plan that incorporates the appropriate inputs, generates optimal production/distribution plans, and enables VCF. Notice that these chapters are structured in a different order than the planning process outlined above because building the initial plan is different than maintaining it, and so the concepts contained within the plan make the most sense when introduced in this order. We also need to address the second action, refreshing the plan on a regular basis and ensuring it is well communicated and adhered to by the organization.

7.6 Clinical Demand and Supply Planning Business Processes

Keeping the demand forecast and supply plan updated is a challenging task for two reasons. First, there are a plethora of inputs and assumptions that need to be updated on a regular basis. Second, these inputs and assumptions come from a wide variety of data sources and stakeholders who help manage the supply chain. To ensure the demand and supply plan is in alignment with both data and stakeholders and to drive alignment across the organization, there are several planning processes that should be implemented. The three most important processes are demand and supply plan creation, demand and supply planning, and clinical demand and operations planning (CD&OP). Within CD&OP, there are two sub-processes: operational CD&OP and strategic CD&OP process. Additional processes include supply plan handover and global supply execution readiness.

These processes share some important characteristics. First, they all rely on a set of data, assumptions, and inputs that are outlined in the chapters listed above. Second, many of them require cross-functional participation and support to successfully execute. And, third, they are all interconnected to each other and reinforce each other. While these processes need to be right sized for the organization's size and maturity level, implementing one strengthens the others, and they work best when all are in place. With this background in mind, we'll transition from the demand and supply planning elements to supply chain reports and these planning processes in the order outlined below:

- *Chapter 14* – Metrics, Summary Reporting, and Budget Forecasting
- *Chapter 15* – Clinical Supply Chain Planning Processes
- *Chapter 16* – A Week in the Life of a Clinical Planner

Collectively, the content contained in these chapters describes how to communicate what's happening within the supply chain and how to keep the plan current and accurate. We spend Chapter 14 discussing how to effectively communicate the supply chain strategy and supply chain performance because this influences the planning processes described in Chapter 15. We'll also spend a week in the life of a clinical planner which brings together all the concepts discussed in the planning section of the book and illustrates the challenges and opportunities that the role brings to bear.

7.7 Alternative Clinical Supply Chain Scenarios

Up to this point, we have largely focused on straightforward clinical trial supply chains in which small-molecule and biologic products are manufactured by the sponsor and delivered to subjects at clinical sites. There are alternative CSCs though, and we'll cover these in the chapters below:

- *Chapter 17* – Comparator and Ancillary Planning
- *Chapter 18* – Just-in-Time Manufacturing and Supply Chain Management
- *Chapter 19* – Direct-to-Patient Supply Chain Planning and Execution

The first case is comparator products that are purchased on the open market and then repurposed for use in clinical trials. There are also "Just-in-Time" (JIT) supply chains which borrow from the "make-to-order" (MTO) production strategy introduced in Chapter 1. The JIT discussion includes both standard small-molecule and biologic JIT labeling as well as cell therapy products that require an end-to-end MTO strategy and therefore operate on a radically different paradigm than more standard drug modalities. The final case is decentralized clinical trial supply chains, otherwise known as a direct-to-patient supply chain. These clinical trial types were introduced in Chapter 2, and we'll discuss supply chain strategy, planning, and execution in greater detail.

7.8 The Role of the Clinical Supply Chain Function

This brings us to the final question we introduced at the beginning of the chapter, "What value proposition does Clinical Supply Chain and the Clinical Planner offer to the organization?" With the clinical planner's objectives and goals in mind, let's consider how the CSC function is organized. Recall from Chapter 3 that the CSC function is typically divided into two halves, planning and execution. The clinical planning function is responsible for delivery of the demand and supply plan and for meeting the objectives discussed above. Meanwhile, the execution function is responsible for oversight of packaging, labeling, distribution, QP certification, and integrated response technology (IRT)-related activities to move the product where it needs to go. In larger organizations, the planning (and execution) function may be further subdivided into more granular responsibilities, but the overall responsibility of the CSC function to deliver a demand and supply plan holds true no matter the organization's size.

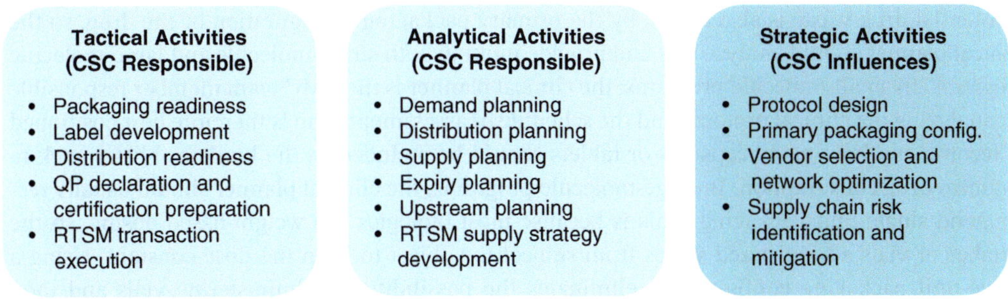

Figure 7.4 Clinical Supply Chain Activities

As discussed in Chapters 2 and 3, clinical planning primarily collaborates with two functions, development and technical operations. Development has subfunctions (e.g. clinical operations, clinical science, and clinical pharmacology) that are responsible for protocol design and execution while technical operations have subfunctions (e.g. DS and DP technical leads, analytical leads, and quality) that are responsible for developing and scaling the manufacturing process. When building and refreshing the plan, clinical planning collects demand inputs from development and supply inputs from technical operations and serves as a bridge between these two groups.

This organizational structure means that planning is often viewed as a purely analytical (not to be confused with analytical operations) function because of how it processes inputs and assumptions into supply plan outputs (e.g. production and distribution plans). And, execution is largely viewed as a tactical function that performs tasks to enable production and distribution of the physical goods. These viewpoints are mostly correct, but they're also incomplete, particularly with respect to planning which is well positioned to act as a strategic partner to both development and technical operations. Figure 7.4 contains a list of tactical and analytical activities which the CSC function is responsible for and a list of strategic activities which the CSC function can influence and support.

The tactical activities listed above were introduced in Chapter 3 (e.g. packaging, labeling, and distribution), Chapter 4 (e.g. QP certification), and Chapter 6 (e.g. RTSM supply strategy development), while the analytical activities listed above will be introduced through the next several chapters. CSC rarely owns the strategic activities listed in the figure, but it potentially wields significant influence over these decisions. Examples of this strategic input are provided below.

With respect to protocol design, the clinical planner may provide analysis of how much additional complexity and/or cost a study will incur by adding additional subjects, cohorts, or countries. While supply chain considerations are rarely the deciding factor in protocol design or regulatory amendment decisions, they are a consideration that can tip the scales one way or another. For example, if a study wants to add a new country, but this would require developing a new label design, executing additional packaging campaigns, updating the RTSM system, and then distributing and storing the product in an expensive location, the team may not consider the addition worthwhile.

A second, and often overlooked, consideration within the protocol is the visit window within the schedule of assessments for take-home products. As we'll see in Chapter 8 – Demand Planning: Basic Concepts, accounting for visit windows means supplying additional IMP to the subject at each visit. As the visit window range increases, so too does the amount of IMP administered and the amount of IMP that's potentially wasted. The clinical planner should analyze and review these consequences with the clinical science and clinical operations teams to determine if tighter visit window ranges that reduce IMP waste are an option within the schedule of assessments.

Potential drug waste is also driven by the primary packaging configuration of the drug, so the clinical planner should analyze this and provide input to both small-molecule and large-molecule programs. In small-molecule programs, the clinical planner is the CMC team member responsible for analyzing the clinical protocol and the schedule of assessments and is therefore best positioned to recommend how many capsules or tablets should be included in the bottle or blister pack to optimize drug consumption. In large-molecule programs, the clinical planner should usually recommend single-unit packaging. This is because most biologics use weight-based dosing, so the number of vials administered varies from subject to subject to keep the dose constant. Using a single-unit packaging configuration eliminates the possibility of administering vials and then throwing them away because the subjects' weight dictated that only a partial number of vials were needed to meet the dose. We'll also look more closely at the packaging configuration in Chapter 8 because it's a primary input for determining individual subject demand.

On the manufacturing side, the production plan informs the CMC team of DS and DP material delivery requirements in terms of quantity and timing. This is important information that feeds the "make-versus-buy" decisions that were discussed in Chapter 1. If the sponsor has internal manufacturing capabilities, but capacity can't meet the demand presented in the supply plan, then a decision must be taken to outsource the material. In this situation, or in a situation where the sponsor relies exclusively on CMOs for manufacturing, the production requirements should be discussed with the CMO to ensure capacity is available and the appropriate equipment trains are on hand to make reasonable batch sizes. End-to-end plans are ideal for this purpose because they show not only where the plan is in the short term (e.g. next 1–2 years), but over the medium–long term (3–5 years). Executing a technical transfer from one CMO to another is extremely costly, so selecting a supplier who can support the entire plan and avoiding that transfer can save a company millions of dollars.

Identifying supply chain risks and recommending mitigations is another strategic activity the clinical planner potentially supports. As we'll soon learn, the supply plan is built on inventory targets and is weeks-of-supply based. That means the planner sees how low (or high) inventory is running at any point in time and at any level of the supply chain. This means the planner knows when supply will become constrained in the future and has the power to inform the CMC team of when this will happen and what actions should be taken to avoid a stockout situation. These actions could be as simple as adjusting the distribution plan or as complex as recommending additional DS campaigns and/or adding another manufacturing site to the mix to meet increasing demand. On the other side of the coin, the planner also knows if the team is planning to produce an excessive and wasteful amount of drugs. In this scenario, the planner should tell the CMC team to delay or reduce production which can also save the sponsor millions of dollars.

This is by no means an exhaustive list of where the clinical planner can influence strategic decisions within the development or the CMC team. It does illustrate the potential power of the demand and supply plan outputs and planning process. To harness this power, the value proposition of the CSC function should be communicated to both sides of the organization – development and technical operations. CSC is tactical in execution, analytical in planning, and a strategic thought partner for development and technical operations.

In some sponsors, CSC reports up through the development function with a dashed line to technical operations, and, in others, the organizational structure is reversed with CSC located in technical operations with a dashed line to development. To be clear, the reporting structure doesn't matter as much as the recognition of the value proposition by the company's leadership within development and technical operations. When the value proposition is recognized in development, the clinical planner is included in discussions about protocol design and execution. And, when the value proposition is recognized in technical operations, the clinical planner is included in discussions about CMO network strategy or supply chain risk identification and mitigations. The complete truth is that CSC underpins the success of both development and technical operations.

7.9 Getting Started

The last concept to introduce before getting into the nuts and bolts of demand and supply planning is identifying when the clinical planner should engage on a new product and when the planner should build the first iteration of the demand and supply plan for upcoming studies. On the first point, it's common for the clinical planner role to formally join the CMC team and/or the team responsible for developing the product 9–12 months before the first Phase I clinical trial. This gives the clinical planner time to support the identification and selection of manufacturers within the supply chain and to influence the development of the clinical protocol. For example, if the initial Phase I study has two options for where the trial may take place, the supply chain role can provide insight into additional activities (e.g. QP certification) that may make one site more desirable than another.

In terms of building the demand and supply plan for a clinical trial, it's recommended to build a plan for all clinical trials within the CDP. In early-stage programs, there may only be assumptions available for the Phase I trial and that's obviously a fair starting point. As the program moves deeper into development and the CDP is populated with planned studies, the clinical planner should have at least a basic demand forecast and supply plan for all studies within a two- to three-year horizon. This is because the end-to-end lead time for manufacturing DS to having FP available for the clinic can take 12–18 months (or longer!). The CMC team needs insight into the overall program requirements to appropriately select the supply chain and plan material production to meet the demand. If there aren't enough assumptions available to make a reasonable demand forecast for studies outside the horizon, then the clinical planner should consider developing some general guesstimates to create a forecast. Even a rough plan is better than no plan at all to give the CMC team something to anchor too as they craft their strategy to support the program.

7.10 Key Takeaways

The objectives of the clinical planner are threefold: to ensure an uninterrupted supply of all ongoing trials within the program; to ensure all other program material demands are met; to minimize the spend required to meet the first two objectives.

1) To meet these objectives, it's important to understand how clinical trial supply chain management is unique when compared to most consumer good supply chains. Almost every demand or supply-related assumption or input within supply chain can change at almost any time.
2) To manage near-constant change, the clinical planner needs to take three actions. First, the clinical planner needs to build a holistic plan that incorporates the inputs, variables, and assumptions that impact demand and supply. Second, the clinical planner needs to refresh the plan on a regular basis. Third, the clinical planner needs to communicate the plan to the appropriate internal and external stakeholders who are responsible for its inputs and execution.
3) There are three mathematical approaches to clinical trial demand and supply planning. Simplistic planning is most appropriate for Phase I studies where total demand for the study

can be calculated exclusively from protocol design elements and resupply production campaigns will not be needed. The deterministic method should be used for all other clinical trials that have demand and/or supply complexity. The stochastic method uses complex statistical analyses to model a range of outcomes and is an excellent supplement to deterministic models if the supply chain is both highly complex and very costly to execute.

4) There are five elements to demand and supply planning: demand planning, distribution planning, FP supply planning, expiry planning, and upstream supply planning. These elements are listed in the same sequence through which the demand and supply plan is built.
5) Three business processes are needed to keep the demand and supply plan current, aligned to operational demand and supply assumptions, and aligned to the overall company strategy. These are demand and supply plan creation, demand and supply planning, and CD&OP. CD&OP can be further subdivided into operational CD&OP and strategy CD&OP.
6) The value proposition of the CSC function is tactical in execution, analytical in planning, and a strategic through partner for development and technical operations. The role of strategic thought partner can take many forms, but some examples include influencing protocol design decisions alongside development or supporting supply chain network optimization strategy within technical operations. It's not so important where the CSC functions sit within the organization; rather it is more important that the CSC value proposition is well understood and applied across both development and manufacturing.
7) The clinical planner should engage in a new program 9–12 months ahead of the target first patient in (FPI) date for the first Phase I study. A demand forecast and supply plan should be created for all clinical studies listed on the clinical development plan. If detailed assumptions aren't available, the clinical planner should work with the development team to make estimates so that a supply plan exists for the CMC team to anchor to as they build the CMC strategy.

8

Finished Product Demand Planning

Basic Concepts

8.1 Introduction to Demand Planning for Clinical Trials

Let's begin with a basic question: What is a clinical trial demand forecast? From the last chapter, we know there are three sources of demand within the trial – subject demand, clinical site stocking demand, and distribution demand. Ultimately, subjects are the final consumers of the finished product, but as we've learned from Chapter 6 – RTSM and its Role in the Clinical Supply Chain, clinical sites receive a baseline amount of supply at the study that may or may not be consumed by subjects. Taking that logic one step further, regional depots need to have supply available to meet clinical site stocking requirements and depot supply may or may not be consumed by site resupplies. Therefore, the demand forecast is the total number of finished product units that are needed to service subject demand, clinical site stocking demand, and depot stocking (e.g. distribution) demand.

Another way to answer that question is that the clinical trial demand forecast is the basis upon which all trial supply plan outputs from finished product through drug substance (DS) are based. We plan supply to meet demand after all. Unfortunately, the demand forecast is nearly always wrong because it's based on a large set of intrinsic and extrinsic trial assumptions that are dynamic and impossible to predict over time. We'll soon learn that the most important assumption driving the demand forecast is the enrollment rate, and the enrollment rate itself is based on a complex set of assumptions around screening criteria, site activations, and historical trends in the therapeutic area and indication of the study. So the question with demand forecasting is not if the forecast is wrong (because it is) but how we can forecast in a way that improves accuracy while also accepting that the forecast will never be perfect.

The answer to this is that we become less wrong and more accurate through the process of demand planning. Demand planning is the process of generating a demand forecast and then maintaining and updating that forecast as assumptions change and demand is realized over time. Demand planning is one of the most complex aspects of the demand and supply planning process because the forecast is built on a combination of mathematical assumptions and qualitative judgments. In other words, creating a demand forecast is both science and art, but there are strategies the clinical planner uses to refine their forecast and improve the accuracy of the plan. After all, the goal of the forecast is to form the basis of a supply plan that ensures uninterrupted supply at all levels – subject, site, and depot – while also minimizing the amount of waste through the supply chain.

We'll spend the rest of this chapter taking a step-by-step approach to building clinical trial demand forecasts. Before we get to the math though, let's recall the primary objective of the clinical planner: "Ensure that *Investigational Medicinal Products* are delivered to *all clinical trial sites and depots* to support *all IMP dispensing visits* when those visits happen and *accomplish this for the*

Supply Chain Planning for Clinical Trials – A Practical Guide, First Edition. Ryan Mills.

least amount of monetary investment possible." The first three underlined components of that phrase are the What (investigational medicinal product (IMP)), Where (sites and depots), and When (visits) of the demand forecast. So we'll begin this chapter by introducing what we're forecasting – the stockkeeping unit (SKU); where we forecast – at the depot level; and the demand forecast period within which the forecast will be calculated over – either weeks or months. With those concepts in hand, we'll turn to building an individual subject demand forecast which incorporates many concepts we've touched on in earlier chapters such as the schedule of assessments (SoA), visit windows, and the packaging configuration of the drug product (DP).

Once the individual demand forecast is set, the next step is to forecast the enrollment curve. The enrollment curve predicts the timing of subject enrollment through the duration of the study. The enrollment curve carries the highest degree of uncertainty within the forecasting process, so we'll spend extra time here discussing not only how to create an enrollment curve but also how to minimize that uncertainty to the greatest extent possible. With subject demand and the enrollment curve in place, we'll then turn to a discussion on overage application. Overages are used to increase the demand forecast to buffer against demand shocks, but if they're not applied carefully, they create the conditions for overproduction and unnecessary waste. We'll finish the chapter by exploring site stocking demand and how to incorporate this into the forecast. Taken together, the individual demand curve, enrollment curve, overage, and site stocking demand quantities all combine to inform the total demand forecast at a particular SKU–depot location.

Through the entirety of this chapter and subsequent chapters, we will present mathematical concepts through written words, formulas, and illustrative figures. The figures are based on a spreadsheet-like format. While many large biopharmaceutical companies use advanced planning systems and software, showing these concepts in this spreadsheet format is the best starting point for those learning how the demand and supply planning process works due to its relative simplicity. We'll also use examples from different trial types and indications to reinforce each concept under a variety of circumstances. While it's impossible to be exhaustive, our goal is to provide enough variety in the examples to reinforce each concept as much as possible.

8.2 The What of Demand Forecasting: The Stockkeeping Unit

In the world of consumer supply chain management, an SKU is a unique type of product that's made available to a consumer, and SKU categorization is the process of distinguishing SKUs from each other. SKU categorizations are made based on product characteristics, and making these distinctions is necessary to enable visibility to customer demand requirements and control over what products to make. To use a simple example, a bicycle store could produce three SKUs – a children's bike, a road bike, and a mountain bike. In the real world, SKUs are segmented based on the underlying attributes of the product (e.g. components, description, color, packaging design, and manufacturing location). This categorization is vital to inventory management within the supply chain because it enables the manufacturer/seller to appropriately organize and manage their inventory. Imagine that the same bike store only having one SKU category – a bike. There would be no mechanism to track how many children's bikes, road bikes, or mountain bikes were sold or manufactured, and figuring out how many units the store wanted to manufacture of each type would become guesswork.

Applying SKU categorization to clinical trial supply chain management is equally important at every level of the supply chain – finished product, primary packaging, DP, and DS. If SKUs are not properly managed, two risks arise with respect to inventory management and regulatory compliance. From an inventory management perspective, there's a risk that either too little or too much of a particular SKU will be made due to lack of visibility into the demand profile of the product. This risk results in underproduction that leads to inventory stockouts or overproduction that leads to waste.

The regulatory compliance risk relates to the concepts introduced in Chapter 5 – Regulatory Considerations. Without proper visibility into specific SKUs, control is lost over finished product batches and their connection to the approved regulatory application. As a quick refresher, module 3 of the regulatory application captures DS and DP characteristics such as manufacturing process, specification, and supply chain. These same characteristics underpin the SKU categorization strategy for DS and DPs. If those characteristics are amended, by definition, a new SKU is created based on the updated characteristics. If these SKUs are not appropriately managed, it creates the risk of shipping noncompliant product to a clinical depot or site.

Within clinical supply chain management, SKU categorization is determined by the attributes listed in Figure 8.1.

At the finished product level, SKUs are categorized by the underlying DP contained within the primary/secondary packaging, the packaging configuration, and by two label characteristics – the label text and the countries that are included on the label. Chapter 3 – Introduction to the Clinical Trial Supply Chain describes the individual elements contained within the label text such as the product description, protocol number, caution statements, and storage condition. To illustrate SKU categorization as it applies to small-molecule finished products, let's start with a product that has the following characteristics:

- *DP*: Wonder Drug 50 mg tablet
- *Packaging Configuration*: 30-count bottle
- *Label Text*: Protocol WD8401, Wonder Drug 50 mg tablet, store below 30°C, etc.
- *Label Countries*: The United States only
- *Secondary Packaging Site*: Clinical drug packaging pros

If any of these attributes change, it would require generating a new SKU. So if the DP is Wonder Drug 100 mg tablet, then we would create a 100 mg SKU and a 50 mg SKU. If the packaging configuration changes from 30-count bottle to 90-count bottle, then we would create 30- and 90-count SKUs. Large-molecule finished products behave similar to small molecules with respect to SKU categorization. If the underlying DP or the packaging configuration change then a new SKU is required. For example, if the original SKU is a single-unit vial, and the clinical planner decides to package two vials together into a single kit, then a new SKU is needed for the two-vial kit configuration (KC).

Finished product

- Underlying drug product
- Packaging configuration
- Label text
- Label countries
- Secondary packaging and/or labeling site

Drug product

- Dosage formulation
- Dosage strength (SM*)/ Fill volume (LM**)
- Container closure
- Manufacturing process
- Specification
- Manufacturing site

Drug substance

- Active pharmaceutical ingredient
- Manufacturing scale (LM**-only)
- Specification
- Manufacturing site

*SM = Small molecule
**LM = Large molecule/Biologic

Figure 8.1 SKU Categorization Attributes

Label text changes are noteworthy because they apply to both small molecules and large molecules, and labels are the only component of the finished product that is included in the regulatory application. If the label text changes, this is a substantial amendment to the application and requires prior approval before the new label can be introduced. Tracking finished product batches with different labels as separate SKUs enables the appropriate degree of control when introducing newly labeled product to the supply chain following regulatory approval. Similarly, if countries are added to the label, this also requires creation of a new SKU because a batch of finished product that does not include a particular country should obviously not be shipped to that country. The last finished product attribute that drives SKU categorization is the secondary packaging site. This makes intuitive sense to maintain visibility and control over where packaging and labeling activities will take place within the supply chain.

At the DP level, there are a few obvious attributes that drive SKU categorization and other less obvious attributes that require careful attention to recognize and account for them. Two of the most obvious attributes are the DP formulation and/or the strength of the DP. Examples of formulation changes include moving from a capsule formulation to a tablet formulation for a small molecule or switching from a liquid to a lyophilisate formulation for a large molecule. Creating a new dosage strength such as the 50 mg tablet example above or changing the fill volume of a large molecule from a 100 mg vial to a 200 mg vial are clear examples of needing to create new SKUs. New container closure systems such as using a new bottle size, deciding to use blisters, or introducing a new vial size (VS) are also obvious examples. Notice that formulation, strength, and container closure are obvious because they are physical aspects of the DP. It's easy to see the difference between a 100 mg and a 200 mg or a bottle and a blister pack.

The more subtle examples of DP attributes that drive SKU creation and categorization are those that compose how the DP is made – the manufacturing process, the specification, and the manufacturing sites. Manufacturing process and specification attributes are particularly tricky because updates can be made to process and specification that do not require creation of a new SKU. How then is it possible to determine when new SKUs are required when the process or specification is updated? The answer lies in Chapter 5 – Regulatory Considerations. If the process or specification change is deemed substantial via the change management process or if the change limits which countries the DP may be released to, then a new SKU is required to appropriately categorize it. This is especially true if a decision is taken to carry multiple regulatory versions of a specification for one DP.

Like finished product, the DP manufacturing site is always a driver for SKU categorization. Even if the formulation, strength, process, and specification are identical but the DP is made at two different sites, then two SKUs are needed to appropriately manage that DP. In the scenario where the DP is identical across sites, then does the finished product made from that DP also needs two SKU categories (e.g. Wonder Drug 50 mg tablet made at DP site 1 and Wonder Drug 50 mg tablet made at DP site 2)? The answer is "No" so long as both DP sites are captured in the regulatory application of each country participating in the trial. If only one manufacturing site is listed though, then the finished product should also have two different SKUs to appropriately control the supply chain.

Many of the attributes that drive DP SKU management also drive DS SKU categorization. The active pharmaceutical ingredient (e.g. the drug itself) is the clearest attribute. It's rather straightforward that if you're making a different drug, it should be a different SKU and planned for independently. Small-molecule DS manufacturing process changes are rarely significant enough in clinical development to warrant creating new SKUs. Even if the process scales from a 1 kg to 100 kg, the underlying materials and process steps are the same. The same is not quite true for biologics where increasing production scale is considered a substantial change and would therefore warrant creating a new SKU. Also like DPs, utilizing multiple versions of the specification for one DS or manufacturing the same DS at different sites requires creating different SKUs to manage the complexity of regulatory compliance.

Appropriately managing SKUs is often overlooked in clinical trial supply chain management for two reasons. First, many attributes that drive SKU complexity are not necessarily transparent to the clinical planner such as the decision to create a country-specific specification. Second, many companies don't have clear guidelines on when to create a new SKU and/or processes in place to ensure the appropriate level of SKU categorization. In other words, SKU identification and categorization require trained people executing an effective process to happen consistently because without SKU management, the clinical planner risks losing visibility and control over the supply chain.

8.3 The Where of Demand Forecasting: Depot Versus Site

The second element the demand forecast should incorporate is the location where demand meets supply. The location indicates where demand meets supply, and when these two elements combine, they become an SKU–location demand forecast. Forecasting against a specific location is important for the same reason forecasting against the appropriate SKU is important – to give the clinical planner visibility into where demand occurs and control over where to place inventory in the supply chain to support demand. To explain this concept, let's revisit the finished product supply chain in Figure 8.2, but this time with the addition of finished product SKUs.

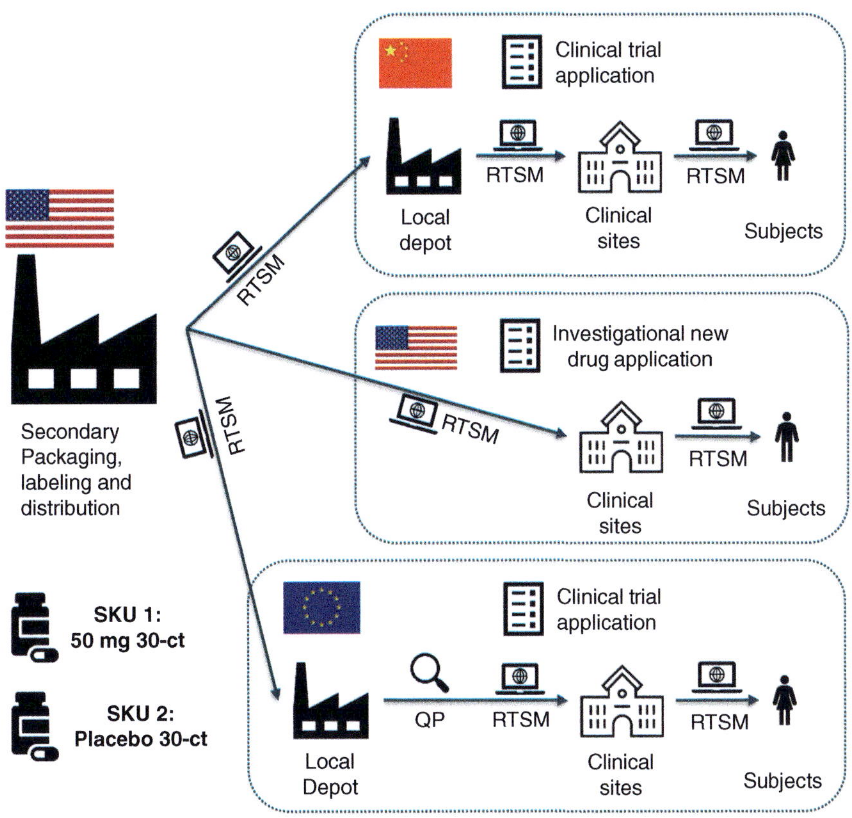

Figure 8.2 The Finished Product Supply Chain with SKU Additions

In our previous supply chain example, the SKU is packaged at a site in the United States. The secondary packaging site serves as a central depot, a depot that supplies other depots, for regional depots in China and the European Union. Regional depots differ from central depots because they only service clinical sites. In this example, the central depot also distributes the SKU directly to clinical sites located in the United States. We therefore have three potential levels of the supply chain where the SKU is either manufactured or distributed that we can choose as locations to forecast SKU demand – the clinical site, the regional depot, and/or the central depot. As we move deeper into the distribution chain from central depot to regional depot to clinical site and add more SKU–location nodes, we gain more visibility into our inventory position, but this comes at a cost. Maintaining a demand forecast and supply plan at each SKU–location requires upkeep, and this takes time and effort to maintain. Thus, deciding between these levels requires a trade-off between visibility and control over the supply chain and the level of resource intensiveness required to maintain the plan.

At first, it seems obvious to forecast at the clinical site level because that's where the subjects are and where the finished product kits will ultimately be dispensed. This forecasting strategy works for single-site studies where all subjects will eventually enroll, but as sites are added, forecasting at the clinical site level becomes increasingly resource intensive for two reasons. First, if clinical trials use multiple sites, the amount of effort to maintain the plan increases, and this is further exacerbated when multiple SKUs are provided in the study. Imagine a Phase II study with 3 SKUs and 20 sites. In this scenario 60(!), SKU–site forecasts need to be maintained if the forecast is kept at the site level. The second reason forecasting at the clinical site level isn't wise is because demand variability increases through the supply chain from central depot to regional depot to clinical site. In other words, the least amount of demand information is certain at the clinical site level. We don't know how many subjects will enroll at a particular site or which arm(s) of the study those subjects will randomize into at a site. This lack of certainty makes forecasting at the site level much more difficult when compared to forecasting at the regional or central depot level. More demand-related information is known at the regional depot level (e.g. the number of clinical sites that will be supported out of that depot) or the central depot level (e.g. total number of subjects who will enroll across the trial).

Instead, the appropriate level to create demand forecasts is at the SKU–depot level for all central and regional depots participating in the study. The SKU–depot combination offers sufficient visibility to control the clinical supply chain without becoming overwhelming. Even the largest global clinical trials usually top out at 20–30 regional depots, and these studies are often Phase III trials with two SKUs (e.g. active and placebo/comparator). It's important to include the regional depot locations because if they are not appropriately segmented in the demand and supply plan, the clinical planner loses too much visibility and control over their inventory management strategy. Clearly, the clinical planner needs to determine how many kits of each SKU to send to the Chinese depot or the EU depot in our example supply chain. Not having the regional depot locations available in the forecast makes this task impossible.

8.4 The Demand Forecast Period: Weeks Versus Months of Supply

The third component of the demand forecast is the demand forecast period, which is the incremental duration of time over which the forecast is calculated. The demand forecast period is measured in days, weeks, or months, and like the forecast location decision there is a trade-off associated between the level of granularity the choice offers and the amount of time it takes a clinical planner to maintain. Forecasting in daily increments offers the highest degree of visibility, but this is a highly resource-intensive approach that implies updating the plan every single day. Additionally, while clinical trial attributes often change quickly, most clinical supply chains aren't nimble enough to react within days. And the good news here is that most stakeholders within clinical trial operations don't plan down to the day either because there are too many moving parts.

So the choice comes down to planning in weekly or monthly periods. Both are workable, and there are advantages and disadvantages to each approach. In consumer good supply chain management, there are arguments for monthly[1] buckets because using this period smooths out demand variability and reduces the burden of the planning cycle. On the other hand, there are arguments for weekly[2] buckets because this approach enables greater accuracy with respect to the demand profile of a good. So which approach is best for clinical trial demand forecasting?

The answer is that either approach can work well, but weekly buckets are a better choice in most cases for four reasons. First, on the demand side, the SoA in many clinical trials calls for weekly rather than monthly visits. In these situations, IMP is dispensed in weekly intervals, and it's therefore easier to create a demand forecast in weekly buckets. If the SoA calls for daily, monthly, or one-time visits and dispensations, then an alternate time interval may be considered. Second, on the supply side, forecasting in weekly buckets enables a greater level of visibility into the supply plan and in communicating when IMP will be available for site shipments. If the demand forecast period is monthly, the output of the supply plan is also measured in months. There is a large difference between communicating that supply will be available in May 2025 and communicating that supply will be available May 28, 2025, ±3 days. It's easy for stakeholders who read the May 2025 date to assume supply will be ready on May 1, 2025, not May 28. Weekly buckets also enable the clinical planner to be much more precise in their production and distribution plan summary reporting, and this protects against harmful miscommunication. Third, forecasting in weekly buckets gives the clinical planner more control over the plan because the level of detail is more granular in a weekly plan. Finally, weekly buckets give the planner more opportunities to do actual planning, and with more practice comes improved skill and performance. If the planner is touching the plan only once a month (or less often), they're not giving themselves as many repetitions as they could or probably should to get better at the job.

One theoretical drawback of forecasting in weekly buckets is the increased time it takes to maintain the plan because more data needs to be captured. This is a valid concern but forecasting in weekly buckets doesn't mean the plan needs to be updated weekly. Instead, the plan can be updated on a less frequent basis at the clinical planner's discretion. We'll explore this point more in Chapter 15 – Clinical Supply Chain Planning Processes. Another argument against planning weekly is that the lead times for clinical trial production are so long that forecasting in weekly buckets is overkill. While that statement is largely true for DS and DP manufacturing, it neglects the fact that other downstream activities such as finished product manufacturing, QP certification, and distribution are better measured in weeks. Planning accurately for these activities helps prevent start-up delays and stockouts.

Ultimately, planning in weekly buckets offers a greater degree of visibility and control than planning in monthly buckets, but unlike SKU categorization and location-based forecasting, this is a matter of degree and not kind. Planning in monthly buckets can still be effective, but it does require additional effort to refine the outputs and ensure alignment to target batch release or product shipment dates. A final practical recommendation when planning in weekly buckets is to always set your forecast to either Sunday or Monday. It doesn't matter which day is selected so long as that day is consistent across all SKU–locations and all levels (e.g. DS, DP, finished product) of the plan. If one level of the plan is on Sundays and the other on Mondays, it becomes confusing to track.

Subject	W1	W2	W3	W4	W5	W6	W7	W8	W9	W10	W11	W12
1												
2												
3												
4												
5												

SKU description: Wonder drug 50 mg, 30-ct bottle, global label
Depot location: US depot

Figure 8.3 Example Demand Forecast Table

8.5 The Demand Forecast Table

In the previous three sections of this chapter, we determined the What (SKU), Where (depot), and time-based periodic (weekly) elements of the demand forecast. Taken together, they translate into the demand forecast table captured in Figure 8.3.

The demand forecast table is designed to eventually capture the total demand forecast. Notice how all three elements are captured in the table. The SKU–location combination is in the bottom left while the forecast itself is quite simple with subject demand organized by row and the weekly buckets organized by columns. Each week begins on a Monday and is intended to capture demand for that week from the Monday through the ending Sunday. This table makes populating the forecast straightforward because the planner can focus on calculating individual subject demand first and then work through the enrollment curve second. With the demand forecast table setup, we now turn to forecasting individual subject demand.

8.6 Forecasting Individual Subject Demand

Building the individual subject forecast is a mathematical exercise that requires calculating how many kits will be dispensed to a subject within each period of the demand forecast over that subject's duration of treatment. The final result of the individual demand forecast is then populated into each subject's row in the demand forecast table. This exercise brings together a wide variety of inputs including the target dose, the formulation strength, the KC, the frequency and duration of treatment, and how often IMP is administered to a subject. There are also modality-specific nuances between small and large molecules that are worth considering. Therefore, we'll break these calculations down as much as possible to simplify the entire process. To calculate the subject demand forecast, follow these steps:

1) Identify the absolute target dose (TD_{ab}) the subject will receive as prescribed in the study protocol.
2) Identify/calculate the total available dose (TAD) within an individual unit of the drug. A single unit of the drug is measured in different ways for large molecules and small molecules. From Chapter 3 – Introduction to the Clinical Trial Supply Chain, large molecules are packaged into vials or prefilled syringes while small molecules are often manufactured as capsules, tablets, or oral suspensions. Each individual unit (e.g. each vial, syringe, capsule, or tablet) has a potential dose strength. Calculating the TAD is usually straightforward for solid-state small molecules (e.g. a 50 mg capsule or a 100 mg tablet), but liquid-based small molecules and large molecules require a simple formula to determine the TAD.

3) Identify the KC. For large-molecule products, this is straightforward when secondary packaging is done on a single-unit basis. For small-molecule products, there are nearly always multiple-unit packaging configurations (e.g. 30-ct bottles and 21-ct blisters).
4) Calculate the number of drug units per dose (NDU). For large molecules, this is measured in number of vials or syringes while for small molecules, this is measured in the number of capsules or tablets.
5) Calculate how many doses a single kit can support or how many kits are required to support a single dose. This calculation is a ratio, and it's often the case for small molecules that a single kit supports many doses where large molecules may require multiple kits to support a single dose.
6) Identify the frequency and duration of treatment from the clinical protocol. Dosing frequency may occur on a daily, weekly, or monthly basis and the duration of treatment can range from a single dose to years of treatment.
7) Identify the frequency with which subjects are administered IMP from the SoA. Note that in some cases, dosing and administration happen at the same time, and this is frequently the case for large molecules that are administered at the clinical site. For many small-molecule programs though, the product is administered by the clinic and then taken home for consumption so enough kits need to be supplied to support all potential doses between IMP administration (DbA) events.
8) Calculate the total number of DbA events. Steps 6 and 7 inform this calculation.
9) Calculate how many kits need to be dispensed at each treatment administration event. Steps 5 and 8 inform this calculation, which is straightforward when dosing and IMP administration overlap (e.g. dosing at site) but becomes more complex when IMP administration events need to cover many doses over time (e.g. dosing at home).
10) Estimate the date of the first subject's first dose. This target typically comes from clinical operations and considers the first site activation (FSA) date and the screening window prescribed within the protocol.
11) Translate the frequency derived from Step 7 and the quantity derived from Step 9 to the demand forecast table. Enter the first value in the demand period identified in Step 10.

We'll walk through each of these steps first with a small-molecule example and then with a large molecule example to highlight the nuances of each modality. Small molecules are simple in terms of identifying the TAD, but they have more complexity when calculating the quantity of kits dispensed (QKD) between administration events (e.g. Step 9) because they are often take-home products. Let's illustrate this process in our first example.

Step 1: Identify the TD_{ab}. For small molecules, the target dose is nearly always prescribed in absolute terms and is taken directly from the clinical protocol. In this example, let's assume an TD_{ab} of 100 mg.

Step 2: Identify the TAD. For solid-state small molecules such as capsules or tablets, this means identifying the dosage form strength. In this example, let's assume a 50 mg capsule as the individual drug unit. For small molecules formulated as an oral suspension, calculating the TAD behaves similar to large molecules, and the process for this is outlined in the second example below.

Step 3: Identify the KC. In this example, let's assume a 30-ct bottle configuration is used, so the KC value = 30.

Step 4: Calculate the NDU with the following formula:

$$\text{NDU}\left(\text{capsules}\right) = \frac{\text{TD}_{\text{ab}}}{\text{TAD}}$$

In this example, the NDU is two capsules per dose (100/50 = 2). Note that for solid-state small molecules, the NDU must always be a wholly divisible ratio. If this is not the case, the TD_{ab} can't be met with the available drug strength.

Step 5: Calculate how many doses a single kit supports. This is known as the available doses per kit (ADpK) and is calculated with the following formula:

$$\text{ADpK}\left(\text{doses}\right) = \frac{\text{KC}}{\text{NDU}}$$

In this example, the ADpK = 15 doses (30/2 = 15).

Step 6: Identify the dosing frequency. In this example, let's assume the drug is dosed twice a day (e.g. BID) every day for 8 weeks.

Step 7: Identify the frequency with which subjects are administered IMP within the SoA. Let's use the SoA listed in Table 8.1 as an example.

Step 8: Calculate the total number of DbA events. From Step 6, the drug is dosed twice daily. From Step 7, there are up to 13 days between administration events, and enough drug needs to be administered to cover this entire period. The number 13 comes from a subject having a 3-day window on either side of the visit. In the most extreme case, a subject could come in as early as three days before their scheduled visit date and then up to 3 days after the next scheduled visit date. For example, a subject could come in as early as day 4 for visit number 4 and then as late as day 17 for visit number 5, which totals 13 days between visits. To complete the calculation, use the following formula:

$$\text{DbA}\left(\text{doses}\right) = \text{doses per day} \times \text{days between administration events}$$

In this example, DbA = 20 doses (2 * 13 = 26).

Step 9: Calculate the QKD at each treatment administration event. This is done by dividing the number of DbA events into the ADpK and rounding up. Rounding up is required because in most cases, kits can't be partially administered and then re-used at a future visit, so the QKD must be rounded up to the nearest whole integer. This results in the following formula:

$$\text{QKD}\left(\text{units}\right) = \text{ROUNDUP}\left(\frac{\text{DbA}}{\text{ADpK}}\right)$$

Table 8.1 Example Schedule of Assessments

			Treatment Period Description (e.g. Blind or Open Label)								
Visit Description	**Screening**	**Baseline**	**Treatment Phase**								**Follow-Up**
Visit Number	1	2	3	4	5	6	7	8	9	10	11
Study Day	−21 to −2	−1	1	7	14	21	28	35	42	49	56
Study Week	−4 to 0	0	1	2	3	4	5	6	7	8	9
Visit Window		0	±3d								
Randomization			X								
Treatment Administration			X	X	X	X	X	X	X	X	

Subject	W1	W2	W3	W4	W5	W6	W7	W8	W9	W10	W11	W12
1				2	2	2	2	2	2	2	2	
2												
3												
4												
5												

SKU description: Wonder drug 50 mg, 30-ct bottle, global label
Depot location: US depot

Figure 8.4 Example Small-Molecule Demand Forecast Table

In this example, QKD = 2 kits (26/15 = 1.733 rounded up to the nearest whole integer of 2 kits). Since the treatment administration events occur weekly, then two kits will be dispensed at each event.

Step 10: Estimate the date of the first subject's first dose. Clinical operations may provide either a first patient in (FPI) target which is the date of the first dosing, or they may provide a FSA target. If the FSA is given, then the screening window (SW) should be added to the FSA date to arrive at the FPI target date:

$$\text{FPI Date} = \text{FSA Date} + \text{SW}\left(\text{days}\right)$$

In this example, let's assume an FSA sate in Week 1 and a screening window of up to 21 days which results in an FPI date in Week 4.

Step 11: Populate the demand forecast table. Taking the calculations from Steps 9 and 10 results in the demand forecast table captured in Figure 8.4.

Notice how the first dispensation visit aligns to the Week 4 FPI date and the QKD is two bottles as calculated in Step 9. Then, two bottles are dispensed at each subsequent weekly visit for eight total visits.

Beyond development of the subject demand forecast, there are two additional points worth noting from this exercise. First, the QKD aligns to the IMP dispensing table in Chapter 6 – RTSM and Its Role in the Clinical Supply Chain. There should be no difference between the values in the randomization and trial supply management (RTSM) specification and the subject demand forecast. Second, it's important to recognize that waste occurs when there is not a 1:1 ratio between the DbA events and the ADpK. In the example above, that ratio is 1.733:1 and that means out of 30 available doses administered at each event, an absolute minimum of 4 doses will be thrown away. The situation is even worse though when the visit window is taken out of the equation. In most cases, the subject will return within one day of the scheduled visit having taken between 14 and 16 doses. That leaves 14–16 doses unused which equates to roughly half of all the drug packaged for this trial going to waste. In early phase studies that are relatively small, 50% waste may be an acceptable figure, but in later-stage trials with hundreds or thousands of subjects consuming tens of thousands of doses, this packaging strategy potentially creates millions of dollars of waste. Therefore, the clinical planner, chemistry, manufacturing, and controls (CMC) team, and clinical operations should closely scrutinize the packaging configuration and the visit windows to maximize the amount of drug taken by the subject and minimize potential waste.

Let's now turn to an example with a biologic product. Large molecules often have more complexity when calculating the target dose (e.g. Step 1) because they're often dosed in relative terms based on the body weight of each subject, but they are more straightforward when calculating the QKD because the dose and the administration event happen simultaneously.

Step 1: Identify the TD_{ab}. While some large molecules are dosed in absolute terms, many have target doses based on the subject's weight to ensure the drug meets necessary exposure levels within the body. A weight-based target dose (TD_{wb}) is typically measured in mg/kg and is used to calculate the TD_{ab} by multiplying it against the estimated average weight of a subject (AWS), as in the formula below:

$$TD_{ab}\left(mg\right) = AWS\left(kg\right) \times TD_{wb}\left(\frac{mg}{kg}\right)$$

Average weight varies based on subjects' age, gender, and geographic location,[3] so research should be done with clinical operations to estimate average patient weight based on the specified participant population in the trial. In this example, let's assume the TD_{wb} is 5 mg/kg, and the estimated average subject weight is 70 kg. The TD_{ab} for this study is 350 mg (70 * 5 = 350).

Step 2: Identify the TAD. For most liquid-based large-molecule programs, each subject is administered drug from a vial that has been filled to a target concentration (TC) such as 10 mg/mL or 5 mg/mL into a standard VS such as a 5 mL or 10 mL vial. Multiplying these variables yields the TAD in a single vial:

$$TAD\left(mg\right) = TC\left(\frac{mg}{mL}\right) \times VS\left(mL\right)$$

Meanwhile, if the biologic is a lyophilisate (e.g. powder form), then the TAD is noted (e.g. 100 mg). In this example, let's assume the concentration is 20 mg/mL and the VS is 5 mL. The TAD is then 100 mg (20 * 5 = 100).

Step 3: Identify the KC. In this example, let's assume a single-unit packaging configuration with one vial per kit.

Step 4: Calculate the NDU. Like small molecules, the number of drug units is calculated by dividing the target dose into the TAD. But, unlike small molecules, the NDU has a ROUNDUP function added to it. This is because a portion of a vial is used to meet the target dose. However, once opened, vials can't be partially consumed and then re-used at a future visit, so the NDU must be rounded up to the nearest whole integer. This results in the following formula:

$$NDU\left(vials\right) = ROUNDUP\left(\frac{TD_{ab}}{TAD}\right)$$

In this example, if the TD_{ab} is 350 mg and the TAD is 100 mg, then the NDU is 4 vials (350/100 = 3.5 rounded up to 4 vials). Note that this logic also applies to liquid-based small molecules (e.g. suspensions) that can be partially administered to meet a target dose. This step is where single-unit packaging simplifies the forecasting process when compared to small molecules or multi-vial large-molecule KCs. With single unit-packaging, the NDU is equal to the QKD for an IMP administration event. This is because the subject only receives one dose between IMP administration events.

Step 5: Calculate how many doses a single kit can support. The ADpK is calculated with the same formula as with small molecules:

$$ADpK\left(doses\right) = \frac{KC}{NDU}$$

Table 8.2 Example Schedule of Assessments

			Treatment Period Description (e.g. Blind or Open Label)								
Visit Description	**Screening**	**Baseline**	**Treatment Phase**								**Follow-Up**
Visit Number	1	2	3	4	5	6	7	8	9	10	11
Study Day	−28 to −2	−1	1	7	14	21	28	35	42	49	56
Study Week	−4 to 0	0	1	2	3	4	5	6	7	8	9
Visit Window		0	±3d								
Randomization			X								
Treatment Administration			X		X		X		X		

In this example, the ADPK = 0.25 (1/4 = 0.25).

Step 6: Identify the dosing frequency. In this example, let's assume the drug is dosed once every other week (e.g. Q2W) for 8 weeks.

Step 7: Identify the frequency subjects will be administered IMP within the SoA. Let's use the SoA in Table 8.2 as an example.

Step 8: Calculate the total number of DbA events. In this example, the drug is administered simultaneously with the subject's visit to the clinic. This means there is only one dose per visit and the DbA = 1.

Step 9: Calculate the QKD at each treatment administration event using the below formula:

$$\text{QKD}\left(\text{units}\right) = \text{ROUNDUP}\left(\frac{\text{DbA}}{\text{ADpK}}\right)$$

In this example, the QKD in this example is four kits (1/0.25 = 4). Note that for large molecules, the QKD will always be the inverse of Step 5 when the DbA is equal to 1. What if the KC contained two vials instead of one vial? In this case, the ADpK would be 0.5 (2/4 = 0.5) and the QKD would be 2 (1/0.5).

Step 10: Estimate the date of the first subject's first dose. This step is agnostic to product modality so we can use the same assumptions from the small-molecule example but with a screening window of 28 days instead of 21 days:

$$\text{FPI Date} = \text{FSA Date} + \text{SW}\left(\text{days}\right)$$

Thus, the FPI date is June 1, 2025 (June 1, 2025 + 28 days = June 29, 2025).

Step 11: Populate the demand forecast table. Taking the information surmised from Steps 9 and 10, the demand forecast table for the first subject is in Figure 8.5.

Note how the first dispensation visit aligns to the Week 5 (e.g. 29-Jun-2025) FPI date and the QKD is four kits every other week. The alternate weeks are populated with zeros to make it clear that the subject is enrolled but no IMP will be administered during those weeks. Also notice the SKU description matches the example details with the drug's concentration, VS, and KC all explicitly listed to make clear what is forecasted in the demand forecast table.

One takeaway from the large-molecule example is the importance of packaging biologics into single-vial KCs for weight-based trials. To illustrate this, let's assume that 10 subjects enroll with

Subject	W1	W2	W3	W4	W5	W6	W7	W8	W9	W10	W11	W12
1					4	0	4	0	4	0	4	0
2												
3												
4												
5												

SKU description: Wonder drug 20 mg/mL, 5 mL vial, 1x kit, global label
Depot location: US depot

Figure 8.5 Example Large-Molecule Demand Forecast Table

the same TD_{ab} of 350 mg. If the lightest subject weighs 50 kg and the heaviest weighs 85 kg, that translates to NDUs of 3 vials and 5 vials, respectively. If this were a two-vial KC, one vial would go unused and then thrown away for each subject at each visit. Imagine a trial with 100 subjects taking 52 weekly doses. That's 5200 visits and if half of those visits involved throwing away an unused vial, that would translate to 2600 wasted vials. This is particularly critical for large molecules because of the relatively high cost of DS and DP manufacturing, which often runs into the millions of dollars for a single batch.

Also note that the individual subject demand forecast should be completed for each arm and/or cohort within the study. Many Phase I and Phase II studies test multiple doses across cohorts, and it's not uncommon for multiple cohorts to be supplied with the same SKU. When that's the case, the clinical planner needs to take care to go through the forecasting exercise for each cohort so as not to under- or over-forecast. Once the individual subject forecast for each of the study arms/cohorts is built, the next step is to forecast when subjects will enroll and populate the remainder of the demand forecast table.

8.7 Forecasting Enrollment

While calculating the individual subject forecast is a straightforward mathematical exercise with discrete inputs, predicting the enrollment curve – when each subject enrolls and begin treatment – relies on a more subjective set of assumptions. That's because forecasting enrollment relies on two inputs that are challenging to judge before the study begins, the enrollment rate and the site activation schedule. The enrollment rate estimates how many patients per clinical site per month (p/s/m) or patients per month (p/m) the trial will enroll. This number is nearly impossible to accurately predict ahead of the trial because it depends on several underlying variables such as the scope of recruitment activities, the inclusion/exclusion criteria of the clinical protocol, and the competitive environment of the trial. How these variables interact with each other to impact study enrollment is complex. In most cases, the enrollment picture doesn't become clear until the trial formally starts, and actual data are collected on recruitment, screening, and enrollment. Enrollment rates may also vary substantially across countries in global clinical trials because recruitment, patient populations, and the competitive environment change across regions.

Meanwhile, the site activation schedule predicts the date each site will activate and begin screening and enrolling subjects. Site activation dates are also difficult to predict because of how many activities must take place before a site is ready. These include negotiating contracts with the site, gaining regulatory approval from both the competent health authority and subject review boards, onboarding the suppliers (e.g. RTSM, assessments, and labs) required to support the study, and training the site staff on how to conduct the trial. Each site has its own process and intricacies that either accelerate or slow the activation process. During the site start up process, the clinical planner should not only document the initial site activation schedule but also work to understand how the activation process unfolds and if sites are experiencing a smooth start-up process or if there are consistent delays that more broadly impact the overall demand forecast for the study.

The bottom line is that both enrollment rate and site activation dates are dynamic targets and that means the clinical planner may need to iterate through enrollment curve forecast many times over the life of the trial. Fortunately, the process of building the enrollment curve is relatively simple and should take the following steps:

1) Collect or calculate the anticipated screen rate (SR), screen-fail rate (SFR), and/or enrollment rate from clinical operations and/or the clinical research organization (CRO) responsible for overseeing trial start-up and execution.
2) Collect the total estimated number of sites by country from clinical operations and/or the CRO.
3) Collect the estimated site activation date for each site by country from clinical operations and/or the CRO.
4) Calculate the discrete and cumulative number of sites activated over the entire study. At this point, the enrollment forecast table begins to take shape.
5) Calculate the discrete and cumulative number of subjects enrolled over the entire study at each country.
6) If applicable, split the total enrollment curve into the applicable arms of the study at each country.
7) If applicable, translate country-level subject enrollment to depot-level subject enrollment. Recall that clinical sites are not supplied on a country-level basis but rather at the depot level. Therefore, the demand forecast at the country level must be translated to the depot level to accurately reflect the supply chain.
8) Translate the enrollment curve forecast to the demand forecast table at each SKU–depot location.
9) If applicable, translate the enrollment curve forecast from monthly buckets to weekly buckets. The site activation forecast provided in Step 3 is most often given in monthly estimates rather than the weekly periodic buckets used in the recommended demand forecast table structure. This is because it's both difficult to predict exact activation dates and because those exact activation dates aren't required to develop a total demand forecast.

Like the subject demand forecast, we'll walk through two examples of how to build an enrollment curve for a clinical trial. The first example assumes site activations occur within a single country, and the second example builds country-level complexity into the calculations. The country-level detail becomes important as the demand forecast extends from a single SKU–depot forecast to the regional SKU–depot combinations required to properly plan for and supply global studies.

Step 1: Collect or calculate the anticipated SR, SFR, and/or enrollment rate. Up to this point, we have discussed the enrollment rate as the primary input to determine the enrollment curve, but we've also noted how difficult this is to accurately assess. In some cases, clinical operations and/or the CRO provide the SR for the study based on feedback from potential sites. The SR is like the enrollment rate in that it is measured in p/s/m, but the SR predicts the rate at which subjects will screen for the study, not how many subjects will enroll in the study. If the SR is provided, then the clinical planner must also request the projected SFR, which is the percentage of subjects who don't pass the inclusion/exclusion criteria and can't proceed into the study. The enrollment rate is then calculated by multiplying the SR into the SFR as below:

$$\text{ER}\left(\text{p/s/m}\right) = \text{SR} \times \text{SFR}$$

In this example, let's assume an SR of 1.0 p/s/m and an SFR of 50% which leads to a projected enrollment rate of 0.5 p/s/m.

Step 2: Collect the total estimated number of sites by country. This information again comes from clinical operations and/or the CRO of the study. In this first example, let's assume 20 participating sites all located within the United States.

Step 3: Collect the estimated site activation date for each site by country. For this example, let's assume clinical operations provide the site activation schedule listed in Table 8.3.

Step 4: Calculate the discrete and cumulative number of sites activated over the entire study. In this example, the discrete and cumulative number of sites are calculated below. Each month, the number of newly activated sites is added to the previous month's total as in Table 8.4.

Step 5: Calculate the discrete and cumulative number of subjects enrolled over the entire study at each country. To perform this calculation, both the enrollment rate and the total sample size (e.g. the total number of subjects who will enroll in the study) are needed to determine the enrollment curve. To determine the discrete number of subjects and where to place each value, follow the below sub-steps:

1) Multiply the cumulative number of sites by the enrollment rate.
2) Round the calculated value up to the nearest integer.
3) Offset each value by the number of months the screening period will last.

Once the discrete number of subjects have been totaled and populated for each month, the final step is to calculate the cumulative number of subjects until the sample size value has been reached.

In Table 8.5 note the following:

- The first month (Month 1) has no subjects enrolled because the first discrete subject value has been offset by the screening period duration of one month. This offset rolls forward into all future months.
- Where applicable (e.g. Month 4), the discrete subject value has been rounded up to the nearest whole integer.

Table 8.3 Example Site Activation Table

Country	Month 1	M2	M3	M4
United States	5	5	5	5

Table 8.4 Enrollment Table with Discrete and Cumulative Site Totals

Enrollment Rate	0.5			
	Month 1	**M2**	**M3**	**M4**
# Disc. Site	5	5	5	5
# Cum. Sites	5	10	15	20

Table 8.5 Enrollment Table with Discrete and Cumulative Subject Totals

Enrollment Rate	**0.5**							
Sample Size	**50**							
	Month 1	**M2**	**M3**	**M4**	**M5**	**M6**	**M7**	**M8**
# Disc. Site	5	5	5	5	0	0	0	0
# Cum. Sites	5	10	15	20	20	20	20	20
# Disc. Subjects	0	3	5	8	10	10	10	4
# Cum. Subjects	0	3	8	16	26	36	46	50

Table 8.6 Enrollment Table with Subject Enrollment Populated

Enrollment Rate	**0.5**								
Sample Size	**50**								
	Month 1	**M2**	**M3**	**M4**	**M5**	**M6**	**M7**	**M8**	**Total**
# Disc. Site	5	5	5	5	0	0	0	0	
# Cum. Sites	5	10	15	20	20	20	20	20	
# Disc. Subjects	0	3	5	8	10	10	10	4	
# Cum. Subjects	0	3	8	16	26	36	46	50	
Active Subjects	0	2	2	4	5	5	5	2	25
Placebo Subjects	0	1	3	4	5	5	5	2	25

- The cut-off at 50 total subjects occurs in Month 8. This table now illustrates the total enrollment curve for the trial with FPI occurring in Month 2 and last patient in occurring in Month 8.
- There is only one row for the cumulative number of subjects because this example assumes the study takes place in a single country.

Step 6: If applicable, split the total enrollment curve into the applicable arms of the study at each country. In this example, let's assume a two-arm design with an active (e.g. Wonder Drug 50 mg) and placebo (e.g. Wonder Drug placebo) with a 1:1 randomization schema. The 1:1 ratio converts to half of the subjects enrolling in the active arm and half in the placebo as in Table 8.6.

Step 7: If applicable, translate country-level subject enrollment to depot-level subject enrollment. There is only one country in this example and therefore only one depot, so this step is not applicable.

Steps 8 and 9: Translate the enrollment curve forecast to the demand forecast table at each SKU–depot location and translate the enrollment curve forecast from monthly buckets to weekly buckets. In this example, there are two SKU–depot locations, the active-US depot and the placebo-US depot. The enrollment curve is populated by taking the total subject forecast and starting it at the appropriate weekly bucket for the predicted number of subjects in that arm. To make the monthly-to-weekly translation, the simplest approach is to place all the subjects in the first weekly

Subject	W1	W2	W3	W4	W5	W6	W7	W8	W9	W10	W11	W12	W13	W14	W15	W16	W17	W18	W19	W20	W21	W22
1						2	2	2	2	2	2	2	2									
2						2	2	2	2	2	2	2	2									
3										2	2	2	2	2	2	2	2					
4										2	2	2	2	2	2	2	2					
5															2	2	2	2	2	2	2	2
6															2	2	2	2	2	2	2	2
7															2	2	2	2	2	2	2	2
8															2	2	2	2	2	2	2	2
9																			2	2	2	2
10																			2	2	2	2
Total subject demand	0	0	0	0	0	4	4	4	4	8	8	8	8	4	12	12	12	8	12	12	12	12

SKU description: Wonder drug 50 mg, 30-ct bottle, global label
Depot location: US depot

Figure 8.6 Example Demand Forecast Table with Subject Demand

bucket of the month. Let's take the individual demand forecast from the small-molecule example above and then plot it against this enrollment curve as in Figure 8.6.

Notice how the populated demand forecast table enables calculating total subject demand (TSD). TSD is the sum of the QKD to each subject within a given demand forecast period (QKD_{DFP}) as in the following formula:

$$TSD(\text{units}) = \Sigma QKD_{DFP}$$

In the second example, we'll build country-level complexity into the forecast. Let's take the same enrollment rate assumption of 0.5 p/s/m from Step 1 of the first example and begin again at Step 2.

Steps 2 and 3: Collect the total estimated number of sites by country and estimated site activation date for each site by country. In this example, let's assume clinical operations has provided the site activation schedule listed in Table 8.7.

Step 4: Calculate the discrete (DSi) and cumulative (CSi) number of sites activated over the entire study as in Table 8.8.

Table 8.7 Example Site Activation Schedule

Country	Month 1	M2	M3	M4
USA	2	3	3	2
France	0	1	1	2
Germany	0	1	2	1
China	0	0	0	2
Total	2	5	6	7

Table 8.8 Enrollment Table with Discrete and Cumulative Site Totals

Enrollment Rate	0.5			
	Month 1	**M2**	**M3**	**M4**
# Disc. Site	2	5	6	7
# Cum. Sites	2	7	13	20

Steps 5 and 6: Calculate the discrete (DSu) and cumulative (CSu) number of subjects enrolled over the entire study at each country:

$$\text{DSu}(\#\text{sub}) = \text{active sites} \times \text{ER}$$

Then split the total enrollment curve into the applicable arms of the study at each country as in Tables 8.9–8.12.

In this example, Table 8.13 sums enrollment across all countries, so the total meets the estimated sample size of 50 subjects, and the split between active and placebo subjects is 1:1 or 25 total subjects enrolled in each arm.

Step 7: If applicable, translate country-level subject enrollment to depot-level subject enrollment. In this example, the United States and China are supplied out of in-country depots, but France and Germany are supplied out of a single depot responsible for all EU-based clinical sites. Therefore, the France and Germany forecasts should be summed to reflect the forecast out of the EU depot as in Table 8.14.

Table 8.9 Enrollment Table with Subject Enrollment Populated in the United States

Enrollment Rate	**0.5**							
Country/Depot	**USA**							
	Month 1	**M2**	**M3**	**M4**	**M5**	**M6**	**M7**	**M8**
# Disc. Site	2	3	3	2	0	0	0	0
# Cum. Sites	2	5	8	10	10	10	10	10
# Disc. Subjects	0	1	3	4	5	5	5	2
# Cum. Subjects	0	1	4	8	13	18	23	25
Active Subjects	0	1	1	2	3	2	3	1
Placebo Subjects	0	0	2	2	2	3	2	1

Table 8.10 Enrollment Table with Subject Enrollment Populated in France

Enrollment Rate	**0.5**							
Country	**France**							
	Month 1	**M2**	**M3**	**M4**	**M5**	**M6**	**M7**	**M8**
# Disc. Site	0	1	1	2	0	0	0	0
# Cum. Sites	0	1	2	4	4	4	4	4
# Disc. Subjects	0	0	1	1	2	2	2	2
# Cum. Subjects	0	0	1	2	4	6	8	10
Active Subjects	0	0	1	0	1	1	1	1
Placebo Subjects	0	0	0	1	1	1	1	1

Table 8.11 Enrollment Table with Subject Enrollment Populated in Germany

Enrollment Rate	**0.5**							
Country	**Germany**							
	Month 1	**M2**	**M3**	**M4**	**M5**	**M6**	**M7**	**M8**
# Disc. Site	0	1	2	1	0	0	0	0
# Cum. Sites	0	1	3	4	4	4	4	4
# Disc. Subjects	0	0	1	2	2	2	2	2
# Cum. Subjects	0	0	1	3	5	7	9	11
Active Subjects	0	0	0	1	1	1	1	1
Placebo Subjects	0	0	1	1	1	1	1	1

Table 8.12 Enrollment Table with Subject Enrollment Populated in China

Enrollment Rate	**0.5**							
Country/Depot	**China**							
	Month 1	**M2**	**M3**	**M4**	**M5**	**M6**	**M7**	**M8**
# Disc. Site	0	0	0	2	0	0	0	0
# Cum. Sites	0	0	0	2	2	2	2	2
# Disc. Subjects	0	0	0	0	1	1	1	1
# Cum. Subjects	0	0	0	0	1	2	3	4
Active Subjects	0	0	0	0	1	0	1	0
Placebo Subjects	0	0	0	0	0	1	0	1

Table 8.13 Enrollment Table with Total Number of Active and Placebo Subjects

	Month 1	**M2**	**M3**	**M4**	**M5**	**M6**	**M7**	**M8**
Total # Cum. Subjects	0	1	6	13	23	33	43	50
Active Subjects	0	1	2	3	6	4	6	3
Placebo Subjects	0	0	3	4	4	6	4	4

Steps 8 and 9: Translate the enrollment curve forecast to the demand forecast table at each SKU–depot location and translate the enrollment curve forecast from monthly buckets to weekly buckets. In this example, there are a total of six SKU–depot locations: active-US, active-EU, active-China, and the respective placebo–depot combinations. Figure 8.7 presents an example of the placebo-EU demand forecast table that's been translated from the enrollment curve listed in Table 8.14.

Table 8.14 Enrollment Table with Subject Enrollment Populated in the European Union

Enrollment Rate	**0.5**							
Depot	**EU**							
	Month 1	**M2**	**M3**	**M4**	**M5**	**M6**	**M7**	**M8**
# Disc. Site	0	2	3	3	0	0	0	0
# Cum. Sites	0	2	5	8	8	8	8	8
# Disc. Subjects	0	0	2	3	4	4	4	4
# Cum. Subjects	0	0	2	5	9	13	17	21
Active Subjects	0	0	1	1	2	2	2	2
Placebo Subjects	0	0	1	2	2	2	2	2

Subject	W1	W2	W3	W4	W5	W6	W7	W8	W9	W10	W11	W12	W13	W14	W15	W16	W17	W18	W19	W20	W21	W22
1										2	2	2	2	2	2	2	2					
2															2	2	2	2	2	2	2	2
3															2	2	2	2	2	2	2	2
4																			2	2	2	2
5																			2	2	2	2
Total subject demand	**0**	**0**	**0**	**0**	**0**	**0**	**0**	**0**	**0**	**2**	**2**	**2**	**2**	**2**	**6**	**6**	**6**	**4**	**8**	**8**	**8**	**8**

SKU description: Wonder drug placebo, 30-ct bottle, global label
Depot location: EU depot

Figure 8.7 Example Demand Forecast Table with Subject Demand

Note how enrollment at the EU depot begins in W10 because the first site isn't activated until Month 2. The forecast table in this example cuts off at W22, but demand continues in this trial through at least W43 when the last patient last visit (LPLV) occurs, and the final treatment is administered.

A more simplistic approach to enrollment forecasting involves assuming a fixed linear rate of enrollment from start-to-finish. In the example above, the clinical planner could assume that 10 subjects/month will enroll from the first month through the last month of enrollment. Clearly, using the proposed methodology of forecasting individual subject demand and then plotting the enrollment curve involves more work than this, so why go through this process? The primary reason for using the recommended approach is that it enables better visibility and control for the clinical planner to decide how much drug needs to be manufactured before study start-up. This is because the enrollment curve not only captures the total amount of drug needed but also when that drug is needed over time. In most clinical trials, less drug is needed when the trial starts and there are fewer sites online and more is needed after all sites are active and enrolling.

Modeling a more accurate enrollment curve provides the clinical planner with the information they need to avoid overproducing drug at the outset of the study. To illustrate this concept, examine Figure 8.8 where the total demand is calculated using a linear approach versus a curved approach. Notice how the curve starts with a more gradual rate of enrollment that increases over time while the linear approach does not change. It's often the case that drugs have less expiry and shorter shelf-life dates early in the study, and overproduction may lead to unnecessary waste. Using a linear approach leads to the potential for increased waste because fewer subjects will enroll early in the study. This risk is especially potent for products that have short expiry periods and will need to be replaced early in the study.

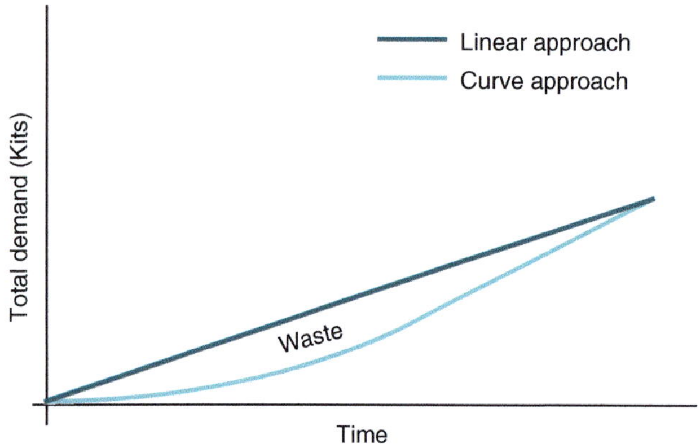

Figure 8.8 The Linear Approach versus the Curve Approach

With this concept in mind, it's also important to remember that clinical trials have a wide variety of potential enrollment curves that depend on many factors such as the therapeutic area, disease indication, and size of the trial. Generally, a broader indication (e.g. Alzheimer's disease and COVID-19) translates to a larger potential patient population, and these studies may enroll much more rapidly. As the disease becomes more rare (e.g. genetic subtypes of cancer) or more targeted, patients become more difficult to find, and enrollment may stretch out for months or years. Similarly, as the trial's sample size increases, the amount of time it takes to enroll the study also increases. While every clinical trial design is different, forecasting enrollment at a granular enough level to account for SKU complexity and depot complexity is necessary to ensure the plan meets total demand for each product in each location. Bypassing enrollment curve forecasting may be appropriate if the trial design is simple (e.g. one SKU and one depot), but as soon as SKU–location complexity increases, these extra steps are warranted.

There is a final concept worth discussion here. Recruitment activities also vary across regions, so clinical operations or the CRO may provide enrollment rate assumptions at a country level rather than at a global level. These assumptions are worked into Step 5 (calculate the discrete and cumulative number of subjects enrolled over the entire study at each country) as needed. However, beware of country-specific enrollment rate projections. They are usually less accurate than an overall study enrollment rate projection because country-specific forecasting attempts to predict both the rate at which subjects will enroll and where they will enroll. Contrast this to the study-level forecast which only attempts to predict the rate at which subjects will enroll and is agnostic to location. There is more uncertainty in a country-level forecast because the country-level forecast attempts to address two variables instead of just one. This attempt at increased precision and granularity leads to increased variability in the outcome versus the original forecast. Fortunately, there is an alternative to managing country-level demand variability in the forecast, aggregate forecasting, which is covered in Chapter 13 – Advanced Concepts in Demand and Supply Planning.

8.8 Overage

The next step in the demand forecasting process is to consider the application of an overage multiplier to the TSD forecast. An overage multiplier is expressed in percentage terms and is used to inflate the demand forecast. Common overage values used in clinical trial forecasting range from 5% to 50%, and the intent of the overage figure is to protect against demand uncertainty and/or variability. When the demand forecast is artificially inflated, additional supply is needed to meet that demand. The extra supply generated by the overage is available to hedge against otherwise unplanned demand. Examples of demand uncertainty include the following:

- Faster-than-expected enrollment.
- More high-enrolling clinical sites than planned in the forecast.
- Broken or damaged kits that need to be replaced.
- Unlimited ability to switch from one treatment (e.g. SKU 1) to another treatment (e.g. SKU 2) in the trial.
- Heavier-than-planned subject populations for weight-based products.

Figure 8.9 presents an example demand forecast table that's been updated with a 40% overage multiplier.

So what overage value should the clinical planner select? Generally, greater uncertainty in the clinical trial design and enrollment means the clinical planner should use a higher overage multiplier, and more uncertainty exists the farther away in time the plan is from study start. Early in study planning (e.g. 6–12 months before a trial begins), there may not be a firm sample size, country list, site activation schedule, or enrollment rate projection. In these situations, an overage of 30–50% may be appropriate in case the sample size increases, the country list expands, or the enrollment rate is increased from the original plan. As more information about the trial design is finalized and the study closes in on dosing the first subject, the overage may be reduced to 15–30% depending on how confident the team is that the study will not enroll faster than the demand forecast calls for. If there is optimism, the study could enroll faster than the forecast, the clinical planner should consider updating the enrollment curve instead of using overage to protect against this outcome.

Once the study begins enrolling, uncertainty continues to decrease as the subject demand picture clarifies, and by the time a study is halfway enrolled, the clinical planner may want to reduce the overage even further to 5–15%. After the study is fully enrolled, the overage should be no higher than 5% at any SKU–depot location because all enrollment demand uncertainty has been removed. The one exception to this is the unlimited ability to switch from one treatment (SKU) to another treatment over the duration of the study. In this scenario, there will always be some degree of demand uncertainty for those SKUs.

The key principle with applying overages is that they are a dynamic value that should be revisited from early in the trial planning process all the way through enrollment completion. If this value isn't regularly updated, the clinical planner risks overproducing material when demand uncertainty to the high side reduces over time and/or fails to materialize altogether.

Subject	W1	W2	W3	W4	W5	W6	W7	W8	W9	W10	W11	W12	W13	W14	W15	W16	W17	W18	W19	W20	W21	W22
1										2	2	2	2	2	2	2	2					
2															2	2	2	2	2	2	2	2
3															2	2	2	2	2	2	2	2
4																			2	2	2	2
5																			2	2	2	2
Total subject demand	**0**	**0**	**0**	**0**	**0**	**0**	**0**	**0**	**0**	**2**	**2**	**2**	**2**	**2**	**6**	**6**	**6**	**4**	**8**	**8**	**8**	**8**
Total subject demand + Overage	**0**	**0**	**0**	**0**	**0**	**0**	**0**	**0**	**0**	**3**	**3**	**3**	**3**	**3**	**8**	**8**	**8**	**6**	**11**	**11**	**11**	**11**

SKU description: Wonder drug placebo, 30-ct bottle, global label
Depot location: EU depot
Overage: 40%

Figure 8.9 Example Demand Forecast Table with Overage

8.9 Site Stocking Demand

There is a second component to the forecast that needs to be included in the demand planning process, site stocking demand. Recall from Chapter 6 – RTSM and Its Role in the Clinical Supply Chain that each site is supplied with IMP before the first subject receives treatment. From a demand perspective, these supply requirements create demand for each SKU at each depot location on top of subject demand and should therefore be included in the forecast as a separate input. For most sites, the initial site stocking may be duplicative with subject demand because when subjects enroll they will consume IMP from these initial shipments, but there are two scenarios where site stocking demand represents a unique source of demand in the plan.

The first scenario occurs when a clinical site is supplied with material and no subject enrolls at that site in the arm for which that material is intended. The initial batch of inventory sits at the site until it expires, or the trial is over. The second situation occurs if the material is labeled with an expiry date and then it expires before it's consumed by a subject. When that initial batch of material expires, the RTSM automatically triggers a resupply shipment from a newer batch with a longer expiry date, creating more demand at the depot level. In this second scenario, the clinical planner may need to create more than one placeholder for site stocking demand to represent replacement shipments for expiring material. This second situation is not uncommon because of the lengthy start-up timelines for global trials and the regulatory requirement to have the expiry date listed on the label.

Site stocking demand is additive to TSD and does not require the application of an overage. This is because site stocking demand is a predetermined value from the RTSM setup parameters or from the clinical planner's discretion if no RTSM will be used in the study. There are three places within the demand forecast that site stocking demand should be added:

1) At the beginning of the study to account for initial supply.
2) If there will be an expiry event and site resupplies need to be accounted for in the plan.
3) At the completion of enrollment to account for any sites that have unused inventory on hand.

The approach for determining the site stocking values for each event above require careful consideration. For the first event, the clinical planner should consider how many clinical sites will be supplied and how active those sites will be. If there are only a few sites and/or sites may not be active out of that depot, the clinical planner should use the full initial supply value for each site out of that depot. This is because those sites may never see a subject enroll for that SKU, and therefore site stocking demand is truly unique from subject demand. For depots supplying many clinical sites or sites that are anticipated to be highly active, a lower site stocking value (e.g. half of the initial supply value) may be used because much of this initial supply will be consumed by patients.

Let's use the same example from Figure 8.9 and assume a low initial supply value of four kits. Given that 11 total placebo subjects are planned to enroll across 8 sites, it makes sense to assume half of the low initial supply value will go to waste as unique demand because subjects are very likely to enroll and consume that supply. The output of this is presented in Figure 8.10.

For the second event, replacing expiring supply at one or many sites, the clinical planner should analyze how many sites have been supplied with IMP, how much IMP is at those sites, and when the expiry resupply shipment will be triggered based on the do not count, do not ship, and do not dispense values in the RTSM. The value of how much supply will go out should be added to the demand forecast table as a total figure. In the above example, if the first five activated sites had been supplied 4 kits each of material that expired during Week 22 and a resupply will be triggered during Week 16, then an additional value of 20 kits should be entered in the site demand section in Week 16.

Subject	W1	W2	W3	W4	W5	W6	W7	W8	W9	W10	W11	W12	W13	W14	W15	W16	W17	W18	W19	W20	W21	W22
1										2	2	2	2	2	2	2	2					
2															2	2	2	2	2	2	2	2
3															2	2	2	2	2	2	2	2
4																			2	2	2	2
5																			2	2	2	2
Total subject demand	0	0	0	0	0	0	0	0	0	2	2	2	2	2	6	6	6	4	8	8	8	8
Total subject demand + Overage	0	0	0	0	0	0	0	0	0	3	3	3	3	3	8	8	8	6	11	11	11	11
# Sites activated							1		1		1	1	1		1	1	1					
Site demand							2		2		2	2	2		2	2	2					

SKU description: Wonder drug placebo, 30-ct bottle, global label
Depot location: EU depot
Overage: 40%

Figure 8.10 Example Demand Forecast Table with Site Demand

For the last event, the clinical planner should analyze how many clinical sites have enrolled subjects and multiply this number by the floor quantity for each site to determine how many extra kits are required on top of subject demand. This only needs to be done for sites that have enrolled subjects because sites that haven't enrolled anyone will have been captured in the first and second event analyses. In nearly all cases, the overage multiplier drives enough supply to cover this last event, but for smaller trials and/or trials that are very tight on supply it makes sense to check and account for this final piece of demand. If the clinical trial only calls for one dose or a short period of dosing (e.g. less than 2 months), this last event may not be necessary to account for in the plan. Instead, the initial site stocking value in the first event should be adjusted to at least the initial supply quantity for all sites even if some sites will enroll many subjects.

8.10 Total Demand and Next Steps

With SKUs, depots, forecast periods, subject demand, enrollment curves, overages, and site stocking demand all accounted for, the final step in building the demand forecast is to total up the subject demand (with overage) and site stocking demand to arrive at the final total demand values:

$$\text{TD}(\text{units}) = \text{TSD} \times \text{OVG} + \text{SSD}$$

Figure 8.11 captures this formula where TSD is multiplied by the overage factor and then added to the site demand to arrive at total demand for the SKU–depot combination. Keep in mind, there are a wide range of scenarios that further complicate the demand forecast such as combination product dosing, dose titration, dose reduction, SKU switching, and open label extensions. These concepts are best covered after introducing the other elements of demand and supply planning, so that is where we'll turn our attention to in the next chapter. Before we turn the page on finished product demand planning, we conclude this chapter with a discussion on the challenge of accurately planning clinical trial demand.

8.11 The Challenge of Accuracy in Clinical Trial Demand Planning

The methodical approach outlined in this chapter oversimplifies demand planning in that it takes the site activation schedule and enrollment rate assumptions at face value. In the real world, these two inputs are extremely challenging to accurately predict because they are based on a highly

Subject	W1	W2	W3	W4	W5	W6	W7	W8	W9	W10	W11	W12	W13	W14	W15	W16	W17	W18	W19	W20	W21	W22
1										2	2	2	2	2	2	2	2					
2															2	2	2	2	2	2	2	2
3															2	2	2	2	2	2	2	2
4																			2	2	2	2
5																			2	2	2	2
Total subject demand	0	0	0	0	0	0	0	0	0	2	2	2	2	2	6	6	6	4	8	8	8	8
Total subject demand + Overage	0	0	0	0	0	0	0	0	0	3	3	3	3	3	8	8	8	6	11	11	11	11
# Sites activated							1		1		1	1	1		1	1	1					
Site demand							2		2		2	2	2		2	2	2					
Total demand	0	0	0	0	0	0	2	0	2	3	5	5	5	3	10	10	10	6	11	11	11	11

SKU description: Wonder drug placebo, 30-ct bottle, global label
Depot location: EU depot
Overage: 40%

Figure 8.11 Example Demand Forecast Table with Total Demand

complex network of interconnected factors that are out of the control of the sponsor to influence. Everything from site staffing levels to contract requirements to inclusion/exclusion criteria to patient outreach efforts influences how quickly sites activate and enroll subjects. To make this problem even more challenging, both clinical operations and the contract research organizations they partner with are incentivized to provide the most optimistic enrollment projections possible. This is because every sponsor wants to enroll their trial as quickly as possible, and they often set aggressive goals toward making this happen.

Consequently, it's far more common for a sponsor to overestimate site activations and enrollment and therefore overestimate the total demand curve for a trial than it is for the sponsor to underestimate demand. Once the trial moves from planning to start-up activities and into enrollment, delays happen. Site activations push out, enrollment pushes out, and the demand curve decrease in the short run and lengthen in time. The real challenge is that nobody knows exactly where site activation delays will happen and/or how much enrollment will be delayed over the duration of the study. Compounding this situation, enrollment may be slower in some regions or countries, while it's accelerated in other countries.

One approach that we'll avoid is to throw our hands up and add excessive overages to the demand forecast. Traditionally, many sponsors have added unrealistic demand assumptions such as assuming a linear, straight-line enrollment curve or that all subjects will receive all kits where dose switching is allowed to protect against forecast inaccuracy. This approach is incorrect and leads to significant amounts of unnecessary waste in the supply chain. We'll return to this topic in more detail in Chapter 13 – Advanced Concepts in Demand and Supply Planning, but it's important context to introduce now before delving into production planning.

8.12 Key Takeaways

1) There are three elements required to create a demand forecast the What (e.g. the SKU), the Where (e.g. the depot level), and the When of the forecast (e.g. the weekly or monthly demand forecast period). These combine to form the demand forecast table through which total demand is captured. To calculate total demand, individual subject demand is calculated, then the enrollment curve is completed which leads to TSD. Then, overage is applied to TSD and site stocking demand is added as a last step to reach total demand at a given SKU–depot location.

2) The SKU is a unique type of product and SKU categorization is fundamental to clinical trial supply chain management because it enables the appropriate amount of visibility to support inventory management and regulatory compliance. SKU categorization is based on the DS and DP characteristics captured in Module 3 of the investigational new drug (IND) or Investigational Medicinal Product Dossier (IMPD) regulatory application. Special attention is required to navigate manufacturing process and specification changes because if these are not deemed substantial, then a new SKU is not required.
3) The depot is the most appropriate planning level to create the demand forecast at because it enables enough visibility for the clinical planner to effectively manage inventory while also not overwhelming the planner with SKU–location combinations. If the clinical site level were used, this could lead to dozens or even hundreds of forecasts that would require maintenance.
4) The demand forecast period is the incremental duration of time (e.g. temporal basis) over which the demand forecast will be calculated. The most common approaches are to use monthly or weekly periods. Either selection is feasible, but the weekly period does offer the clinical planner a greater degree of visibility within the plan and is therefore the recommended approach.
5) Forecasting individual subject demand is an 11-step process that incorporates the target dose, formulation strength, KC, and SoA. There are unique considerations for small-molecule (e.g. visit windows) and large-molecule (e.g. KC design) modalities that impact how much potential waste is generated in the supply chain.
6) The enrollment curve is based on two assumptions, the enrollment rate and site activation schedule, both of which are notoriously difficult to accurately forecast. Calculating the enrollment curve requires a nine-step process that gives the clinical planner the information needed to create a demand forecast table at each relevant SKU–depot location. Using this process is superior to assuming a fixed linear enrollment curve because in most studies, the actual enrollment rate increases over time rather than staying at a fixed rate because site activations are staggered over weeks and months.
7) Overage is a multiplier applied to TSD, and its purpose is to protect the plan against demand uncertainty. Selecting an overage value depends on how far away trial start is and how much uncertainty there is in the overall forecast. Over time, overage should be reduced as enrollment materializes and uncertainty reduces.
8) Site stocking demand is demand created from having to supply each clinical site with inventory before participants screen and enroll in the study. It's therefore unique from subject demand if no subject enrolls at a given site or if that inventory expires before a subject consumes it. Site stocking demand is additive to TSD and taken together these two elements form the total demand forecast at a given SKU–depot location.
9) Accurately planning clinical trial demand is extremely challenging because of the unpredictability of site activations and enrollment rates. However, it's important to keep the forecast as accurate as possible because excessive demand increases potential waste in the supply chain.

Notes

1 https://insidesmallbusiness.com.au/latest-news/why-monthly-is-better-than-weekly-in-inventory-forecasting#:~:text=Forecasts%20by%20month%20are%20almost,not%20subject%20to%20over%2Dreactivity.
2 https://www.linkedin.com/pulse/should-you-forecast-monthly-weekly-stefan-de-kok/.
3 https://www.ncbi.nlm.nih.gov/pmc/articles/PMC3408371/.

9

Finished Product Supply Planning

The Production Plan

9.1 Introduction to Supply Planning

After the demand forecast is built, the next step in the planning process is to translate that forecast into a finished product supply plan. At the finished product level, the supply plan takes two forms, a production plan that describes the manufacturing strategy for a particular stockkeeping unit (SKU) and a distribution plan that describes how a SKU will move through the supply chain from one depot to another. These plans are composed of production orders and distribution orders, respectively. The supply plan and its individual production and distribution orders answer the questions of when, where, and how much of a given SKU needs to be made or distributed.

The simplest approach to determining a production order quantity (POQ) is to sum up all demand at a given SKU–depot combination and then ensure the production order is available when demand calls for it. For example, if the total demand for a given SKU–depot combination is 100 units over a 1-year period, why not plan a production order to supply 100 units for use 1 month prior to the first demand signal? In this overly simplistic example, a single batch may in fact cover total demand, but most clinical trial designs and supply chains are significantly more complex. Here are some questions to consider:

- What happens if the amount of total demand changes over time?
- What happens if the timepoint when supply is needed changes?
- What if the expiry date of the first batch doesn't last as long as demand is forecasted?
- What if the capacity of the drug product (DP) plant is less than 100 units, and therefore multiple batches are needed?

Answering these questions may require planning multiple production orders to satisfy demand, and when multiple batches are necessary, the goal is to find the most economical production strategy possible. Therefore, a systematic method to supply planning is necessary to account for capacity constraints, expiry, and other considerations that go into the plan. One common, relatively simple strategy is to focus only on a 6-month horizon where only enough finished product material is made to cover 6 months of demand. Unfortunately, this approach has three drawbacks. First, it doesn't inform the clinical planner when resupply events need to occur, which reduces visibility and increases the risk of a stockout. Second, there may be a more efficient packaging strategy involving larger campaigns that saves money and reduces resource burden. Third, it doesn't enable the clinical planner to make optimal decisions about how to plan upstream DP and drug substance (DS) campaigns.

Supply Chain Planning for Clinical Trials – A Practical Guide, First Edition. Ryan Mills.

In this chapter, we'll cover the concepts that are needed to create production orders at the finished product level and to develop a robust production plan. Ultimately, these concepts translate to both production and distribution planning, but it's more logical to start with production planning when building a supply plan for the first time. We'll begin with an introduction to the minimum–maximum (min–max) inventory management methodology and how it creates order decisions (ODs) that drive order quantities and timing. Then, we'll introduce future demand as the basis for calculating the min–max quantities that drive production order management.

Earlier in Chapter 1 – Supply Chain Management, we shared the concept of weeks-of-supply-based planning, and we'll expand on that to show how weeks-of-supply combined with the min–max methodology creates a more dynamic basis for production planning that's ideal for clinical trial demand profiles. We'll then discover how the demand forecast interacts with a set of supply planning parameters to create a requirements-based production plan. Like demand forecasting, requirements-based plans are rooted entirely in mathematical concepts and give the clinical planner a starting point to make informed and optimized decisions about when and how much to produce. Requirements-based production planning also enables the clinical planner to make updates as the demand profile for a SKU–depot combination changes. If demand decreases or increases, the production requirements adjust to reflect this and help the clinical planner see what actions need to be taken to ensure demand is satisfied while also reducing waste.

Once the initial production requirements are gathered, we'll discuss the variables – expiry, cost, and capacity – that influence the strategy behind setting the min–max parameters that generate production order requirements. We'll examine each of these three considerations individually and then show how the clinical planner should think about them collectively when making decisions about setting safety stock and order target (OT) values. Then, we'll take a deeper look at the difference between production order requirements and actual production orders and highlight important considerations for quantity and timing when turning the production plan into actual production orders. Finally, we'll conclude the chapter with the output of the planning process, the production plan summary and by highlighting the importance of production order readiness to effectively transition production orders into execution.

9.2 Min–Max Inventory Management

Min–max inventory management models use minimum and maximum threshold values to drive how much inventory is held in the supply chain. The min–max inventory management methodology succeeds because it provides a framework to make production planning decisions that focus on delivering the appropriate quantity of inventory at the right time, two of the key objectives of the clinical planner. It does this by establishing three values that generate production order requirements which are important for successfully managing on-hand inventory (OHI) levels over time:

- *Value 1*: Safety stock value that inventory levels should not fall below at any point in the planning process. The safety stock quantity (SSQ) also communicates the resupply order requirement date (RORD) when a new batch of material must be available for use. The SSQ influences both the timing and the quantity of resupply orders.
- *Value 2*: "Minimum (Min)" level of inventory that triggers a resupply order to begin so that inventory is replenished before the safety stock inventory level is breached. The Min value influences the quantity and timing of resupply order requirements.
- *Value 3*: "Maximum (Max)" level of inventory that should be manufactured in that resupply order. The Max value also influences the quantity and timing of resupply order requirements.

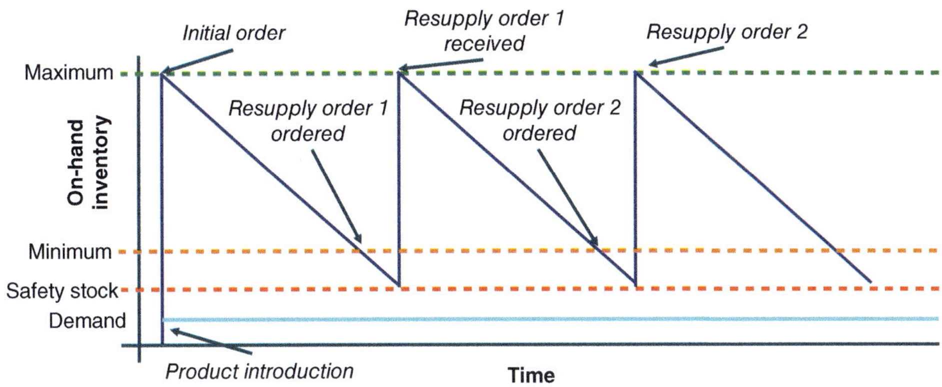

Figure 9.1 The Min–Max Inventory Management Methodology

Production order requirements are suggested production orders that are solely determined by the safety stock, min, and max values. These requirements tell the clinical planner when inventory is needed and how much is needed to replenish OHI. We'll demonstrate that two of the values listed above, safety stock and the maximum, are determined by the clinical planner, and these values provide a great deal of control over the production planning strategy. When implemented, the min–max method methodology creates a production order strategy like the one outlined in Figure 9.1.

At product launch, an initial order is generated to produce enough supply to cover demand for a predetermined period (e.g. up to the maximum quantity). Over time, demand consumes OHI until a predetermined threshold (e.g. the minimum) is reached at which point the first resupply order is triggered. From the point, the order is triggered until the order is received, demand consumes inventory week by week until the safety stock level is reached. The goal is to have the first resupply order received and ready to ship before the SSQ is breached. This pattern then repeats itself for the life of the product.

Before using the min–max methodology, it's important to understand the respective inputs – demand, OHI, the safety stock value, the minimum value, and the maximum value – and how they all work together. The first input is demand, which we have already covered extensively in Chapter 8 – Finished Product Demand Planning: Basic Concepts. Recall that the example forecasts generated in that chapter are not derived from historical analyses or other common forecasting techniques like time series, regression analysis, or seasonality assessments. This is because total demand for clinical trial material is calculated based on inputs from the protocol and study design that translate to subject demand, the enrollment curve, overage estimations, and site stocking demand. These culminate in the total demand curve for a specific SKU at a specific location.

When production planning for clinical trials, only future demand should be used as opposed to using historical (e.g. actual) demand. Future demand is defined as demand occurring in future weeks beginning at the first future week from the current week of the demand and supply plan. There are two reasons to use future demand. First, it wouldn't be logical to use any other approach because past demand has already happened, and the other statistical techniques mentioned above such as time series or regression analyses aren't applicable to clinical trial forecasting. Second, using future demand enables the min–max inputs (e.g. safety stock and maximum threshold) to create dynamic production order requirement suggestions (e.g. order timing and quantity). This is critical for clinical trial production planning management because the demand forecast values change constantly from week to week. We'll elaborate on this point further after introducing the other concepts that make up the min–max methodology. Also note that we assume future demand is measured in the same weekly periods used in Chapter 8 because the unit of measure for the safety stock calculation needs to be constant across safety stock, min, and max values. Using monthly buckets for min–max values is also appropriate as long as this period is consistent with the forecast period for demand.

The second input to the min–max methodology is OHI. OHI is simply the total amount of available inventory of a particular SKU–depot combination available for shipment in a specific planning period. It's calculated by taking OHI from the previous week (OHI_{CW-1}) subtracting total demand in the current week (TD_{CW}) and adding any production order quantities that are released in the current week (POQ_{CW}). This calculation is expressed in the below formula:

$$OHI_{CW}\left(\text{units}\right) = OHI_{CW-1} - TD_{CW} + POQ_{CW}$$

In clinical trial planning, OHI_{CW} is also thought of as the sum of all available released lots that may be shipped from a depot to a site or from a clinical site to a patient during a specific weekly/monthly planning period.

There are a few points of emphasis with respect to OHI_{CW} for clinical trial planning. First, it's the sum of all available inventory across both clinical depot and clinical sites because the forecast is maintained at the depot level, not at the site level. If the demand forecast was constructed at the clinical site level, then OHI would be the total amount of available inventory at that site. Second, OHI_{CW} is the sum of ALL available lots, not just an individual batch. Finally, OHI_{CW} needs to account for expiry. We'll return to this last concept in the next chapter, Chapter 10 – Finished Product Expiry Planning.

The next input to determine is the SSQ. To calculate safety stock, sum future demand over the lead time required to replenish (FD_{LTR}) the finished product. The lead time required to replenish the finished product should account for all production order readiness, manufacturing, and batch release lead times. This results in the formula listed below:

$$SSQ = \Sigma\left(FD_{LTR}\right)$$

The reason to use this approach is to protect against an unexpected supply shock. If there is a supply-related issue (e.g. a resupply order is lost due to a quality issue), there is enough time to start a new production event and replenish inventory levels before demand fully consumes the safety stock inventory. Therefore, the SSQ drives the **OD** of whether a production order should be created or not. The OD takes the following logic:

$$\text{OD is "Yes"} = \text{IF}\left(OHI_{CW} \le SSQ\right)$$

$$\text{OD is "No"} = \text{IF}\left(OHI_{CW} > SSQ\right)$$

If OHI_{CW} is less than or equal to the SSQ, a production order should be created. If OHI_{CW} is greater than the SSQ, then no production order is needed. To illustrate the SSQ and OD concepts, let's use an example with the following assumptions:

- Lead time to replenish (LT_r expressed in number of weeks) = 9
- Sum of future demand over LT_r (expressed in number of units) = 90
- SSQ (expressed in number of units) = 90
- $OHI_{CW+10} = 90$ units

In this example, the OD is "Yes" because the SSQ is equal to the OHI in the current week. If OHI_{CW} were 91 units or higher, then the OD would be "No."

There are a few important insights from the safety stock calculation. First, the safety stock value drives the RORD, which is the date a resupply batch of material must be available to replenish OHI. In the example above, the OD is "Yes," triggering a resupply order with a requirement date of Week 10.

Second, the SSQ reacts dynamically when the demand forecast changes. If demand increases, this pulls in the timing of the production order requirement date because demand consumes OHI faster. If demand decreases, this pushes out the timing of the production order requirement date because demand consumes inventory more slowly. If the demand forecast changes so that average weekly future demand increases to 11 units, then the SSQ rises to 99 units and the RORD pulls into Week 9. If the average weekly future demand reduces to 8 units, then the SSQ comes down to 72 units and the RORD pushes out to Week 13. Table 9.1 and Figure 9.2 clearly illustrate that as the SSQ increases, more OHI needs to be held at the depot before a resupply arrives. This need for additional inventory also brings in the desired resupply date to an earlier point.

The next insight is that the second variable influencing the SSQ, the lead time to replenish supply of that specific SKU, is also dynamic. From Chapter 3 – Introduction to the Clinical Supply Chain, we learned that the lead times for production events vary, sometimes dramatically. At the finished product level, the manufacturing lead time for labeling and packaging ranges from 1 to

Table 9.1 Scenarios for Safety Stock Quantities and Resupply Order Requirement Dates

Current Week		**Week 1**	
Lead Time to Replenish	**9**	**9**	**9**
Average Weekly Future Demand	11	10	8
Safety Stock Quantity	99	90	72
On-Hand Inventory on Week 10	90	90	90
Order Decisions	Yes	Yes	No
Resupply Order Requirement Date	Week 9	Week 10	Week 13

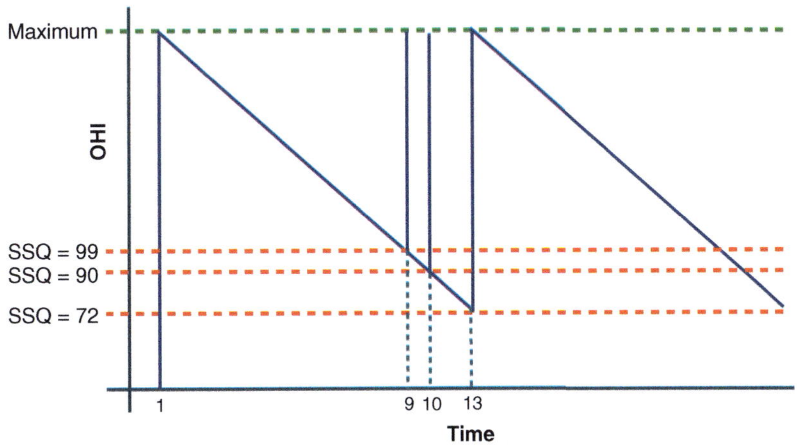

Figure 9.2 Scenarios for Safety Stock Quantities and Resupply Order Requirement Dates

4 weeks and the batch release lead time also ranges from 1 to 4 weeks leading to a total lead time of anywhere from 2 to 8 weeks. However, these lead times don't account for the time it takes to complete preproduction activities such as booking a production slot, ordering components, and/or creating batch records. After adding production readiness, the total lead time to replenish may take anywhere from 10 to 16 weeks or longer depending on the product and where it's being packaged and/or labeled. The important point is that the lead time to replenish input should be analyzed and adjusted as needed on a case-by-case basis for each SKU–depot combination.

The last insight with respect to the SSQ is that it can be expressed in two ways, either as a quantity (e.g. 99 units) or in the unit of measure of the lead time to replenish (e.g. 9 weeks). Describing safety stock levels in terms of weeks or months is the superior approach for finished product planning in clinical trials. This is because demand changes constantly before and during the trial, which in turn leads to changes in the SSQ. Recall from Chapter 1 – Supply Chain Management that OHI levels should always be measured relative to demand rather than in absolute terms because absolute measurements are often misleading and don't give the proper context for how long supply will last. Going forward, we'll describe the SSQ in number of weeks.

There is another implication to describing safety stock in time-based units rather than absolute quantities. It gives the clinical planner more control over how to manage the plan because the planner can change the safety stock value when replenishment lead times change. When there are issues that slow manufacturing and release, the safety stock value should be increased from 9 weeks to 13 weeks or 18 weeks, for example. Increasing the number of weeks of safety stock is a more conservative approach because, as we've shown through the safety stock formula, it increases the total quantity of inventory that needs to be held in OHI before a resupply order is triggered. On the other hand, a more aggressive approach may be taken if manufacturing and batch release happen faster than expected. The safety stock value can then be reduced, which in turn decreases the amount of OHI required before a resupply order is triggered.

Interestingly, these insights about SSQ and the lead time to replenish also mean that the minimum threshold value does not need to be calculated. This is because the only purpose of the min value is to inform the clinical planner of the resupply order start date (ROSD), or when that resupply order needs to begin to be ready for use by the timepoint the SSQ is breached. The ROSD can be inferred directly from the RORD by subtracting the lead time to replenish:

$$\mathrm{ROSD} = \mathrm{RORD} - \mathrm{LT_R}$$

Table 9.2 illustrates this formula.

To summarize the safety stock and OD discussion, the total amount of future demand over the lead time to replenish is equivalent to the SSQ. Safety stock values are most appropriately described in weeks, which are a relative measurement, than in total units, which are an absolute measurement. The point in time when OHI falls below the SSQ is the point when a production order requirement is generated, and this time point represents the RORD for a new batch of inventory.

Table 9.2 Calculating Resupply Order Start Date from Resupply Order Requirement Date and Lead Time to Replenish

Resupply Order Requirement Date	Week 10	Week 11	Week 13
Lead Time to Replenish (weeks)	9	9	9
Resupply Order Start Date	Week 1	Week 2	Week 4

The ROSD is then inferred from the RORD by deducting the lead time to replenish. Now, we know when action needs to be taken for each production order in terms of timing. The next logical step is to derive the quantity for each requirement.

Determining the POQ is the purpose of the "Max" quantity in the min–max methodology. Another term for the max is OT, a description we'll take forward from here. Like the safety stock calculation, the OT also incorporates future demand, but unlike safety stock, it's based on a predefined number of weeks of future demand (e.g. 13, 26, and 52) that should be covered by a particular production order:

$$\text{POQ} = \Sigma\left(\text{FD}_{\text{OT}}\right)$$

Let's use Table 9.3 as an example.

In Figure 9.3, the clinical planner could select an OT anywhere in the range of 2–26 weeks' worth of demand. Let's use three example OT values of 13 weeks, 18 weeks, and 26 weeks. These values generate production order quantities of 170 kits (the sum of demand from Weeks 1 to 13), 270 kits (the sum of demand from Weeks 1 to 18), and 430 kits (the sum of demand from Weeks 1 to 26), respectively. Notice that like the safety stock value, the OT value is also measured in weeks as both safety stock and OT are determined by the number of weeks of future demand they are trying to cover. Safety stock sets the floor of desired OHI, and the OT sets the ceiling.

Now that we have the main building blocks of the min–max methodology in place, we can see how the inputs work together to create production order requirements both in terms of timing when the safety stock level is breached, and in terms of quantity to resupply based on how many weeks of demand need to be covered by the OT. In Figure 9.4, the initial order covers the first 26 weeks of demand. Seventeen weeks later, the safety stock of 9 weeks is hit, and the min–max method calls for the first resupply order for another 26 weeks-of-supply. It's straightforward to see that an order will be generated every 17 weeks when these values are selected. Selecting optimal safety stock and OT values is covered in subsequent sections of this chapter.

Table 9.3 Setting Order Targets Based on Expiry Period and Supply Chain Lead Times

A	B	C	D	E	F = B+C+D+E	G = A − F	H
Expiry Period	Drug Product LT	Pack and Label LT	Distribution LT	RTSM LT	Total LT	Remaining Expiry = Max Order Target	Suggested Order Target
78 Weeks (18 mo.)	18 weeks	9 weeks	4 weeks	4 weeks	35 weeks	43 weeks	31–36 weeks
104 Weeks (24 mo.)	26 weeks	13 weeks	4 weeks	2 weeks	45 weeks	59 weeks	39–48 weeks

	W1	W2	W3	W4	W5	W6	W7	W8	W9	W10	W11	W12	W13	W14	W15	W16	W17	W18	W19	W20	W21	W22	W23	W24	W25	W26
Total demand	0	2	4	6	8	10	12	14	16	18	20	20	20	20	20	20	20	20	20	20	20	20	20	20	20	20

SKU description: Wonder drug placebo, 30-ct bottle, global label
Depot location: EU depot

Figure 9.3 Example Total Demand Forecast from Week 1 Through Week 26

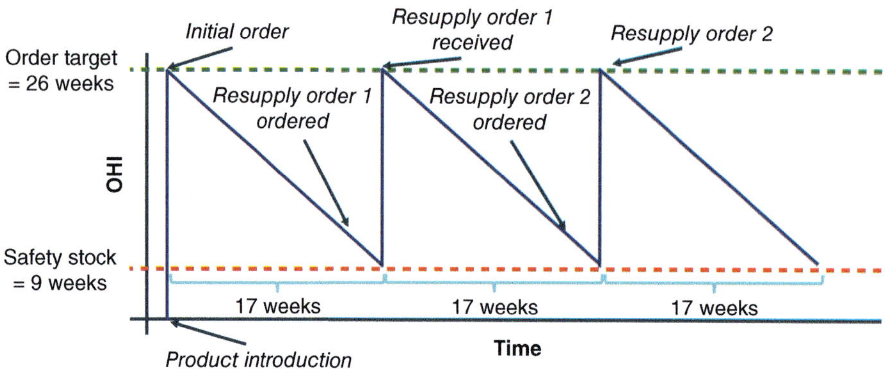

Figure 9.4 The Min–Max Methodology with Safety Stock and Order Target Quantities

9.3 Weeks-of-Supply

Both safety stock and OT are inputs to the min–max method, and, to this point, we've described them in terms of how many weeks of demand each parameter covers. We've also seen how the min–max method generates outputs in the form of production order requirements that tell the clinical planner when an order needs to be made available (e.g. RORD), when the order should begin production (e.g. ROSD), and the desired order quantity.

In Chapter 1, weeks-of-supply is defined as a description of how much OHI is available within a specific forecast period. We now expand on that definition to state that weeks-of-supply measures how much OHI is available to service future demand at any given timepoint within the forecast. Weeks-of-supply is the most important metric in production planning because it tells the clinical planner exactly where their inventory position is relative to demand. Weeks-of-supply also responds dynamically as the demand forecast changes precisely because it's a function of future demand.

Weeks-of-supply and future demand have an inverse relationship. As future demand increases, available weeks-of-supply decreases, and, as future demand decreases, then available weeks-of-supply increases. For example, if there are 10 weeks-of-supply of OHI and the safety stock is set to 9 weeks, the min–max method will not call for a resupply order. But, if the planner increases the demand forecast (e.g. sites activating or subjects enrolling earlier than expected), this pushes weeks-of-supply below the safety stock threshold and triggers the clinical planner to begin ordering material earlier than they originally would have. This makes intuitive sense because if demand increases, inventory is consumed faster, and resupplies should happen earlier. Conversely, if the demand forecast decreases (e.g. a subject unexpectedly drops out), this increases the weeks-of-supply of OHI and will delay a production order requirement or eliminate it altogether.

The weeks-of-supply calculation is made by summing the total number of future weeks of demand that OHI can cover:

$$\text{WOS} = \text{Number of Weeks of future demand covered by on} - \text{hand inventory}$$

This formula applies to the OHI total in any given planning period, and the metric adjusts automatically as updates to future demand or OHI are made from week to week. Within the clinical supply chain, the formula applies specifically to finished product inventory located at a depot that services subject demand. There are more complex weeks-of-supply calculations for aggregating finished product weeks-of-supply across all depots for a particular SKU and for calculating weeks-of-supply for upstream material. These formulas are covered in Chapter 11 – Finished Product Supply Planning: The Distribution Plan and Chapter 12 – Upstream Supply Planning: Primary Packaging, Drug Product, and Drug Substance, respectively.

9.4 The Production Plan Table

In this section, we'll put future demand, OHI, safety stock, the OD, and OT values together to create production order requirements, and we'll also tie weeks-of-supply in as the metric the clinical planner uses to make production planning decisions. The framework we'll use for this is the production plan table. The first step to building the table is to create the respective rows for the inputs listed above as in Figure 9.5.

In this example, total demand has already been calculated, the safety stock has been set to 9 weeks, and the OT has been set to 18 weeks. There is an order of operations the min–max methodology takes in the current week (e.g. Week 1 in Figure 9.5), which is the present week in which the order of operations will run:

1) Make the OD in the current week: yes or no?
   a) If the OD is yes, create a production order requirement in the production plan row that covers demand over the number of weeks specified in the OT row.
   b) If the OD is no, do not create a production order.
2) Calculate OHI in the current week.
3) Calculate weeks-of-supply in the current week.

This sequence of calculations is then executed in current Week +1, then current Week +2, and so on until the entire table is populated. Assuming Week 1 is the current week, the following calculations are made and the values presented in Figure 9.6 are the result.

1) OD is "Yes".
2) Production order is created covering demand from W2 to W19 which is equal to 94 units.
3) OHI is calculated to be 94 units = 0 $(OHI_{CW\text{-}1})$ − 0 $(Demand_{CW})$ + 94 (Prod. Order Qty_{CW}).

The order of operations is then rolled through each subsequent week. The result of these calculations is presented in Figure 9.7.

The first production order requirement is created in Week 1 (e.g. W1) and is intended to cover future demand from Weeks 2 to 19 based on the 18-week OT value. When the initial order is made, it's based on the OT which covers 18 weeks of demand and is therefore equal to 18 weeks-of-supply.

	CW W1	W2	W3	W4	W5	W6	W7	W8	W9	W10	W11	W12	W13	W14	W15	W16	W17	W18	W19	W20	W21	W22	W23	W24	W25	W26
Total demand	0	2	4	6	4	6	4	8	10	2	6	8	4	6	4	8	4	6	2	4	6	8	2	4	8	6
OHI																										
Prod. plan																										
WOS																										
Safety stock	9	9	9	9	9	9	9	9	9	9	9	9	9	9	9	9	9	9	9	9	9	9	9	9	9	9
Order target	18	18	18	18	18	18	18	18	18	18	18	18	18	18	18	18	18	18	18	18	18	18	18	18	18	18

SKU description: Wonder drug placebo, 30-ct bottle, global label
Depot location: United States depot

Figure 9.5 Production Plan Table with Total Demand

CW	W1	W2	W3	W4	W5	W6	W7	W8	W9	W10	W11	W12	W13	W14	W15	W16	W17	W18	W19	W20	W21	W22	W23	W24	W25	W26
Total demand	0	2	4	6	4	6	4	8	10	2	6	8	4	6	4	8	4	6	2	4	6	8	2	4	8	6
OHI	94																									
Prod. plan	94																									
WOS	18																									
Safety stock	9	9	9	9	9	9	9	9	9	9	9	9	9	9	9	9	9	9	9	9	9	9	9	9	9	9
Order target	18	18	18	18	18	18	18	18	18	18	18	18	18	18	18	18	18	18	18	18	18	18	18	18	18	18

SKU description: Wonder drug placebo, 30-ct bottle, global label
Depot location: United States depot

Figure 9.6 Base Case Production Plan Table Through Week 1

CW	W1	W2	W3	W4	W5	W6	W7	W8	W9	W10	W11	W12	W13	W14	W15	W16	W17	W18	W19	W20	W21	W22	W23	W24	W25	W26
Total demand	0	2	4	6	4	6	4	8	10	2	6	8	4	6	4	8	4	6	2	4	6	8	2	4	8	6
OHI	94	92	88	82	78	72	68	60	50	100	94	86	82	76	72	64	60	54	90	86	80	72	70	66	58	52
Prod. plan	94	0	0	0	0	0	0	0	0	52	0	0	0	0	0	0	0	0	38	0	0	0	0	0	0	0
WOS	18	17	16	15	14	13	12	11	10	18	17	16	15	14	13	12	11	10	18	17	16	15	14	13	12	11
Safety stock	9	9	9	9	9	9	9	9	9	9	9	9	9	9	9	9	9	9	9	9	9	9	9	9	9	9
Order target	18	18	18	18	18	18	18	18	18	18	18	18	18	18	18	18	18	18	18	18	18	18	18	18	18	18

SKU description: Wonder drug placebo, 30-ct bottle, global label
Depot location: United States depot

Figure 9.7 Base Case Production Plan Table Through Week 26

Demand consumes inventory until Week 10 when the SSQ of 9 weeks is reached, and a second production order requirement is created for 52 units. The total of 52 units is calculated by summing future demand from Weeks 20 to 28. Keep in mind that calculation starts at week 20 because demand from weeks 2 to 19 has already been included in the first POQ and only extends to Week 28 because the OT is set to 18 weeks.

Following the first resupply order, OHI increases from 50 units to 100 units in Week 10 and weeks-of-supply improves from 10 weeks to 18 weeks. The safety stock already covers nine weeks of future demand (e.g. from Weeks 11 to 19) so the resupply order covers the next nine weeks of future demand (e.g. from Weeks 20 to 28) to bring total OHI to 18 weeks-of-supply. The pattern repeats itself in Week 19, when the safety stock value is hit, and a second resupply order is created for 38 units. Notice how the weeks-of-supply oscillate from 18 down to 9, when another delivery is scheduled to bring weeks-of-supply back to 18 weeks.

9.5 The Impact of Future Demand on Production Order Requirements

The expression of the min–max methodology in the production plan table gives the clinical planner a detailed framework for understanding how changes to future demand impact production order requirements. We demonstrated earlier in this chapter that increases in demand increase the SSQ and pull in production order requirements, while decreasing demand reduces the SSQ and therefore delays the need for resupply orders. The production plan table enables the clinical planner to measure the impact of demand changes on production order requirements in terms of both

	W1	CW W2	W3	W4	W5	W6	W7	W8	W9	W10	W11	W12	W13	W14	W15	W16	W17	W18	W19	W20	W21	W22	W23	W24	W25	W26
Total demand	0	2	4	6	6	8	6	10	12	4	8	10	6	8	6	10	6	8	4	6	8	10	4	6	10	8
OHI	94	92	88	82	76	130	124	114	102	98	90	80	74	66	130	120	114	106	102	96	88	78	74	94	84	76
Prod. plan	94	0	0	0	0	62	0	0	0	0	0	0	0	0	70	0	0	0	0	0	0	0	0	26	0	0
WOS	14	13	12	11	10	18	17	16	15	14	13	12	11	10	18	17	16	15	14	13	12	11	10	18	17	16
Safety stock	9	9	9	9	9	9	9	9	9	9	9	9	9	9	9	9	9	9	9	9	9	9	9	9	9	9
Order target	18	18	18	18	18	18	18	18	18	18	18	18	18	18	18	18	18	18	18	18	18	18	18	18	18	18

SKU description: Wonder drug placebo, 30-ct bottle, global label
Depot location: United States depot

Figure 9.8 Production Plan Table with Increased Demand over Base Case

quantity and timing. In the example below, let's assume it's Week 2, and we have learned that an additional subject will enroll in China in Week 5 adding two units of demand per week to the total demand forecast as in Figure 9.8.

The first impact is the change to weeks-of-supply in Week 1, which has dropped from 18 weeks to 14 weeks. The weeks-of-supply value adjusts because the 94 units of inventory received in the first production order now cover fewer weeks of demand (e.g. Weeks 1–14) because demand has increased beginning in Week 5. This increase in demand also impacts the production plan because the safety stock value now hits four weeks earlier than the base case example, creating a production order requirement in Week 6 rather than in Week 10. The resupply is also for a larger 62-unit order quantity than the order for 52 units in the base case. Finally, note how there are three total resupply requirements in Weeks 6, 15, and 24 rather than two resupply order requirements in the base case.

The power and practicality of the min–max method combined with the production plan table are evident in this example. The clinical planner has made a change to the demand forecast at this SKU–depot combination, and they can now see the impact of that change reflected tangibly in the production plan row where production order requirements have also changed in terms of timing and quantity. This gives the clinical planner visibility to the impact of demand changes and enables better decision-making and control over how to mitigate the impact of updates to demand.

In the final example, let's assume the forecast is revised downward after a planned site activation fails to materialize leading to a reduction of two units of demand per week from the base case from Week 5 onward. The resulting production plan is presented in Figure 9.9.

	W1	CW W2	W3	W4	W5	W6	W7	W8	W9	W10	W11	W12	W13	W14	W15	W16	W17	W18	W19	W20	W21	W22	W23	W24	W25	W26
Total demand	0	2	4	6	2	4	2	6	8	0	4	6	2	4	2	6	2	4	0	2	4	6	0	2	6	4
OHI	94	92	88	82	80	76	74	68	60	60	56	50	48	44	42	36	34	56	54	50	44	44	42	36	32	30
Prod. plan	94	0	0	0	0	0	0	0	0	0	0	0	0	0	0	0	0	26	0	0	0	0	0	0	0	0
WOS	26	25	24	23	22	21	20	19	18	17	16	15	14	13	12	11	10	18	17	16	15	14	13	12	11	10
Safety stock	9	9	9	9	9	9	9	9	9	9	9	9	9	9	9	9	9	9	9	9	9	9	9	9	9	9
Order target	18	18	18	18	18	18	18	18	18	18	18	18	18	18	18	18	18	18	18	18	18	18	18	18	18	18

SKU description: Wonder drug placebo, 30-ct bottle, global label
Depot location: United States depot

Figure 9.9 Production Plan Table with Reduced Demand from Base Case

In this example, the reduction in demand increases the weeks-of-supply available in Week 1 from 18 weeks to 26 weeks. As a result, the need for a resupply order pushes out from Week 10 to Week 18, and the production order requirement also decreases to 26 units. The relationship between future demand, OHI, weeks-of-supply, and production order requirements is clear at this point. As demand increases, OHI and weeks-of-supply decrease, which in turn leads to larger production order requirements that need to be delivered earlier in time. As demand decreases, OHI and weeks-of-supply increase, which in turn leads to smaller production order requirements that need to be delivered later in time.

9.6 Selecting Supply Planning Parameters and Their Impact on Production Order Requirements

The other levers the clinical planner has at their disposal to influence production order requirements are the safety stock and OT values. Taken collectively, these two values are the supply planning parameters. Selecting optimal supply planning parameter values requires assessing a handful factors embedded within the supply chain. Safety stock values for production planning should be driven by the lead time to replenish secondary packaging and labeling production orders. However, this approach assumes that DP is available if needed to support packaging campaigns. If this is not the case, and DP supply is constrained or otherwise unreliable, it makes sense for the clinical planner to consider adding weeks of safety stock at the finished product level that equal the amount of time it takes to replenish supply from the point DP manufacturing begins. A finished product safety stock value of 18 or even 26 weeks may be appropriate in extreme cases if DP manufacturing is highly constrained. This more conservative approach protects against a situation where a DP resupply batch is delayed or lost altogether and allows the supply chain sufficient time to recover while preventing a stockout in a worst-case scenario.

Defining OT values is less straightforward because there are more considerations at play, and because the OT can be treated more dynamically over the course of the trial than the safety stock value. The first consideration is the stage at which the production plan is run – before the trial begins, as the study enrolls, or after enrollment completes. Before the study begins, it makes the most sense to select an OT value within 18–39 weeks depending on how quickly the study may enroll. If the projection is for the study to enroll quickly, a longer OT of 26–39 weeks is more appropriate because demand is more likely to consume supply over this longer horizon. On the other hand, if the study will take greater than a year to enroll, it makes sense to start with a lower OT of 18–26. Setting the OT at this lower range creates smaller initial order requirements and therefore reduces the amount of drug that will be wasted if site and/or subject demand fail to materialize. We know from Chapter 8 – Finished Product Demand Planning: Basic Concepts that it's very common to over-forecast demand early in the study and then encounter delays in enrollment, so this approach protects against unnecessarily throwing expiring drug away. As the study enrolls and demand solidifies, the OT may be increased to a larger value such as 39–78 weeks because we're now certain the supply produced in those larger orders will be consumed.

A second variable to consider when setting the OT is the expiry period of the product. The expiry period of the product establishes the maximum value the OT should be set at because if the expiry period is exceeded, then it's guaranteed that some of the order will expire. Recall that expiry is based on the date of manufacture of the DP, and it usually takes anywhere from 4 to 8 months to complete DP production, secondary packaging and labeling, and have the finished product released for distribution out of the central depot via the Randomization and Trial Supply Management (RTSM) platform. There may also be distribution lead times (e.g. do not ship and do not dispense values) to consider if the product needs to be shipped from the central depot to regional depots. Table 9.3 illustrates suggested order target based on two example product expiry periods and supply chain lead times.

Note that in these two examples, the suggested OT is lower than the max OT because of the potential likelihood that demand will push out. If the max OT is used and demand fails to materialize, then there will be wasted product from those batches.

A third consideration is the DP batch size and capacity constraints within the DP manufacturing process. It's highly undesirable to mix two different DP lots into one finished product batch because if a quality issue is detected in the finished product, it implicates both DP lots which then result in a larger, lengthier stockout situation. Thus, the maximum finished product batch size is equal to the maximum DP batch size. It therefore does not make sense to increase the OT value to cover more demand than can be provided in a single DP batch.

The final consideration in setting the OT is the cost to package and label the product. If packaging and labeling operations are relatively cheap, then running more frequent campaigns with smaller batch sizes is appropriate, and a lower OT is preferable. If packaging and labeling operations are relatively expensive though, then the goal shifts to minimizing the number of campaigns and batches while also trying to minimize potential waste. In this situation, a higher OT is better. Table 9.4 summarizes the considerations discussed above and how they influence setting the OT.

Selecting the safety stock and OT values requires assessing the supply chain in terms of lead times, expiry, and cost. It's also important to remember that these variables may change over time, so it makes sense to review and monitor the safety stock or OT values at regular intervals during the trial. When these values are updated, it impacts the production plan and suggested production order requirements as shown in Figures 9.10 and 9.11.

Increasing the safety stock value creates more frequent production order requirements with smaller batch sizes, while decreasing safety stock values translates to less frequent production order requirements with larger batch sizes. In the example above, the safety stock value of 9 weeks will create a production order requirement every 17 weeks rather than every 21 weeks if it were at 5 weeks. Therefore, it's usually best to start with a more conservative safety stock target and then reduce it if supply chain stability and consistency increase. As a rule of thumb, it's easier to push production orders out in time than to bring them in.

The same concepts are at work when adjusting the OT quantity. Increasing the OT results larger, less frequent batches while decreasing the OT results in smaller, more frequent batches. This further supports the case for beginning a study with a lower OT and then increasing it over time because it's easier to push production orders out in time than to bring them in. Table 9.5 summarizes how toggling the supply planning parameters influences OHI levels and production order requirement quantity and timing.

Table 9.4 Order Target Selection Considerations

Order Target Value	Study Phase	Expiry Period	DP Batch Size	Cost
Lower Order Target (18–26 Weeks)	Planning to study start	12–18 mo	Small	Inexpensive
Medium Order Target (26–52 Weeks)	Study start to <50% enrolled	18–24 mo	Neutral	Expensive
High Order Target (52–78 Weeks)	>50% Enrolled to 100% enrolled	24 mo+	Large	Very expensive

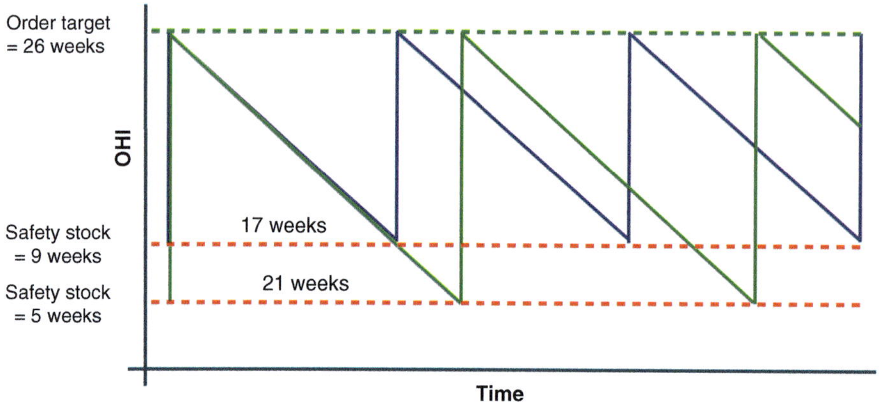

Figure 9.10 Safety Stock Value Impact on Production Order Requirements

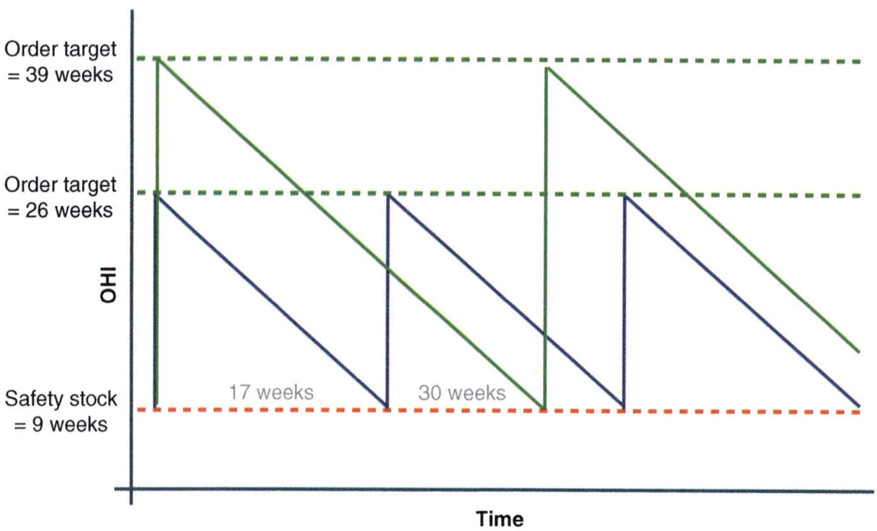

Figure 9.11 Order Target Value Impact on Production Order Requirements

Table 9.5 Supply Planning Parameter Impact on On-Hand Inventory and Production Order Requirements

Supply Planning Parameter	On-Hand Inventory	Requirement Quantity	Requirement Timing
Safety Stock Increases	Increases	Decreases	Sooner
Safety Stock Decreases	Decreases	Increases	Later
Order Target Increases	Increases	Increases	Later
Order Target Decreases	Decreases	Decreases	Sooner

9.7 From Requirements to Orders

Up to this point, we've explored how the min–max methodology generates production order requirements based on the interaction between future demand and the supply planning parameters. It's important to re-emphasize that production order requirements are not actual production orders. Requirements are named as such because they are the amount of inventory required at a specific time point based on future demand's interaction with the supply planning parameters. To convert a production order requirement into an actual production order, the clinical planner needs to confirm the quantity and timing of the order within the plan. To accomplish this in the production plan table, an additional row, "Production Actual," should be added. When a production actual value is populated, it automatically supersedes the suggested production order requirement listed in the "Production Plan" row.

Implementing the min–max methodology to generate production order requirements gives the clinical planner visibility into what's needed to maintain inventory levels. Implementing the "Production Actual" row gives the planner control over production planning decisions to make sure they're realistic and achievable. At the finished product level, a realistic production plan means that sufficient DP is available at the time manufacturing of the finished product production order begins. After all, without the underlying DP, it's impossible to make the finished product. In most cases, the production order requirement will need to be adjusted up or down in quantity based on the size of the DP batch that's available for consumption, and, in some cases, the timing of the order may need to be pulled in or moved out depending on the availability of the DP batch.

Let's revisit the base case example with the following production order requirements:

- Initial supply order requirement: 94 units in Week 1
- Resupply order requirement: 52 units in Week 10
- Resupply order requirement: 38 units in Week 19

Let's assume the following DP lots are available:

- DP Lot 1: 100 units available for packaging to begin eight weeks before Week 1
- DP Lot 2: 50 units available for packaging to begin in Week 5
- DP Lot 3: 50 units available for packaging to begin in Week 13

And, finally, let's assume a 6-week lead time from packaging start to packaging release. An updated production plan table is provided in Figure 9.12.

CW

	W1	W2	W3	W4	W5	W6	W7	W8	W9	W10	W11	W12	W13	W14	W15	W16	W17	W18	W19	W20	W21	W22	W23	W24	W25	W26
Total demand	0	2	4	6	4	6	4	8	10	2	6	8	4	6	4	8	4	6	2	4	6	8	2	4	8	6
OHI	100	98	94	88	84	78	74	66	56	54	98	90	86	80	76	68	64	58	56	102	96	88	86	82	74	68
Prod. actual	100	0	0	0	0	0	0	0	0	0	50	0	0	0	0	0	0	0	0	50	0	0	0	0	0	0
Prod. plan	94	0	0	0	0	0	0	0	0	52	0	0	0	0	0	0	0	0	38	0	0	0	0	0	0	0
WOS	18	17	16	15	14	13	12	11	10	9	18	17	16	15	14	13	12	11	18	18	17	16	15	14	13	12
Safety stock	9	9	9	9	9	9	9	9	9	9	9	9	9	9	9	9	9	9	9	9	9	9	9	9	9	9
Order target	18	18	18	18	18	18	18	18	18	18	18	18	18	18	18	18	18	18	18	18	18	18	18	18	18	18

SKU description: Wonder drug placebo, 30-ct bottle, global label
Depot location: United States depot

Figure 9.12 Production Plan Table with Production Actuals

Let's first examine the timing of orders in this plan. The initial supply of 100 units is packaged by Week 1 and available to service demand beginning in Week 2 because the first DP lot is available 8 weeks before the requirement date, and the packaging and labeling lead time are only 6 weeks. However, the first resupply lot isn't available for packaging until Week 5. With a 6-week lead time for packaging to complete, the resupply won't be available until Week 11. This causes the plan to run below the safety stock target for 1 week, but the 50 units of supply available in the second DP lot are sufficient to bring weeks-of-supply back up to the target of 18 weeks. The second resupply lot is not available for packaging until Week 13 and therefore must also be planned for availability in Week 19 instead of the original Week 18 delivery target.

From a quantity perspective, the initial supply and first resupply orders are close enough to the production requirements that it makes sense to pack all 100 and 50 units, respectively. It wouldn't make sense to leave a handful of kits behind that would otherwise go to waste. With respect to the last order, the planner has a decision to make. The plan only calls for 38 units, but there are 50 units available in the third DP lot. It usually makes sense to package an entire lot if there are <100 units of difference between the requirement and what's available to package. This is because running another packaging campaign for such a small number of kits is usually financially wasteful due to incremental pre- and postproduction costs associated with each batch.

Let's look at one more example with a different set of DP availability assumptions:

- DP Lot 1: 60 units available for packaging to begin eight weeks before Week 1
- DP Lot 2: 120 units available for packaging to begin in Week 3
- DP Lot 3: 50 units available for packaging to begin in Week 9

One possible production plan with this set of assumptions is provided in Figure 9.13.

In this scenario, DP supply is relatively constrained at the outset of the study. Sixty kits represent only 11 weeks-of-supply, and the first resupply batch is not available for release until Week 9, at which point, inventory has dropped below 4 weeks-of-supply. Notice that the production plan row has created a requirement for 44 additional kits in Week 3 because the safety stock level has hit 9 weeks-of-supply. This would have brought inventory up to 98 kits, which would have represented the OT of 18 weeks-of-supply, but the DP inventory is not available so the clinical planner must wait until Week 9 before inventory can be replenished.

Fortunately, this second packaging campaign contains 120 kits, and this replenishes inventory levels up to 24 weeks-of-supply. The most interesting decision for the clinical planner in this scenario is how to manage the timing of the third packaging campaign. In Figure 9.13, the third

CW

	W1	W2	W3	W4	W5	W6	W7	W8	W9	W10	W11	W12	W13	W14	W15	W16	W17	W18	W19	W20	W21	W22	W23	W24	W25	W26
Total demand	0	2	4	6	4	6	4	8	10	2	6	8	4	6	4	8	4	6	2	4	6	8	2	4	8	6
OHI	60	58	54	48	44	38	34	26	136	134	128	120	116	110	106	98	94	88	86	82	126	118	116	112	104	98
Prod. actual	60	0	0	0	0	0	0	0	120	0	0	0	0	0	0	0	0	0	0	0	50	0	0	0	0	0
Prod. plan	94	0	44	0	0	0	0	0	0	0	0	0	0	0	0	0	0	0	0	0	0	0	0	4	0	0
WOS	11	10	9	8	7	6	5	4	24	23	22	21	20	19	18	17	16	15	14	13	18	17	16	15	14	13

1

Safety stock	9	9	9	9	9	9	9	9	9	9	9	9	9	9	9	9	9	9	9	9	9	9	9	9	9	9
Order target	18	18	18	18	18	18	18	18	18	18	18	18	18	18	18	18	18	18	18	18	18	18	18	18	18	18

SKU description: Wonder drug placebo, 30-ct bottle, global label
Depot location: United States depot

Figure 9.13 Updated Production Plan Table with Production Actuals

campaign is packaged in Week 21 when safety stock levels fall below 13 weeks-of-supply. There are two competing principles here. The first principle is to pack as late as possible so that more of the total demand picture is known. On the other hand, there is also a principle to minimize risk within the supply chain, and if it's possible to package and release product before the safety stock is met, this is a wise decision. Here, the clinical planner splits the difference, choosing to package the product 6 weeks after it's available and 3 weeks before the next requirement would have been hit in Week 24.

There is a final point to make on finished product production planning. We know that earlier in the study, demand is more uncertain because enrollment is still theoretical and that over time demand becomes more certain as subjects enroll and the forecast becomes reality. This concept is why OT values are set lower at the outset of a study and should then be increased over time. The same logic applies to converting production order requirements to actual production orders. Initial finished product production orders should cover a relatively low weeks-of-supply (e.g. 18–26 weeks) because there's a greater chance that demand fails to materialize and those initial batches go to waste. As subjects enroll, and there is greater certainty in the forecast, larger packaging jobs are more appropriate because that inventory is guaranteed to be consumed by subject demand.

Below is a summary of the finished product production planning strategies discussed so far:

- *Quantity and Timing*: It's impossible to package DP that is not available. DP must be available in sufficient quantities and with sufficient lead time to complete the secondary packaging and labeling jobs in the plan.
- *Quantity*: Create smaller production orders early in the study when demand is uncertain. As subjects enroll consider increasing the production order size.
- *Quantity*: Avoid leaving small portions (e.g. <100 units) of DP lots unpackaged. It's inefficient to run many, small production campaigns if that DP lot is allocated for the same study, and the extra units may come in handy if clinical site resupply shipments are damaged.
- *Timing*: Find the sweet spot. Delay the timing of resupply production orders to confirm as much real subject demand as possible. But, don't delay production so much that safety stock levels are breached.

There are two additional factors to consider when planning finished product production, the first of which is to account for the blinding strategy of the study. Typically, blinding strategies dictate that the appearance of the active IMP and placebo or comparator IMP must be identical to the subjects including the product itself, the packaging, and the labeling. In Chapter 3 – Introduction to the Clinical Trial Supply Chain, we learned that clinical labels must include the expiration date of the product. Practically speaking, this means that in blinded trials a finished product batch of the active drug should be accompanied by a finished product batch of the placebo drug. The reason for this is so the expiry dates of the active and placebo batches stay the same across the two labels. If an older batch of either active or placebo needs to be paired with the other, that means the older expiry date needs to be applied to the newer product so the same expiry date can be printed on both labels. Therefore, to get the most shelf life out of each SKU, it typically makes sense to run active and placebo finished product batch manufacturing together.

This discussion on blinding and expiry considerations leads to the second factor, which is incorporating the expiry period of the DP and the finished product shelf-life assignment into the plan. This step is critical to successful planning, but it's also complex enough to warrant an entire chapter, Chapter 10 – Finished Product Expiry Planning, which we'll cover next.

9.8 The Production Plan Summary

Before we move on to expiry management though, we'll close this chapter by introducing two more concepts, the production plan summary and the production order readiness process. At the completion of each planning cycle, the clinical planner shares the output of the demand and supply planning process, the **production plan summary**. The summary is a detailed report of when, where, and how much of each SKU will be produced to support the clinical trial. At minimum, the production plan summary should capture the following elements:

1) Protocol number
2) SKU description
3) Manufacturing site
4) Type of manufacturing (e.g. packaging, labeling, and shelf-life extension)
5) Batch number
6) Quantity produced in the batch
7) Unit of measure
8) Production start date
9) Production release date
10) Expiry date
11) Process version (if maintained)

The production plan summary captures the entire production plan for each SKU for the duration of the trial and serves many important uses. An example is shown in Table 9.6. First, the summary is an important touchstone when making updates to the plan, something we'll cover extensively in Chapter 15 – Clinical Supply Chain Planning Processes. This table is also a mechanism for communicating the plan to internal stakeholders such as the chemistry, manufacturing, and controls (CMCs) team and external stakeholders such as the manufacturing sites responsible for executing the plan. And, lastly, the summary is the basis for many other reports and analyses such as distribution planning, budgeting, and supply chain analytics. No matter which planning system or tool is used (e.g. spreadsheets and enterprise resource planning (ERP) systems), the production plan summary is essential to the clinical planner to communicate the planned finished product production events required to support the trial.

There is a final note to make on the production plan summary. In this example, the production start date is equivalent to the date of manufacture of the finished product batch, and the production release date is equivalent to the date the batch is fully released by the sponsor's quality unit for

Table 9.6 The Production Plan Summary

Protocol	SKU	Mfg. Site	Type	Batch No.	Qty	UoM	Start Date	Release Date	Expiry Date	Process Version
Study 1	100 mg Active	US CMO	Packaging	A1A	500	Kits	31-May-2025	15-July-2025	31-Dec-2025	DS1.0–DP1.0
Study 1	100 mg Active	US CMO	SLE	A1A	200	Kits	31-Oct-2025	30-Nov-2025	30-Jun-2026	DS1.0–DP1.0
Study 1	100 mg Active	EU CMO	SLE	A1A	100	Kits	31-Oct-2025	30-Nov-2025	30-Jun-2026	DS1.0–DP1.0

use in the clinic. It's important to recognize though that a host of activities must take place before the batch goes into the secondary packaging and/or labeling process, which is the final topic of discussion in this chapter.

9.9 Production Order Readiness

The last aspect of production planning to consider is production order readiness. In the context of clinical supply chain management, production order readiness is defined as being prepared and able to execute the production order so that it meets the desired release date target. From a timing perspective, order readiness activities need to begin at least 6 months before the desired release date of a batch. This relatively lengthy readiness lead time is meant to account for all the activities that need to take place for production to occur and complete. Within the physical supply chain, components need to be available before the production event begins including receipt of the underlying DP at the packaging site, receipt of all other packaging materials, and printing and receipt of the clinical labels. From a documentation perspective, the master batch record (MBR) for the packaging and/or labeling jobs needs to be written and approved before production can begin, and when the batch completes production, the executed batch record must be reviewed and approved before the batch is released for use in the clinic. Lastly, the packaging site needs to confirm both capacity and labor are available to meet the order requirements. It's prudent to give the manufacturing site as much time as possible to schedule finished product production orders to avoid any potential constraints.

To accomplish all these activities, multiple stakeholders are needed from within the sponsor including clinical supply chain, quality, and the manufacturing team if the packaging and/or labeling operation is carried out internally. If the packaging and/or labeling jobs are outsourced to a third party, then the contract manufacturing organization (CMO) needs to be included as well. Within clinical supply chain, the clinical planner is responsible for the development of the production plan and for communicating the plan to the manufacturing site. In mid-large size biopharmaceutical companies, clinical supply chain may be split into a planning function and execution-related subfunctions who are responsible for oversight of production and distribution order completion. Quality must also be included so they can prepare for and schedule document reviews and batch release activities into their own queue. When manufacturing is outsourced to a supplier, the clinical supply chain function must work closely with the supplier to understand their internal business processes and timelines for setting up packaging and labeling activities.

Given the scope and cross-functional nature of these activities, it's necessary to have a production order readiness business process in place for the clinical planner to communicate the production plan to internal and external stakeholders within an agreed upon time horizon. The production order readiness process typically involves a weekly or bi-weekly meeting within the clinical supply chain and quality functions to review existing production orders as they move toward their production date as well as sharing new orders entering the planning horizon. The planning horizon is the amount of time prior to a desired production order release date that the production order needs to be shared with the manufacturing site to guarantee order readiness activities will be completed in time to support the release date. The finished product planning horizon is usually around 6 months. If manufacturing is internal, then they attend as well, while if manufacturing is outsourced, a separate meeting takes place with the supplier to monitor order progress. Activity trackers for each order are an excellent too to document which activities have been completed, which are yet to complete, and due dates for each activity listed. Establishing this process ensures that the good work done in the demand and supply planning process doesn't fall apart when it's time to execute.

9.10 Key Takeaways

1) Finished product production planning should rely on a minimum–maximum inventory management methodology. Three values underpin the min–max methodology, the safety stock value that inventory should not fall below, the min value that triggers the initiation of a resupply order, and the max value that drives how much inventory should be manufactured in a given production order. These three values work together to generate production order requirements.
2) Production planning for clinical trials is based on future demand, which is demand occurring in future weeks beginning at the first future week from the current week of the demand and supply plan. Future demand is the basis for production planning because clinical trial demand forecasting doesn't rely on assessing historical demand to determine future demand. Using future demand also enables the min–max methodology to react dynamically to changes in the demand forecast from week to week.
3) The safety stock level is calculated by summing future demand over the lead time to replenish the product in the event of an unexpected supply shock. The safety stock value drives the OD of whether a production order requirement is needed or not. The date when the OD is "Yes" is known as the RORD. Safety stock values fluctuate because both future demand and the lead time to replenish are dynamic. Given this dynamism, it makes sense to express safety stock levels in terms of how much time (e.g. number of weeks) safety stock values cover rather than absolute quantities.
4) The max level is also known as the OT, and this input determines production order requirement quantities. The OT is derived by summing future demand over the desired number of weeks selected by the clinical planner.
5) Together, the safety stock and OT values are known as supply planning parameters. Changing these parameters directly impacts (e.g. increases or decreases) OHI levels and production order requirement outputs.
6) Weeks-of-supply is the most important metric in production planning and measures how much OHI is available to service future demand at any given timepoint within the forecast. Weeks-of-supply automatically updates as changes are made to the demand forecast or to production order requirements.
7) The production plan table brings together each of the key inputs, demand, OHI, and the supply planning parameters to generate the production order requirement outputs. The production plan table makes clear how changes to the demand forecast impact production order requirements. If demand increases, production order requirements pull in and increase in size. If demand decreases, production order requirements push out and reduce in size.
8) Selecting an optimal safety stock value is based on the lead time to replenish and the amount of risk in the supply chain. Selecting an optimal OT value is based on the study phase, DP expiry period, DP batch size, and the cost of finished product production.
9) The min–max methodology produces production order requirements, but these are only suggested production orders. The clinical planner must confirm production order quantities and timing within the plan. Important considerations for finished product production planning include assessing DP inventory availability, DP batch size, and balancing the timing of production to reduce stockout risk and reduce the potential for waste.

10) The production plan summary captures the details of when, where, and how much of each SKU will be made over the duration of the clinical trial. The summary is a mechanism for communicating to internal and external stakeholders and serves as the basis for the distribution plan, supply chain budgeting, and analytics.
11) Production order readiness is defined as being prepared and able to execute a production order so that it meets the desired release date target. Production order readiness encompasses a wide range of tasks that multiple stakeholders have responsibility for including clinical supply chain, quality, and manufacturing. Effective production order readiness relies on a business process that brings these functions together to review order status and timing.

10

Finished Product Expiry Planning

10.1 Expiry Management and the Planning Process

In Chapter 3 – Introduction to the Clinical Trial Supply Chain, we devoted an entire section to stability studies and the process of establishing and assigning the expiry period to the drug product (DP). Let's quickly recap a few key points from that discussion. First, every DP planned for use in a clinical trial must have samples go up on a stability study that is used to establish the expiry period of the drug. Over time, data is collected at predetermined points (e.g. at time = 0 months, time = 1 month, and time = 2 months), and the expiry period is extrapolated from analysis of the product's degradation over those durations.

Second, the expiry period describes the amount of time a batch of DP is considered good for use from its date of manufacture. The shelf-life date is then assigned to the finished product, and this date is determined by taking the date of manufacture of the DP and adding the expiry period duration to it. For example, a DP may have an assigned expiry period of 24 months. If that DP is manufactured in June 2025, the finished product batch that was packaged and/or labeled off that DP lot would have a shelf-life date of June 30, 2027. The shelf-life date is nearly always printed on the label of the finished product kit. The one exception to this rule of thumb is when the finished product is labeled for use in the United States only, where expiry dates are not required on clinical labels.

Third, the expiry period assignment is unique to each DP formulation/primary packaging combination. This means that a 50 mg capsule packaged in a bottle has a different stability profile and expiry period assignment than a 100 mg capsule packaged in a blister pack. Occasionally, comparability can be established between different formulation strengths within the same primary packaging configuration, but even this isn't always possible.

Finally, expiry period assignment updates don't happen immediately after a stability time point is reached in a stability study. It can take several weeks or months to analyze the data and issue an updated expiry period assignment. This means the clinical planner needs to work closely with their analytical operations and quality counterparts to understand the stability strategy, expiry period assignment dates, and how much confidence there is in the stability of the DP that will be supplied to the clinic. Each of these points is important to the demand and supply planning process.

Successfully managing expiry periods and shelf life within the finished product supply plan is a challenge in clinical trial supply chain management for multiple reasons. The expiry period assignment is usually not static over the course of a clinical trial. This situation is especially common in Phase I and Phase II clinical trials where new formulations and packaging configurations are

Supply Chain Planning for Clinical Trials – A Practical Guide, First Edition. Ryan Mills.

typically introduced. It may be the case in Phase III trials that the same DP and packaging configuration are carried forward, but even then the expiry period assignment may still have room to change. The potential for the expiry period to lengthen as the trial progresses means the clinical planner must understand the structure and timing of the stability studies to accurately plan which shelf life will be assigned to the finished product.

The clinical planner must also thoroughly plan for expiry events when a batch or several batches of finished product reach their shelf-life date and expire. When this happens, replacement inventory must be available at both the clinical site and the supplying clinical depot to ensure continuity in patient dosing. Making sure enough inventory is available to cover these situations is imperative, and the clinical planner nearly always has a few options at their disposal to consider. Like so many other decisions in the planning process, selecting the best plan to manage expiry events requires assessing trade-offs between capability, risk, and cost.

There is another potential outcome the clinical planner must consider with respect to expiry period assignment; what happens if the expiry period is unexpectedly cut short due to product degradation over the course of the stability study? This situation is not uncommon, particularly for liquid-based biologic products which tend to be less stable than their small molecule or solid-state biologic counterparts. When this happens, the clinical planner must revisit the entire finished product production plan to determine the impact of the reduced expiry period on inventory levels and production order requirements.

To fully cover the topic of expiry planning, we'll begin with a discussion of the first in first out (FIFO) and last in first out (LIFO) inventory management methodologies and how these concepts are applicable to managing expiry within the clinical supply chain. Then, we'll cover the importance of breaking aggregate on-hand inventory into individual batches so that expiry can be appropriately analyzed within the finished product production plan. These topics set the stage for how to incorporate expiry into the production plan and how to forecast expiry events. We'll conclude the chapter with guidance on how to choose a strategy for extending or replacing expiring lots and how to manage the dreaded expiry limitation.

10.2 First In First Out and Last In First Out Inventory Management

The first concept to understand in expiry management is how inventory flows through the supply chain. There are two primary strategies in the clinical supply chain world, FIFO and LIFO which are both illustrated in Figure 10.1. In the standard consumer good world, FIFO and LIFO are inventory valuation methods used to determine the cost of goods (COGs) of inventory, but FIFO and LIFO also happen to describe how inventory physically moves through the supply chain. When the FIFO methodology is used, older inventory (e.g. inventory received first) is sent to the end consumer first while the LIFO methodology flips that and sends the newest inventory (e.g. inventory received last) to the consumer first. The FIFO methodology is more often used in accounting, but it also makes intuitive sense to use this approach in physical inventory management for perishable products like biopharmaceuticals. This is because older batches typically have a shorter expiration date and therefore expire sooner while newer batches typically have a lengthier expiration date and can therefore sit on the shelf for longer before they're potentially lost.

Clinical trial material is perishable, so it may seem obvious that FIFO is the inventory methodology of choice when it comes to material flow through the supply chain. However, this is not always the case. In clinical trial supply chains, there are five node-to-node transfers where FIFO/LIFO methodologies are applicable. The most distal node in the supply chain is distribution from a

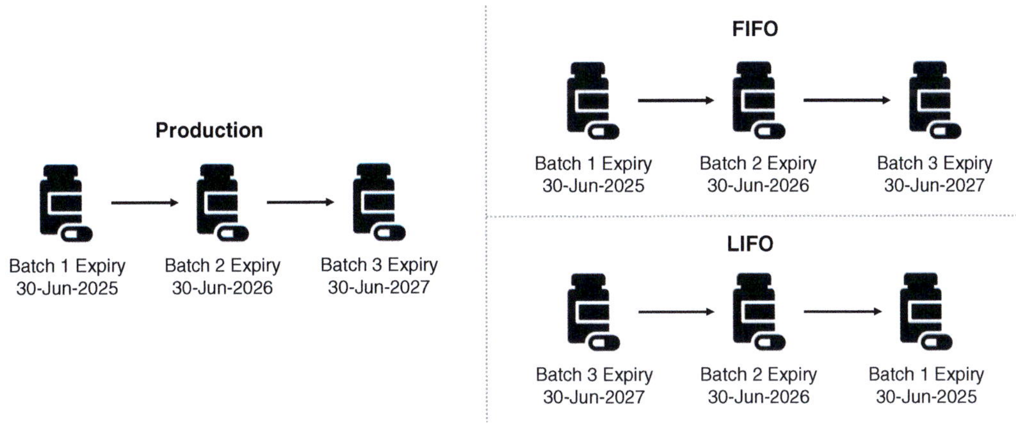

Figure 10.1 FIFO and LIFO Inventory Management Methodologies

clinical site to a subject, and this transfer should always follow the FIFO methodology to minimize drug waste at the clinical site. The second node is distribution from a regional depot to resupply a clinical site, and this should also always follow FIFO for the same reason, to minimize waste at the supplying depot.

The FIFO/LIFO decision becomes more challenging at the third node, distribution from a central depot to a regional depot. At this location in the supply chain, there may be multiple lots of material available, and it may not always make sense to resupply a regional depot with an older lot that expires sooner instead of a newer lot that expires later. This is because there may not be subject or site demand to consume the older lot before it reaches its expiration date, so applying a FIFO strategy would only lead to more waste. We'll explore how to forecast and plan for these scenarios in Chapter 11 – Finished Product Supply Planning: The Distribution Plan.

As the supply chain moves further upstream to the DP level, applying FIFO or LIFO remains murky for the same reason. Older DP lots may not have enough available expiry to justify packaging them and shipping them through the supply chain if there's not enough demand to consume them before lot ultimately expires. We'll discuss how to manage DP expiry and make appropriate allocation decisions in Chapter 12 – Upstream Supply Planning: Primary Packaging, Drug Product, and Drug Substance.

Interestingly, the most upstream node of the supply chain, drug substance (DS) manufacturing and its allocation to DP production, should always follow a FIFO approach. This is because the expiry period of the DS is completely independent from the expiry period of the DP. It's therefore always wise to consume older DS batches first so they do not expire. This is especially true because DS is the most expensive manufacturing step within the supply chain, and it's financially painful to throw out DS because of a poor allocation decision.

10.3 The Importance of Batch Visibility

Applying FIFO and/or LIFO methodologies to inventory management is predicated on having visibility into batch-specific inventory levels. While this may seem obvious, it's not uncommon to see planning models that look at on-hand inventory from an aggregate perspective where batch-specific information such as quantities and expiry dates are missing. Without this batch-specific visibility, it's impossible to make sound decisions about how to move inventory through the supply chain. This is particularly true for clinical trial supply chain management because of the two issues identified at the opening of the chapter – the dynamism of expiry period assignments and the need to appropriately plan for expiry events.

On the dynamism point, recall that while the DP has a standard expiry period, each DP batch has its own unique shelf-life date based on the batch's date of manufacture. Consequently, each finished product lot also has a unique shelf-life date. When the expiry period is extended for a DP stockkeeping unit (SKU), each finished product batch that was packaged off that DP needs to be identified so that a strategy can be developed around how to apply that extension to each batch. There are also situations where an expiry period may be limited within a particular country by its national regulatory body. When this happens, the clinical planner must create a strategy to supply clinical sites with inventory that meets the approved shelf life of that country.

With respect to expiry events, the clinical planner must see which batch is expiring, when the batch is expiring, where the batch is expiring (e.g. site, depot, or both), and how much of that batch is expiring to develop a coherent strategy for extending the shelf life of that batch or to replace that batch if a shelf-life extension is not possible. And, if that batch must be replaced, the planner also needs visibility into which upstream batches are available, the location of those batches in the supply chain, and how much inventory is available within each batch. Without batch-specific visibility, none of these analyses could happen.

Fortunately, the demand forecast table and production plan table provide much of the structure needed to add the lot-specific information required to effectively manage expiry. The temporal component is captured in the demand forecast period, and the weekly bucket gives the planner greater precision when incorporating randomization and trial supply management (RTSM) expiry management parameters that shorten a batch's expiry date. The SKU–depot structure of the demand forecast table also tells the planner which finished product is expiring and where in the supply chain action needs to be taken. Finally, the production plan table shows the potential impact expiring inventory has on weeks-of-supply. The final step to incorporating expiry into the finished product supply plan is to take on-hand inventory from an aggregate level and identify the individual batches that are available at the clinical site and depot levels of the supply chain.

10.4 Incorporating Expiry into the Finished Product Plan

The first step to incorporate expiry into the finished product production planning process is to create a new section in the plan dedicated to lot-specific inventory. Within this section, each row represents an individual batch, and the batch number and shelf-life date are both visible. Recall that in most cases, the date the product is no longer available for dispensation to a subject or shipment to a site is not the same date as the expiry (e.g. shelf life) date printed on the label. When an RTSM system is used to control investigational medicinal product (IMP) dispensation the system short-dates the shelf life to ensure expired product is not taken by a subject, and the date when supply is no longer available to dispense to subjects is known as the expiry short date (ESD).

In Chapter 6 – RTSM and Its Role in the Clinical Supply Chain, we identified four different expiry management parameters that need to be considered when calculating the ESD of each batch. These parameters are the do not dispense (DND), do not ship (DNS), do not include (DNI), and do not count (DNC) settings. The formula for identifying the ESD is below:

$$\mathrm{ESD} = \mathrm{ED} - \mathrm{DND} - \mathrm{DNS} - \mathrm{DNC} - \mathrm{DNI}$$

CW	W1	W2	W3	W4	W5	W6	W7	W8	W9	W10	W11	W12	W13	W14	W15	W16	W17	W18	W19	W20	W21	W22	W23	W24	W25	W26
Total demand	0	2	4	6	4	6	4	8	10	2	6	8	4	6	4	8	4	6	2	4	6	8	2	4	8	6
OHI	60	58	54	48	44	38	34	26	136	134	128	120	116	110	106	98	94	88	86	82	126	118	116	112	104	98
Prod. actual	60	0	0	0	0	0	0	0	120	0	0	0	0	0	0	0	0	0	0	0	50	0	0	0	0	0
Prod. plan	94	0	44	0	0	0	0	0	0	0	0	0	0	0	0	0	0	0	0	0	0	0	0	4	0	0
WOS	11	10	9	8	7	6	5	4	24	23	22	21	20	19	18	17	16	15	14	13	18	17	16	15	14	13
Safety stock	9	9	9	9	9	9	9	9	9	9	9	9	9	9	9	9	9	9	9	9	9	9	9	9	9	9
Order target	18	18	18	18	18	18	18	18	18	18	18	18	18	18	18	18	18	18	18	18	18	18	18	18	18	18
BATCH: A1A EXP: W13 ESD: W10																										
BATCH: B2B EXP: W26 ESD: W23																										
BATCH: C3C EXP: W39 ESD: W36																										

SKU description: Wonder drug placebo, 30-ct bottle, global label
Depot location: United States depot
DND = 14 days; DNS = 7 days

Figure 10.2 Production Plan Table with Batch Visibility

Note that in some cases, DNC parameters set at the depot level may overlap with the DND parameter, so only the DNC setting at the site may be applicable. The planner needs to understand how the parameters are configured and how they behave relative to the ESD calculation. After the individual rows for each batch are added, the RTSM parameters have been identified, and the ESD has been calculated, the production plan table looks like Figure 10.2.

In this example, we've carried over the production plan table that is captured in Figure 9.13. There are three finished product production orders planned for release in Week 1, Week 9, and Week 21 in the production actual row so three rows have been added as a separate section in the table. Each row represents a batch, and the batch number, expiry week (EXP), and ESD week are identified within the row. In this example, the ESD is calculated by subtracting three weeks from the expiry date (EXP). Individual batch on-hand inventory (BOHI) values are calculated the same way the on-hand inventory row was calculated in the previous chapter, by subtracting total demand in the current week from the on-hand inventory in the previous week. In the week the batch is released for use, on-hand inventory is 0 in the previous week, so the below formula still applies at the individual batch level:

$$\mathrm{BOHI}_{\mathrm{CW}} = \mathrm{BOHI}_{\mathrm{CW}-1} - \mathrm{Demand}_{\mathrm{CW}} + \mathrm{POQ}_{\mathrm{CW}}$$

Then, when the batch hits the ESD week inventory for that batch goes to zero. The aggregate on-hand inventory then becomes a total of each batch's on-hand inventory in that week, as in the formula below:

$$\mathrm{TOHI}_{\mathrm{CW}} = \Sigma\left(\mathrm{BOHI}_{\mathrm{CW}}\right)$$

When these formulas are applied to the production plan table above, it results in Figure 10.3.

In this example, populating the individual batch data changes the total on-hand inventory and weeks-of-supply values. When Batch A1A's ESD is reached, 16 kits of inventory are removed from on-hand inventory, bringing the total down from 136 kits to 118 kits in Week 10. Similarly, there are 52 units kits of Batch B2B available when it hits the short date in Week 23 and so on-hand inventory falls from 102 kits to 48 kits. The delta of 54 kits between these values is equal to the number of kits from Batch B2B that expired (e.g. 52 kits) plus total demand in that week (e.g. 2 kits).

CW

	W1	W2	W3	W4	W5	W6	W7	W8	W9	W10	W11	W12	W13	W14	W15	W16	W17	W18	W19	W20	W21	W22	W23	W24	W25	W26
Total demand	0	2	4	6	4	6	4	8	10	2	6	8	4	6	4	8	4	6	2	4	6	8	2	4	8	6
OHI	60	58	54	48	44	38	34	26	136	118	112	104	100	94	90	82	78	72	70	66	110	102	48	44	36	30
Prod. actual	60	0	0	0	0	0	0	0	120	0	0	0	0	0	0	0	0	0	0	0	50	0	0	0	0	0
Prod. plan	94	0	44	0	0	0	0	0	0	0	0	0	0	0	0	0	0	0	0	0	0	0	0	4	0	0
WOS	8	7	6	5	4	3	2	1	14	13	12	11	10	9	8	7	6	5	4	3	13	12	11	10	9	8
Safety stock	9	9	9	9	9	9	9	9	9	9	9	9	9	9	9	9	9	9	9	9	9	9	9	9	9	9
Order target	18	18	18	18	18	18	18	18	18	18	18	18	18	18	18	18	18	18	18	18	18	18	18	18	18	18
BATCH: A1A EXP: W13 ESD: W10	60	58	54	48	44	38	34	26	16	0	0	0	0	0	0	0	0	0	0	0	0	0	0	0	0	0
BATCH: B2B EXP: W26 ESD: W23	0	0	0	0	0	0	0	0	120	118	112	104	100	94	90	82	78	72	70	66	60	52	0	0	0	0
BATCH: C3C EXP: W39 ESD: W36	0	0	0	0	0	0	0	0	0	0	0	0	0	0	0	0	0	0	0	0	50	50	48	44	36	30

SKU description: Wonder drug placebo, 30-ct bottle, global label
Depot location: United States depot
DND = 14 days; DNS = 7 days

Figure 10.3 Production Plan Table with Batch Data Populated

Weeks-of-supply is tied to on-hand inventory, so the removal of these kits is reflected in this row as well. When Batch A1A hits the short date, the 16 kits represent three weeks of future demand (e.g. Weeks 10–12) that are no longer available in inventory. This reduces weeks-of-supply by three weeks from Week 1 onward, and on-hand inventory is all the way down to 1 week by the time Batch B2B is available as a resupply. The logic continues through the release of Batch B2B which only replenishes on-hand inventory to 14 weeks-of-supply because there is only enough total on-hand inventory to cover Week 9 through Week 22 inclusive of the start and ending weeks.

This specific example is a highly risky situation! On-hand inventory runs all the way down to just 1 week left before the first resupply batch is available, and this highlights exactly why planning inventory to the lot level and incorporating expiry management is so critical. The supply chain above is 1 week away from a potential stockout, and now the clinical planner has visibility to this risk and can act accordingly to bring in the timing of Batch B2B from Week 9 to an earlier release date to mitigate that risk.

It's also worth noting that this production plan table does not distinguish between inventory held at the depot and inventory held at the clinical site. Instead, the total is aggregated across all sites supplied by the depot and the depot itself. Generally, splitting site-level inventory from depot level inventory is not needed because this approach has site-level inventory hitting the ESD earlier than it will in the real world. In other words, this is a more conservative approach that forces the clinical planner to account for expiry events at site and depot level without risking a stockout due to expiry. The advantage to this approach is that the clinical planner does not have to maintain site-level inventory updates which would be a much more resource intensive approach than aggregating site and depot together. We'll return to this concept in Chapter 13 – Advanced Concepts in Demand and Supply Planning to show how to break out site-level inventory for greater accuracy within the plan if needed.

Incorporating expiry into the finished product production plan is about more than identifying individual batches, their on-hand inventory quantities, and expiry dates, though. It's also about understanding how expiry impacts the production plan strategy with respect to translating production order requirements into production orders. Here, we return to the concepts of quantity and timing that were introduced in the previous chapter.

The clinical planner must understand the overall stability strategy and whether the expiry period will remain stable or whether expiry period extensions will occur during the trial. If the expiry period remains stable with no further extensions planned, it simplifies the strategy for the clinical planner. From a quantity perspective, only enough kits should be packaged to satisfy total demand over the period of time before the finished product ESD is reached. If demand is highly variable, then the planner may consider packaging extra kits in case of an increase in subject demand. Remember though that in most cases, demand pushes out, so if the clinical planner packages extra units and demand fails to materialize those additional kits will go to waste. From a timing perspective, the clinical planner should heed the advice offered in the previous chapter. Find the sweet spot where safety stock levels aren't put at risk by packing too late, but delay production as much as possible to confirm demand.

The equation becomes more challenging when expiry period extensions are a part of the packaging strategy. With respect to production order quantities, the clinical planner has the option to create larger orders so long as three conditions are met. The first condition is demand certainty. If demand is highly uncertain, it makes sense to label smaller quantities anyways, and this is especially true if the expiry period and resulting shelf life are relatively short because that situation increases the risk that a significant portion of the batch goes to waste. Second, there should be a high degree of confidence the expiry period extension will not fail due to poor stability data. If there's concern the expiry period may not be extended, then the clinical planner should only label enough material to cover demand through the period already assigned to the DP. With this approach, product isn't wasted if the expiry extension doesn't materialize. Finally, the clinical planner should analyze if shelf-life extension relabeling activities make more economical sense than replacing expiring inventory. Shelf-life extensions of finished product batches are logistically challenging and costly as we'll discover later in this chapter. Even if an expiry extension is guaranteed, it may still make sense to package smaller lot sizes and plan to replace expiring inventory with more recently manufactured DP batches.

Potential DP expiry extensions also influence the timing of production orders because it's always desirable to have more shelf life on a finished product batch than less. This is one of the few absolutes in clinical planning, but in this case, it's true. If a finished product batch has more shelf life, then it's available for longer in the event demand pushes out. In other words, there is no downside to printing a longer shelf-life date on the label. If it's possible for the clinical planner to push out a labeling production order to wait for a better shelf-life assignment without jeopardizing safety stock levels, they should do it.

If finished product safety stock levels run too tight (e.g. <9 weeks-of-supply) to wait, then the clinical planner should work with analytical operations and quality to analyze the possibility of labeling the campaign "at risk" before the stability data is available to justify the longer shelf-life date. This concept was introduced in Chapter 4 – Quality Considerations, and it's particularly valuable in this situation. The only downside to labeling at risk is if the stability data do not justify further extending the batch. In this event, the batch would need to be relabeled again with the shorter expiry date, a costly and time-consuming scenario. This situation is why analytical operations needs to weigh in on the confidence level of receiving that expiry period extension and quality needs to give their approval before moving forward with this strategy.

In a scenario where the stability data no longer justify further expiry extensions, then the clinical planner needs to revisit the production plan in its entirety. The order target may need to be shortened, and this results in smaller, more frequent packaging campaigns. The need for smaller, more frequent finished product orders means that smaller, more frequent DP campaigns are required. This situation places a greater emphasis on accurately forecasting expiry events at both the site and the depot level, which is covered in the next section.

Before moving on, let's update the production planning guidelines begun in the previous chapter:

- *Quantity and Timing*: It's impossible to package DP that is not available. DP must be available in sufficient quantities and with sufficient lead time to complete the secondary packaging and labeling jobs in the plan.
- *Quantity*: Create smaller production orders early in the study when demand is uncertain. As subjects enroll, consider increasing the production order size, but only to the limit of a given batch's shelf-life date.
- *Quantity*: If expiry extensions are not expected, don't label more kits than demand can cover before the batch's shelf-life date is reached.
- *Quantity*: If expiry extensions are expected, the finished product batch size may be increased to cover demand over a longer period than justified by the current assigned expiry period and corresponding shelf-life date so long as three conditions are met:
  - Demand is relatively certain.
  - There is a high degree of confidence the expiry period will be extended.
  - Shelf-life extension relabeling activities make economic sense when compared to supplying replacement batches.
- *Quantity*: Avoid leaving small portions (e.g. <100 units) of DP lots unpackaged. It's inefficient to run many, small production campaigns if that DP lot is allocated for the same study, and the extra units may come in handy if clinical site resupply shipments are damaged.
- *Timing*: Find the sweet spot. Delay the timing of resupply production orders to confirm as much real subject demand as possible, but don't delay so much that safety stock levels are breached.
- *Timing*: If it's possible to delay a production order until an expiry period extension is received, it's wise to do so because more shelf life is always preferable to less shelf life. Consider labeling "at risk" before the expiry extension is finalized only if there is a high degree of confidence the expiry period will be extended.

Remember that these are only guidelines. The clinical planner's judgment must be applied to each production order decision!

10.5 Forecasting Expiry Events

While the expiry period and shelf-life assignments impact the finished product production order strategy, it's also necessary to forecast the batch expiry events within the plan for another reason. When the expiry short-date is reached for a batch, available on-hand inventory of that batch immediately goes to zero. This results in the RTSM generating a resupply order to restock all impacted clinical sites with a newer batch of inventory that has a longer shelf-life date so that subject dosing may continue. There must be sufficient inventory at the supplying depot to cover all resupply shipments of finished product to impacted clinical sites. For studies that have multiple active clinical sites, the clinical planner needs to accurately forecast expiry events, or they risk not having enough inventory with sufficient shelf life in the right place at the right time to avoid site resupply shipment failures. It's a horror story in clinical trial supply chain management when an expiry event is missed, and then depots stockout of replacement inventory which in turn impacts the ability to continue dispensing drug to patients. This scenario takes weeks to recover from and can significantly delay the overall trial timelines.

To avoid this situation, the clinical planner needs to forecast for expiry events, and the most direct way to do this is through use of the demand forecast and production plan tables. From a timing perspective, forecasting expiry events should begin at least 6–8 months ahead of the actual expiry event. The reason behind starting so early is to ensure sufficient replacement inventory is available. If a new DP batch needs to be manufactured, that activity may need to start several months ahead of the actual expiry event. Forecasting the expiry event itself follows a step-by-step process:

1) Identify the finished product batch and its ESD, which we've already done in the previous section.
2) Forecast the number of clinical sites that will have supply of expiring inventory from the impacted batch at the point the expiry period is reached. Keep in mind that when the initial expiry forecast is created, this may be largely guesswork if not all clinical sites have activated and/or begun screening and enrolling subjects. Communication with clinical operations is key to understanding how many potential sites will be supplied with inventory at the point the expiry event is reached.
3) Forecast the amount of inventory that will be stocked at each clinical site at the point the expiry period is reached. This can be even more challenging to forecast because the amount of inventory held at each site is dependent on that site's RTSM site supply strategy (e.g. low, medium, or high), the floor quantity set in the predictive threshold resupply model (PTRM), and the number of subjects enrolled at the site. When forecasting these quantities, it makes sense to err on the side of more inventory expiring than less. It's almost always better to have more replacement kits available than too few available.
4) Sum the total amount of projected inventory that will expire at the clinical site and enter this value in the site demand field of the demand forecast table as in Figure 10.4.

   In this example, initial site demand was 16 units spread across four site activation events from Weeks 2 to 5. The planner has entered site demand of 18 units to account for an expiry event in Week 10, the ESD of the initial batch supplied to the sites. The amount of expiring kits is an estimate based on the assumption that 3 sites require at least 4 kits on hand and the fourth site requires at least 6 kits on hand for a total of 18 kits. From Chapter 9, we know that increasing demand leads to increased production order requirements. In other words, adding site demand increases the production order requirement at the supplying depot, so this step creates a signal to the planner that more supply of a newer batch is needed to replace the expiring batch.
5) Continually refine the forecast as the expiry event draws closer. From Chapter 6 – RTSM and Its Role in the Clinical Supply Chain, we know that the RTSM is capable of generating site inventory reports that tell the clinical planner which sites have been stocked with inventory and how

	W1	W2	W3	W4	W5	W6	W7	W8	W9	W10
Total subject demand					4	8	8	8	8	10
# Sites activated		1	1	1	1					
Site demand		4	4	4	4					18

Figure 10.4 Example Expiry Event Demand

	W1	W2	W3	W4	W5	W6	W7	W8	W9	W10	W11	W12	W13	W14	W15	W16	W17	W18	W19	W20	W21	W22
Total subject demand					4	8	8	8	8	10	10	10	10	10	12	12	12	12	12	14	14	14
# Sites activated		1	1	1	1																	
Site demand		4	4	4	4					18												
Total demand	0	4	4	4	8	8	8	8	8	28	10	10	10	10	12	12	12	12	12	14	14	14
OHI	80	76	72	368	360	352	344	336	328	300	290	280	270	260	248	236	224	212	200	186	172	158
Prod. actual	80	0	0	300	0	0	0	0	0	0	0	0	0	0	0	0	0	0	0	0	0	0
Prod. plan	62	168	0	0	0	0	0	0	0	0	0	0	0	0	0	0	0	0	0	0	0	0
WOS	9	8	7	28	27	26	25	24	23	22	21	20	19	18	17	16	15	14	13	12	11	10
Safety stock	9	9	9	9	9	9	9	9	9	9	9	9	9	9	9	9	9	9	9	9	9	9
Order target	18	18	18	18	18	18	18	18	18	18	18	18	18	18	18	18	18	18	18	18	18	18
BATCH: A1A EXP: W13 ESD: W10	80	76	72	68	60	52	44	36	28	0	0	0	0	0	0	0	0	0	0	0	0	0
BATCH: B2B EXP: W39 ESD: W36	0	0	0	300	300	300	300	300	300	300	290	280	270	260	248	236	224	212	200	186	172	158

SKU description: Wonder drug placebo, 30-ct bottle, global label
Depot location: United States depot
DND = 14 days; DNS = 7 days

Figure 10.5 Production Plan Table with Expiry Event Forecasted and Completed

much inventory is at each site. This is valuable information to analyze when updating expiry event forecasts.

6) Plan replacement production orders based on the logic described in the previous section.

To illustrate how this process plays out, consider Figure 10.5.

This production plan table builds on the expiry event in the previous figure and on the production planning concepts we've discussed so far. First notice that Batch A1A's ESD is calculated for Week 10. In Week 1, the plan created a production order requirement of 62 kits based on the total amount of future demand from Week 2 to Week 10, the point at which the batch expires. The clinical planner decided to create an initial production order of 80 kits, greater than the suggested order quantity of 62 kits. This decision was made to cover a potential up-side demand scenario and because there were only 80 kits of DP available to package.

Due to the ESD of the initial batch, only 9 weeks-of-supply are available on the first batch, so the production plan calls for an immediate resupply of 168 units based on the 9-week safety stock and 18-week order target values. However, finished product Batch B2B isn't available for distribution to the clinic until Week 4 due to DP unavailability. The good news in this plan is that the underlying DP batch has a much better ESD of Week 36, so the clinical planner creates a much larger production order of 300 units to cover future demand for 28 weeks from Week 4 to Week 31 (Weeks 23–31 not pictured).

Then, in Week 10, the expiry event takes place. Entering that week, there are 28 kits of Batch A1A left, and of those 28, 10 kits are consumed by subject demand while the remaining 18 kits expire at the clinical site and go to waste. The amount of available inventory from Batch A1A goes from 28 kits to 0, and demand in future weeks eats into inventory from Batch B2B. Also note that on-hand inventory of Batch B2B remains at 300 kits in Week 10. That's because the model assumes the 10 kits of total subject demand are still satisfied from Batch A1A and because this model aggregates depot and site inventory together. In other words, there are still 300 kits of Batch B2B spread between the depot and site levels of the supply chain from Week 10 onward.

There are a few takeaways from this example. First, if the expiry event was not included in the forecast in this example, the clinical planner would still be OK because there was so little of the DP packaged to begin with, and this is by design! This example illustrates why packaging smaller orders is wise early in the study when demand is uncertain and when expiry is poor. That strategy limits the potential exposure to waste and stockouts from expiry events because there's less inventory to replace in the supply chain. Second, the weeks-of-supply calculation highlights the risk inherent to the expiry event, and immediately calls for a resupply order requirement in the plan. Finally, the individual batch section gives the clinical planner the visibility and control they need to appropriately manage the expiry event when multiple batches are available at the same time. While that wasn't illustrated in this example, it's not hard to imagine a world where multiple batches are expiring at the same time and multiple batches are available to replace those expiring batches. Keeping track of that complexity requires separating total on-hand inventory into individual batch inventory levels. Without this granularity, the clinical planner will struggle to choose how best to replace the expiring batch, via a new batch of DP or through a shelf-life extension activity, a decision we'll cover in the next section.

10.6 Shelf-Life Extensions Versus Replacement Inventory

When inventory expires, the clinical planner may opt to replace the expiring batch with a batch of finished product made from a new DP batch, or they can execute a shelf-life extension on the expiring batch. In a shelf-life extension operation, the expiration date printed on the label is crossed out, and a sticker is added to the label that lists the batch number and the new expiry date. Shelf-life extensions take place at any central or regional depot that has been qualified to perform good manufacturing practice (GMP) labeling operations, and they are usually a simple and straightforward option for most small-molecule and biologic products.

However, there are potential constraints that limit the feasibility of executing a shelf-life extension. If the primary packaging is labeled with an expiry date and it's difficult to access because it's been further packaged in a secondary container (e.g. a sachet in a carton), then it might not be possible to perform the shelf-life extension operation without damaging the entire kit. Or if the product must be kept frozen, then the location where the shelf-life extension takes place must have a frozen labeling room, and capability may not be available at a particular depot. Or if the study is blinded, and the expiry period for the active or placebo is extended for one and not the other, it may not be possible to perform the shelf-life extension without compromising the blinded status of the IMP.

Assuming that a shelf-life extension is possible, the clinical planner should take the following variables into consideration when selecting an expiry management strategy:

- *Confidence Level in the Stability Profile of the DP*: If the formulation development and analytical operations teams have concerns about the potential stability profile of the drug, the clinical planner should default to planning replacement batches of DP rather than planning shelf-life extensions. It's simply too risky not to plan a new batch and then have the expiry period limited without a replacement batch on hand.
- *Amount of Time Left in the Stability Study*: If there is relatively little time left on the stability study, that means a relatively lengthy expiry period of at least 24 months has already been established. In that situation, it typically makes sense to plan replacement DP inventory because the finished product batches already cover an extensive period of demand. If the stability study is

relatively new and the established expiry period is less than 18 months, then shelf-life extensions make more sense to increase the amount of demand a single finished product batch covers.

- *Cost of the DS and DP to Manufacture*: The more expensive the DS and DP is to manufacture, the more sense it makes to make to limit the amount of DS and DP campaigns required to support a study. As DS and/or DP costs increase, the argument in favor of executing shelf-life extensions also increases.
- *Cost of an Individual Shelf-life Operation*: Most shelf-life extensions are relatively cheap when compared to making a new DP batch, but this isn't always the case, particularly for complex packaging configurations and/or cold chain products. The cost of executing a shelf-life extension strategy may also increase exponentially when the next factor is taken into the analysis.
- *Number of Depots at Which the Operation Could be Performed*: In global clinical trials with large depot networks, multiple shelf-life extensions on a single batch may need to be performed at once. This increases the complexity and cost of the shelf-life extension strategy considerably. For example, if a single shelf-life operation costs $8000 to execute, but an active and a placebo SKU need to be extended across 10 depots, that's 20 unique operations which brings the total cost to $160,000. That cost may indeed exceed a new DP campaign. As the number of SKUs and depots increase, it typically makes more sense to plan replacement DP inventory.
- *Number of Batches of On-hand Inventory at the Time of the Potential Shelf-life Extension*: Having more than one batch of on-hand inventory helps considerably when executing shelf-life extensions. When there's only one batch of inventory available for extension, that batch must be split into two because some inventory needs to be available to support continued dosing. This means the clinical planner must estimate how many kits are needed to cover demand while the first extension takes place. If two batches are available, the planner can extend one batch followed by extending the second batch, a much more straightforward approach.
- *Amount of Inventory Within Each Available Batch at the Time of the Potential Shelf-life Extension*: Carrying the last bullet point forward, if there is only one batch available for a shelf-life extension, it's beneficial if that batch has at least 9 weeks-of-supply available at the time of the extension. Shelf-life extensions can take up to a month to execute after accounting for batch release activities, so having a relatively large amount of inventory available in the batch provides more room for error when splitting the batch into two parts.
- *Capability of the Depot Network*: There's an old saying that goes, "Just because you can doesn't mean you should." Even if the depot network has the capability to perform a shelf-life extension, when the product is sensitive to light, temperature, or humidity, then it may make sense to simply plan for replacement inventory and not risk losing a batch.
- *Capacity Within the Depot Network*: This consideration is especially important to consider when shelf-life extensions need to be turned around in a tight timeframe. If the depot network has limited capacity to perform labeling jobs, then it may not be able to perform the shelf-life extension within the production requirement window.

Unfortunately, there aren't simple answers for when to perform shelf-life extensions and when to provide replacement DP inventory. Each situation needs to be analyzed on a case-by-case basis, and the clinical planner should consider each of the points made above before selecting one strategy or the other. There are even times when both strategies make sense within a single plan. For example, a shelf-life extension strategy may make sense early in the study if the expiry period is short, there's high confidence in the stability profile, and the DP is relatively expensive to make, but may no longer make sense as the study enrolls, the expiry period is relatively long, and batches are spread out across a wide range of depots and sites.

If a decision is taken to perform shelf-life extensions, then quantity and timing of the operation must also be carefully thought through. From a timing perspective, it's best to execute a shelf-life extension as soon as the underlying expiry period extension is in place. In some cases, this may be only a few weeks before the ESD of the first batches, so careful planning with the depot and quality may be needed to expedite the process.

In terms of calculating shelf-life extension order quantities, when a second batch of material is available with enough supply to cover the shelf-life extension lead time, simply extend the entire expiring batch. If this isn't the case, and the only batch available is the one that needs to be extended, more analysis is necessary. The clinical planner needs to calculate the total amount of demand over the shelf-life extension lead time and subtract that value from the on-hand quantity of the batch at the time the extension is scheduled to take place. For example, if the lead time to execute the shelf-life extension is 4 weeks, there are 200 kits on hand at the time of the extension, and average weekly demand is 10 kits, then the clinical planner needs to leave at least 40 kits of supply on hand to cover demand over the shelf-life extension lead time (e.g. 10 kits/week * 4 weeks). However, it's also prudent for the planner to assume some uncertainty in demand and/or a delay in the extension process, so they may consider leaving more kits of supply on hand to cover these possibilities. In this example, a smart strategy would be to extend up to 140 kits in one operation and then extend the remaining kits if it's financially advisable to do so.

10.7 Strategies for Managing Expiry Limitations

A final challenge with respect to expiry management is confronting expiry period limitations on the DP. There are four situations where a limitation can occur, two related to the underlying stability of the DP and two related to interactions with the competent health authority (CHA). The two stability-related scenarios occur:

1) At the start of a clinical study with a new formulation and/or packaging configuration that has a limited stability profile.
2) When stability data fails unexpectedly, limiting the expiry period across all studies and countries.

In the first situation, the clinical planner has the most flexibility, and there are a few strategies used to manage through tight (e.g. between 6 and 12 months) expiry period windows. One common strategy is to package small, frequent batches to minimize potential waste. This is generally acceptable because demand uncertainty is high at the beginning of the study so the plan should be to package smaller quantities anyways. If there is high confidence the DP will receive longer expiry periods, larger batches may be planned and then updated via shelf-life extensions.

A second commonly employed strategy for clinical trials that begin in the United States is to create a separate finished product SKU with a US-only label. This is because there is no legal or regulatory requirement to print the shelf-life date on the label within the United States. Creating a US-only SKU in a global study creates additional supply chain complexity, but the advantages are significant. Since there is no shelf life printed on the label, a batch's shelf life can be updated via the RTSM, and no shelf-life operation needs to be performed.

In the second situation, the clinical planner needs to reassess the entire supply plan to determine the impact of the expiry limitation. Actions may include:

- Analyzing the distribution planning strategy to determine if more frequent depot resupplies are required to replace expiring batches. Chapter 11 – Finished Product Supply Planning: The Distribution Plan covers these strategies in detail.

- Planning smaller, more frequent finished product batches because each batch covers a smaller period of demand.
- Planning additional DP campaigns to support those finished product batches because expiry extensions are no longer available as an option.
- Analyzing the potential for introducing a new DP formulation that has a stronger stability profile for future clinical trials.

The two CHA-related scenarios occur when a CHA:

1) Requires prior notification before a lengthier expiry period may be applied to the DP.
2) Restricts the expiry period to a shorter duration than that applied for in the regulatory application.

In Chapter 4 – Regulatory Considerations, we explored the regulatory application components, submission process, and commitments in detail. The applicable section of the regulatory application in this discussion is P.8.1 – Stability Data. In many countries including the United States, the United Kingdom, and those in the European Union, shelf-life extension activities are allowed with approval of the application package. However, in some cases, a specific country may ask to receive prior notification of expiry period extensions before product labeled with the lengthier expiry period is introduced to that country. When this happens, it's the sponsor's responsibility to provide a notification to the CHA containing the updated stability data and justification for the new expiry period. This can be an easy detail to lose track of during the trial and further justifies use of a regulatory status tracker tool to capture this type of commitment.

Another concept discussed in Chapter 4 are restrictions, which occur when a CHA limits or restricts the criteria contained within the regulatory application. With this context in mind, it's possible for a CHA to place a restriction or limitation on the DP expiry period if there are concerns with the stability data package. When this happens, it's usually because the sponsor is attempting to leverage stability data from a comparable product or formulation, and the CHA is not willing to accept the comparison. In this situation, the clinical planner has three options to manage the restriction within the finished product production strategy:

1) Label a batch of finished product with the lengthier shelf life granted in some countries and another batch with the restricted shelf life granted in other countries. This is an appropriate strategy if it's important to supply drug to the restricted country right away. Labeling one batch with longer shelf life reduces the amount of potential waste from that batch because it covers a longer period of demand, while labeling another batch with restricted shelf life enables site start-up and enrollment in the restricted country as soon as possible.
2) Label all batches of the finished product with the restricted shelf life. This strategy is appropriate when the restricted shelf life is greater than 18 months and there is pressure to begin enrollment in the restricted country as soon as possible. Consolidating production into a single batch reduces cost and shelf-life extensions can be performed in depots/countries that have not been restricted to further reduce waste if demand pushes out or the underlying DP is expensive.
3) Label all batches with the lengthier shelf life granted in some countries and wait until the restriction is lifted in the applicable countries before supplying those batches. This strategy is appropriate if it's not critical to start that country right away or when the restriction limits shelf life to less than 18 months.

Keep in mind that replacement inventory must be planned for the restricted country no matter which strategy is chosen until the restriction is lifted. In some cases, the restriction may be permanent, but in other cases, an amendment to the CHA containing additional stability data may be sufficient to lift the restriction. When restrictions occur, the clinical planner should work closely with analytical operations, quality, and regulatory affairs to develop a strategy for managing the restriction and to ensure the finished product production plan aligns with the chosen strategy.

10.8 Key Takeaways

1) Every DP used in a clinical trial has an expiry period assigned to it that is determined through stability studies that are conducted on samples of the DP. The expiry period describes the amount of time a batch of DP is considered good for use from its date of manufacture. Finished product batches are assigned a shelf-life date based on the date of manufacture of the underlying DP and the expiry period of the DP.
2) Expiry management is often difficult in clinical trials because the expiry period may not be static over the duration of the study and because expiry period assignments aren't guaranteed. Successfully managing expiry requires incorporating expiry into the demand and supply plan as well as collaborating closely with key CMC team members such as analytical operations, quality, and regulatory affairs.
3) FIFO and LIFO are two inventory management systems. In clinical trial supply chain management, FIFO is always utilized when distributing product to a subject or to a clinical site from a depot. Upstream in the supply chain, the FIFO/LIFO strategy becomes more complex as expiry management is introduced into the equation.
4) Building individual batch inventory levels into the plan is necessary so the clinical planner has visibility to when each batch expires, how much of each batch will expire, and where a batch will expire. Without this information, it's impossible to control the supply chain by accurately forecasting and preparing for expiry events.
5) To incorporate finished product expiry into the plan, the ESD is calculated using the RTSM DND, DNS, and DNC parameters. Then, on-hand inventory is disaggregated into individual batch inventory. With these elements in place, the clinical planner can use the updated production plan table to forecast expiry events and create production orders to replace expiring inventory.
6) Forecasting expiry events follows a step-by-step process that includes identification of the finished product batch, forecasting the number of clinical sites that will have expiring inventory, forecasting the amount of inventory at each site that will expire, and adding that quantity to the demand forecast in the week the batch expires. This process creates demand for newer inventory at the depot level to replenish impacted clinical sites.
7) There are two approaches to replacing expiring inventory. The first is to execute a shelf-life extension, which is an operation to update the expiry date on the label of the expiring batch. The second option is to replace the expiring inventory with a new batch of material. Selecting the optimal strategy requires considering several variables including confidence level in the stability profile of the underlying DP, cost of the DP, cost of the shelf-life extension operation, and capability/capacity of the depot network to execute a broad shelf-life extension strategy.
8) Managing expiry limitations is another potential challenge. Expiry limitations come from three sources, a limited stability profile, a regulatory restriction, or an unexpected failure of stability data. In each situation, the clinical planner has multiple strategies at their disposal to work around the limitation.

11

Finished Product Supply Planning

The Distribution Plan

11.1 Introduction to Distribution Planning

Distribution planning is the process of determining how much inventory to ship to a regional depot, when to ship that inventory, and which batch of inventory is most appropriate to ship. The result of the distribution planning process is the distribution plan itself which complements the production plan. Together these two plans compose the entire supply plan for the trial. Recall from Chapter 7 – Supply Chain Planning for Clinical Trials that a network of depots within the supply chain for the trial is responsible for moving finished product from the packaging and/or labeling site to participating clinical sites around the world. Distribution planning brings together many of the planning concepts discussed so far including demand forecasting, production planning, and expiry management.

Identifying the country list and the depot network that will supply the clinical sites within those countries is one of the first actions a clinical planner takes on a newly planned trial. The output of this work is a map of the distribution supply chain, how inventory flows from one depot to another, and how long it takes inventory to move from one depot to the next. Once the distribution supply chain is mapped, the next step is to understand the various requirements of importing product into each country. All participating countries require competent health authority (CHA) approval before investigational medicinal product (IMP) can be imported, but each country has myriad requirements that must be accounted for in the planning process.

Documenting the physical supply chain is a prerequisite to incorporating the depot network into the supply plan. Fortunately, effective distribution planning is built on many of the concepts introduced in Chapter 9 – Finished Product Supply Planning: The Production Plan and Chapter 10 – Finished Product Expiry Planning:

- The stockkeeping unit (SKU) –depot setup applies to SKU–regional depot combinations as well as it does to the SKU–central depot combination.
- The demand forecast table and demand calculations are the same at regional depots and central depots. Therefore, the demand planning process is the same at regional depots as at central depots.
- The production plan table structure, inputs, and parameters are all the same. Instead of production order requirements though, distribution order requirements are generated from application of the min–max methodology and supply planning parameters at regional depots. The clinical planner's objective in the distribution planning process is to take distribution order requirements and convert them to actual distribution orders.

Supply Chain Planning for Clinical Trials – A Practical Guide, First Edition. Ryan Mills.

- Weeks-of-supply (WoS) is used in distribution planning the same way as in production planning.
- The guidelines governing production planning with respect to order timing and quantity apply equally well to distribution planning and distribution order timing and quantity.
- Expiry considerations are also the same whether planning production orders at a central depot or distribution orders at a regional depot.

There are real differences between production planning and distribution planning though. Most notably, the strategy for supplying regional depots tilts heavily toward accounting for site demand and incorporating expiry management strategies into the plan. These differences sometimes mean adjusting inputs in the demand forecast table or in the production plan table. The other notable concept with distribution planning is determination of the allocation strategy for each batch. In demand and supply planning, allocation is the process of selecting a batch to fulfill a production or distribution order. We'll return to allocation in Chapter 12 – Upstream Supply Planning: Primary Packaging, Drug Product, and Drug Substance, where allocation decisions are made on which drug product (DP) batch should fulfill a finished product production order and which drug substance (DS) batch should fulfill a DP batch.

We begin this chapter with an overview of common clinical distribution networks and considerations when importing to various countries around the world. This culminates in an overview of how to think about and calculate distribution lead times, one of the key inputs in the distribution planning process. Then, we turn our attention to how to set up a distribution network within the demand forecast and production plan table framework. Next, we revisit demand and supply inputs that require reassessment when distribution planning versus production planning before moving into the four components of the distribution planning process. These strategies include determination of initial depot stocking order, allocation of finished product batches to order, balancing supply across depots as the study progresses, and managing regulatory constraints within the distribution plan. We'll also introduce the WoS summary, a useful data set for identifying supply chain imbalances. We'll conclude the chapter by summarizing the output of the distribution planning process and the distribution plan summary and discussing the distribution order readiness business process.

11.2 Mapping the Clinical Packaging and Distribution Network

The country list is the basis of the distribution strategy for every clinical trial. Typically, in early-stage or smaller studies such as Phase I or clinical pharmacology studies, it's most common for the trial to take place in only one country. This is because these studies rely on healthy volunteers or a low (e.g. <100) number of subjects that can be recruited from a smaller geographic footprint. As clinical development progresses, the country list for late-phase studies typically expands both to enroll the study as quickly as possible and to collect data from subjects within a variety of countries to support potential commercial applications. Increased competition for subjects within a specific disease indication also plays a role in trials entering more countries than in the past. It's not uncommon for a large Phase III study to take place in several dozen countries spanning almost every continent.

Once the country list takes shape, typically 9–12 months before the start of a global clinical trial, the clinical planner should identify the network of distribution depots that will supply each country. Many of the largest biopharmaceutical companies have a wholly owned network of depots, while smaller companies usually rely on third-party contract manufacturers and distributors that have established depots to ship their products internationally. Whether using an internal or external network though, it's important to recognize that every distribution network is unique, and therefore the distribution depot strategy for each trial needs to be mapped on case-by-case basis.

The first depot the clinical planner needs to identify is the central depot, which is typically the finished product manufacturing site and has the responsibility to supply other regional depots in the study. While central depots primarily supply regional depots, in some very large studies, there may be multiple central depots that perform finished product manufacturing who can also supply each other. And, while it's typical for central depots to perform packaging and/or labeling of finished product for a study, this is not always the case. In most global packaging and distribution networks, there are multiple packaging site and/or central depot options to choose from. Common geographic regions for clinical packaging hubs include the United States, the United Kingdom, the European Union, and Asia–Pacific. It's also typical for central depots to ship product directly to clinical sites, especially, when the central depot sits in the United States or the European Union. Selecting an optimal packaging network is beyond the scope of this chapter, but we'll return to this concept in Chapter 20 – Contract Organization Management and Oversight. For now, it's sufficient to say that a majority of trials with a global footprint operate with a single packaging site that also acts as the central depot responsible for shipping IMP to one or multiple regional depots for in-country distribution.

After the central depot has been selected, the next step is to identify the regional depots who are responsible for shipping finished product directly to clinical sites. A regional depot is defined as a depot that only supplies clinical sites and is not capable of manufacturing the finished product. While every distribution network has nuances, there are two general approaches to supplying clinical sites within a country. Either a country requires an in-country regional depot or a country can be serviced from a hub-based regional depot that serves many countries in a particular geographic area. Some countries have legal requirements to service clinical sites from an in-country depot, while other countries use in-country depots to reduce costs. Similarly, hub-based regional depots are used to reduce supply chain complexity and depot costs in specific regions. Below is a list of widely used countries in clinical trials and the type of regional depot that supplies those regions:

- *United States*: Sites are typically supplied from an in-country regional depot to minimize shipping costs.
- *European Union (EU)*: Sites within EU countries are typically supplied from a hub-based regional depot to minimize supply chain complexity when shipping finished product across the numerous countries in the EU.
- *United Kingdom*: Sites must be supplied from an in-country depot due to Brexit and regulatory requirements for UK-based qualified person (QP) certification.
- *Eastern Europe including Russia and Ukraine*: Sites are typically supplied from in-country regional depots.
- *China and Japan*: Sites are typically supplied from an in-country regional depot.
- *Asia–Pacific including Australia, Singapore, South Korea, and SE Asia*: Sites are typically supplied from a single in-country regional depot to minimize supply chain complexity.
- *Latin America Including Argentina, Brazil, Chile, Colombia, Mexico, and Peru*: Sites are typically supplied from in-country regional depots due to local regulations and import license requirements.

A final reason to identify and map the distribution network of a new study is that clinical trial regulations and import requirements are constantly evolving. The strategy used in a previous study may no longer apply or may no longer be the optimal network for a new study.

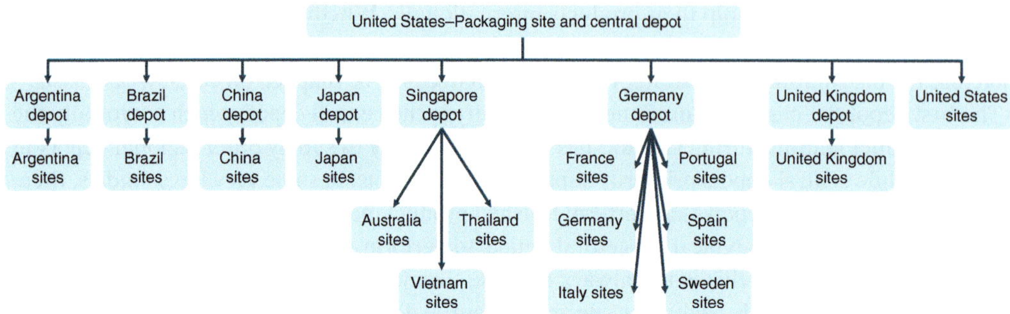

Figure 11.1 Example Clinical Trial Distribution Network

An example of a fully mapped distribution network for a global study is captured in Figure 11.1. In this example trial, the finished product packaging site is located in the United States. This site is responsible for supplying drug directly to clinical sites within the United States, and it acts as the central depot to seven regional depots. Five of these depots – Argentina, Brazil, China, Japan, and the United Kingdom – are in-country regional depots. There are also two hub-based regional depots in Singapore and Germany that service multiple Asian and EU-based countries, respectively. For the sake of this example, let's also assume this supply chain supports a Phase III study testing a 50 mg dose versus placebo, and the two finished product SKUs are a 50 mg capsule of the active drug and a placebo drug. Building out a supply plan for this trial requires creating 8 SKU–depot combinations for the 50 mg capsule and 8 SKU–depot combinations for the placebo, for a total of 16 SKU–depot combinations. Two of those SKU–depot combinations, the 50 mg-US depot and placebo-US depot, are the SKU–central depot combinations. In the next section, we'll show how to link the in-country and hub-based regional depots back to the supplying depot in the supply plan.

Before taking that step though, the clinical planner has more information to gather about each of the regional depots. Specifically, the process and lead times associated with importing product into each country need to be captured. All international shipments of biopharmaceutical products require a standard sequence of operations:

- Preparation of shipment documentation including the proforma invoice.
- Review of shipment documentation and customs preclearance.
- Acquisition of the appropriate shipping containers and temperature monitors.
- Scheduling of shipment pickup.
- Transit to the port of entry at the destination country.
- Customs clearance of the goods.
- Transit to the regional depot.
- Receipt at the regional depot.

Moving biopharmaceutical products internationally is time consuming. Typically, each of the steps listed above takes days or weeks to complete. While some steps may take place in parallel (e.g. review of shipment documentation and acquisition of shipping containers), others must be performed sequentially (e.g. transit, customs clearance, and delivery to the regional depot). Most challenging of all, the amount of time these activities take varies from country to country and from shipment to shipment, even when those shipments are executed within the same depot network. With that context in mind, the single most important variable in distribution planning is the depot-to-depot lead time because this lead time drives the timing of when distribution orders must be initiated to arrive on time at the receiving depot to service demand. The depot-to-depot lead time captures the amount of time it takes to complete all activities required to set up and execute a distribution order from the supplying depot to the receiving depot. Analyzing, understanding, and documenting lead times for each depot-to-depot transfer are a prerequisite to linking regional depots to the central depot within the plan.

In Chapter 3 – Introduction to the Clinical Trial Supply Chain, we introduced two additional country-specific activities on top of the general steps listed above, import license requirements and the QP certification process. Import licenses are documents issued by national governments that grant authority to import products for a specified purpose. Another way to say this is that in countries requiring an import license, it's impossible to import product without an approved license. Import licenses fall into one of the two categories, umbrella licenses and shipment-specific licenses. Umbrella licenses cover the entire quantity of IMP for the trial, while shipment-specific licenses must be applied for with each individual shipment. Some countries require both an umbrella import license and subsequent shipment-specific licenses. If the sponsor needs to ship additional inventory above what's listed in an umbrella license, then the license needs to be amended and approved. Obtaining an import license takes anywhere from one to several weeks depending on the country, so it's an important factor to incorporate when analyzing distribution lead times.

QP certifications were covered extensively in Chapter 4 – Quality Considerations and typically take 2–3 weeks to execute, although they can take longer if there are questions and follow-up activities needed to obtain the certification. Below is a list of common countries that require an import license or QP certificate:

- *Umbrella Import Licenses*: Argentina, Chile, Hong Kong, India, Israel, Peru, Russia, Taiwan, and Turkey
- *Shipment-specific Import Licenses*: Argentina, China, Colombia, Egypt, Malaysia, Mexico, and Vietnam
- *QP Certification*: All European Union countries, Israel, and the United Kingdom

The final output of the distribution lead time analysis is a table of countries that will participate in the trial, the supplying depot for each country, and the total depot-to-depot lead time from shipment initiation through shipment receipt and QP certification (if applicable). A representative table of the example clinical trial supply chain is presented in Table 11.1.

With the supply chain mapped, the depot network identified, and the distribution lead times analyzed and documented, the next step is to build out the SKU–regional depot demand forecast and production plan tables and link up the SKU–regional depot tables to the SKU–central depot table.

11.3 Connecting Regional Depots to the Central Depot

There are two steps needed to connect regional depots to central depots in the supply plan. The first step is to update the production plan table to a distribution plan table. To do this requires a number of sub-steps starting with renaming the production plan table to the distribution plan table at each SKU–regional depot combination. This is because regional depots don't manufacture finished product; they receive finished product from the central depot. Within the table, the production plan row becomes the distribution plan row, and instead of the plan generating production order requirements, the model generates distribution order requirements. Distribution order requirements are then turned into actual distribution orders through the same process that production order requirements are turned into production orders via use of

Table 11.1 Table of Depot-to-Depot Lead Times

Central Depot	Country	Regional Depot	Depot-to-Depot Lead Time	Notes
United States	United States	N/A	N/A	None
	Argentina	Argentina	8 weeks	Umbrella and shipment licenses
	Brazil	Brazil	10 weeks	Umbrella license
	China	China	12 weeks	Shipment license
	Japan	Japan	3 weeks	None
	Australia	Singapore	3 weeks	Extra week for depot-to-site shipments
	Thailand			
	Vietnam			
	France	Germany	3 weeks	QP certification
	Germany			
	Italy			
	Portugal			
	Spain			
	Sweden			
	United Kingdom	United Kingdom	3 weeks	QP certification

	W1	W2	W3	W4	W5	W6	W7	W8	W9	W10	W11	W12	W13	W14	W15	W16	W17	W18	W19	W20	W21	W22
Total subject demand												2	2	2	2	2	2	2	2	2	2	2
# Sites activated							1		1													
Site demand							4		4													
Total demand	0	0	0	0	0	0	4	0	4	0	0	2	2	2	2	2	2	2	2	2	2	2
OHI	0	0	0	0	0	40	36	36	32	32	32	30	28	26	24	22	20	58	56	54	52	50
Dist. actual	0	0	0	0	0	40	0	0	0	0	0	0	0	0	0	0	0	40	0	0	0	0
Dist. plan	0	0	0	0	0	34	0	0	0	0	0	0	0	0	0	0	0	18	0	0	0	0
WOS	0	0	0	0	0	21	20	19	18	17	16	15	14	13	12	11	10	29	28	27	26	25
Safety stock	0	0	0	0	0	9	9	9	9	9	9	9	9	9	9	9	9	9	9	9	9	9
Order target	0	0	0	0	0	18	18	18	18	18	18	18	18	18	18	18	18	18	18	18	18	18
BATCH: A1A EXP: W30 ESD: W27	0	0	0	0	0	40	36	36	32	32	32	30	28	26	24	22	20	18	16	14	12	10
BATCH: B2B EXP: W50 ESD: W47	0	0	0	0	0	0	0	0	0	0	0	0	0	0	0	0	0	40	40	40	40	40

SKU description: Wonder drug placebo, 30-ct bottle, global label
DND = 14 days; DNS = 7 days

Depot location: United Kingdom depot
Depot-to-depot lead time: 3 weeks

Figure 11.2 Regional Depot Distribution Plan Table

the distribution actual row. Figure 11.2 illustrates these updates in a regional depot distribution plan table.

This is a relatively straightforward supply plan. The subject demand forecast begins in Week 12 and continues with a weekly dispensing total of two kits. There are two clinical sites forecasted to activate in Week 7 and Week 9, with each site receiving a total of four kits. The supply planning parameters drive a suggested distribution order requirement of 34 kits in Week 6, but the clinical planner opts to increase this to 40 kits. Then, the plan suggests a resupply of 18 kits in Week 18 based on the 18-week order target, but again the planner increases this quantity to 40 kits. Note how the production actual and production plan rows are now listed as distribution actual and distribution plan rows and note how the assumptions listed in the bottom right corner of the figure capture the depot-to-depot lead time of 3 weeks.

The second step is to link the regional depot distribution plan table to central depot production plan table. When this happens, distribution orders supplied to a regional depot create demand at the central depot for finished product to complete those orders. To capture this demand–supply relationship, a new section needs to be created in the demand forecast table at the central depot to capture distribution demand from each of the regional depots. Adding a distribution demand section to the demand forecast at the central depot gives the clinical planner visibility into their distribution strategy because each depot location is included as its own row in a separate distribution demand section. In this example, we are connecting the UK regional depot as well as the other depots listed in Table 11.1 to the central depot. This allows the planner to see each distribution order requirement at every depot as illustrated in Figure 11.3.

In this updated central depot demand forecast table, seven rows of regional depot distribution demand have been added, one each for the regions identified in Table 11.1. Notice the UK depot row has demand for 40 units in Week 3 and another 40 units of Week 15. These two orders coincide with the distribution order receipts (e.g. distribution actual row) in Week 6 and Week 18 of the UK regional depot production plan table. The 3-week offset is equal to the depot-to-depot lead time of 3 weeks from the US central depot to the UK depot. That same logic is applicable to each of the regional depot distribution demand rows. Argentina has an 8-week lead time, meaning that the 80-unit order falling in Week 11 needs to arrive to the Argentina regional depot by Week 19 and the 80-unit order falling in Week 20 needs to arrive by Week 28.

It's clear from this example that regional depot supply requirements create additional demand for finished product material at the central depot, so the clinical planner must have a clear strategy for how they intend to supply each of the regional depots. In the example above, there are 188 units of subject demand, 16 units of site stocking demand, and 1500 units of distribution demand! This is an extreme example, but it's also not uncommon for distribution demand to account for a

	W1	W2	W3	W4	W5	W6	W7	W8	W9	W10	W11	W12	W13	W14	W15	W16	W17	W18	W19	W20	W21	W22
Total subject demand					4	8	8	8	8	10	10	10	10	10	12	12	12	12	12	14	14	14
# Sites activated		1	1	1	1																	
Site demand		4	4	4	4					10												
Argentina depot											80									80		
Brazil depot													100									100
China depot																		120				
Japan depot																				100		
Singapore depot									120									120				
Germany depot			200									200									200	
Uk depot			40												40							
Total demand	0	4	244	4	8	8	8	8	128	10	90	210	110	10	52	12	12	252	12	194	214	114

SKU description: Wonder drug placebo, 30-ct bottle, global label
DND = 14 days; DNS = 7 days
Depot location: United States depot

Figure 11.3 Central Depot Demand Forecast Table with Distribution Demand

majority of total study demand. Before we analyze distribution planning strategies though, we'll explore how to reconsider demand and supply planning parameters at the regional depots, as changes to these inputs ultimately impact distribution order requirements at the central depot.

11.4 Demand and Supply Planning Parameters Revisited

Planning total demand at a regional depot follows the exact step-by-step logic described in Chapter 8 – Finished Product Demand Planning: Basic Concepts. Individual subject demand is determined by the target dose, formulation strength, kit configuration, and intrinsic aspects of the clinical protocol (e.g. frequency and duration of treatment). Site activation and enrollment rates are provided by clinical operations and/or the contract research organization (CRO), and, then, an overage is applied to calculate total demand. With respect to using country-specific enrollment rates to project when subjects begin dosing, Chapter 8 advises the clinical planner to take caution with this approach. These are often inaccurate and lead the clinical planner toward undersupplying or oversupplying a regional depot. Instead, the clinical planner can hold all subject demand at the central depot until subjects are identified and ready to screen and enroll at the regional depot. We'll return to this aggregate forecasting strategy in Chapter 13 – Advanced Concepts in Demand and Supply Planning.

The other demand-related input to analyze is the overage multiplier. If a relatively high (e.g. 20–30%) overage is applied at the central depot level, then the clinical planner should consider reducing overage at the regional depot level to 10–15%, especially if a significant percentage (e.g. <80%+) of subject demand is planned at other depots. Using large overages leads to over-stocking depots and increases waste because it is almost never real demand. Instead of using the overage input to generate higher depot stocking requirements, the clinical planner should rely on the supply planning parameters and distribution planning strategies to drive suggested and actual distribution orders, respectively.

On the supply side, the clinical planner should carefully assess their safety stock and order target supply planning parameters at each regional depot. The first rule of thumb is that the safety stock value should never be set lower than the depot-to-depot lead time for that regional depot. For example, if the regional depot in China has a lead time of 12 weeks to resupply, the safety stock value should be set to more than 12 weeks. This ensures that if a shipment is lost to a temperature excursion or damage, there is time to initiate a resupply shipment before the depot stocks out. For regional depots with lengthy depot-to-depot lead times (e.g. >9 weeks), it's typically unwise to set a safety stock target much higher than the lead time because then the risk of waste increases if subjects drop out of the study and demand falls.

For regional depots with a distribution lead time that's less than the finished product production lead time, it makes sense to continue using the finished product production lead time as a safety stock value. That's because finished product production and release lead times are rarely longer than 13 weeks and are often as short as 6–9 weeks. Aligning the safety stock value to the finished product replenishment lead time ensures there won't be a stockout if there is a delay in a finished product resupply campaign and prevents over-stocking the depot by setting the safety stock target too high.

With respect to the order target value, the same logic used to set an order target value at a central depot also applies at the regional depot. Recall from Chapter 9 – Finished Product Supply Planning: The Production Plan that the point in time when the study is planned greatly influences the order target value and that it makes the most sense to set a shorter order target value when demand is uncertain. Uncertainty is heightened at a regional depot where there are a smaller number of sites than there are when forecasting at a study-wide aggregate level. Therefore, it makes sense to stick with a lower order target value of 18–26 weeks. For regional depots with long distribution lead times, a higher order target value of 26 weeks makes more sense since depot resupply order requirements are generated by the difference between the safety stock value and the order target value. If the safety stock value is relatively high (e.g. 10–13 weeks) and the order target value is low (e.g. 18–22 weeks), this creates resupply order requirements every 5–10 weeks, which is too often. Setting a higher order target value of 26 weeks means resupply order requirements are generated at minimum every 3 months, a more reasonable interval from a logistics perspective. As the regional depot enrolls subjects, the order target value can be gradually increased based on the other considerations such as the expiry period of the DP and the other supply chain lead times analyzed in Chapter 9.

11.5 Distribution Planning Strategies

Up to this point, the clinical planner has mapped the supply chain, identified the depot lead times, built out the relevant SKU-depot combinations, linked the regional depot locations to the central depot location so distribution orders at the regional depot become distribution demand at the central depot, and tweaked the demand and supply planning parameters at each regional depot as needed. We are now ready to begin the distribution planning process which has four elements:

- Determining the initial depot stocking orders with respect to both timing and quantity.
- Allocating finished product batches to fulfill distribution order requirements.
- Monitoring enrollment and adjusting the distribution plan as needed to balance supply across depots.
- Managing regulatory commitments and restrictions as needed in the distribution plan.

Let's begin with determination of the initial depot stocking strategy. From a timing perspective, there are two events the clinical planner needs to keep in mind. The first is the timing of the first site activation (FSA) within that country and the second is the anticipated CHA approval date, which is necessary for shipments to in-country regional depots. Hub-based regional depots, on the other hand, may receive supply without CHA approval, but they do need CHA approvals for a country before shipments to clinical sites may begin. For example, drug may be sent to a hub depot in the EU prior to CHA approval in any EU country, but then shipment of drug to a clinical site is gated to QP certification, a process that also requires CHA approval to proceed. Similarly, a shipment of finished product may be sent to a hub in Asia (e.g. Singapore and Hong Kong), but that batch can't be shipped from the hub into another country without the CHA approval.

The CHA approval therefore sets the earliest timepoint the initial batch can be shipped to an in-country depot or to a clinical site from a hub-based depot, while the FSA sets the latest date the batch should be received at the depot for shipment. If there is a relatively limited amount of time between the CHA approval date and the FSA date, the clinical planner should work closely with their depot network counterparts to move shipments as quickly as possible to clinical sites. On the other hand, if there's a lengthy amount of time between CHA approval and FSA, the clinical planner should delay sending the initial batch until it's relatively close (e.g. within a few weeks) to the FSA date. Waiting to send the first shipment allows for two things to happen. First, the demand profile becomes more certain over time and, second, there may be a newer batch of material with better expiry to use as the initial shipment to that depot.

To determine the initial depot stocking quantities, a combination of mathematical analysis and qualitative judgment should be applied. On the mathematical side, the order target value drives a suggested distribution order requirement for the clinical planner to consider. From Chapter 10 – Finished Product Supply Planning: The Production Plan, we know this requirement is based on a combination of projected subject demand and site seeding demand over the duration of the order target value. With respect to subject demand, we know that it's more difficult to accurately predict enrollment at the regional level than at the total study level. We also know from Chapter 6 – RTSM and Its Role in the Clinical Supply Chain that site seeding demand is driven by the site supply strategy and initial site stocking quantities in the randomization and trial supply management (RTSM) design. Given these variables, it's critical for the clinical planner to understand the following inputs at each regional depot:

- Estimated number of active sites and the timing of each site activation.
- RTSM initial supply value for each active site.
- Estimated number of potential subjects that will screen over the first two to three months across all sites.
- Estimated screen failure rate.
- Randomization schema of the study (e.g. 1:1, 3:2, and 2:1:1).

Note that the number of potential subjects screening is directly related to the initial supply strategy. If sites within a country are expected to enroll a small number of subjects (e.g. <5 total) over the duration of the study, then it makes sense to have all sites set up with a low-supply strategy. Similarly, if many subjects are expected to screen and enroll at the beginning of the study, multiple sites may be set to medium- or high-supply strategies, and the site seeding demand needs to reflect this in the demand forecast table. Doing this increases demand at the regional depot and creates a larger initial distribution order to meet that increased site demand.

Also note that the estimated screen failure rate and randomization schema of the study both potentially limit the number of subjects enrolling onto a particular arm of the study, and this should also be analyzed as part of the forecasting process. For example, if 10 subjects are expected to screen at a site, but there is a 60% screen fail rate, then only four subjects are likely to randomize. And, if there are two arms in the study with a 1:1 randomization schema, those four subjects are likely to split two into one arm of the study and two into the other. Of course, there is a possibility of unbalanced enrollment or more subjects then expected may pass screening. We'll return to how to manage the unpredictable nature of demand in Chapter 13 – Advanced Concepts in Demand and Supply Planning.

Ultimately, the clinical planner needs to digest the inputs provided by clinical operations and the CRO to estimate if the initial distribution order requirement suggested by the plan is sufficient or if it should be reduced or increased. Additional variables to consider in this analysis include:

- Confidence level in the demand forecast and amount of potential subject demand: When confidence in subject demand meeting the forecast is high, and the relative amount of subject demand is high, then stocking up a depot with more initial supply makes sense. This is usually the case for regional depots supplying many countries and sites versus in-country regional depots that may be supplying a limited number of sites. When confidence in demand is low, then sending a relatively small initial quantity makes more sense.
- Cost of the drug to manufacture or purchase: If the product is expensive to manufacture or purchase, then the clinical planner may consider reducing the initial depot quantity to avoid waste. Conversely, if the product is relatively cheap and/or plentiful, then increasing the initial supply quantity is more appropriate.

- Shelf life of the finished product batch: The last variable to consider is also one of the most important. The clinical planner must consider the expiry date of the finished product batch that will be sent in the initial shipment. All the principles discussed in Chapter 10 – Finished Product Expiry Planning hold as true for distribution planning as they do for production planning. If the initial batch of material has a short expiry date, then the clinical planner should reduce the quantity of the initial shipment so as not to waste product due to expiry. In a scenario where shelf-life extensions are available, this also means more inventory is held at the central depot to include in an extension for future consumption in the study.

Analyzing expiry within the context of setting an initial supply quantity is an excellent segue to the next component of distribution planning, allocation. Allocation is the process of selecting a batch to fulfill a production or distribution order, and it's central to the distribution planning process. This is because it's rare in global trials to produce a single finished product batch to satisfy demand at every depot and for every patient. Of course, if this is the case, it makes the allocation process easy! Typically, multiple batches of DP are necessary over the duration of a global trial, and the clinical planner is left with a choice of which available finished product batch to allocate to a distribution order.

To make these decisions, the clinical planner must have visibility to available inventory at the individual batch level within the plan, a concept introduced in Chapter 10 – Finished Product Expiry Planning. Without this visibility, it's impossible to allocate the appropriate batch to the depot. Another concept introduced in Chapter 10 were the first in first out (FIFO) and last in first out (LIFO) inventory management methodologies. With respect to distribution order management, FIFO is not an optimal strategy because it often does not make sense to ship an older batch of material that expires sooner than a newer batch with better shelf life if demand is not certain to consume that supply. When demand fails to materialize, a greater quantity of that older batch expires and then needs to be replaced by the newer batch. This is a simple mistake to make that's mitigated by incorporating distribution demand into the demand forecast table and then planning for it in the production plan table.

Let's illustrate how the clinical planner makes allocation decisions by continuing with the example presented in Figure 11.3 earlier in this chapter.

In Figure 11.4, the full scope of demand forecasting, production planning, and distribution planning becomes clear. Subject demand, site demand, and distribution demand sum together in the

OHI	400	396	152	148	140	132	124	716	588	578	488	278	168	1758	1706	1694	1682	1430	1418	1224	1010	896
Prod. actual	400	0	0	0	0	0	0	600	0	0	0	0	0	1600	0	0	0	0	0	0	0	0
Prod. plan	1182	0	904	0	0	0	0	0	704	0	0	0	0	0	0	0	0	0	0	0	0	0
WOS	8	7	6	5	4	3	2	10	9	8	7	6	5	18	17	16	15	14	13	12	11	10
Safety stock	9	9	9	9	9	9	9	9	9	9	9	9	9	9	9	9	9	9	9	9	9	9
Order target	18	18	18	18	18	18	18	18	18	18	18	18	18	18	18	18	18	18	18	18	18	18
Batch: A1A EXP: W22 ESD: W19	400	396	152	148	140	132	124	116	108	98	88	78	68	58	46	34	22	10	0	0	0	0
Batch: B2B EXP: W28 ESD: W25	0	0	0	0	0	0	0	600	480	480	400	200	100	100	100	100	100	100	98	84	70	56
Batch: C3C EXP: W39 ESD: W36	0	0	0	0	0	0	0	0	0	0	0	0	0	1600	1560	1560	1560	1320	1320	1140	940	840

SKU description: Wonder drug placebo, 30-ct bottle, global label
DND = 14 days; DNS = 7 days

Depot location: United States depot

Figure 11.4 Central Depot Demand Forecast and Production Plan Table

total demand row. Then, the production plan row suggests production order requirement quantities and dates to fulfill demand based on the WoS, safety stock, and order target values in each column. The clinical planner confirms the actual production orders, and the last section at the bottom of the table captures batch-specific information including on-hand inventory (OHI) quantities for each batch during each week. Within the batch section, the clinical planner makes allocation decisions, and this provides both control over the plan and enables a flexible strategy. Control comes from the ability to select which batch should be allocated to which order, and flexibility is driven by the ability to make changes to the distribution plan on a weekly basis as demand- and supply-related inputs change. Let's review this example in detail to illustrate these points.

The first batch produced, Batch A1A, only contains 400 kits, but that's because it has a relatively short expiry date in Week 22 that's further reduced by the do not dispense (DND) and do not ship (DNS) Randomization and Trial Supply Management (RTSM) parameters to Week 19. In this plan, the first distribution orders occur in Week 3 with 2 requirements to send 200 kits to the Germany regional depot and 40 kits to the UK regional depot. At this point, the planner only has Batch A1A available, so this batch is allocated to fulfill these two production orders. OHI in Week 3 reduces from 396 kits to 152 kits accordingly, and the plan moves forward to the next production event, which takes place in Week 8 when Batch B2B is released for use in the clinic increasing OHI up to 716 kits.

There is a distribution order requirement for 120 kits to the Singapore depot in Week 9, and now the clinical planner has a choice. They can either fulfill this order entirely from Batch B2B using the LIFO methodology or allocate the remainder of Batch A1A and use a partial quantity of Batch B2B to fulfill the order using the FIFO methodology. Here, the clinical planner uses the LIFO methodology and fulfills the order entirely from Batch B2B. The justification for this decision is that Batch A1A's expiry short date is Week 19, and, by the time the batch arrives in Singapore, it would only have seven weeks of effective expiry left. Sending batch B2B improves the odds that less material from this initial shipment goes to waste because there is subject demand at the central depot to consume batch A1A, and there is more time at the Singapore depot for Batch B2B to go to use.

Between Weeks 9 and 13, there are a total of four distribution order requirements fulfilled by Batch B2B to Singapore, Argentina, Germany, and Brazil. Then, in Week 14, the third production order, Batch C3C, is received. This batch has even better shelf life, with the expiry short date not occurring until Week 36 so the clinical planner decides to pack 1600 kits of inventory for this job. This new inventory is then shipped to every available depot between Weeks 15 and 22 so that all depots are resupplied with material that has better expiry than batch A1A or batch B2B. Meanwhile, notice how batch A1A is consumed by subject demand all the way through Week 19, the exact date it expires, which is further justification for the initial production order quantity because those 400 kits are just enough to satisfy demand before the batch's expiry short date. Once batch A1A is depleted, subject demand then consumes batch B2B through the remaining duration of the plan shown. Only after batch B2B is consumed does subject demand begin to consume batch C3C. In this plan, subject demand consumes inventory via a FIFO methodology while distribution demand consumes inventory via a LIFO methodology.

Beyond allocation, the clinical planner has another lever to consider within the plan which is the timing of the production order resupplies. If the release of batch B2B or batch C3C is delayed and prevents fulfillment of distribution orders in subsequent weeks, the planner may consider moving the distribution order requirements out by a week (or several weeks) until the batch finishes production and is available to complete the desired distribution order. This can be done by pushing out the distribution order receipt at the regional depot level. There is a trade-off to using this strategy though, which is that WoS levels at the regional depot may run lower than the desired safety stock level. Therefore, the clinical planner needs to consider how much demand uncertainty there is at the regional depot level before moving orders out. If subjects are already enrolled, then resupply orders should be moved out only if absolutely necessary. Conversely, if batches release earlier, then distribution orders can be pulled in, a favorable situation for the planner.

Before moving on, let's review the production strategy in the example above. Notice how the WoS value in Week 7 falls to what appears to be a dangerously low level of only 2 weeks. This is somewhat misleading because WoS at the central depot encompasses both subject demand and distribution demand. If only subject demand is considered, there are still 11 WoS available in Week 7 so if the release of batch B2B were delayed, it wouldn't affect subject dosing until Week 19. The primary consequence of such a delay would instead be its impact on the timing of distribution order fulfillment. In that scenario, the clinical planner should work with clinical operations and regulatory affairs to understand the impact of delaying those initial distribution orders. Countries and regions that have confirmed site activations and potential subjects identified would need to be prioritized.

With the initial depot stocking strategy set and an allocation strategy for planned resupply batches in place, the next component of the distribution planning strategy is to monitor subject enrollment and OHI levels across the depot network. As sites activate and subjects enroll over time, inventory usually becomes imbalanced across the SKU–depot combinations. This is because subjects typically enroll more on one arm than the other at the depot level, so WoS for the SKU supplying that arm drops more quickly than anticipated in the plan. A consequence of this is that fewer subjects than planned enroll on that same arm in a different region or country causing WoS to increase at that depot. The planner needs to monitor these changes and adjust the distribution plan accordingly to keep WoS balanced across all depots. Depots that have high WoS need to have their resupply orders reduced, delayed, or eliminated entirely while depots that have low WoS need to have their resupply orders increased, brought in to arrive sooner or both depending on the severity of the imbalance. The next section introduces the WoS summary, which helps facilitate this rebalancing process.

The final component of the distribution plan is to manage regulatory commitments and restrictions. In Chapter 5 – Regulatory Considerations, we learned about restrictions, which limit the importation of a batch that fails to meet criteria specified by the CHA in their response to a clinical trial application. To effectively manage restrictions, we also introduced the concept of process version management (PVM), which creates differentiation between DS and DP batches that have unique regulatory attributes. Implementing PVM gives the clinical planner, quality, and regulatory affairs visibility into which batches meet or fail to meet restrictions from CHAs. This is because each finished product batch is associated to the underlying DS and DP versions, and this gives the entire team line of sight to which batches have been filed and approved in each country where the trial takes place. Control over distribution of material to hubs or in-country regional depots that have restrictions in place is fully realized within the distribution planning strategy.

There are two potential points when a restriction is placed on a product, either before the trial starts following a review of the initial application or following review of an amendment to the application. Managing a restriction before the trial starts is the more straightforward scenario. Enrollment and dosing can't begin until a finished product batch is available that meets the criteria of the restriction. Unfortunately, when subject dosing has already begun and a restriction is placed on new batches of material, the analysis becomes significantly more complex. When this happens, the clinical planner needs to identify:

- How much inventory of the approved process version (e.g. Process Version DS1.0 – DP1.0) is available within the entire supply chain.

- The production order strategy of the restricted process version (e.g. Process Version DS2.0 – DP2.0) in terms of quantity and timing.
- Which countries have approved the restricted process version, and which have not.

Following this assessment, the clinical planner then needs to:

- Allocate inventory of the approved process version (e.g. Process Version DS1.0 – DP1.0) to those countries that don't accept the restricted process version (e.g. Process Version DS2.0 – DP2.0).
- Allocate inventory of the restricted process version (e.g. Process Version DS2.0 – DP2.0) to those countries that will accept the restricted process version (e.g. Process Version DS2.0 – DP2.0).
- Plan for this allocation strategy to continue until the restriction is lifted.
- If the restriction won't be lifted, then unique finished product SKUs need to be created for the two different process versions. Without separating the SKUs, it becomes more difficult to distinguish which batches fall under the approved process version and which fall under the restricted process version.

Over the course of this section, we've covered the four components of distribution planning. These are determining initial depot stocking orders, making allocation decisions, balancing depot inventory levels over time, and managing regulatory restrictions within the plan. As enrollment completes and the study winds down, the clinical planner also needs to assess the final depot stocking shipment quantity and timing. Like planning the last finished product production event, the clinical planner needs to account for the minimum inventory quantities held at each participating site in addition to subject demand. This data can be gleaned from an understanding the RTSM's resupply methodology and RTSM-generated reports that show site inventory levels. Applying a small overage factor of 3–5% also makes sense in case final site shipments are damaged or lost. Carefully managing these final depot stocking distribution orders helps prevent over-production at the central depot and reduces cost and waste within the plan.

11.6 Aggregate Finished Product Weeks-of-Supply

When a clinical supply chain requires multiple depots to service clinical sites, finished product inventory is supplied across the entire network. From an inventory management perspective, the clinical planner needs to understand how many WoS of inventory are available at each SKU–depot location to appropriately plan resupply shipments. But it's also important to understand how much finished product inventory of each SKU is available at the study level relative to the total demand forecast across all depots. This information is useful for reporting on the overall inventory position of the trial and for creating a balanced allocation strategy for upstream materials to finished product SKUs. Another name for describing the sum total of inventory across all depots is the aggregate finished product WoS (AFPWoS). The AFPWoS value also plays a role in determining WoS for upstream materials which will be covered in Chapter 12 – Upstream Supply Planning: Primary Packaging, Drug Product, and Drug Substance.

The AFPWoS is a weighted average of the individual central and regional depot WoS values. Calculating the AFPWoS is a multi-step process:

1) Identify the OHI at each central and regional depot and then sum the total OHI across all depots. Any central or regional depot that has OHI and no subjects enrolled should be removed from the total OHI figure.

a) Central depots which do not service clinical sites are treated separately in this analysis. See Step 5 below for how to include supply at these depots in the AFPWoS.
b) Regional depots which have no subjects enrolled are removed because that supply goes to waste and is not available for future subject demand if no subjects enroll. As soon as a subject enrolls at a regional depot, the supply at that depot should be counted as part of the sum.

2) Identify each central and regional depot's WoS.
3) Calculate each central and regional depot's weighted WoS by dividing the regional depot OHI into the total OHI across all depots and multiplying that value by the depot's WoS.
4) Sum the weighted WoS across all depots to arrive at the AFPWoS.
5) If there is a central depot that does not service clinical sites and subjects, the next step is to calculate the implicit total run rate (ITRR) of the study. The ITRR is taken by dividing the total OHI into the AFPWoS. The ITRR describes the average weekly demand over the period of future demand which the OHI in that week is available to cover. This number can then be used to identify how many additional WoS are available at the central depot as well as for upstream materials.
6) Identify the OHI at the central depot that does not service clinical sites or subjects. Divide the OHI into the ITRR to determine how many WoS are available at the central depot. Add the central depot WoS to the weighted WoS across all depots to arrive at the AFPWoS.

Figure 11.5 illustrates how AFPWoS are calculated including a regional depot with no subject demand and a central depot that does not service clinical sites and subjects. In the example, there are four total regional depots, the first three of which are servicing subjects. Regional depot 4 does not have any subjects so total OHI at that depot has been removed. The weighted WoS calculation includes values from regional depots 1 to 3, which are then summed into the regional depot AFPWoS. In Week 1, there are 3641 total kits of OHI leading to a regional depot AFPWoS value of 33 weeks. Note the 33 WoS here does not yet include the central depot supply. The ITRR is then calculated by dividing the 3641 kits into the 33 WoS to arrive at the 107 kits per week value in Week 1. Then, the central depot implied WoS is calculated by dividing its OHI into the ITRR. In the last row, the AFPWoS sums the regional depot AFPWoS with the implied central depot WoS to arrive at the total AFPWoS figure for the SKU.

AFPWoS can also be calculated across studies for finished product SKUs that consume the same underlying DP and DS. To do this, take the OHI of each SKU at the total study level and the

	W1	W2	W3	W4	W5	W6	W7	W8	W9	W10	W11	W12	W13	W14	W15	W16	W17	W18
OHI RD1	3357	3313	3095	3047	2948	2894	2840	2779	2718	2650	2582	2448	2377	2303	2226	2160	1988	1915
OHI RD2	164	325	306	287	271	255	239	223	207	191	164	148	132	116	100	84	75	66
OHI RD3	120	110	100	90	250	240	230	220	210	200	190	180	170	160	150	140	130	120
OHI RD4	0	0	0	0	0	0	0	0	0	0	0	0	0	0	0	0	0	0
TOTAL OHI	3641	3748	3501	3424	3469	3389	3309	3222	3135	3041	2936	2776	2679	2579	2476	2384	2193	2101
WoS RD1	36	35	34	33	32	31	30	29	28	27	26	25	24	23	22	21	20	19
WoS RD2	10	25	24	23	22	21	20	19	18	17	16	15	14	13	12	11	10	9
WoS RD3	12	11	10	9	25	24	23	22	21	20	19	18	17	16	15	14	13	12
WoS RD4	27	26	25	24	23	22	21	20	19	18	17	16	15	14	13	12	11	10
Weight. WoS RD1	33	31	30	29	27	26	26	25	24	24	23	22	21	21	20	19	18	17
Weight. WoS RD2	0	2	2	2	2	2	1	1	1	1	1	1	1	1	0	0	0	0
Weight. WoS RD3	0	0	0	0	2	2	2	2	1	1	1	1	1	1	1	1	1	1
Weight. WoS RD4	0	0	0	0	0	0	0	0	0	0	0	0	0	0	0	0	0	0
RD AFPWoS	33	33	32	31	31	30	29	28	27	26	25	24	23	22	21	20	19	18
ITRR	107	112	108	109	113	114	115	116	117	117	117	116	116	117	117	118	114	115
OHI CD1	300	300	300	300	300	300	300	300	300	300	300	300	300	300	300	300	300	300
Implied WoS CD1	3	3	3	3	3	3	3	3	3	3	3	3	3	3	3	3	3	3
Total AFPWoS	**37**	**36**	**35**	**34**	**33**	**32**	**31**	**30**	**29**	**28**	**28**	**27**	**26**	**25**	**24**	**23**	**22**	**21**

Figure 11.5 Aggregate Finished Product Weeks-of-Supply

AFPWoS of each SKU and use those to calculate the weighted WoS of each study. These values are then summed into an all-study AFPWoS. Eventually, this value can be combined with upstream implicit WoS to inform the total level of resiliency across every level of the supply chain.

11.7 The Weeks-of-Supply Summary

Another data set that's especially useful for distribution planning is the amount of inventory and how many WoS are available at each level of the supply chain. The WoS summary is an inventory report that captures current available OHI and WoS values at every finished product SKU–depot location in the study. Like the production plan and distribution plan summaries, the finished product data contained within the WoS summary is available in the production plan tables covered in Chapter 9 – Finished Product Supply Planning: The Production Plan. Upstream SKUs can also be added to the summary. Chapter 12 – Upstream Supply Planning: Primary Packaging, Drug Product, and Drug Substance covers how to plan for upstream materials and how to build WoS into an upstream production plan table for inclusion in the WoS summary. At minimum, the WoS summary contains the following data elements:

1) Current date
2) Protocol number
3) Supply chain level
4) SKU
5) Depot location/manufacturing site
6) Target WoS
7) WoS available
8) Total OHI available
9) Unit of measure

This summary is a powerful tool to use in the distribution planning process because it clearly highlights inventory imbalances across the finished product supply chain as shown in Table 11.2. As enrollment occurs, the demand profile at each SKU–depot varies from the original plan. More active subjects than planned may enroll at sites supported by a single regional depot while fewer active subjects enroll at sites supported by another depot and vice versa for placebo subjects. The WoS summary highlights these imbalances for the clinical planner. This summary tells the planner where to focus during the planning process to ensure these imbalances are managed appropriately to avoid stockout risk and minimize waste. The WoS summary is also useful for communicating the immediate status of the upstream supply chain. We'll revisit how to incorporate the WoS summary into the demand and supply planning process in Chapter 15 – Clinical Supply Chain Planning Processes.

11.8 The Distribution Plan Summary

Like its counterpart the production plan summary, the distribution plan summary is an output of the demand and supply planning process that summarizes the key details of each order within the distribution plan. At minimum, the distribution plan summary contains the following elements:

1) Protocol number
2) SKU description

Table 11.2 Example Weeks-of-Supply Summary

Current Date	W78

Protocol	Level	SKU	Depot	Target WoS	Available WoS	On-hand Inventory	UOM
Study 1	Finished product	100 mg active	Aggregate	18/52	24	472	Kits
		100 mg placebo			26	410	
		100 mg active	USA	18/52	26	416	
		100 mg placebo			24	336	
		100 mg active	Brazil	18/26	14	56	
		100 mg placebo			37	74	
N/A	Drug product	100 mg tablet	USA	0	0	0	Bottles
N/A	Drug substance	Molecule 1	India	52	52	20	kg

3) Supplying depot
4) Receiving depot
5) Batch number
6) Quantity to ship
7) Unit of measure
8) Shipment initiation date
9) Shipment receipt date
10) Expiry date of the batch
11) Process version (if maintained)

The distribution plan summary shown in Table 11.3 serves many of the same purposes as the production plan summary. It's an important reference point during the demand and supply planning process for the clinical planner to review how the distribution plan is coming together and whether the planned orders are logical. Where the production plan summary provides information to the chemistry, manufacturing, and controls (CMC) team and packaging sites regarding upcoming manufacturing events, the distribution plan summary provides information to the distribution depots responsible for preparing and receiving each shipment. It also gives the clinical operations and regulatory affairs functions insight into when IMP will be ready to ship within a specific country, and this helps guide site and regulatory strategy. The

Table 11.3 Example Distribution Plan Summary

Protocol	SKU	Supplying Depot	Receiving Depot	Batch No.	Qty	UoM	Initiation Date	Receipt Date	Expiry Date	Process Version
Study 1	100 mg active	US depot	Germany depot	A1A	200	Kits	1-Jun-2025	22-Jun-2025	31-Aug-2025	DS1.0–DP1.0
Study 1	100 mg active	US depot	Germany depot	B2B	200	Kits	27-Jul-2025	17-Aug-2025	31-Oct-2025	DS1.0–DP1.0
Study 1	100 mg active	US depot	UK depot	A1A	40	Kits	1-Jun-2025	22-Jun-2025	31-Aug-2025	DS1.0–DP1.0

process version column also gives the clinical planner a mechanism to communicate how regulatory restrictions are controlled as different versions of the product are introduced to the supply chain. Combined, the production plan and distribution plan summaries represent the entire supply plan for the trial.

11.9 Distribution Order Readiness

Distribution planning shares another commonality with production planning, both require a tremendous amount of activity and coordination between cross-functional stakeholders to translate the planning outputs into execution. Interestingly, more functions and groups are required to support distribution order readiness and execution than production order readiness. Where distribution planning is focused on the strategy of determining distribution order quantities, timing, and allocation, distribution order readiness is defined as being prepared and able to execute a distribution order so that it meets the desired target receipt date and is released for shipment to a clinical site. Distribution order readiness typically requires coordinating several activities from preparation and review of shipment documentation to acquisition of shipping containers and materials (e.g. temperature monitors and global position system [GPS] monitors) to physical pickup and delivery of the shipment. And, don't forget about the import license application process in countries that require those and the QP certification process for countries within the EU, the United Kingdom, and Israel.

Executing distribution orders requires bringing together clinical supply chain, the supplying depot, the receiving depot, and another often overlooked party, the importer of record (IoR) listed on the proforma invoice. The IoR is important to highlight for two reasons. First, the IoR often needs to have a physical location in the country where the import takes place, and, second, the IoR is typically responsible for filing the import license application in countries that require those. At large biopharmaceutical companies that have their own distribution network of depots, the company itself may act as the IoR because they have a presence established in these countries. But smaller biopharma companies who don't have their own depot network must rely on a third party to play the IoR role, and this responsibility is usually taken by either the packaging and distribution contract manufacturer or by the CRO. The person responsible for overseeing distribution order execution within the clinical supply chain function needs to understand who will play the IoR role to ensure that the appropriate documentation and import licenses are in place to enable shipments into those countries. The international logistics process and stakeholders are covered in much more detail in Chapter 21 – International Logistics for Clinical Biopharmaceutical Products.

Given the complexity of international logistics for biopharmaceutical products, a distribution order readiness business process is also required to communicate the distribution plan to internal and external stakeholders. This process should include the clinical supply chain functions (e.g. planning and logistics) responsible for planning and executing distribution orders, and the supplying and receiving depots responsible for coordinating the shipments. Quality also needs to be involved when QP certifications are needed. The planning horizon for new distribution orders is at least three to 4 weeks before the shipment initiation date to ensure sufficient time for readiness activities to complete before the batch is ready to ship to its destination. The order readiness process should also collect data and metrics on lead times to ship from one depot to another. Global supply chains and distribution networks are highly dynamic and any changes to lead time assumptions need to be fed back into the planning models, so distribution orders are planned for accurately.

Table 11.4 Production Order Readiness and Distribution Order Readiness Comparison

	Production Order Readiness	Distribution Order Readiness
New Order Horizon	Six months before order start date which is equivalent to the date of manufacture	One month before order initiation date which encompasses all activities required to prepare and ship the order
Key Stakeholders	Clinical supply chain, quality, manufacturing (internal or external)	Clinical supply chain, quality, supplying depot, receiving depot, qualified person (if necessary)
Meeting Frequency	Weekly or bi-weekly as needed	Weekly or bi-weekly as needed
Process Outputs	Activity tracker for all production orders Feedback on production order lead times	Activity tracker for all distribution orders Feedback on depot-to-depot lead times

Finally, it's necessary to align with the other stakeholders in the supply chain on distribution order lead times and which activities need to occur between the distribution order initiation date and distribution order receipt date. For example, if umbrella import license applications (or shipment-specific import licenses for that matter) are included in the scope, then additional distribution lead time should be added to allow for this activity to take place. Similarly, if the sponsor is expecting the QP certification process is expected to complete by the shipment receipt date, which they absolutely should expect, then quality, the receiving depot, and the QP need to be on the same page with the desired release date. If this coordination and alignment doesn't happen, then orders end up starting too late or aren't released in time to support site shipments when they're needed.

Table 11.4 provides a comparison of production and distribution order readiness attributes.

11.10 Key Takeaways

1) Distribution planning shares many commonalities with production planning. The SKU–depot setup, demand forecast table, production plan table structure, expiry considerations, and WoS values are all the same. Instead of producing production order requirements, the distribution plan generates distribution order requirements that are then converted into distribution orders.
2) The first step in the distribution planning process is to identify the country list of the trial and the network of distribution depots that will supply each country. There are two types of depots, central depots and regional depots. Central depots are depots that supply other depots, and they often perform finished product manufacturing in addition to distribution activities. In most cases, central depots may service site and subject demand directly, while in more rare cases central depots may service other central depots.
3) Regional depots are responsible for shipping finished product directly to clinical sites and are typically not capable of manufacturing finished product. Regional depots are either in-country meaning they are located within the country where the trial takes place or hub-based depots meaning they service an entire region of countries.

4) A standard sequence of operations is required to execute international shipments of biopharmaceutical products include preparation and review of shipment documentation, acquisition of shipment components, and the physical transit process. In addition to these activities, some countries also require import licenses to ship product into their countries, while others require a QP certificate to enable shipment of a batch from the depot to a clinical site.
5) Setting up a distribution network in the demand and supply plan requires linking regional depots to the central depot location. This is done by adding a section for each regional depot in the demand forecast table of the central depot. Distribution orders generated at the regional depot feed directly into the central depot location and are offset by the depot-to-depot lead time, which captures all activities required to set up and execute a distribution order from the supplying depot to the receiving depot.
6) The clinical planner should analyze the demand and supply planning parameters at each regional depot location in the plan. Specifically, the overage calculation should not be set too high, and the safety stock and order target values deserve scrutiny. The safety stock value should never be set lower than the depot-to-depot lead time.
7) Four elements compose the distribution planning process. These are determination of the initial depot stocking order, allocation of finished product batches to distribution orders, adjustment of distribution orders to balance supply across the depot network, and management of regulatory commitments and restrictions within the distribution plan.
8) AFPWoS measures how much inventory of each SKU is available across all regional and central depots in the study. AFPWoS is useful to reporting on the overall inventory position of the study and for creating a balanced allocation strategy in upstream planning. The ITRR describes the average weekly demand over the period of future demand which OHI in that week is available to cover.
9) The WoS summary captures the current amount of WoS at each SKU–depot location within a study. This information is especially useful in the distribution planning process to identify inventory imbalances across the finished product supply chain.
10) The distribution plan summary is an output of the demand and supply planning process that summarizes the key details of all distribution orders within the plan. It's an important reference during the demand and supply planning process and serves as a communication tool to internal and external stakeholders responsible for conduct of the trial and execution of the supply chain.
11) Distribution order readiness is being prepared and able to execute a distribution order so that it meets the desired target receipt date and is released for shipment to a clinical site. Given the complexity of international shipments, implementing a distribution order readiness meeting to effectively transition from planning to execution is necessary. It's also critical to gain alignment on which order readiness activities are included in the depot-to-depot lead time to avoid late shipments.

12

Upstream Supply Planning

Primary Packaging, Drug Product, and Drug Substance

12.1 Planning for Upstream Materials

In the clinical trial supply chain, the terms upstream manufacturing and downstream manufacturing are used to describe the relative order of operations in the chain. Events happening upstream come before those happening downstream. Thus, in clinical trial supply chain management, upstream supply planning encompasses primary packaging (PP), drug product (DP), and drug substance (DS) production because each of these steps precedes finished product (FP) manufacturing. PP is upstream of the FP and downstream of the DP and DS. DP is upstream of PP and downstream from DS, and DS manufacturing is the most upstream good manufacturing practice (GMP) step. Recall from Chapter 3 – Introduction to the Clinical Trial Supply Chain that in most solid-state small-molecule supply chains, PP is a distinct GMP manufacturing step from bulk DP manufacturing. Meanwhile, for biologics and liquid-state small molecules, bulk DP manufacturing and PP take place in one continuous GMP operation.

Supply planning for upstream materials requires a different approach than planning at the FP level for a few reasons. First, all upstream materials are, by definition, components of a downstream manufacturing step – DS is a component of the DP operation; DP is a component of the PP operation in small molecules; DP (in biologics) or PP (in small molecules) is a component of the FP. This means the assumptions used in upstream production planning reflect those used in component planning for widgets. Inputs such as the bill of materials (BOMs), production yields, production lead times, and capacity considerations are common to all upstream materials no matter the modality (e.g. small or large molecule) or type of material (e.g. PP, DP, or DS).

The second difference relates to the nature of demand on FPs versus the nature of demand on upstream materials. Demand for FPs is driven primarily by subject demand, which is random and uncontrollable. This is known as independent demand because it occurs independently of any other source or material. Consequently, if a batch of FP is unavailable to meet demand, then that consumption opportunity is lost and can't be recovered. In clinical trials that translates to missing a subject dosing event. Conversely, demand on upstream materials is dependent on production requirements for the next downstream step in the manufacturing process. In other words, demand on primary packaged units for small molecules and DP for biologics is driven by FP production orders, and demand on DS is driven by DP production orders. This is known as dependent demand, which is demand on a component that is driven by a downstream production requirement. If a batch of DP or DS is not available to meet the downstream production requirement, then the production event is delayed, putting supply for an entire trial at risk.

Supply Chain Planning for Clinical Trials – A Practical Guide, First Edition. Ryan Mills.

The third area where upstream production planning differs from FP planning is cost. DP manufacturing is typically more expensive than FP manufacturing and DS manufacturing is often significantly more expensive than DP manufacturing, especially for biologics. The expense of producing DS and DP adds another layer to the planning process and the overall supply planning strategy for a program. If an FP batch goes to waste, the sponsor loses out on tens of thousands of dollars, but if a DS batch goes to waste, that represents hundreds of thousands or even millions of dollars. At the same time, running out of DP or DS when its needed is also not an option because the consequences of stocking out of DS or DP are so severe. This is the balance of upstream production planning in a nutshell, to minimize the risk of running out of drug while simultaneously minimizing the investment needed to maintain uninterrupted supply.

Fortunately, upstream production planning does share similarities with FP supply planning. The min–max inventory management methodology is applicable to upstream products, although changes to the safety stock and order target values are necessary to appropriately reflect the desired production and inventory management strategy. Similarly, weeks-of-supply is useable as it relates to measuring on-hand inventory quantities at the PP, DP, or DS level. Expiry management considerations are also largely applicable to upstream planning because the expiry period is assigned at the DP level, and because DS batches also have retest dates that must be accounted for in the plan. While shelf-life extension strategies are not applicable to upstream planning because DS and DP don't have clinical labels, planning for expiry events is necessary to ensure replacement batches are on hand when needed.

We'll begin this chapter by introducing a general approach to forecasting demand on upstream materials and a general approach to supply planning for upstream materials. Within these sections, the upstream production planning table is used to incorporate the various demand and supply planning assumptions that enable upstream planning. The supply planning section also touches on how to use supply planning parameters to create upstream production and inventory strategies that align with the overall program strategy. Given the wide variety of upstream manufacturing processes between small and large molecules and PP, DP, and DS manufacturing steps, we'll provide examples and specific considerations for each modality and stage of manufacturing. The chapter concludes with the upstream production summary report and a discussion of production order readiness considerations.

12.2 General Approach to Demand Planning for Upstream Materials

Planning for any material starts with the demand forecast, and upstream production planning is no exception. We began Chapter 8 – Finished Product Demand Planning: Basic Concepts by introducing the concept of stockkeeping units (SKUs) and how SKU categorization forms the basis of what is forecasted within the plan. The first step with upstream production planning is to categorize the PP (if applicable), DP, and DS SKUs that need to be forecasted. Recall from Chapter 8 that DPs may be distinguished by dosage formulation, dosage strength, fill volume for large molecules, container closure systems, manufacturing process, specification, or manufacturing site. DSs may be distinguished by the active pharmaceutical ingredient (API), manufacturing scale for large molecules, specification, or manufacturing site. Each DP and DS SKU needs to have an individual demand and supply plan to maintain visibility of all materials and control over planning decisions that impact each SKU.

After SKU categorization has been completed, the next step is to identify the sources of demand against the upstream material. Typically, the most significant source of demand is to satisfy downstream production requirements for which the upstream material is a component. However, there are other sources of demand on upstream materials that also need to be accounted for in the plan. Upstream materials are commonly needed for nonroutine sampling events to support stability studies and for additional process development activities. In Chapter 3 – Introduction to the Clinical Trial Supply Chain, we discussed process development at both the DP and DS levels. Material from DS batches is often used for both DP and DS manufacturing process development. Similarly, material from DP batches may be used to improve the process for subsequent DP manufacturing campaigns.

There are five inputs required to build a demand forecast for an upstream material:

1) Downstream production orders that require the upstream SKU as a component in the operation. These orders are fixed by the clinical planner during the production planning process for the downstream material and both the quantity and timing of the order are included in the upstream production planning process.
2) The production lead time (LT_P) of the downstream material which captures all manufacturing and batch release activities at the upstream step before a batch enters the manufacturing process at the downstream step. Recall from Chapter 4 – Quality Considerations that production lead times may or may not include quality release activities depending on the nature of the supply chain. Manufacturing and batch release may take place in one of the two ways. Either the batch moves "sequentially" from an upstream node to a downstream node only after both manufacturing and batch release activities take place or the batch moves "at risk" from an upstream node to a downstream node immediately after manufacturing takes place but before batch release activities complete. It's also possible, and relatively common, for a batch of the same SKU to be manufactured "at risk" and then a subsequent batch to be manufactured with a full quality release.
3) The BOM quantity of the upstream material required to complete a downstream production order. The BOM quantities for upstream materials are provided by the relevant chemistry, manufacturing, and controls (CMC) team members. Clinical supply chain typically owns the FP BOM which dictates how many primary packaged or DP units are needed while the formulation development/DP lead owns the DP BOM which dictates how much DS is needed to support a batch. The master batch records (MBRs) also contain BOM information and are a source of truth for these quantities.
4) The production yield of the manufacturing step, which is the percentage of component productivity expressed as the number of units produced/number of units entered in the manufacturing process. Production yields are estimated prior to the first manufacturing campaign of a batch. Then, as batches are manufactured and yield data from each batch becomes available, yield estimates are refined. Like BOM quantities, the relevant CMC team member owns the yield calculations so clinical supply chain owns FP yield estimates, the DP lead owns DP yield estimates, and the DS lead owns DS yield estimates. Yields are also provided on executed batch records (EBRs) postproduction.
5) Other sources of demand on the upstream material include routing sampling, nonroutine sampling, and process development requirements. Again, clinical supply chain needs to work closely with the CMC team to identify and document any nonroutine demand on an upstream material. Routine sampling requirements are typically documented in the MBR while nonroutine sampling requirements may be captured in the EBR or in other documents (e.g. the stability protocol for the SKU).

With these inputs in hand, calculating the demand forecast follows the step-by-step process outlined below:

1) Identify and capture downstream production orders that require the upstream material as a component. For example, in a scenario where the DP of a biologic is used in FP manufacturing, all FP production orders that consume the DP need to be captured. Both order quantity and timing are required.
2) Translate the downstream production order finish dates to downstream production order start dates. This translation is determined by the downstream production lead time. The downstream production order start date is also known as the upstream material's requirement date because the upstream material needs to be available to meet the start date of the downstream manufacturing job.
3) Convert the downstream production order quantity into the quantity of upstream material required to support the production event through this formula:

$$\text{Upstream requirement conversion} = \text{Qty of downstream order} \times \frac{\text{BOM qty}}{\text{production yield}}$$

For example, if a downstream production order calls for 500 units of material, the BOM Qty is 2 units of upstream material for each unit of downstream material, and the production yield is 50%, then the total upstream requirement is 500 * 2/0.5 = 2000 units of materials.

4) Identify and add additional sources of demand on the upstream material to the plan such as nonroutine stability sampling and/or process development requirements.

Planning demand and supply for upstream materials requires a similar approach to that used in planning FP materials – start with the demand forecast and then develop a supply plan to meet that demand. Building an upstream production plan table with the same general structure as the FP production plan creates a solid framework to work through this analysis. The upstream production plan table begins with calculating the demand forecast based on the inputs and steps outlined above. Figure 12.1 captures an example demand forecast for a large-molecule DP SKU.

First note how the columns in the upstream production plan table match the weekly cadence used in the FP production plan table examples from earlier chapters. Whether weekly or monthly buckets are used, the time periods between downstream and upstream materials must match. If different periods are used, then it's difficult to link the supply plan of the FP to its upstream

		W1	W2	W3	W4	W5	W6	W7	W8	W9	W10	W11	W12	W13	W14	W15	W16	W17	W18	W19	W20	W21	W22	W23	W24	W25	W26
FP	FP SKU 1 Order finish														500												
FP	FP SKU 2 Order finish																			200							
FP	FP SKU 1 Order start								500																		
FP	FP SKU 2 Order start													200													
DP	FP SKU 1 Order requirement								500																		
DP	FP SKU 2 Order requirement													200													
DP	Other Dp requirements								1					1				50									
DP	Total DP requirements								501					201				50									

SKU description: Wonder drug 50 mg/mL in a 10 mL vial
FP production lead time: 6 weeks

DP:FP bill of material quantity: 1 unit
FP production yield: 100%

Figure 12.1 Upstream Production Plan Table: Demand Requirements

component. Confusion may also result around the timing of events between downstream and upstream activities. This is especially true when the supply chain is constrained, and events need to be tightly coordinated to deliver supply on time and avoid a stockout.

In the above example, each of the inputs is identified, and the forecasting steps are followed in sequence. There are two FP SKUs that consume the DP, and downstream production order finish dates for each are captured in the top two rows, respectively. The next two rows offset the production order finish dates to production order start dates based on the production order lead time of six weeks that's identified in the assumptions at the bottom of the table. Then, the quantities of the FP production orders are converted into requirements for DP material to satisfy each FP production order start date. This is a simple example where the BOM quantity is 1 unit of DP, and the production yield is 100% so the production order of 500 kits for FP SKU 1 in Week 8 converts to a requirement of 500 vials of DP while the production order of 200 kits for FP SKU 2 in Week 13 converts to a requirement of 200 vials of DP.

Finally, there are three additional requirements captured in the other DP requirements row captured. Each of the FP production orders requires a single retention sample which is a regulatory requirement in the United States and the European Union and is common practice for FP manufacturing around the world. This brings the total DP requirement in Weeks 8 and 13 to 501 and 201 kits, respectively. There is also a requirement entered for 50 vials in Week 17, a request from the DP formulation development to ship samples of the DP to a testing lab for additional method development activities. This brings the total DP requirement quantity to 752 vials from Weeks 1 to 26. The 752-vial requirement is equivalent to demand of 752 vials on DP supply over this period.

Before moving to the supply planning side of the equation, let's reflect on the five inputs that drive demand for upstream materials and how much variance there is within each. The amount of variance matters because this impacts how closely the supply plan matches reality. For example, if an input has a wide potential variance, then the plan is more likely to be wrong and misrepresent the physical supply chain. If multiple inputs have a wide variance, the likelihood and magnitude of how wrong the plan is increases, so let's walk through each input ordered from least to most variance.

The input with the lowest variance is the BOM quantity which should have no variance at all. If the BOM quantity changes, that triggers the need for a new SKU because the DP formulation or the packaging configuration has changed. The second input is downstream production orders. Typically, the clinical planner has some degree of control over changing downstream order quantity and timing, but ultimately those orders will need to be delivered, so the variance on this input is relatively low. Third are the nonroutine sampling requirements which vary from batch to batch and are sometimes not related to a batch at all (e.g. process development requirements). The good news with nonroutine sampling requirements is that the clinical planner usually has visibility to these. Batch sampling quantities are determined before production begins, and process development requirements are also determined as part of the CMC strategy, which the clinical supply chain function should be plugged into.

The fourth input is the production yield, and here variance increases making this harder to predict. Production yield variances correlate to the complexity of the manufacturing process. Relative to upstream manufacturing, FP production is a routine operation, and yields are usually in the 98–100% range. DP manufacturing is significantly more complex, so yields vary from 70% to 95% in most cases. DS manufacturing is highly complex, but less so for small molecules where production yields become relatively predictable over time. Biologics DS manufacturing, on the other hand, is the most complex operation and is generally unpredictable from batch to batch in clinical development where process experience is relatively limited. Yields may vary widely with one batch yielding less than 50% of what was expected and the next yielding more than 125% of expectation. Generally, as development progresses and more batches are made, experience increases and yields tend to stabilize over time, but this is often a challenging input to predict from batch to batch within the DP and DS production plans.

Finally, the input with the most variance is the production lead time. Like production yields, production lead times increase in variance as the supply chain moves upstream and the manufacturing processes become more complex. FP production lead times typically have the lowest variance from batch to batch with a lead time standard deviation that's measured in weeks. This is largely driven by the simplicity of the manufacturing process and the fact that FPs do not undergo analytical testing like their upstream counterparts. DP manufacturing lead times typically vary more than FP lead times, and standard deviations for DP manufacturing typically also measure in weeks. DSs nearly always have the widest lead time variation and the standard deviation measures in months in some cases. The variance for each production lead time is exacerbated further when production is outsourced to a contract manufacturer because there is a two-step batch release process when using a contract manufacturing partner. Both the contract manufacturer's quality unit and the sponsor's quality unit must sign off on the batch, and this adds time to the process. And, if a deviation occurs during a manufacturing campaign, a not uncommon event for DP and DS manufacturing, this adds even more time and variance to the process. This also don't account for whether production lead times include the batch release process or are performed "at risk," which is another factor to consider when projecting the production lead time input.

At a passing glance, predicting demand against upstream materials seems straightforward, simply identify the downstream production requirements and calculate how many units of the upstream material are necessary to meet that demand at the desired timepoint. But the challenges in forecasting production yields and lead times mean that careful attention must be paid to how closely the plan matches reality. Adjustments need to be made to these inputs on a regular basis to maintain the plan as an accurate representation of the physical supply chain, a subject we'll return to in Chapter 15 – Clinical Supply Chain Planning Processes.

12.3 General Approach to Supply Planning for Upstream Materials

With a complete requirement forecast for the upstream material in place, we now turn our attention to supply planning. Fortunately, the same min–max inventory management methodology and supply planning parameters used in FP production and distribution planning are applicable to supply planning for PP, DP, and DS materials. However, there are two new elements unique to upstream manufacturing and a twist on an existing element, weeks-of-supply, that we'll incorporate into the upstream production planning table.

The first element to consider is equipment train constraints within the manufacturing process. At the FP level, there is almost never a constraint on how many kits can be made in a single batch because the equipment used in production does not have a predefined limit on input quantities. The same is not true for some upstream manufacturing processes which are run on pieces of equipment that have limited processing capacity. For example, clinical-scale biologics DS production occurs in bioreactor tanks with a specific size (e.g. 500 L, 1000 L, or 2000 L). This means that batch sizes are capped at an amount based on the expected production yield within a particular tank. Likewise, some small-molecule DP manufacturing processes utilize equipment that limits the input amount of DS, which in turn limits the DP batch size. It's important for the upstream supply plan to account for these equipment constraints when downstream requirements have greater quantities than a single upstream batch can produce. When this happens, multiple upstream batches are needed to meet the full requirement, or the clinical planner needs to adjust the downstream requirement quantity down to fit the available batch size. We'll look at this process more closely in Chapter 15 as well.

The second element to incorporate with upstream production planning is the expiry period management. Unlike FP planning, upstream materials do not have shelf-life extensions performed on them because they don't have a full clinical label printed on them. When the expiry period is extended for an upstream batch, the inventory management system is updated at locations where the batch is held, and the batch is considered good for use until the new expiry date is reached. However, DP and primary packaged batches should not be placed into FP production within four to 6 months of their listed expiry date if no further expiry period extensions are planned. Remember that the FP expiry date is based on the underlying DP, so by the time FP production completes and the batch is released for distribution into the clinic, the expiry date is already near. Also recall from Chapter 6 – RTSM and Its Role in the Clinical Supply Chain that the RTSM do not ship and do not dispense parameters further eat into the amount of remaining shelf life on the FP.

To protect against allocating DP batches nearing their expiry date to FP batches, the upstream production plan needs to include an expiry window assumption. The expiry window acts similar to the RTSM parameters in that it short-dates the DP expiry within the plan to remove it from available supply prior to its expiration date. DS batches do not require an expiry window application because they can be consumed into DP manufacturing campaigns right up to the last day of their assigned retest date.

Production planning for upstream materials also follows a step-by-step process that's similar to the process for FP production planning:

1) The upstream production plan table generates suggested production order requirements based on the interaction between the total downstream requirements on the upstream material, the safety stock value, and the order target value. The methodology is exactly like that described in Chapter 9 – Finished Product Supply Planning: The Production Plan.
2) The clinical planner converts the suggested order requirements into actual production orders and incorporates applicable capacity constraints in this conversion process.
3) The upstream production plan table translates the production orders into individual batch information including quantities and expiry dates, and then the expiry window value short-dates PP and DP SKUs where applicable. The methodology here is exactly like that described in Chapter 10 – Finished Product Expiry Planning.
4) The clinical planner allocates production orders to downstream requirements. It's often the case that multiple upstream production batches are available to service downstream production orders, and allocation decisions are required to optimize the plan. In most cases, a first in first out (FIFO) inventory management methodology makes sense, particularly when long expiry periods are assigned to the upstream material. However, in some cases, last in first out (LIFO) is more appropriate when a newer batch with better expiry dating is available and the older batch would not be consumed by subject demand before reaching its expiration date.

Figure 12.2 shows an example upstream production planning table with a complete production plan.

In this example, the same three downstream demand requirements from Figure 12.1 are documented in the top rows of the table. The safety stock value is set to 1 week while the order target value is set to 26 weeks, and this combination results in a suggested production order requirement of 752 kits in Week 7. Week 7 is 1 week before the first upstream demand requirement

		W1	W2	W3	W4	W5	W6	W7	W8	W9	W10	W11	W12	W13	W14	W15	W16	W17	W18	W19	W20	W21	W22	W23	W24	W25	W26
DP	FP SKU 1 order requirement								500																		
DP	FP SKU 2 order requirement													200													
DP	Other DP requirements								1					1				50									
DP	Total DP requirements								501					201				50									
DP	OHI	0	0	0	0	0	0	600	99	99	99	99	399	198	198	198	198	148	148	148	148	148	148	99	99	99	99
DP	Production actual	0	0	0	0	0	0	600	0	0	0	0	300	0	0	0	0	0	0	0	0	0	0	0	0	0	0
DP	Production plan	0	0	0	0	0	0	752	0	0	0	0	0	0	0	0	0	0	0	0	0	0	0	0	0	0	0
DP	Safety stock value	1	1	1	1	1	1	1	1	1	1	1	1	1	1	1	1	1	1	1	1	1	1	1	1	1	1
DP	Order target value	26	26	26	26	26	26	26	26	26	26	26	26	26	26	26	26	26	26	26	26	26	26	26	26	26	26
DP	Batch: DP1A EXP: W41 ESD: W23	0	0	0	0	0	0	600	99	99	99	99	99	99	99	99	99	49	49	49	49	49	49	0	0	0	0
DP	Batch: DP2B EXP: W46 ESD: W28	0	0	0	0	0	0	0	0	0	0	0	300	99	99	99	99	99	99	99	99	99	99	99	99	99	99

SKU description: Wonder drug 50 mg/mL in a 10 mL vial
FP production lead time: 6 weeks
DP capacity constraint: 600 units

DP:FP bill of material quantity: 1 unit
FP production yield: 100%
DP expiry window: 18 weeks

Figure 12.2 Upstream Production Planning Table: Production Plan

located in Week 8 (e.g. equal to the safety stock value of 1 week). The suggested quantity of 752 kits covers the next 26 weeks of upstream demand requirements (e.g. 501 + 201 + 50).

Notice in the assumptions on the bottom left of the table that the DP has a capacity constraint of 600 units which means the actual production order can't exceed this quantity, so the clinical planner has entered an order of exactly 600 units. Then, in Week 12, another order is added for 300 units, half the constraint, to supply the remaining requirements. The second production order is set to 300 units and not 152 units because of the potential yield variance described in Section 12.2. If the clinical planner aims for exactly 152 units and less are yielded, then another costly DP run may need to be scheduled to meet demand. While this approach may yield some waste, that's less costly than having to execute a third DP campaign.

Individual batch data is captured in the bottom two rows. The initial DP batch, DP1A, has an assigned expiry date of Week 41 which is then short-dated to Week 23 because of the 18-week expiry window input. Also note that the expiry date for the DP batch is based on the date of manufacture which is not shown in this upstream production plan table. The second DP batch, DP2B, has an assigned expiry date of Week 46 which is then short-dated to Week 28.

It's also possible to see the allocation strategy the clinical planner selected through the individual batch data. The first downstream production requirement of 501 kits is clearly supplied from the DP batch DP1A, and the second downstream production requirement of 201 kits is also clearly supplied by DP batch DP2B. But notice how the final downstream production requirement of 50 kits could be supplied by other DP1A or DP2B. The clinical planner has allocated DP1A because this material will be used for process development, so the shorter expiry date is not impactful. If that last DP requirement had been a downstream production order intended for the clinic, the clinical planner would allocate material from batch DP2B to get better shelf life on the FP label.

Note how weeks-of-supply are missing in the table above. Weeks-of-supply for upstream materials can be calculated in two different ways. The first approach, known as standard weeks-of-supply, counts the number of weeks of future downstream demand for that material that is covered by on-hand inventory of the upstream material. Another way of saying this is that standard weeks-of-supply measures the number of weeks between upstream production and downstream consumption of the upstream material. This is a useful approach to understand how much time there is before on-hand inventory will be consumed. However, standard weeks-of-supply is not helpful for understanding the overall production strategy because it only looks at upstream inventory through the lens of immediate downstream demand. Standard weeks-of-supply does not describe whether the supply chain can absorb a supply shock such as a lost batch because it does not consider the total amount of inventory across the supply chain that's in place to protect against a lost batch or an unexpected plant closure.

To address these questions, there is an alternative weeks-of-supply approach that measures the total amount of on-hand inventory available to service future demand at the FP level. This second approach is known as implicit weeks-of-supply because it measures upstream on-hand inventory directly against future FP demand. Implicit weeks-of-supply contextualizes how much FP demand on-hand upstream inventory is able to support. Measurements of upstream implicit weeks-of-supply can also be added to aggregate finished product weeks-of-supply (AFPWoS) to provide a complete picture of inventory and quantify the amount of resiliency in the entire supply chain. To measure implicit weeks-of-supply, take the following steps:

1) Follow Steps 1–5 in the AFPWoS section of Chapter 11 – Finished Product Supply Planning: The Distribution Plan to determine the implicit total run rate (ITRR) of each Finished Product SKU that consumes the upstream material. The ITRR is the appropriate input because it measures the average FP weekly demand over the period of future demand which on-hand inventory in that week is available to cover across all depots in the study.
2) Sum the ITRR's across all SKU's that consume the upstream material to arrive at the Total ITRR.
3) Convert the total ITRR into the upstream unit of measure using the BOM quantity outlined earlier in this chapter. Translating FP demand all the way back to the DS input quantity requires identifying both the FP to DP BOM ratio and the DP to DS BOM ratio.
4) Divide total on-hand inventory of the upstream material into the total ITRR offset by the lead time required to turn the upstream material into FP, and this gives the weeks-of-supply calculation. The offset provides a more accurate picture of when upstream inventory will be available to service that FP demand. For DP (or Primary Packaged) materials, the offset should be set to the FP production lead time, and for DS materials, the offset should be set to the sum of the DP and FP production lead times.

The process to calculate implicit weeks-of-supply is illustrated in Figure 12.3.

In this example, there are two FP production orders in Week 7 and Week 18 which are captured at the top of the table. To service this demand, the clinical planner has scheduled three DP orders

		W1	W2	W3	W4	W5	W6	W7	W8	W9	W10	W11	W12	W13	W14	W15	W16	W17	W18	W19	W20	W21	W22	W23	W24	W25	W26
DP	Total DP requirements							600											600								
FP	FP SKU 1 ITRR	0	0	0	0	0	0	0	0	0	0	0	0	10	10	10	12	12	12	14	14	14	16	16	16	18	18
FP	FP SKU 2 ITRR	0	0	0	0	0	0	0	0	0	0	0	0	8	8	8	8	8	10	10	10	12	12	12	14	14	14
FP	Total ITRR	0	0	0	0	0	0	0	0	0	0	0	0	18	18	18	20	20	22	24	24	26	28	28	30	32	32
DP	FP : DP Conversion	0	0	0	0	0	0	0	0	0	0	0	0	18	18	18	20	20	22	24	24	26	28	28	30	32	32
DP	Standard DP WOS	0	0	4	3	2	1	11	10	9	8	7	6	5	24	23	22	21	20	19	18	17	16	15	14	13	12
DP	Implicit DP WOS	0	0	0	0	0	0	33	33	33	30	30	27	25	50	46	43	43	20	19	19	19	19	19	19	19	19
DP	OHI	0	0	600	600	600	600	600	600	600	600	600	600	600	1200	1200	1200	1200	600	600	600	600	600	600	600	600	600
DP	Production actual	0	0	600	0	0	0	600	0	0	0	0	0	0	600	0	0	0	0	0	0	0	0	0	0	0	0

SKU description: Wonder drug 50 mg/mL in a 10 mL vial
FP production lead time: 6 weeks
DP capacity constraint: 600 units

DP:FP bill of material quantity: 1 unit
FP production yield: 100%
DP expiry window: 18 weeks

Figure 12.3 Upstream Weeks-of-Supply Calculation Example

of 600 units each for delivery in Weeks 3, 7, and 14, which are captured in the bottom two rows of the table. In the middle section, rows have been added to capture the ITRRs for the two SKUs that consume the DP. The total ITRR is summed and then converted to the DP units using the DP:FP BOM quantity of 1 unit. Then, the DP on-hand inventory is divided into the FP:DP conversion and offset 6 weeks into the future to account for the FP production lead time. Meanwhile, the standard weeks-of-supply counts how many weeks before the DP batch is consumed into downstream production.

12.4 Upstream Supply Planning Strategies

Upstream production planning goes beyond the mechanics outlined so far in this section though. There are numerous strategic drivers for the clinical planner to incorporate into their decision-making process for converting suggested production requirements into actual production orders from both a quantity and timing perspective. The key principle here is to find the appropriate balance between building a desirable amount of inventory and resiliency within the upstream supply chain while minimizing the cost to do so and the potential for wasted inventory. Strategic drivers impacting the quantity and timing of production order placement include the following:

- *Quantity – Level of Uncertainty in the Demand Forecast*: When there is a low degree of uncertainty in the demand forecast (e.g. clinical studies are >50% enrolled), DP and DS inventory should be built up to cover a longer period (e.g. 18–36 months-of-supply) of future demand. When there is a high degree of uncertainty in the forecast, the clinical planner should consider either building less inventory to avoid excess waste if demand fails to materialize or building more inventory to avoid a stockout if enrollment happens faster than expected. The strategic clinical demand and operations planning process is designed to manage a high degree of FP demand uncertainty in upstream supply planning. This process is covered in Chapter 15 – Clinical Supply Chain Planning Processes.
- *Quantity – Expiry Period of the Upstream Material*: When the expiry period of the upstream material is relatively long (e.g. >24 months), inventory should be built to cover future demand over the expiry period, but the planner should proceed with caution here. When there is high uncertainty in the demand forecast, the planner may only want to build 12 months-of-supply to avoid excess waste. When the expiry period is relatively short, on the other hand, smaller, more frequent batches of upstream manufacturing are appropriate until a longer expiry period is established.
- *Quantity – Cost of the Upstream Material*: If the upstream material is relatively expensive to manufacture (e.g. large-molecule DS), the more cautious the clinical planner needs to be with building out inventory. If the product is relatively cheap to manufacture (e.g. small-molecule DP manufacturing that consumes an inexpensive DS), then producing extra inventory may be worthwhile to protect against demand uncertainty, even if some of the product ends up going to waste.
- *Quantity – Clinical Program Phase*: Earlier in development, it makes more sense to build smaller batches of inventory because there's not a guarantee the program will progress to the next stage of development. By late-stage development, larger quantities of inventory are needed to support clinical demand and to eventually reach commercial scale manufacturing.
- *Timing – Level of Supply Chain Agility*: Agility is defined as how quickly the supply chain can react to a change in the supply plan. A particular manufacturing site's agility level is dictated by

raw material, capacity, and labor availability. As supply chain agility increases, the amount of on-hand inventory required to buffer against a supply shock reduces. This is because the supply chain can react to changes in demand requirements more quickly, so less inventory is needed to maintain supply chain performance. Unfortunately, most upstream materials have relatively long production lead times and may also have long raw material sourcing lead times, especially for biologic products. It's also important to consider whether manufacturing is insourced or outsourced when assessing agility. Manufacturing that takes place at internal sites tends to be more agile and flexible than outsourced manufacturing because there is more control over the production schedule. Contract manufacturers are often less agile and usually need several months of lead time to schedule an upstream production order into their plant. To combat this, some sponsors consider dual sourcing which increases supply chain agility because there are two manufacturing options which increase total supply chain capacity.

- *Quantity and Timing – Level of Supply Chain Reliability*: Reliability is defined as how consistently the manufacturing site delivers a quality batch within the estimated lead time. Both elements, quality and lead time, are important in biopharmaceutical manufacturing because missing on either may result in a stockout situation. Like agility, as reliability increases, the optimal level of on-hand inventory decreases. If the supply chain is not reliable, however, carrying extra on-hand inventory makes sense because the likelihood of a failed batch is higher.
- *Quantity and Timing – Level of Supply Chain Resiliency*: Supply chain resiliency is defined as the supply chain's ability to withstand a supply shock and not miss demand. By this definition, supply chain resiliency means carrying sufficient on-hand inventory so that if production is delayed, a batch is lost, or a plant is closed, the supply chain can perform until the impacted supply can be replaced or restarted.

One methodology to assess these factors and develop a coherent upstream production and inventory management strategy is through the lens of risk management. The first point on demand uncertainty can be reframed as the level of demand risk, which is then subcategorized into upside and downside demand risk. Upside demand risk is the potential for subject enrollment to happen faster than expected increasing stockout risk, while downside demand risk is the potential for subject enrollment to happen more slowly than expected increasing excess waste risk. Meanwhile, supply risk is the potential inability of the supply chain to deliver material consistently to the clinic. Both supply chain agility and supply chain reliability encompass the level of supply risk in the upstream supply chain. Demand and supply risks then need to be weighed against the expiry period and the cost of the DP and DS. Table 12.1 lays out these factors and their impact on the production and inventory management strategy.

Each of the factors listed above needs to be considered as well as the overall strategy of the program and the company. For established, commercial-stage biopharmaceutical companies, it may be worth the extra cost to invest in an extra batch of DS or DP because it represents a fraction of the overall program budget. For early-stage companies without revenue, carrying fewer weeks-of-supply may be more desirable because dollars are scarce, and they need to be allocated in other areas of clinical development. Unfortunately, there is no magic formula when it comes to selecting the optimal level of upstream inventory to build and carry as safety stock, but there are techniques the clinical planner can use. We'll return to a few of these in Chapter 13 – Advanced Concepts in Demand and Supply Planning.

Ultimately, the desired level of supply chain resiliency needs to be reflected in the plan via the safety stock and order target values of each upstream material. For example, if the team agrees that a minimum of 6 months of DP inventory needs to be held, the safety stock value should be set to

Table 12.1 Demand Risk, Supply Risk, and Cost Considerations in Production and Inventory Management

Factor and Description	Impact to Production and Inventory Management Strategy
High Upside Demand Risk: high degree of uncertainty in the demand forecast with potential for enrollment to be faster than anticipated.	Increase on-hand inventory levels to protect against stockout risk. Counterbalanced by a short expiry period and high cost, both of which increase excess waste risk.
High Downside Demand Risk: high degree of uncertainty in the demand forecast with potential for enrollment to be slower than anticipated.	Reduce production and on-hand inventory levels to protect against downside demand. May be counterbalanced by a long expiry period and low cost, both of which reduce excess waste risk.
Low-demand Risk: low degree of uncertainty in the demand forecast.	No change to the production strategy. Build inventory to cover demand over expiry period.
High-supply Risk: high degree of uncertainty in supply chain performance. Low levels of supply chain agility and/or reliability.	Increase on-hand inventory levels at the level downstream of the supply chain risk to protect against a lost batch at the risk source.
Low-supply Risk: low degree of uncertainty in supply chain performance. High levels of supply chain agility and/or reliability.	No change to the production strategy. Build inventory to cover demand over expiry period.
High Cost: relatively expensive product to procure or manufacture (e.g. >$1M total per batch).	Reduce production and on-hand inventory levels to protect against excess waste risk. Counterbalanced by a long expiry period and low-demand risk.
Low Cost: relatively inexpensive product to procure or manufacture (e.g. <$1M total per batch).	No change to the production strategy. Build inventory to cover demand over expiry period.

26 weeks as a starting point for planning. Likewise, if 1 year of DS inventory needs to be on hand, then the safety stock value at the DS level should be set to 52 weeks. The order target value should also be adjusted based on the factors identified above that influence the desired quantity of each order: the confidence level in the FP demand forecast, the expiry period, and the cost of the upstream material. Keep in mind though that the safety stock and order target values only provide suggested production order requirements. The clinical planner always needs to apply their judgment and account for the elements listed above to make optimal production order decisions.

With respect to safety stock values and how the safety stock value at the FP level interacts with the safety stock level at upstream levels. If the upstream plan incorporates some level of supply chain resiliency, and a minimum supply of DP is always kept on hand, then the FP safety stock values at the central depot may be lowered to the FP production lead time. This reduces unnecessary inventory at the FP level and creates more flexibility within the supply plan. It's a supply chain planning rule of thumb that it's always better to hold material as far upstream as possible to maximize allocation flexibility across SKUs and studies. That way, the supply chain can better react to changes in program or study demand because there is upstream material available to respond to potential changes in demand.

A final point to consider when developing and implementing a supply chain resiliency strategy is the need for alignment within the CMC team and other internal stakeholders such as clinical development and finance. The costs, risks, and trade-offs of the upstream production strategy need to be made clear to all team members so if things go wrong, everyone understands why key decisions were made and what the path to recovery entails. We'll return to these concepts in

Chapter 15 – Clinical Supply Chain Planning Processes, where we'll review the business processes for gaining alignment on the upstream production strategy with the appropriate stakeholders.

Next, we'll turn to more granular considerations for planning upstream materials by modality (e.g. small molecule and biologics) and planning level (e.g. PP, DP, and DS). Each of these sections briefly reintroduces the manufacturing process and how that commonly influences planning inputs such as BOM quantities, production yields, sampling requirements, and capacity constraints. Average production lead times by modality and level are covered in Chapter 3 – Introduction to the Clinical Trial Supply Chain. Then, an example production table is shown followed by a discussion of strategic considerations for that modality-level combination.

12.5 Small-Molecule Primary Packaging Planning

Small-molecule PP operations are a relatively simple manufacturing process where individual units of bulk DP are packaged into bottles, blister packs, or sachets. With respect to BOM quantities, SKUs that use bottles often have a 1:1 BOM ratio (e.g. one bottle per kit), but it's also possible for multiple bottles to be packaged into a final kit. On the other hand, multiple blister packs are usually combined into a single FP kit to achieve the desired dose, but a 1:1 configuration is also possible for kits containing blisters.

Production yields are also relatively high with PP SKUs due to relative process simplicity, and these can range anywhere from 85% to 99% depending on the operation. Sampling requirements vary from batch to batch because the stability profile of the DP is based on the primary packaged presentation. The first GMP batch of a primary packaged SKU almost always has samples taken from it to support stability, and subsequent batches may also have samples taken if process changes take place. The clinical planner should always consult with the formulation and analytical operations leads to document sampling requirements from each primary packaged batch. Finally, PP operations almost never have a capacity constraint. One exception to this is if a raw material such as a foil has a finite amount of stock on hand at the manufacturing site that caps the potential quantity of a batch.

In Figure 12.4, the first elements to call out are the planning inputs at the bottom of the table. The SKU is described as a 1×7 blister strip with a PP: FP BOM quantity of three strips, which means that three blisters are needed to manufacture one FP kit. There is a 6-week lead time to manufacture the FP, the production yield is estimated at 98%, and there is no capacity constraint in this process. These assumptions converge to translate the two FP production orders into PP requirements. First, each FP order is offset by the 6-week lead time, and then the FP order quantities are turned into the PP input quantities through the upstream requirement conversion formula described earlier in this chapter. Note that stability samples are taken from the initial PP batch so 300 extra units have been added to the requirements forecast in Week 8. Without a capacity constraint in this scenario, the clinical planner elects to plan a single production order of 2500 units to satisfy the FP production orders and the stability requirement.

There is another element to consider when planning PP SKUs. It's only advisable to combine multiple batches of primary packaged material into a single FP kit if a single underlying DP batch has been used to manufacture the PP batches. This is because if multiple DP batches are used in a single FP batch, and that FP batch has a quality issue, each underlying DP batch is implicated and may need to be quarantined from use in the clinic. In other words, combining multiple DP batches into a single FP batch puts each of those DP batches potentially at risk. While this is usually a low probability risk, the impact could be catastrophic for a study and program.

		W1	W2	W3	W4	W5	W6	W7	W8	W9	W10	W11	W12	W13	W14	W15	W16	W17	W18	W19	W20
FP	FP SKU 1 Order finish (kits)														500						
FP	FP SKU 2 Order finish (kits)																			200	
PP	FP SKU 1 Order requirement (strips)								1531												
PP	FP SKU 2 Order requirement (strips)													613							
PP	Other PP requirements (strips)								300					1							
PP	Total PP requirements (strips)								1831					614							
PP	OHI	0	0	0	0	0	0	2500	669	669	669	669	669	55	55	55	55	55	55	55	55
PP	Production actual	0	0	0	0	0	0	2500	0	0	0	0	0	0	0	0	0	0	0	0	0
PP	Production plan	0	0	0	0	0	0	2445	0	0	0	0	0	0	0	0	0	0	0	0	0
PP	Safety stock value	1	1	1	1	1	1	1	1	1	1	1	1	1	1	1	1	1	1	1	1
PP	Order target value	26	26	26	26	26	26	26	26	26	26	26	26	26	26	26	26	26	26	26	26
PP	BATCH: PP1A EXP: W41 ESD: W23	0	0	0	0	0	0	2500	669	669	669	669	669	55	55	55	55	55	55	55	55

SKU description: Wonder drug 50 mg 1 x 7 blister strip
FP production lead time: 6 weeks
PP capacity constraint: N/A

PP:FP bill of material quantity: 3 strips
FP production yield: 98%
PP expiry window: 18 weeks

Figure 12.4 Small-Molecule Primary Packaging Production Plan Table

12.6 Small-Molecule Drug Product Planning

Small-molecule bulk DP manufacturing is significantly more complex than PP and involves combining excipients with the DS to create an ingestible version of the drug. There are myriad processes depending on the formulation type – tablets, capsules, and oral solutions are some of the most common – and, as a result, BOM quantities vary from a few units to hundreds of units in a single container. Chapter 3 covers some of the key considerations when selecting a packaging-count configuration such as aligning the DP strength to the intended dose and to the visit schedule of the study. The most common capsule or table configurations in clinical trials are 30-count, 60-count, or 90-count containers. Oral solutions, on the other hand, behave more like biologic DPs because they are manufactured and packaged at the same time, and they are usually packaged in a 1:1 configuration from DP to FP.

DP production yields also have a wider variance than PP operations. This is especially true early in clinical development when batches are often manufactured at smaller scale, and there's a lack of process experience. As clinical demand increases over time, two forces converge that typically lead to significant improvements in yields. First, process improvements are identified and implemented with each new batch. Second, as clinical demand increases over time, batch sizes increase, and economies of scale are recognized. It's not uncommon for the first DP batches to have yields in the 60–80% range and then see those yields improve to 85%+ by late-stage manufacturing.

Routine sampling requirements are usually consistent with bulk DP manufacturing and are typically limited to in-process control samples which can be incorporated into the production yield calculations. However, there are often requests to take nonroutine samples from batches to support manufacturing process and analytical improvement efforts so these need to be accounted for in the plan. Equipment-based capacity constraints are also common with small-molecule DP manufacturing, and these constraints need to be accounted for in the plan.

Figure 12.5 takes the PP production order from the previous example and works through the same sequence of operations with planning inputs specific to the DP. The PP order quantity is offset by the 2-week production lead time for the blisters, then multiplied by the DP:PP BOM quantity of 7 units and then divided by the PP production yield of 95% to arrive at the DP order

		W1	W2	W3	W4	W5	W6	W7	W8	W9	W10	W11	W12	W13	W14	W15	W16	W17	W18	W19	W20
PP	PP SKU 1 Order finish (strips)							2500													
DP	DP SKU 1 Order requirement (tablets)					18421															
DP	Other DP requirements (tablets)					350															
DP	Total DP requirements (tablets)					18771															
DP	OHI	0	0	0	20000	1229	1229	1229	1229	1229	1229	1229	1229	1229	1229	1229	1229	1229	1229	1229	1229
DP	Production actual	0	0	0	20000	0	0	0	0	0	0	0	0	0	0	0	0	0	0	0	0
DP	Production plan	0	0	0	18771	0	0	0	0	0	0	0	0	0	0	0	0	0	0	0	0
DP	Safety stock value	1	1	1	1	1	1	1	1	1	1	1	1	1	1	1	1	1	1	1	1
DP	Order target value	26	26	26	26	26	26	26	26	26	26	26	26	26	26	26	26	26	26	26	26
DP	BATCH: DP1A EXP: W41 ESD: W23	0	0	0	20000	1229	1229	1229	1229	1229	1229	1229	1229	1229	1229	1229	1229	1229	1229	1229	1229

SKU description: Wonder drug 50 mg tablet
PP production lead time: 2 weeks
DP capacity constraint: 30,000

DP:PP bill of material quantity: 7 tablets
PP production yield: 95%
DP expiry window: 18 weeks

Figure 12.5 Small-Molecule Drug Product Production Plan Table

requirement of 18,421 tablets by Week 5. Then, in-process sampling requirements are added to arrive at the total DP quantity required to support each of the PP orders. There is a capacity constraint of 30,000 tablets, but given the short expiry period, a single production order of 20,000 tablets is appropriate.

With small-molecule DP manufacturing, SKU categorization is particularly important because it's easy to confound materials and batches without this. The first step in categorization is to identify the physical products and delineate between formulation strengths, packaging configurations, and manufacturing sites. Equally important is to understand when bulk DP and PP operations are separate events with unique batch records and to make sure these are identified as separate SKUs. It's not always the case that the bulk DP is immediately packaged into the packaging container, and keeping visibility into where the inventory is and how it will be allocated is critical to effective planning.

12.7 Small-Molecule Drug Substance Planning

Manufacturing small-molecule DSs occurs through a series of chemical reactions where intermediate compounds are combined with reagents, solvents, and other materials to eventually yield the desired API. Chemical manufacturing is relatively complex and time consuming to execute with the physical manufacture of the product often taking weeks (or months) to complete depending on how many steps are required to generate the compound. Fortunately, determining the DS BOM quantity for solid-state DP batches is simple and based on the strength of the DP SKU and the production yield of the DP:

$$\text{DS BOM quantity}(\text{kg}) = \frac{\text{DP strength}(\text{kg})}{\text{DP production yield}(\%)}$$

For example, if the DP strength is 50 mg and the DP production yield is 80%, then the DS BOM quantity is 0.00005 kg/0.8 = 0.0000625 kg.

DS manufacturing yields are relatively predictable because there are individual production yield calculations for each synthetic step within the overall process. However, it's not uncommon for

		W1	W2	W3	W4	W5	W6	W7	W8	W9	W10	W11	W12	W13	W14	W15	W16	W17	W18	W19	W20
DP	DP SKU 1 Order finish (tablets)				20000																
DS	DS SKU 1 Order requirement (kg)		1.25																		
DS	Other DS requirements (kg)					0.5	0.5														
DS	Total DS requirements (kg)		1.25			0.5	0.5														
DS	OHI	3	4.75	4.75	4.75	4.25	3.75	3.75	3.75	3.75	3.75	3.75	3.75	3.75	3.75	3.75	3.75	3.75	3.75	3.75	3.75
DS	Production actual	3	3	0	0	0	0	0	0	0	0	0	0	0	0	0	0	0	0	0	0
DS	Production plan	2.25	0	0	0	0	0	0	0	0	0	0	0	0	0	0	0	0	0	0	0
DS	Safety stock value	1	1	1	1	1	1	1	1	1	1	1	1	1	1	1	1	1	1	1	1
DS	Order target value	26	26	26	26	26	26	26	26	26	26	26	26	26	26	26	26	26	26	26	26
DS	Batch: DS1A Retest date: W53	3	1.75	1.75	1.75	1.75	1.75	1.75	1.75	1.75	1.75	1.75	1.75	1.75	1.75	1.75	1.75	1.75	1.75	1.75	1.75
DS	Batch: DS1A Retest date: W54	3	3	3	3	2.5	2	2	2	2	2	2	2	2	2	2	2	2	2	2	2

SKU description: Wonder drug 50 mg tablet
DP production lead time: 2 weeks
DS capacity constraint: N/A

DS:DP bill of material quantity: 0.0000625 kg
DP production yield: 80%
DS expiry window: N/A

Figure 12.6 Small-Molecule Drug Substance Production Plan Table

issues to develop as manufacturing scales up. Close collaboration with the DS lead on the CMC team is required to capture estimated DS batch yields on a batch-by-batch basis. It's also rare for small-molecule production to have a true capacity constraint, but there are cases where equipment trains limit estimated yields to either a minimum or maximum amount of DS per batch.

In Figure 12.6, the DP SKU production order finish date is set for Week 4. Based on a DP production lead time of 2 weeks and a BOM quantity of 0.0000625 kg, a DS requirement of 1.25 kg is generated in Week 2 to satisfy the DP production order requirement. Notice there are two other DS requirements of 0.5 kg each in Weeks 5 and 6 to support process development work which add to the total DS requirement. On the supply side, two DS batches of 3 kg each are planned for release in Weeks 1 and 2 to provide supply for the DP run as well as the process development requirements. The first batch is allocated to the DP requirement, and the second batch is allocated to the process development activities. One justification for this could be that the first batch had a higher purity level than the second batch and would therefore be better suited for the clinic. Finally, note how capacity constraints and the expiry window are not applicable to the DS, and each DS batch has a retest date assignment rather than an expiry date.

There are two considerations unique to small-molecule DS planning, determining the batch scale and accounting for raw materials. Selecting the batch scale (e.g. the desired batch size) has strategic implications for the program. Ideally, the intended batch sizes are sufficient to cover all clinical demand requirements, to provide a healthy amount of DS for process development at both the DS and DP levels, and to cover potentially unexpected events (e.g. changes to the clinical development plan). It's also typically a smart strategy in clinical development to plan for at least two batches of DS within each production campaign. This is to reduce both inventory and financial exposure in the event one of the batches fails during the manufacturing process. If a batch fails, then there are still raw materials available to make the second batch after identifying and remediating the cause of failure. One common strategy is to start with a relatively small scale (e.g. 5–10 kg) during early-phase development to save on costs and then move up to larger batch sizes to meet increasing demand and to drive the cost of goods down by achieving economies of scale. With that said, it's always wise to have at least some DS on hand during clinical development. As data emerges from ongoing trials and manufacturing process information is gained with each DS and DP batch, demand requirements on DS change all the time. Given the lengthy lead times to source raw materials (more on those soon), manufacture the DS, and then get it released for upstream production, planning for at least some extra inventory within each campaign makes sense.

The second consideration unique to small-molecule DS planning is to account for the raw materials used in the DS manufacturing process. Raw materials often take several months to source, and they typically represent the most significant cost component of small-molecule DS manufacturing. Ensuring their availability to support planned DS campaigns is essential. Generally, the same principles used in planning DSs apply to planning raw material availability. BOM quantities, lead times, and cost are all drivers for determining how much to order and when to order raw materials to support DS production. Like DSs, raw materials also have economies of scale so it can be a smart strategy to stockpile raw material inventories to two or three times the required amount to support the next scheduled DS manufacturing campaign. In some organizations, clinical supply chain is responsible for planning raw materials, while in others, this responsibility lies with the DS lead to plan and organize. In either situation, the responsibilities for raw material planning and sourcing need to be clear so raw materials do not become a bottleneck to the DS plan.

12.8 Large-Molecule Drug Product Planning

DP manufacturing for biologics is also known as a fill-finish operation because the DS is combined with excipients, sterilized, and filled into a glass container such as a vial or ampoule. Secondary packaging and labeling operations strive for a 99–100% production yield of DP to FP. It's usually safe to assume a very high yield for the secondary packaging step because of its simplicity. DP BOM quantities for secondary packaging are also quite straightforward with a 1:1 kit configuration as the recommended option for weight-based dosing to reduce product waste as much as possible. Other common configurations are 2:1 (e.g. 2 vials per FP kit) or 3:1 (e.g. 3 vial per FP kit), and these often make sense when planning for a total desired dose.

Upstream, DP production yields are highly variable and influenced by the biochemical properties of the DS (e.g. level of product viscosity) and sterilization/filtration steps within the manufacturing process. Losses during fill-finish operations are known as line losses, a reference to the manufacturing lines upon which DP manufacturing takes place. Line loss predictions can have wide ranges depending on the product characteristics and equipment trains used, so it's important to track these via the DP lead on the program from batch to batch as GMP manufacturing experience is accumulated.

In addition to line losses, analytical testing requirements often consume many units of the completed DP batch. Samples needed for analytical release testing and for stability take hundreds of vials/ampoules/syringes away from the final batch quantity. Analytical samples are often consistent from batch to batch, while stability samples may or may not be consistent. Even when DP manufacturing inputs are well documented, unexpected events may also occur during manufacturing that lead to unexpected line losses and/or to larger-than-expected sample pulls. It's therefore important for the clinical planner to document each of these planning assumptions both prior to production and following completion of each batch to determine if changes to the planning inputs need to be made.

Fortunately, large-molecule filling lines rarely have capacity constraints and can fill as many vials as needed based on the amount of DS available to fill.

In Figure 12.7, the same FP production orders for 500 kits in Week 14 and 200 kits in Week 19 generate a set of DP order requirements for 506 vials in Week 8 and 203 vials in Week 13 based on the FP production yield of 99% and the one kit sample retain needed as part of each FP batch. There is an

		W1	W2	W3	W4	W5	W6	W7	W8	W9	W10	W11	W12	W13	W14	W15	W16	W17	W18	W19	W20
FP	FP SKU 1 Order finish (kits)														500						
FP	FP SKU 2 Order finish (kits)																			200	
DP	FP SKU 1 Order requirement (vials)								506												P
DP	FP SKU 2 Order requirement (vials)													203							
DP	Other DP requirements (vials)								500												
DP	Total DP requirements (vials)								1006					203							
DP	OHI	0	0	0	0	0	0	1500	494	494	494	494	494	291	291	291	291	291	291	291	291
DP	Production actual	0	0	0	0	0	0	1500	0	0	0	0	0	0	0	0	0	0	0	0	0
DP	Production plan	0	0	0	0	0	0	1209	0	0	0	0	0	0	0	0	0	0	0	0	0
DP	Safety stock value	1	1	1	1	1	1	1	1	1	1	1	1	1	1	1	1	1	1	1	1
DP	Order target value	26	26	26	26	26	26	26	26	26	26	26	26	26	26	26	26	26	26	26	26
DP	Batch: DP1A EXP: W41 ESD: W23	0	0	0	0	0	0	1500	494	494	494	494	494	291	291	291	291	291	291	291	291

SKU description: Wonder drug 20 mg/mL, 5 mL vial
FP production lead time: 6 weeks
DP capacity constraint: N/A

DP:FP bill of material quantity: 1 vial
FP production yield: 99%
DP expiry window: 18 weeks

Figure 12.7 Large-Molecule Drug Product Production Plan Table

additional requirement of 500 vials in Week 8 to account for analytical release testing and stability samples. A DP production order of 1500 vials has been scheduled in Week 7 leaving 291 vials left over. The production order provides this 291-vial cushion in case of unexpected line losses or in the event a higher number of analytical samples are required due to issues in the manufacturing process.

There is another important consideration to keep in mind when planning DP material which is the principle of comparability. Comparability was alluded to in Chapter 5 – Regulatory Considerations and is defined as demonstrating a product's equivalence to previous batches after a process change. Comparability is especially important in large-molecule DP manufacturing because seemingly minor changes to the manufacturing process, equipment, or container closure system can impact the potential expiry period assignment of the DP. When competent health authority bodies analyze and interpret comparability assessments that are included as part of regulatory amendments, they can limit the expiry period of the new product/process if they feel comparability has not been suitably established. If comparability is not deemed appropriate, the clinical planner needs to incorporate this into the FP supply plan for both production and distribution and at the DP level. Accounting for comparability enables appropriate and compliant DP allocation decisions by the planner. For example, if the original expiry period is 24 months, and a CHA decides comparability has not been established and limits the expiry period to 18 months for the newer batch, this 18-month limitation needs to be reflected at the DP level to avoid assigning a 24-month shelf life to a batch destined for that country.

12.9 Large-Molecule Drug Substance Planning

Large-molecule DS manufacturing (e.g. bioprocessing) is broken up into two stages, upstream processing where cells are grown with the intent of expressing the protein of interest, and downstream processing where the cell culture is purified to extract the DS of interest. After production finishes, the bulk DS is frozen to maintain it for future DP manufacturing campaigns. The entire process is highly complex, takes months to complete, and production yields can vary dramatically. Therefore, careful analysis is required to determine the BOM quantity for DP manufacturing and to estimate production yields from the DS manufacturing process.

Calculating the DS BOM quantity for DP manufacturing is based on two variables, the DS concentration and the container closure volume size. The DS concentration is the amount of DS mass (e.g. milligrams) contained within a specified volume (e.g. milliliter) of liquid, and it's measured by dividing the mass into the volume (e.g. 100 mg/1 mL). Container closure sizes are typically measured in milliliter fill volume, so to obtain the BOM quantity for a single vial of DP, the following formula is used:

$$\text{DP BOM quantity}\left(\text{mg/unit}\right) = \text{DS conc.}\left(\text{mg/mL}\right) \times \frac{\text{DP fill volume}\left(\text{mL/unit}\right)}{\text{DP yield}\left(\%\right)}$$

This formula converts the concentration and fill volume into a mass-based output equivalent to the DS unit of measure (e.g. grams). For example, if the DS concentration is 10 mg/mL and the fill volume is 10 mL/vial, then the DS per vial quantity before accounting for line losses in the DP process is 100 mg/vial which is equal to .1 g/vial of DS. Let's assume a DP yield of 75% which brings the final DP BOM quantity to 0.1333 g/vial.

Estimating the production yield for a DS campaign also requires multiple inputs. Where small-molecule DS manufacturing is relatively predictable, bioprocessing campaigns tend to produce a wider range of yields because there are more variables at play. Small-molecule production is a sequence of chemical reactions with precise quantities of raw materials, reagents, and solvents used as ingredients to the operation. Large-molecule production relies on growing living cells that excrete proteins in a media composed of amino acids, vitamins, and other ingredients designed to facilitate cellular growth. Then, the media goes through multiple filtration steps to remove cell fragments, aggregates, host proteins, and viruses. There are more elements within bioprocessing that can go awry leading to unexpected outcomes in cellular production, protein expression, and/or recovery yields post-purification.

To calculate the production yield, the following inputs are needed:

- *Bioreactor Size (L)*: Total size of the bioreactor as measured in liters. Common clinical-scale bioreactor sizes are 500 L, 1000 L, and 2000 L. The largest commercial scale reactors have a 25,000 L scale.
- *Product Titer (g/L)*: Measures how much DS is produced within a single liter. Products with excellent expression have titers that range from >10 g/L, while poorly expressing products have a titer <1 g/L.
- *Harvest Yield (%)*: The percentage of DS retained from the cell culture and harvest upstream processes. Typical harvest yields range from 80% to 95%.
- *Purification Yield (%)*: The percentage of DS retained from the purification process. Typical recovery yields range from 20% to 50%.

Multiplying these four inputs together gives the estimated production yield for a DS batch:

$$\text{DS yield}\left(\text{g}\right) = \text{reactor size}\ \left(\text{L}\right) \times \text{titer}\left(\frac{\text{g}}{\text{L}}\right) \times \text{harvest yield}\left(\%\right) \times \text{purification yield}\left(\%\right)$$

For example, if a production process takes place in a 1000 L tank, has a 2 g/L titer, a 95% harvest yield, and has a 40% purification yield, then the estimated DS batch size is 760 g of DS. This formula can be taken a step further to estimate the number of DP vials yielded from a DS batch:

$$\text{DP yield}\left(\text{Number of units}\right) = \frac{\text{DS yield}\left(\text{g}\right)}{\text{DP BOM quantity}\left(\text{g/unit}\right)}$$

With an estimated DS batch size of 760 g and a DP BOM quantity of 0.1 g/vial, then the DS batch is estimated to yield 7600 vials. What happens though if the titer of a batch is only 1.7 g/L and the purification yield is 30%? Then, the DS batch size reduces to 485 g and a total DP yield of 4850 vials, an almost 40% reduction from the previous estimate. These large swings in DP yields are not uncommon in clinical phase DS manufacturing, so this is a compelling reason to closely monitor all the inputs listed above on a per batch basis. As more batches are manufactured and process understanding improves, yields tend to become more predictable. But, even this assumption is complicated by manufacturing process development activities and changes to equipment or scale, all of which can impact DS and DP yields.

Sampling is also required in biologics DS manufacturing. Both routine samples for analytical testing and nonroutine samples for stability testing need to be incorporated into the plan, and other one-off sampling requests for DS material are also common. However, due to manufacturing process complexity, DS sampling often varies significantly from batch to batch, especially when issues develop in production that require adjustments from the MBR which in turn require additional samples to be pulled for in-process testing. As an alternative to predicting specific sampling requirements, sampling losses can be incorporated into the harvest and purification yield assumptions. For example, if the harvest yield is estimated at 95% before sampling is accounted for, then it can be reduced to 90% to account for in-process cell culture sampling. Or if the purification yield is estimated at 40%, it could be reduced to 35% to account for stability sampling.

DS manufacturing is constrained to the bioreactor size in which it takes place, but this is captured in the DS yield filed above, so it does not need to be called out as a separate assumption like it does in small-molecule planning.

The upstream production plan table in Figure 12.8 incorporates the DP production lead time offset and the DP BOM quantity calculation to arrive at the first DS order requirement of 200 g in Week 5. The DP order finish date of Week 7 is offset to a Week 2 production start date and the quantity is translated from 1500 vials to 200 g of DS. The DS:DP BOM quantity is calculated by multiplying the DS concentration (20 mg/mL) into the DP fill volume (5 mL) and then dividing by the DP production yield percentage (75%) to arrive at the 0.1333 g quantity. This quantity is then multiplied into the 1500 vial requirement which equals the 200 g requirement.

		W1	W2	W3	W4	W5	W6	W7	W8	W9	W10	W11	W12	W13	W14	W15	W16	W17	W18	W19	W20
DP	DP SKU 1 Order finish (vials)							1500													
DS	DP SKU 1 Order requirement (G)		200																		
DS	Other DS requirements (g)		0																		
DS	Total DS requirements (g)		200																		
DS	OHI	1575	1375	1375	1375	1375	1375	1375	1375	1375	1375	1375	1375	1375	1375	1375	1375	1375	1375	1375	1375
DS	Production actual	1575	0	0	0	0	0	0	0	0	0	0	0	0	0	0	0	0	0	0	0
DS	Production plan	200	0	0	0	0	0	0	0	0	0	0	0	0	0	0	0	0	0	0	0
DS	Safety stock value	1	1	1	1	1	1	1	1	1	1	1	1	1	1	1	1	1	1	1	1
DS	Order target value	26	26	26	26	26	26	26	26	26	26	26	26	26	26	26	26	26	26	26	26
DS	Batch: DS1A retest date: W92	1575	1375	1375	1375	1375	1375	1375	1375	1375	1375	1375	1375	1375	1375	1375	1375	1375	1375	1375	1375

DS SKU description: Wonder drug drug substance

DS requirement assumptions:
DP production lead time: 5 weeks
DP formulation: 20 mg/mL, 5 mL vial
DP production yield: 75%
DS:DP bill of material quantity: 0.1333 g

DS yield assumptions:
DS bioreactor capacity: 1000 L
DS titer: 5 g/L
Harvest yield: 90%
Purification yield: 35%
DS expiry window: N/A

Figure 12.8 Large-Molecule Drug Substance Production Plan Table

On the supply side, the DS bioreactor size (1000 L), titer (5 g/L), harvest yield (90%), and purification yield (35%) are multiplied together to arrive at the estimated DS batch size of 1575 g. The bioreactor size represents the capacity limitation, so the production order reflects this predicted batch size. Also note how there is no expiry window. While the DS production lead time isn't listed in the table, the DS production start date is derived by offsetting the DS production order delivery in Week 1 by that lead time. If the expiry period is 24 months (e.g. 104 weeks), and the retest date is Week 92, this implies a DS production lead time of 12 weeks.

One area where biologics DS supply planning differs from small-molecule planning is the ability to pool (e.g. combine) more than one DS batch into a single DP batch. Several factors combine to make this strategy attractive. First, large-molecule DSs are very expensive to manufacture so wasting even a small portion of a batch is undesirable. Second, while DP manufacturing lines typically don't have capacity constraints, it's not economical to make small batches because line losses and analytical requirements eat up a higher percentage of the batch as the DS input size decreases. In other words, it's highly desirable to make larger batches where possible. Finally, DS production is highly unpredictable in clinical development, especially during early-phase production, when a handful of batches or fewer have been made and process knowledge is relatively low. All these reasons combine to produce situations where multiple DS batches may be available to support DP manufacturing. Instead of making small batches where one DS batch goes to one DP batch though, it makes more sense to pool multiple DS batches into a single DP batch.

12.10 The Upstream Production Plan Summary

The upstream production plan summary is an output of the demand and supply planning process and contains similar information to that found in the FP production plan summary. However, there are a few new elements specific to upstream planning. At minimum, the upstream production plan summary contains the following:

1) SKU description
2) Manufacturing site
3) Batch number
4) Quantity
5) Unit of measure
6) Date of manufacture
7) Further processing date (if applicable)
8) Quality release date
9) Downstream allocation (if applicable)
10) Upstream allocation (if applicable)
11) Process version (if maintained)

The upstream production plan summary in Table 12.2 contains a wealth of valuable information about the upstream supply plan strategy. First, the table captures basic batch information including SKU descriptions, batch numbers, production sites, quantities, and manufacturing dates. But, it also tells a story about the manufacturing lead time (e.g. the difference between the date of manufacture and the further processing date) and the total production lead time (e.g. the difference between the date of manufacture and the full quality release date). As a reminder, if a batch needs to be shipped under quarantine for further processing prior to full quality release, the clinical planner must work with the CMC team to identify the activities required and lead time between the date of manufacture and the further processing date.

Table 12.2 Example Upstream Production Plan Summary

DP SKU	Mfg. Site	Batch No.	Qty	UoM	DoM	Further Process Date	Quality Release Date	Downst. Allocation	Upst. Allocation	Process Version
50 mg Tablet	DP CMO 1	DP1A	20,000	Tablets	1-Jun-2025	15-Jul-2025	15-Aug-2025	PP1A PP2B	DS1A	DS1.0 – DP1.0
50 mg Tablet	DP CMO 1	DP2B	20,000	Tablets	1-Dec-2025	15-Jan-2026	15-Feb-2026	PP3C	DS1A	DS1.0 – DP1.0
50 mg Tablet	DP CMO 1	DP3C	20,000	Tablets	1-Jun-2026	15-Jul-2026	15-Aug-2026	PP4D	DS1A	DS1.0 – DP1.0

The downstream and upstream allocation strategies are also critical when sharing the production plan with the CMC team and the manufacturing site. It's important for quality, regulatory, and the technical leads to understand which batches are planned for use in downstream batches so there is traceability to the FP level, and it's important for the manufacturing site to know which upstream batches are intended as inputs for the SKU in question. Lastly, the process version documents how each batch is categorized based on the regulatory attributes of that batch.

One field missing in this report is the expiry period and/or retest date assignment of the SKU in question. This is an optional field because the expiry period for DP and DS materials is usually a moving target. Some teams may find it helpful to include this field in the table, but it's not required to enable DP or DS manufacturing, and there are usually other quality-controlled documents (e.g. specifications) that list the most current expiry period. Keep in mind that whether expiry or retest dates are included in the summary table or not, monitoring and updating batch-specific expiry dates remain a critical component of the planning strategy and the upstream production plan tables.

12.11 Upstream Production Order Readiness

Like FP SKUs, upstream production order readiness is necessary to ensure the supply plan is executed on time and in full. DP and DS manufacturing campaigns require more lead time to prepare than FPs do because raw materials often take a long time to source (e.g. up to 12–18 months for some biologics DS components) and because production capacity is often booked out months in advance. Therefore, an appropriate planning horizon for upstream materials is at least a year ahead of the desired release date. The beauty of taking an end-to-end approach in clinical planning is that the entire supply plan through the duration of the clinical program is available to share which means manufacturing capacity can be secured far in advance of when it's needed.

Given these long order readiness lead times, upstream manufacturing sites are also much less agile and flexible than FP manufacturing sites. From a timing perspective, if a FP order is delayed due to DP availability, a component sourcing issue, or a labor shortage, those jobs can usually be rescheduled within weeks. If similar delays occur for upstream jobs, then the order may not be rescheduled for several months. And, from a quantity perspective, upstream production quantities often can't be changed after a specific point in time because of the raw material lead times mentioned in the previous paragraph. It's therefore imperative for the clinical planner to understand and abide by the upstream planning time fence. The planning time fence is the amount of time from the production order release date past which the quantity and timing of the order should not change. If changes are made to increase order quantities or pull in the desired release date of the batch within the planning time fence, then the manufacturing site will likely not be able to accommodate them. FP time fences are often measured in weeks while upstream time fences are measured in months.

When upstream manufacturing is outsourced, aligning on which party will source and provide raw materials is another key to success. In DP manufacturing, it's common for the contract development and manufacturing organization (CDMO) to source excipients and packaging materials, so it's advisable for the sponsor to not only confirm which materials are planned for use but also confirm those materials have been secured in sufficient quantities to support the upcoming campaign. Unfortunately, raw material sourcing can also be unpredictable which in turn impacts the quantity and/or timing of the upstream batch. This is the primary reason why building upstream inventory well ahead of need is a smart strategy.

Order readiness and alignment is also necessary with internal stakeholders. The entire CMC team should understand the manufacturing, analytical, and quality release strategy for each batch. If a batch needs further processing under quarantine (e.g. prior to full release of the underlying DS or DP), time is usually critical. The manufacturing site and quality team typically have a minimal set of activities that must be completed to release the batch for further processing. Standard batch release activities and timelines are also important to document and monitor to ensure the upstream supply chain flows smoothly. An effective and efficient business process is needed to ensure production order readiness and supply plan alignment for all internal and external stakeholders. We'll return to this topic in Chapter 15 – Clinical Supply Chain Planning Processes.

12.12 Key Takeaways

1) Upstream supply planning encompasses PP, DP, and DS materials. This type of planning is fundamentally different from FP supply planning because upstream materials are components of a downstream material, and therefore demand on upstream materials is dependent on the production plan of the corresponding downstream material. However, the approach to planning upstream materials also shares similarities with FP planning including use of the min–max inventory management methodology and incorporation of expiry planning principles.
2) Demand planning for upstream materials starts with SKU categorization. Five inputs are needed to build the upstream demand forecast including downstream production orders for which the upstream material is a component, the production lead time of the downstream material, the BOM quantity of the upstream material, the production yield of the upstream material, and any other sources of demand on the upstream material. Production yields and production lead times are the inputs with the highest variances, so these require careful monitoring to keep them current within the plan.
3) Supply planning for upstream materials relies on the same min–max inventory management methodology used in FP supply planning. Capacity constraints and expiry management are two additional cornerstones of upstream supply planning. Production planning follows a step-by-step process that includes generation of production order requirements, conversion of production order requirements into actual production orders, translation of orders into batch-specific information, and allocation of upstream batches to downstream production orders.

4) There are two approaches to measuring weeks-of-supply for upstream materials. Standard weeks-of-supply measures the number of weeks until the upstream batch is consumed in downstream production while implicit weeks-of-supply measures the amount of upstream inventory against FP demand via the FP implicit total run rate. Standard weeks-of-supply measurements are most relevant when no safety stock is carried while the implicit weeks-of-supply measurement is most appropriate when upstream safety stocks are held to protect against supply loss.
5) Developing an optimal upstream supply plan requires a balance between holding inventory in case of a supply shock (e.g. supply chain resiliency) and minimizing the amount of cost investment involved to do so. Strategic drivers for this balancing process include confidence in the demand forecast, the expiry period of the upstream material, the cost of the upstream material, the level of supply chain agility, the level of supply chain reliability, and the desired level of resiliency.
6) Small-molecule PP is a relatively simple manufacturing process. Production yields are often high, but sampling requirements are especially important to account for because PP units are required to support stability studies. PP batches should also never be combined into a single FP batch because batch-specific accountability is lost leading to multiple batches getting implicated in a quality event.
7) Small-molecule DP manufacturing is a relatively complex manufacturing process. Production yields typically range from 60% to 95%, but they often improve over time as process experience is gained. Capacity constraints are also routine in small-molecule DP manufacturing. Both PP and DP batches also have an expiry window where batches should no longer be allocated to FP batches because there isn't enough time left on expiry for the product to be consumed after FP manufacturing completes and RTSM parameters are applied.
8) Small-molecule DS manufacturing is a complex manufacturing process. Fortunately, yields tend to become predictable with process experience. Determining the optimal scale for DS manufacturing is a key planning consideration so that production both meets clinical needs and leaves some inventory on hand to service other needs. Raw materials planning is also unique to DSs, and ensuring raw material availability at a reasonable cost is another essential element of managing the supply chain.
9) Large-molecule DP manufacturing is a complex manufacturing process. Line losses impact the eventual production yield as do sampling requirements for analytical testing and stability studies. Establishing comparability between DP lots is another important consideration when process changes at the DS or DP level have been implemented between production campaign.
10) Large-molecule DS manufacturing is the most complex manufacturing process. Determining production yields requires multiplying the bioreactor size, product titer, harvest yield, and purification yield together. Sampling requirements can vary significantly from batch to batch as well. However, unlike their small-molecule counterparts, it's common practice to combine or pool multiple DS lots into one DP lot to maximize the use of the material in the clinic.
11) The upstream production plan summary captures the supply plan for an upstream material including a description of the SKU, the manufacturing site, key batch information, manufacturing and release dates, and the allocation strategy. This is important information to align both internal and external stakeholders around the supply plan.
12) Upstream production order readiness is necessary to ensure raw materials and plant capacity are available to service the supply plan. Abiding by planning time fences, the amount of time from the production order release date past which the quantity and timing of the order should not change is important to ensure planned orders are delivered by the manufacturing site.

13

Advanced Concepts in Demand and Supply Planning

13.1 Advanced Concepts in Demand and Supply Planning

So far, we've introduced the basic elements of clinical trial demand and supply planning: finished product (FP) demand planning, FP supply planning, expiry planning, distribution planning, and upstream drug product (DP) and drug substance (DS) planning. These concepts are applicable to almost every clinical program. However, clinical trial designs are rich with diversity. Therefore, a variety of more advanced planning concepts and techniques are needed to navigate complex studies. Chapter 2 – Introduction to Clinical Trials presents a summary of clinical trial types and their design elements and describes the trend toward increasing trial design complexity and flexibility within to meet a study's desired endpoints. Elements such as variable treatment durations (e.g. not fixing how long a subject will remain on treatment to a specific number of weeks), switching from one treatment arm to another, dose titrations (e.g. increasing the dose), dose reductions (e.g. reducing the dose), and offering treatment extension options have become more common in therapeutic areas such as oncology, virology, and neurology.

Each of these design elements makes the demand forecasting process more uncertain and less predictable. If it can't be intuited from the protocol how long a subject will take a drug or even which drug they'll take, then precisely estimating individual subject demand becomes much more difficult. The natural reaction when there is significant uncertainty is to increase the demand forecast at each stockkeeping unit (SKU)–depot combination. Simply predict that each subject always doses on each available treatment arm, and the uncertainty problem is solved. Unfortunately, reducing uncertainty by artificially increasing demand leads to a bloated supply chain full of excess inventory because in reality, each subject takes only one treatment at a time. Therein lies the challenge for the clinical planner when it comes to managing highly complex study designs – to protect against demand uncertainty and to prevent stockouts while also minimizing excess inventory and waste.

We'll examine this challenge from both the demand and supply sides of the equation through the remainder of this chapter. There are two general approaches the clinical planner can take when faced with demand uncertainty. Either the plan attempts to model demand accurately as it will unfold in the real world or it includes unrealistic assumptions that inflate demand to drive greater amounts of inventory to hedge against potential stockouts. Traditionally, the industry has favored the latter, which is a more conservative approach to demand forecasting that features relatively large overages (e.g. 20–50%) and/or planning subjects to dose on two arms at once to ensure the forecast drives supply for multiple treatments. We'll argue that taking the former approach and focusing on accuracy is a more beneficial approach. Following this discussion of accuracy versus

Supply Chain Planning for Clinical Trials – A Practical Guide, First Edition. Ryan Mills.

conservatism in the forecast, we'll cover two additional techniques the clinical planner can use to improve forecast accuracy, aggregate forecasting and incorporating attrition into the plan.

Notice that the most complex forecasting scenarios such as variable durations of treatment and SKU switching (e.g. dose titration and reduction) are not explicitly identified from a pure demand management perspective. That's because the best way to manage demand uncertainty is not through increasing the forecast, but via a holistic inventory management strategy that includes randomization and trial supply management (RTSM) settings, depot stock, and upstream supply. An important concept for the clinical planner to know is that trial design complexity and flexibility (e.g. variable durations of treatment; SKU switching) come at a price, namely waste. It's not possible to have a 100% service level and manage waste levels down to near zero when multiple treatment options need to be available to all subjects at any point during the trial. Rather, the goal is to reduce waste as much as possible, and this is achieved by actively managing inventory levels as the trial and program progress.

After a deeper analysis of how to use the supply plan to manage demand forecast uncertainty, we'll introduce demand and supply planning techniques for the following scenarios:

- Variable durations of treatment
- SKU switches
- Extension phase studies

We'll conclude the chapter by looking at how to segregate site inventory from depot inventory for better visibility in the supply plan, by discussing inventory pooling as a supply strategy, and by introducing the low-case–high-case forecasting technique. Low-case–high-case planning gives the clinical planner a tool to assess the FP production plan under multiple demand scenarios and enables the planner to optimize the plan so that it reduces waste in a low-demand scenario while simultaneously avoiding stockouts in a high-demand scenario. Low-case–high-case planning also gives the clinical planner a tool to set optimal safety stock and order target values through each level of the supply chain. In the final section of the chapter, we'll briefly review deterministic and stochastic modeling techniques and the benefits of each approach.

13.2 Accuracy or Conservatism in Demand Planning

As introduced in the previous section, there are two general strategies in clinical trial demand planning, and these can be thought of as forecasting accurately and forecasting conservatively. First, it must be said that attempting to forecast accurately does not mean the forecast will be accurate! In Chapter 8 – Finished Product Demand Planning: Basic Concepts, the dynamism of the demand forecast for clinical trials was covered extensively. The demand forecast is never 100% correct at predicting when and where subjects enroll, and which arm they ultimately randomize into. Instead, attempting to forecast accurately means that realistic assumptions about how demand will play out over the course of the trial are used in the forecasting process. Table 13.1 lists some of the assumptions and strategies a clinical planner may employ to plan either accurately or conservatively when setting up their forecast:

The goal when taking an accuracy-based approach is to predict how many FP kits will be consumed by subjects (e.g. total subject demand), and how many kits will remain at clinical sites (e.g. total site demand) and at clinical depots (e.g. total study demand) at the conclusion of the study. This goal matters because the supply plan then produces and distributes the minimum number of kits required to meet the total demand forecast. Conversely, when demand planning takes a

Table 13.1 Accuracy Versus Conservatism in Clinical Trial Demand Planning

Accuracy in Demand Planning	Conservatism in Demand Planning
• Plan in weekly buckets and reflect the exact number of kits dispensed at each visit to most accurately represent how investigational medicinal product (IMP) will be dispensed at each patient visit. • In weight-based studies, take an average subject weight to forecast the number of kits dispensed per visit for all subjects. • In studies with dose titration or reduction schemas, only forecast a realistic fraction (e.g. <25%) of subjects to dose up or dose down unless feedback from clinical science or clinical operations indicates the percentage should be higher. • In studies with SKU-switching schemes, forecast a small fraction (e.g. <25%) of subjects to switch from one SKU to another SKU unless feedback from clinical science or clinical operations indicates the percentage should be higher. • In studies with dose titration, reduction, or SKU-switching schemas, only forecast subjects to take one treatment at each visit rather than multiple treatments at each visit. • Monitor the demand overage multiplier closely over time and reduce overages as subjects enroll in the study. For slow-enrolling studies, consider setting a very low overage (e.g. <10%) or not using an overage multiplier at all. • Actualize demand on at least a monthly basis to assess trends and adjust all aspects of the demand forecast (e.g. subject enrollment, dispensing, titration, and reduction) as needed. This last point is covered extensively in Chapter 15 – Clinical Supply Chain Planning Processes.	• Plan in monthly buckets and round up the number of kits dispensed to the nearest integer to represent how IMP will be dispensed across patient visits within that month. • In weight-based studies, take the highest possible subject weight to forecast the number of kits dispensed per visit for all subjects. • In studies with dose titration or reduction schemas, forecast a large fraction (e.g. >50%) of subjects to dose up or dose down without consulting clinical science or clinical operations on a realistic percentage. • In studies with SKU-switching schemas, forecast a large fraction (e.g. >50%) of subjects to dose up or dose down without consulting clinical science or clinical operations on a realistic percentage. • In studies with dose titration, reduction, or SKU-switching schemas, forecast subjects to take multiple treatments simultaneously. • Set a relatively high-demand overage multiplier (e.g. >20%) and do not update the multiplier as the study enrolls. • Actualize demand infrequently and only adjust subject enrollment. Continue to forecast subjects taking multiple treatments simultaneously.

conservative-based approach, the supply plan automatically produces and distributes a surplus of kits that won't be consumed during the trial because the demand forecast drives that overestimation. The amount of surplus inventory is then further exacerbated when multiple conservative techniques are utilized within the same forecast (e.g. rounding up demand and using higher overages). It's then exacerbated again by the tendency of the contract research organization (CRO) and clinical operations to take an aggressive stance on the enrollment rate projections. Taken together, conservative techniques have a multiplicative effect that inflates demand far beyond what will be consumed during the trial.

Inflating the demand forecast also increases the potential for the bullwhip effect within the supply chain. The bullwhip effect is a phenomenon where changes in demand at the FP level lead to amplified demand signals for upstream materials. This amplification increases with each upstream step in the chain. In other words, a small change in demand at the FP level leads to a larger change in the production order strategy to fulfill that change in demand. The bullwhip effect is most pronounced in clinical trial supply chain management during the study planning phase where conservatism at the FP level drives larger than necessary DP and DS orders. Over time as demand fails to materialize, excess inventory goes to waste, and replacing expiring material creates more demand for DP and DS.

From the supply side of the equation, it's also been a hallmark of clinical trial supply chain management to artificially increase inventory beyond what the supply plan calls for to meet demand. This buffering approach can occur at every level of the supply chain, from overseeding clinical sites to maintaining large pools of FP, DP, and DS inventory. When conservative demand planning meets conservative supply planning, the result is huge unnecessary waste at the end of the trial or program. For this reason, attempting to forecast demand as accurately as possible and then implementing a smart, holistic inventory strategy reduces waste without compromising sufficient on-hand inventory to prevent stockouts.

13.3 Aggregate Forecasting

One of the best techniques to employ to forecast demand accurately and reduce the bullwhip effect is to place all subject demand at the depot location with the highest expected subject enrollment for as long as reasonably possible. This approach is known as aggregate forecasting. Aggregate forecasting utilizes the demand forecasting process from Chapter 8 –Finished Product Demand Planning: Basic Concepts. Then, the clinical planner identifies which central or regional depot is likely to enroll the most subject demand over the course of the study and populates all subject demand at that SKU–depot location. This is usually the central depot location, but it could also be at a high-enrolling regional depot. Meanwhile, the site activation forecast for each depot is populated, and the initial depot stocking strategies introduced in Chapter 11 – Finished Product Supply Planning: The Distribution Plan are applied to calculate initial depot stocking quantities at each SKU–depot combination. Only when subjects begin screening at active sites within the SKU–depot location should the clinical planner begin moving subject demand from the central depot location to the appropriate regional depot location.

This approach differs from attempting to forecast country-level enrollment and randomization before clinical sites are active within a country or region. It's often the case that the CRO responsible for conducting the study and/or clinical operations provide country-level estimates, and it's tempting for the clinical planner to take these and start populating country or regional subject demand into the forecast. However, forecasting depot-level subject demand without any activity at the sites or depots is less effective than aggregate forecasting because forecasting at the country or regional level is inherently more unpredictable than consolidating demand out of a single depot location.

Forecasting at the country level requires predicting three variables: when the subject will enroll, which arm of the study they will enroll into, and where the subject will enroll. Consolidating demand out of one depot removes one of these variables, the location of enrollment, and this results in lower demand fluctuations within the forecast from week-to-week. Reducing these fluctuations is valuable because this reduces the bullwhip effect from regional depot to central depot. If subject demand is included at the regional depot early in the study start-up process, the demand forecast out of that depot increases or decreases as subject enrollment changes, which in turn leads to changes in the distribution plan. These changes in the distribution plan then cascade back to the FP production plan. For example, if 10 subjects are populated at a SKU–depot location, and then clinical operations increase the forecast to 20 subjects, this creates an increased requirement out of the central depot, which then leads to an increased production order requirement to satisfy the larger depot stocking requirement. However, total demand for the study has not changed, and it's possible that the forecast may be revised downward in future weeks, and then the additional inventory sent to the regional depot ends up as waste.

There are two other advantages to aggregate forecasting. First, forecasting demand out of a single depot enables greater flexibility within the supply plan because FP inventory is held at the central depot for longer. Once material is sent to a clinical site, it cannot be returned to the regional depot, and once material is sent to a regional depot, it is difficult (and wasteful) to return it to a central depot. Holding inventory at the central depot maximizes flexibility within the plan because it has not yet been committed to a downstream node (e.g. regional depot or clinical site). Second, maintaining subject demand forecast in one SKU–depot location is far more resource efficient than maintaining it in several SKU–depot locations before enrollment begins.

Aggregate forecasting should be adhered to for as long as possible within the demand plan. A few factors influence the timing of when to move subjects to their actual depot:

- *Number of Subjects in Screening*: If a high number of subjects enter screening simultaneously at sites supplied from a regional depot, then some subject demand should be moved to that depot location ahead of randomization.
- *Screening Window Duration*: If the screening window is relatively short (e.g. <14 days), then the clinical planner should consider moving subjects over as soon as they enter screening.
- *Screen Failure Rate*: If the screen failure rate is relatively low (e.g. <30%), then demand can be moved over during the screening or prescreening process, but if screen failure rates are high, then waiting for subject randomization is more sensical.
- *RTSM Strategy and Depot Stocking Strategy*: If depots are stocked with sufficient inventory to cover all site activations and additional inventory is available on top of this to protect against speedy screening and enrollment, then holding demand until subjects are in screening or even randomizing makes sense.

Ultimately, the clinical planner needs to use their best judgment on the best time to move subject demand from the aggregate depot to the appropriate regional depot.

Another scenario the clinical planner may encounter is building a forecast when multiple central depot locations are utilized to manufacture and distribute FP. In this situation, aggregate forecasting may still be employed, but the subject demand forecast should be split 50/50 across the two central depot locations. Each regional depot location needs to be connected to the appropriate central depot location, and site activation demand and depot stocking requirements should flow back to a single central depot location. If a regional depot location can be supplied by either central depot location, the clinical planner must take care to only count depot stocking requirements against a single central depot to avoid double counting. This strategy ensures the production plan out of each central depot is sufficient to meet total subject demand for the study plus site and depot stocking requirements from each regional depot.

13.4 Forecasting Subject Attrition and Subject Replacement

Attrition in the context of clinical trial demand forecasting is defined as the annual rate at which subjects drop out of the study over time. While most clinical trials experience some degree of subject attrition, the rate at which this occurs varies widely depending on the nature of the trial. Studies that are relatively short in duration (e.g. Phase I or Ib) or have a relatively short duration of treatment (e.g. single dose) or take place in nonterminal disease may see little to no attrition. Other studies that are relatively long in duration (e.g. Phase II or III) or have a long duration of treatment (e.g. months to years) or take place in life-threatening diseases typically see higher attrition rates.

As the attrition rate increases, the greater the impact it has on the demand forecast via lowering actual demand against the forecast. Thus, it's important to incorporate the attrition rate into the clinical trial forecast where high attrition rates (e.g. >10% per year) are expected. To estimate the attrition rate for a study, the clinical planner should first look for historical data from past trials in similar indications and of similar duration. This analysis provides a starting point from which the attrition rate can be further refined. Consulting with the clinical science and clinical operations on the dropout assumptions used in their planning is also recommended both for their respective experience and to see if any attrition mitigation strategies will be utilized in the study. Finally, like the enrollment rate, the attrition rate needs to be carefully monitored through the duration of the trial, and the clinical planner should adjust the rate up or down as actual attrition data comes through from the trial.

There are two general approaches to implementing an attrition rate into the forecast. The first is to attempt to guess at the individual level which subjects will drop out and when the drop out will occur. This is not recommended because it's pure guesswork, and, in larger studies, it's time consuming to continually make updates at the subject level. A better way to incorporate attrition is to apply it at the total subject demand level. To do this, rows are added to the demand forecast table introduced in Chapter 8 for the year number, the attrition rate, and the updated total subject demand after accounting for attrition. The year number should begin its count when the first subject is dosed on that SKU. An example demand forecast table with attrition is presented in Figure 13.1.

In this example, it's assumed that no subjects drop out in the first year of the study, but from Year 2 on, attrition compounds annually by the 15% annual rate. Total subject demand with attrition (TSD_A) is calculated via the formula below:

$$\text{Subject demand with attrition} = \text{subject demand} \times (1 - \text{attrition rate})\,^\wedge\,\text{year}$$

Note that taking an exponential of the year number against the inverse of the attrition rate creates the desired compounding effect. An alternative to this approach is to apply the attrition rate to the first year as well, but this could lead to under-forecasting demand. However, this approach is OK when sufficient FP safety stock exists at the depot to mitigate this potential risk.

There is another attrition-related study design element to consider which is whether the trial calls for replacement subjects if a subject drops out before enough data is captured to include them in the study analysis. In the event, replacement subjects are included in the study design, the clinical planner should forecast no attrition within the study. When a subject attrits, that subject's

	W50	W51	W52	W53	W54	W55	...	W104	W105	W106	...	W156	W157	W158	...	W208	W209	W210	...	W260	W261	W262
Total subject demand	80	80	80	90	90	90		100	100	100		110	110	110		120	120	120		130	130	130
Year	1	1	1	2	2	2		3	3	3		4	4	4		5	5	5		6	6	6
Attrition rate	0%	0%	0%	15%	15%	15%		28%	28%	28%		39%	39%	39%		48%	48%	48%		56%	56%	56%
Total subject demand with attrition	80	80	80	77	77	77		72	72	72		67	67	67		62	62	62		57	57	57

SKU description: Wonder drug 50 mg, 30-ct bottle, global label
Depot location: US depot
Attrition rate: 15%

Figure 13.1 Demand Forecast Table with Attrition

individual demand forecast should be closed out, and a new subject added at a future date within the forecast. The supply planning parameters will pick up the new subject's demand as part of future production order requirements to ensure sufficient FP inventory availability.

13.5 End-to-End Supply Planning Strategy

Previous chapters discussed each of the levels within the clinical trial supply chain as discrete steps with their own production and management strategies. We've worked our way from the most downstream node of the chain, the clinical site, to the most upstream node of the chain, DS manufacturing:

- Clinical site = Chapter 6 – RTSM and Its Role in the Clinical Supply Chain
- Regional depot = Chapter 11 – Finished Product Supply Planning: The Distribution Plan
- Central depot = Chapters 9 and 10 – Finished Product Supply Planning: The Production Plan and Finished Product Expiry Planning
- Primary packaging, DP, and DS = Chapter 12 – Upstream Supply Planning: Primary Packaging, Drug Product, and Drug Substance

Within each chapter, strategies have been introduced for how to design optimal RTSM site supply parameters, how to develop depot stocking strategies at regional depots, how to set safety stock values and order targets at the FP level, and how much inventory to hold at the DP and DS levels. But, we have not yet taken a holistic view of how to manage end-to-end inventory. In Chapter 12, the concepts of demand and supply risk were applied to upstream materials. Here, we extend that analysis to the FP level.

As a refresher, upside demand risk is the potential for subject enrollment to happen faster than expected increasing stockout risk while downside demand risk is the potential for subject enrollment to happen more slowly than expected increasing excess waste risk. Supply risk is the potential inability of the supply chain to deliver material consistently to the clinic. As upside demand risk increases, more inventory needs to be held at each FP node of the chain, and, as supply risk increases, more inventory needs to be held at upstream nodes. To capture this, Table 13.2 summarizes demand and supply risks at each node and describes safety stock strategies to prevent these risks from leading to missed doses. Bear in mind that as safety stock values increase, so too does excess waste risk. The clinical planner needs to assess how much risk exists at each of these levels and incorporate that into the production and inventory management strategy accordingly.

Starting at the top of the table, notice how the clinical site is the first line of defense against upside demand risk-related stockouts and how the RTSM predictive threshold resupply model (PTRM) parameters are the key to this protection. In Chapter 6 – RTSM and Its Role in the Clinical Supply Chain, we referenced the demand profile as a key consideration for setting the minimum amount of inventory available at a clinical site. Table 13.3 lists those drivers.

As the sample size and speed of enrollment increase, and as the screening window shortens, the PTRM minimum values should increase relative to the baseline estimates set in Chapter 6. Recall that those baseline values were primarily driven by the number of kits dispensed at each visit and the maximum number of subjects expected to enroll at a site simultaneously. The baseline values need to increase precisely because these variables – increased sample size, increased speed of enrollment, short screening windows – increase the number of subjects who could enroll simultaneously. If SKU switching is unpredictable, this means that subjects can dose titrate, dose reduce, or switch to a new SKU at any time, so minimum levels also need to increase for all available SKUs

Table 13.2 Stockout Risk, Safety Stock Strategy, and Waste Risk within the Clinical Supply Chain

Node and/or Modality	Stockout Risk Category	Stockout Risk Description	Safety Stock Strategy	Excess Waste Risk
Clinical Site	Upside demand risk	Multiple subjects enrolling into a single arm simultaneously depletes inventory Dose titration or dose reduction with insufficient inventory SKU switching with insufficient inventory	Set minimum thresholds for each SKU to protect against a pre-defined number of subjects randomizing or switching into each SKU	Inventory at clinical site expires or goes unused
Regional Depot	Upside demand risk	Multiple sites activating simultaneously and/or misalignment on site supply strategies depletes depot inventory	Set initial depot stocking quantity to cover agreed upon site supply strategies	Initial depot stocking inventory expires or goes unused
		Fast or unbalanced enrollment at clinical sites depletes depot inventory	Set safety stock target at regional depot to cover replenishment lead time to depot	Inventory at regional depot expires or goes unused
		Expiry events at clinical sites depletes depot inventory	Forecast expiry events within the demand plan to drive resupply	
Central Depot	Upside demand risk	Fast or unbalanced enrollment at sites supplied depletes depot inventory	Utilize aggregate forecasting and incorporate depot stocking demand	Inventory at central depot expires or goes unused
		Fast or unbalanced enrollment at sites supplied out of regional depots depletes depot inventory	Set safety stock target to cover replenishment lead time of FP production	
Primary Pack (SM)	Upside demand risk	Fast enrollment at the study level depletes FP inventory	Set inventory strategy based on desired minimum level of weeks-of-supply	Primary pack or DP inventory expires or goes unused
OR	Supply risk	Unexpected loss of a FP or DP resupply batch depletes PP or DP inventory		
Drug Product (LM)				
Drug Substance	Supply risk	Unexpected loss of a DP or DS resupply batch depletes DS inventory	Set inventory strategy based on desired minimum level of weeks-of-supply	DS inventory expires or goes unused

to ensure supply availability for that scenario. An alternative to increasing the minimum quantities is to place more sites at a medium- or high-supply setting in the RTSM, but either way, more inventory is held at clinical sites across the study.

To summarize, as demand speeds up and/or becomes more uncertain, minimum inventory levels at the site must increase to maintain service levels and avoid stockouts. This has the unfortunate consequence of increasing potential excess waste at the conclusion of the study, but this is the

Table 13.3 Clinical Trial Demand Profile and Impact to RTSM Minimum Values

Demand Characteristic	Demand Level	Impact to PTRM Minimum
Total Sample Size of the Study	High	Increases
	Low	Decreases
Speed of Enrollment	Fast	Increases
	Slow	Decreases
Screening Window	Short (<1 week)	Increases
	Long (>1 week)	Decreases
SKU Switching (e.g. Dose Titration and Reduction)	Unpredictable	Increases
	Predictable	Decreases

price of maintaining service levels in the face of demand variability. But, many clinical trials do not fit this profile! Clinical trials are often slow enrolling, and they also tend to enroll a small number of subjects simultaneously at most participating sites within a trial. This is typically due to resource constraints at sites or when the indication is a highly targeted subset of a broader patient population. When trials are slow enrolling, or fewer subjects are expected, or SKU switching is highly predictable, then baseline minimum levels in the RTSM make the most sense.

The same risks that affect how much inventory needs to be held at the site also impact the amount of inventory that needs to be held at regional depots. If sites are at risk of stocking out due to aggressive or unbalanced enrollment, then the regional depot needs to have sufficient supply available to resupply the sites. To accomplish this, the initial depot stocking strategy described in Chapter 11 – Finished Product Supply Planning: The Distribution Plan should be adhered to. Stock the regional depot with enough supply to cover initial site activations and then add supply on top of this to cover anticipated patient demand through the first several weeks of the trial. If the study has an aggressive enrollment profile, then aggregate forecasting may need to be abandoned in favor of adding subject demand directly to the regional depots ahead of patient enrollment to ensure sufficient regional depot inventory is on hand to protect against fast enrolling sites. However, it's wise to use the aggregate forecasting technique when demand is low or slow enrolling to avoid overstocking the regional depot. As subjects enroll and the forecast becomes certain, the supply planning parameters take over and drive the appropriate level of safety stock at the regional depot without the planner needing to add further inventory.

At the central depot level, safety stock levels protect against two outcomes, demand uncertainty from subject demand supplied out of that depot and demand uncertainty at the regional depot level. With respect to subject demand, the safety stock supply planning parameter is designed precisely to protect against this risk. As for regional depot uncertainty, keep in mind that those depots should already have some amount of additional inventory in place early in the trial to protect against demand upside. Therefore, more safety stock does not need to be added on top of the safety stock already in place. When subject demand and distribution demand are accounted for at the central depot level, the clinical planner should trust their safety stock to hold.

The other consideration the clinical planner should incorporate into their inventory management strategy at the FP level is the amount of expiry available. Chapter 10 – Finished Product Expiry Planning introduced the concept of building larger FP batches as expiry improves. This is true when demand is relatively certain (e.g. later in the study when enrollment is locked in) or when the study is expected to be fast enrolling. However, when demand is highly uncertain and the study is expected to be slow enrolling, it makes sense to continue building smaller batches of inventory and stocking depots with relatively small amounts of inventory. This protects against excess waste. Only when demand solidifies should the planner feel confident in increasing the batch size.

Trusting the RTSM thresholds and FP safety stock levels without adding more supply is further supported by the upstream production and inventory management strategy. It's usually wise to carry a minimum of 26–52 weeks-of-supply of primary pack inventory for small molecules or DP inventory for large molecules as insurance against upside demand risk that consumes FP safety stock or against a supply risk such as a lost FP batch or lost DP resupply. In a situation where only downside demand risk exists and the supply chain has low risk, the clinical planner may decide to carry less than 26 weeks of upstream safety stock. Managing DS inventory follows a similar story. As supply chain reliability increases, the amount of DS that needs to be held may decrease, and this is especially true when DS is expensive to make.

No matter the scenario, the clinical planner needs to maintain a flexible approach and regularly assess how much inventory is held at each node of the supply chain to protect against demand uncertainty and supply risk. Always keep in mind that the demand profile changes over time. Typically, more inventory needs to be held at sites and depots early in the study when demand is most uncertain, and some expiry-induced waste due to this strategy may be inevitable. Over time as demand solidifies, the clinical planner should review everything from the demand forecast to the RTSM thresholds to the safety stock levels at the FP, DP, and DS levels to ensure as much supply as possible makes it to the subject and as little as possible ends in the garbage can.

13.6 Demand and Supply Planning for Variable Durations of Treatment

The duration of treatment is the period over which a subject consumes drug during the trial, and it's a foundational element of the subject demand forecast. In many trial designs, all subjects are on drug for a discrete number of days, weeks, or months, which simplifies the demand forecasting process because all subjects are on treatment for the same amount of time. There is another type of design known as an event-driven study where the duration of treatment is not fixed on an individual basis but instead is tied to the observation of a prespecified number of events within the overall study. For example, an oncology trial may define an event as tumor growth within a subject and call for 100 events to occur before the trial concludes. If that trial had a sample size of 200 subjects, then the study would end when 100 of them experienced tumor growth.

From a demand planning perspective, the first step the clinical planner should take when setting up the demand plan for the study is to consult with biostatistics about the estimated point in time at which the number of events will be reached. While it's not possible to predict exactly when this will happen, biostatisticians typically base the sample size of the study on when it's expected the required number of events will take place based on subject enrollment and time on treatment.[1] Additionally, events are tracked as the study enrolls and data is collected from participating subjects. One challenge with predicting when the number of events will be reached in the demand plan is that events require subjects to be enrolled and on treatment, and the timing of this is inherently unpredictable. Holding a regular meeting with the biostatistics lead for the study through the duration of the trial to check on both enrollment and number of events observed is strongly recommended. This feedback helps the clinical planner further hone when the prespecified number of events will be reached and refine the duration of treatment for all remaining subjects.

From a supply planning perspective, the clinical planner should consider the following strategies at each node:

- *Clinical Site*: No change to RTSM thresholds. These don't need to be adjusted upward because the variable duration of treatment does not impact how many subjects will enroll simultaneously.
- *Regional Depot*: After the study is fully enrolled, increase safety stock at each regional depot to double the replenishment lead time for that depot. This ensures that sufficient inventory is on hand if the demand forecast for all remaining subjects is extended following an update from biostatistics.
- *Central Depot*: No change to safety stock levels is needed. Demand from regional depots will drive increased production requirements for FP.
- *Upstream*: Hold a minimum of 6 months of primary pack or DP inventory to support extended subject dosing. Additional FP resupply campaigns siphon off this inventory as needed.

Study designs with variable durations of treatment have more potential for waste than their fixed duration counterparts because of increased uncertainty around when treatment stops. Through collaboration with biostatistics, taking care not to overinflate demand, and managing regional depot supplies carefully, extra waste can be minimized.

13.7 Demand and Supply Planning for Dose Titrations, Reductions, and SKU Switches

Another common clinical trial design element is to call for or allow dose titrations (e.g. increases in dose) and/or dose reductions. Dose titrations and reductions come in two flavors, planned and unplanned. Planned changes in dose are documented in the schedule of assessments, and these are easily accounted for in the individual subject demand forecast. Unplanned dose changes present a more difficult challenge. Designs with dose changes are further complicated when those changes require different FP SKUs to meet the desired doses. In large molecule studies, this SKU complexity is mitigated by planning a single vial per kit design, but in small molecule studies, it's typically the case that higher doses require a higher formulation strength to avoid pill burden. Pill burden occurs when a subject must take more than five capsules or tablets in a single dose.

Demand planning for unplanned dose changes is not easy. The planner should consider forecasting at least some percentage of subjects to increase or decrease per the flexibility prescribed in the protocol, but the difficulty lies in predicting which percentage of subjects will do so. Zero is too low, and 100% is almost certainly too high, but it can feel like anywhere in between is fair game. One recommended approach is to consult with the clinical science team to better understand why dose titrations and reductions are allowed in the study and what their expectations are for how many subjects will follow this path. In some studies, dose titrations are encouraged for all subjects who can tolerate the drug to extract the maximum possible benefit. For a study like this, forecasting a high percentage of subjects to increase their dose makes sense. In other studies, dose reductions are allowed only if a subject can't tolerate the desired dose. If safety data exists from previous studies, this can be used to extrapolate a potential percentage of subjects who won't tolerate the drug and could potentially dose reduce. Historical data of any kind is especially useful in predicting titration or reduction probabilities.

Within the demand plan for the study, careful attention to both the percentage of subjects dosing up or down and the SKU strategy is required. In a scenario where only one SKU is

Subject	W1	W2	W3	W4	W5	W6	W7	W8	W9	W10	W11	W12	W13	W14	W15	W16	W17	W18	W19	W20	W21	W22
1		1	1	1	1	1	2	2	2	2	2	3	3	3	3	3	3	3	3	3	3	3
2				1	1	1	1	1	1	1	1	1	1	1	1	1	1	1	1	1	1	1
3						1	1	1	1	1	2	2	2	2	2	3	3	3	3	3	3	3
4								1	1	1	1	1	2	2	2	2	2	3	3	3	3	3
5										1	1	1	1	1	2	2	2	2	2	2	2	2
6										1	1	1	1	1	2	2	2	2	2	3	3	3
7											1	1	1	1	1	2	2	2	2	2	3	3
8											1	1	1	1	1	2	2	2	2	2	2	2
9												1	1	1	1	1	2	2	2	2	2	3
10												1	1	1	1	1	1	1	1	1	1	1
Total subject demand	**0**	**1**	**1**	**2**	**2**	**3**	**4**	**5**	**5**	**7**	**10**	**13**	**14**	**14**	**16**	**19**	**20**	**21**	**21**	**22**	**23**	**24**
Total subject demand + Overage	**0**	**1**	**1**	**2**	**2**	**4**	**5**	**6**	**6**	**8**	**12**	**16**	**17**	**17**	**19**	**23**	**24**	**25**	**25**	**26**	**28**	**29**

SKU description: Wonder drug 50 mg/mL in a 10 mL vial, global label
Depot location: EU depot
Overage: 20%

Figure 13.2 Subject Demand Forecast Table with Dose Titration and no SKU Switch

required to support all doses, the subject demand forecast should make a best attempt to accurately reflect the percentage of subjects expected to dose up or down. For example, if there are three potential doses requiring one vial (e.g. one kit), two vials (e.g. two kits), and three vials (e.g. three kits), respectively, and 80% of subjects are expected to reach the two-vial dose while 60% of subjects are expected to reach the three-vial dose, then the forecast should present like Figure 13.2.

In this figure, 6 out of 10 subjects reach the three-kit threshold after 10 weeks, while 2 subjects reach the 2-kit threshold, and 2 subjects remain at the 1-kit threshold. Notice too how the overage is maintained at 20%, even after the study has fully enrolled. Unplanned dose titrations are a scenario where maintaining a slightly higher overage (e.g. 15–20%) in the demand plan makes sense to ensure additional supply availability at the depot level is available.

On the supply planning side, the clinical planner should consider the following strategies at each node:

- *Clinical Site*: Monitor enrollment activity across active sites. As sites enroll multiple subjects, consider increasing the RTSM supply strategy as needed from low enrolling to medium or high enrolling. This strategy increases the minimum inventory held at the site to protect against multiple subjects titrating up or dosing down simultaneously.
- *Regional Depot*: Use the demand overage input to increase inventory held at the regional depot levels.
- *Central Depot*: No change to safety stock levels is needed. Demand from regional depots will drive increased production requirements for FP.
- *Upstream*: No change to upstream inventory levels is needed because increased FP production requirements drive DP resupply quantities.

In a scenario where unique SKUs are required to achieve higher or lower doses, a demand forecast for each SKU needs to be generated with enough demand to ensure sufficient supply at both the site and depot. From an individual subject demand perspective, the planner should take the titration and/or reduction assumptions discussed earlier in the chapter and forecast subject demand at each SKU. In the previous example, if the highest dose required a unique SKU, then the subject demand forecast for that SKU should be populated with projected subject visits and demand. The site stocking strategy also takes on added importance in this situation. Every site needs to be supplied with inventory of each SKU that a subject may need as their dose changes through the trial. As sites enroll more patients, the minimum quantity required of each SKU also goes up because multiple subjects may need to dose titrate or dose reduce simultaneously.

Studies with unplanned dose changes into and out of multiple SKUs generate more excess waste than studies that can get away with using a single SKU, or easier yet that have planned dose changes. This can be somewhat mitigated through the following supply planning strategies:

- *Clinical Site*: Set RTSM minimum thresholds for each required SKU in the study. As sites enroll multiple subjects, consider increasing the RTSM supply strategy as needed from low enrolling to medium or high enrolling. This strategy increases minimum inventory held at the site across all SKUs to protect against multiple subjects titrating up or dosing down simultaneously.
- *Regional Depot*: Use the demand overage input at each SKU–depot location to increase inventory held at the regional depot levels.
- *Central Depot*: No change to safety stock levels is needed.
- *Upstream*: No change to upstream inventory levels is needed.

13.8 Demand and Supply Planning for Extensions

Many clinical trial designs create an opportunity for subjects to remain on study drug after the initial treatment period of the study concludes. This additional segment of the study is called an extension phase. Extension phases give subjects a chance to receive treatment over a longer period or, if the study is blinded, to guarantee that subjects will receive the active drug if they were randomized to placebo during the double-blind portion of the study. Extensions also give the sponsor an opportunity to collect longer-term safety and/or exploratory data from participants in the trial.

Accurately forecasting extension phase demand presents a different challenge than forecasting the double-blind portion of the study. First, the good news. Two demand planning variables are typically known, the timing of subject enrollment and the depot location of the subject. With these two pieces of information, the demand plan for the extension phase is straightforward to create. There are various types of extension scenarios, though. The most basic involves an open label design that has a prescribed period of treatment and dose followed by an extension phase on the same dose. In this scenario, the same SKU used to support the first period of treatment is also used for the extension, so the individual subject demand forecast for each subject enrolled at the SKU–depot location can simply be extended for as long as necessary. When the study is blinded and subjects transition from a blinded FP SKU to an open label SKU, the demand forecast at each applicable open label SKU–depot location needs to capture both the timing of transition and the individual subject demand curve of the open label design. No matter the scenario, attrition forecasting is almost always a core component of extension phase demand planning. The clinical planner should use the approach outlined earlier in this chapter to apply attrition rates to the extension phase forecast.

There is a third extension phase scenario that combines variable durations of treatment with SKU switching. This happens when a double-blind study is event driven, and the protocol then calls for an open label extension phase.

When this happens, subject rollover to the extension phase is gated to a minimum number of events observed, which is inherently unpredictable. As a result, the clinical planner needs to ensure open label supply is available at the earliest possible point the minimum number of events is reached and blinded supply is available until the latest possible point the minimum number of events is reached.

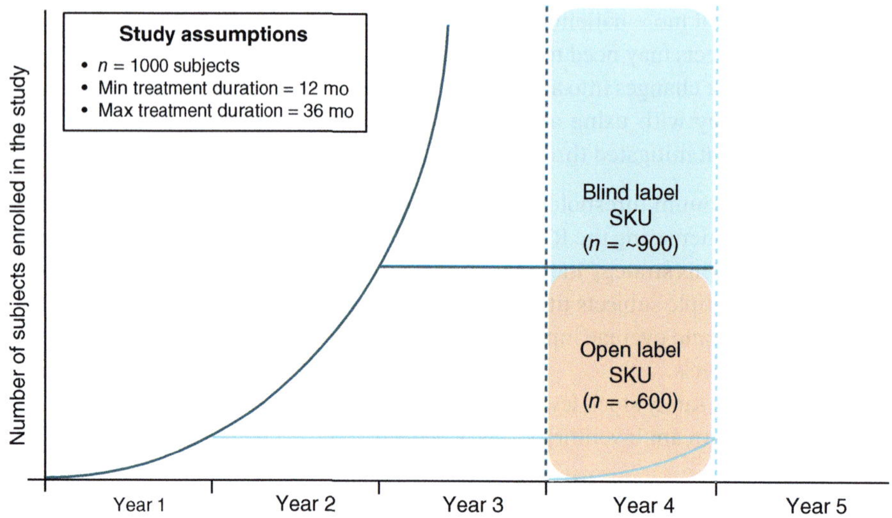

Figure 13.3 Extension Phase with Variable Duration of Treatment and SKU Switch

Figure 13.3 illustrates what planning extension phases with a variable duration of treatment leading into a SKU switch looks like. In this example, the sample size is 1000 subjects, and the protocol calls for a minimum treatment duration of 12 months and a maximum treatment duration of 36 months. The study has been designed to reach the minimum number of events at the end of Year 3, at which point, approximately 60% of subjects will have hit the 12-month treatment minimum and reached open label eligibility. Thus, the open label supply plan calls for sufficient inventory to support those 600 subjects. But, what if reaching that minimum number of events is delayed, keeping subjects on the blind label SKU for longer than expected? The figure estimates the latest possible date this could happen to be at the end of Year 4. By the time the end of Year 4 is reached, only 10% of subjects have reached the 36-month maximum treatment, so blind supply must be available to support the remaining 900 subjects as they continue blinded treatment.

This scenario exemplifies what happens when multiple points of uncertainty combine to force the plan into carrying large amounts of excess inventory. The clinical planner has a few options available to them to reduce waste in this circumstance. First, the goal should be to narrow the range of time between the earliest and latest date the minimum number of events will be reached. Close collaboration with biostatistics to monitor events and adjust the FP plan as the study unfolds is necessary to achieve this.

From a supply planning standpoint, the clinical planner should also consider the following strategies:

- *Clinical Site*: No change to RTSM parameters is required as the number of subjects enrolled at each site is known for both the blind and open label SKUs.
- *Regional Depot*: For the blind SKUs, do not add surplus inventory at the regional depot, and consider reducing demand overage to near zero as soon as the study is fully enrolled. For the open label SKUs, focus on stocking regional depots with sufficient inventory to cover subject demand over lead time for all subjects enrolled out of that depot over the lead time required to resupply that depot. More initial depot stocking may be added if the product has a strong expiry profile.

- *Central Depot*: For both the blind and open label SKUs, consider scheduling smaller, more frequent batches leading up to the potential window when the minimum number of events could be reached. The planner should also work closely with the FP manufacturing site to build as much agility into the production strategy as possible so orders can be changed over from blind to open label as soon as the event threshold is passed.
- *Upstream*: Planning for both blind and open label FP inventory for overlapping periods naturally creates more demand for DP and DS. As soon as the rollover to the extension phase happens, further resupply of the blind material is no longer necessary, and this may create a surplus of upstream inventory. The clinical planner should monitor upstream production orders in the year leading up to the window to avoid overproduction during this time.

13.9 Segregating Site Inventory from Depot Inventory

In the FP production plan table examples that were provided throughout Chapters 9–11, on-hand inventory is shown as an aggregate of depot-level and clinical site-level inventory. In most cases, combining inventory between the two levels is manageable because the ratio of depot inventory to site inventory is weighted more toward the depot, meaning that most inventory is truly available to service subject demand at the site level. However, there are scenarios where more inventory is held at the clinical site level than at the depot level. This happens when a central or regional depot services a large number (e.g. >15) of clinical sites and production/distribution orders supplying the depot are relatively small. This situation becomes problematic when clinical sites hold a large amount of on-hand inventory, but subject demand is concentrated at a small number of those sites. In this circumstance, the inventory held at sites without subjects is not actually available to service subject demand, and the result is that on-hand inventory is overstated in the production plan table. This can give the clinical planner a false sense of security when on the ground depot inventory can run dangerously low creating a stockout risk.

The solution to this problem is to segregate depot inventory from site inventory in the production plan table as in Figure 13.4.

In this example, 60 kits are produced in Week 1, and 10 of those are immediately shipped to a clinical site to service subject demand in Week 2. Then, in Week 2, another 10 kits are shipped to the clinical site bringing site on-hand inventory up to 18 kits and depot on-hand inventory down to

CW

	W1	W2	W3	W4	W5	W6	W7	W8	W9	W10	W11	W12	W13	W14	W15	W16	W17	W18	W19	W20	W21	W22	W23	W24	W25	W26
Total demand	0	2	4	6	4	6	4	8	10	2	6	8	4	6	4	8	4	6	2	4	6	8	2	4	8	6
OHI	60	58	54	48	44	38	34	26	136	134	128	120	116	110	106	98	94	88	86	82	126	118	116	112	104	98
Site OHI	10	18	14	18	14	18	14	16	16	14	18	20	16	20	16	18	14	18	16	22	16	18	16	22	14	18
Depot OHI	50	40	40	30	30	20	20	10	120	120	110	100	100	90	90	80	80	70	70	60	110	100	100	90	90	80
Prod. actual	60	0	0	0	0	0	0	0	120	0	0	0	0	0	0	0	0	0	0	0	50	0	0	0	0	0
Prod. plan	94	0	44	0	0	0	0	0	0	0	0	0	0	0	0	0	0	0	0	0	0	0	0	4	0	0
WOS	11	10	9	8	7	6	5	4	24	23	22	21	20	19	18	17	16	15	14	13	18	17	16	15	14	13

1

	W1	W2	W3	W4	W5	W6	W7	W8	W9	W10	W11	W12	W13	W14	W15	W16	W17	W18	W19	W20	W21	W22	W23	W24	W25	W26
Safety stock	9	9	9	9	9	9	9	9	9	9	9	9	9	9	9	9	9	9	9	9	9	9	9	9	9	9
Order target	18	18	18	18	18	18	18	18	18	18	18	18	18	18	18	18	18	18	18	18	18	18	18	18	18	18

SKU description: Wonder drug placebo, 30-ct bottle, global label
Depot location: European Union depot

Figure 13.4 Example Production Plan Table Segregating Site Inventory from Depot Inventory

40 kits. This continues through to Week 8 when depot on-hand inventory falls all the way down to just 10 kits. If another site needed to be activated prior to Week 8, the plan may not be able to accommodate this change. Without segregating site inventory from depot inventory, the clinical planner would not have the visibility they need to analyze this and make subsequent changes to the plan.

Fortunately, populating and maintaining site on-hand inventory and depot on-hand inventory can be done through utilization of the RTSM inventory reports. Weeks-of-supply measurements at the site and depot level can also be added to the plan by forecasting how many weeks of future subject demand each level of inventory supports. Breaking out this detail doesn't need to be done at every SKU–depot combination because it does add time to the demand and supply planning process, but it is useful to monitor at those depots servicing a large number of clinical sites where supply may be relatively constrained.

13.10 Inventory Pooling

Inventory pooling in consumer good supply chain management is the act of serving two or more markets using a common SKU.[2] When inventory pooling is utilized, it reduces uncertainty within the demand forecast due to aggregation, and this in turn lowers the amount of inventory the manufacturer needs to hold to service that demand profile. This also happens to be the same benefit that aggregate forecasting creates within the demand plan when subject demand is held entirely out of one depot for as long as possible. In clinical trial supply chain management, inventory pooling creates the same advantages that it does in consumer supply chains. There are two opportunities to implement inventory pooling at the FP level:

- *FP at Clinical Trial Sites*: If multiple trials studying the same drug are underway at a single site, then unlabeled inventory can be sent to the site for use in each respective trial. Labeling can then take place at the clinical site on an as-needed basis.
- *FP at Central and Regional Depots*: Inventory pooling at the regional and central depot levels is possible, but it requires a unique, just-in-time supply chain configuration to properly execute.

Successfully implementing inventory pooling requires a paradigm shift within the supply chain from a make-to-stock to a make-to-order strategy. This is not how clinical trial supply chains have traditionally operated, and for good reason. To make this change requires reorganizing how the supply chain functions, so we'll return to this topic in Chapter 18 – Make-to-Order and Just-in-Time Supply Chains in Clinical Trials with a more complete overview of the make-to-stock versus make-to-order debate and a discussion of when and how it's best applied in the clinical trial space.

13.11 Low-Case–High-Case Demand and Supply Planning

There is another demand and supply planning technique the clinical planner should utilize, low-case–high-case planning. Low-case–high-case is a form of scenario planning that's most helpful when there is a high degree of uncertainty around the enrollment rate and the total demand forecast for a study. Instead of using a single demand forecast upon which to base the FP and upstream production plans, low-case–high-case planning creates two scenarios, a low case where enrollment is projected to be slow, and a high case where enrollment is projected to go fast. These

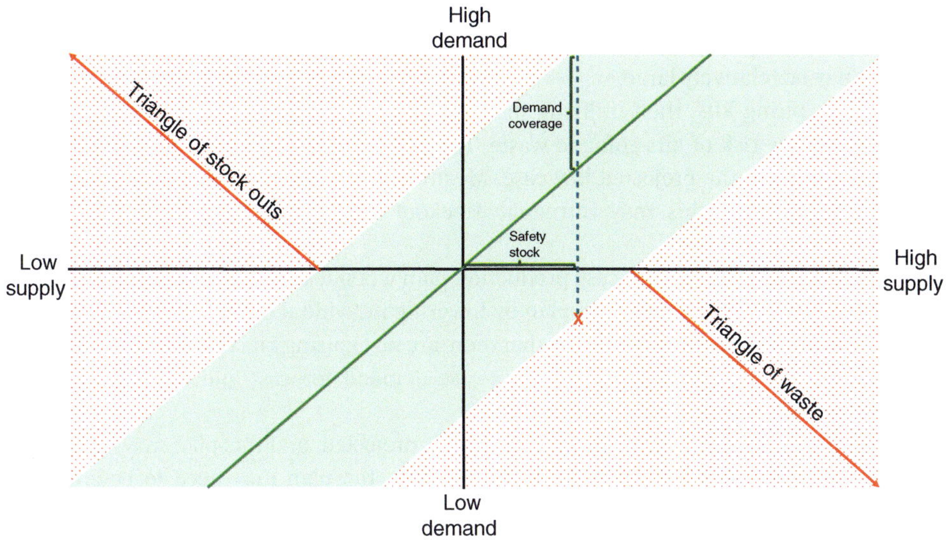

Figure 13.5 The Low-Case–High-Case Quadrant

two scenarios create two poles against which the FP supply plan can be analyzed and then updated to reduce waste from expiry in the event low-case demand materializes while also mitigating stock-out risk if high-case demand comes through.

Figure 13.5 presents the low-case–high-case quadrant which plots supply on the *x*-axis versus demand on the *y*-axis. The green line intersecting the center point of the axes represents a "perfect" plan where demand and supply are completely balanced after accounting for safety stock inventory held at each node of the supply chain. The distance along the *x*-axis from the origin point to the point where the triangle of waste is breached represents the maximum amount of safety stock to build. The distance along the *y*-axis from the safety stock point represents the maximum amount of demand that can be covered by the safety stock before the triangle of stockouts is breached. In Figure 13.5, the safety stock provides enough demand coverage to meet the highest demand case. This comes at a cost though. If the low-demand case materializes below the red "X," then drug goes to waste. The low-case–high-case analysis is a tool the clinical planner can use to quantify the safety stock, the demand coverage, and the excess waste risk against the low-case and high-case demand scenarios.

Building a low-case–high=case framework involves the following steps:

1) Build a study demand forecast with the fastest projected enrollment rate. This is the high-case demand forecast.
2) Set the safety stock and order target supply planning parameters to a reasonable level (e.g. 9–18 weeks for safety stock and 39–78 weeks for order target) based on the study and product characteristics discussed in Chapter 9 – Finished Product Supply Planning: The Production Plan.
3) Allow the safety stock and order target parameters to create a suggested production plan for the high-case enrollment rate and do not make any updates to that plan.
4) Build a second study demand forecast with the slowest projected enrollment rate. This is the low-case demand forecast.
5) Plot the high-case production plan against the low-case demand forecast and analyze how many go unconsumed due to shelf life in this scenario. If the expiry period of the DP is

relatively long (e.g. >24 months), the quantity of expiring kits may be relatively small or even non-existent. If the expiry period is relatively short (e.g. <24 months), then the amount of expiring kits may be relatively large.

6) If there are no expiring kits, then utilizing the high-case production plan as is makes sense because there is a low risk of kits going to waste. This only happens if demand materializes at an even lower rate than the projected low case. Meanwhile, if the production plan is reduced from the high-case plan, this may introduce stockout risk if the high-demand scenario materializes.
7) If there are expiring kits in the high-case production plan when it's plotted against the low-case demand forecast, reduce the production plan by lowering individual orders or by lowering the safety stock and/or order target values such that there are no expiring kits remaining. Then, plot this updated production plan against the high-case demand forecast and analyze on-hand inventory and weeks-of-supply levels in the plan.
8) If on-hand inventory stays positive, then no stockout is projected, and this plan may be acceptable, but if the weeks-of-supply are too low for comfort, the plan may need to be adjusted upward via increasing a single order quantity and/or revising the safety stock parameter upward, even if this creates some excess waste risk in the other direction.
9) If on-hand inventory goes negative, then a stockout is projected to happen, and the plan may need to be increased if the team does not want to risk a stockout in the event high-case demand materializes.

If the enrollment rate gap between the low-case and the high-case scenarios is relatively small (e.g. an enrollment rate of 0.3 p/s/m and an enrollment rate of 0.4 p/s/m) and the expiry period of the DP is relatively long, then a single iteration through this process may be sufficient to confirm the production plan is optimal. But, as the gap between the low-case and high-case enrollment rates increases and, as the expiry profile shortens, the number of iterations to find an optimal plan also goes up. This makes intuitive sense because as demand uncertainty increases (e.g. the range of possible demand outcomes is wider) so too does stockout and/or excess waste risk in the supply chain.

Low-case–high-case planning can be applied at any point in the clinical planning process, but it's most useful in two situations. The first is the point when production for initial DP and DS batches needs to be made to support a study (or studies). This is typically well before any demand information is known, and other variables such as the estimated sample size or the duration of treatment may influence the low-case or high-case demand forecast, further broadening the gap between the two cases. Low-case–high-case planning provides the most value when the DP and/or DS are very expensive, and real trade-offs need to be considered in balancing service levels and cost.

As adjustments are made to the FP production plan to avoid waste (e.g. reducing the quantity or pushing out the timing of orders) or to avoid stockouts (e.g. increasing the quantity or pulling in the timing of orders), the same logic can be applied to the DP plan:

1) Follow steps 1–4 above.
2) For both the high-case and low-case production plans, plan the amount of DP required to support each scenario with the desired inventory management strategy. For example, in the high case, the inventory management strategy may be relatively conservative (e.g. hold at least 26 weeks-of-supply of DP at all time), while in the low case, the inventory management strategy may be relatively aggressive (e.g. package DP immediately upon receipt).

3) Plot the high-case FP production plan against the low-case DP production plan. Assess the plan to determine if DP production is sufficient to meet the FP requirements. It's likely that it won't be and that DP orders may need to be adjusted upward or pulled in to avoid stockouts.
4) Plot the low-case FP production plan against the high-case DP production plan. Assess the plan to determine if DP batches go to waste. It's likely that some quantity or number of batches will go to waste, and DP orders may need to be adjusted downward or pushed out to avoid waste.
5) Alternatively, iterate through Steps 1–9 above and take the final desired FP plan and use this to drive the DP plan. Each scenario may be compared to the others to identify an optimal path that balances stockout risk and excess waste risk and minimizes the cost required to achieve the desired strategy.

Performing this analysis sheds light on the upstream inventory management strategy, and it illustrates how resilient the supply chain is when faced with a high-demand scenario.

The second point occurs when a simplified version of low-case–high-case planning can be applied to determine the initial FP production order quantity and the timing of the first resupply order. In this situation, the clinical planner runs the high-case and low-case demand forecasts, and then assesses:

1) How much FP can be manufactured in the first batch before it expires in the low case? This answer sets the quantity of the initial packaging order.
2) How long will this inventory last if high-case demand materializes? This answer sets the timing of the first resupply order. Figure 13.6 illustrates this concept.

This analysis complements the FP production planning process outlined in Chapter 9 and provides a sanity check on the supply planning parameters used at the FP level. If the low-case–high-case result suggests a much lower or higher quantity than the production plan, the clinical planner should consider adjusting the order target levels down or up, respectively, because they are likely out of step with the expiry profile of the drug.

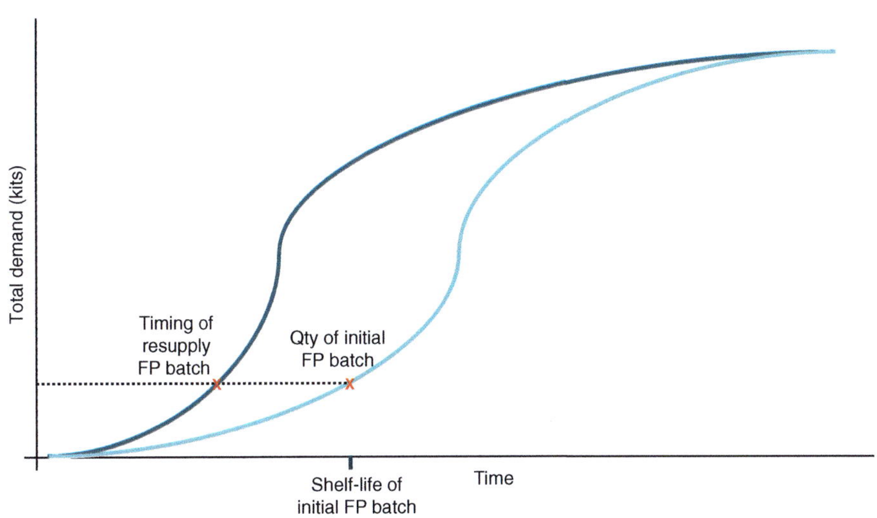

Figure 13.6 Low-Case–High-Case Finished Product Planning

13.12 Deterministic Versus Stochastic Demand and Supply Planning

Throughout our discussion of demand and supply planning concepts and techniques, we've used a deterministic framework and approach. Deterministic models assume that outcomes are discrete and predictable, and, as a result, they use linear logic and programming. All the formulas described in earlier sections of this book fit the deterministic profile. There is an alternative methodology that's known as stochastic planning. Stochastic models assume that outcomes are random and unpredictable, so these models use probabilistic statistical techniques to generate a range of potential outcomes with associated likelihoods. In the real world, clinical trial supply chains behave stochastically with randomness and uncertainty at every corner.

From these descriptions, stochastic modeling seems like a superior fit for clinical trial demand and supply planning, yet this book has focused exclusively on deterministic formulas and frameworks. The primary reason for this is that building, maintaining, and interpreting stochastic models are a significantly more complex enterprise than using deterministic models, so it takes much more time and money to use a stochastic model than a deterministic one. One example where stochastic modeling makes sense is for large, global clinical trials that have a highly uncertain demand forecast and that utilize a very expensive DS or comparator product. In this situation, there is a high risk of excess waste due to demand uncertainty, and the sponsor may feel the risk of reduced service level is worth the tradeoff of saving millions of dollars. Stochastic modeling can quantify how much risk the plan takes on by producing and distributing fewer kits of drug to depots and sites. On the other hand, stochastic modeling is typically not appropriate for relatively simple clinical trial designs or studies where the product is inexpensive to procure or manufacture.

Stochastic models do share something in common with deterministic models. Both approaches are only as good as the assumptions and inputs that go into them. Stochastic models are often highly sensitive to their inputs so when changes are made to these the model may produce wildly different outputs. This has two implications. First, the clinical planner must be highly attuned to the inputs and assumptions they're using whether the model is deterministic or stochastic. Second, all models produce outputs that require further analysis and scrutiny before finalizing them as the plan. No model's outputs should be taken as gospel truth without a person checking that they're feasible and fit in with the broader strategy of the team and the company. Fortunately, several companies offer stochastic modeling solutions dedicated to clinical trial supply chain management and experts-for-hire who can tailor these models to the specific nuances of clinical trials.

13.13 Key Takeaways

Clinical trial protocols often contain design elements that introduce uncertainty to the demand forecast such as variable durations of treatment, dose titrations and reductions, and extension phases. These complex elements create a challenge for the clinical planner because they increase demand uncertainty and further complicate the balance between maintaining service levels while avoiding excessive waste.

1) The first instinct to manage demand uncertainty is to artificially increase the forecast to ensure the plan drives extra inventory. This is a conservative approach to demand planning. Striving

for accuracy within the demand forecast is a superior strategy because when the forecast is over-inflated, it increases the potential for the bullwhip effect within the supply chain. The bullwhip effect is a phenomenon where changes in demand at the FP level lead to amplified demand signals for upstream materials. The primary consequence of the bullwhip effect is to create excessive inventory and waste. This is especially pernicious in clinical trial supply chain management where subject demand changes constantly.

2) Aggregate forecasting for clinical trials is a technique where all subject demand is planned out of a single depot location for as long as reasonably possible. This is typically done out of the depot that projects to have the highest percentage of subjects enrolled in the study. Aggregate forecasting reduces demand uncertainty and the bullwhip effect, and it also creates a more flexible plan because it helps prevent the clinical planner from overcommitting inventory to regional depots.
3) Attrition is the annual rate at which subjects drop out of a clinical trial, and it's a common feature of clinical trials that are long in duration and/or study a serious disease. Incorporating attrition into the forecast is best done by applying the attrition rate to total subject demand rather than at the individual subject level.
4) Building an effective supply planning strategy for a trial or program requires analyzing the demand profile of that study and then developing an inventory management strategy at each level of the supply chain to match that profile. As upside demand risk increases, RTSM minimum thresholds and regional depot levels may need to be increased to protect against simultaneous enrollment or SKU switches. But, those changes don't necessarily mean that central depot or upstream inventory levels also need to be adjusted upward.
5) Variable durations of treatment most often occur in event-driven studies where the duration of treatment is not fixed on an individual subject basis but is instead tied to the observation of a pre-specified number of events within the overall study. Collaboration with the biostatistics function is necessary to predict when the number of events will be reached.
6) Dose titrations and dose reductions may be planned or unplanned. Unplanned changes in dose are more complex to forecast, and this challenge is exacerbated when different doses are supported by unique SKUs. Careful attention must be paid to the RTSM parameters and to depot stocking quantities. As enrollment progresses, assumptions around dose changes should be updated as needed.
7) After the initial treatment period of a study completes, subjects may be offered an opportunity to continue taking the drug offered in a clinical trial. This is known as an extension phase. In most cases, extension phase planning is straightforward because subject enrollment is known. When it's an event-driven study though, planning the timing of the extension phase start is difficult.
8) Low-case–high-case planning is a type of scenario planning that involves creating a production plan based on the slowest potential enrollment rate (e.g. the low case) and a production plan with the fastest potential enrollment rate (e.g. the high case) and then comparing each plan against the opposite profile. When the high-case plan is plotted against the low-case demand forecast and vice versa, the production plan can be optimized to reduce potential waste and protect against stockouts.
9) There are two approaches to modeling clinical trial demand and supply, deterministic and stochastic. Deterministic models assume discrete, predictable outcomes and use linear formulas to provide singular outputs while stochastic models assume random, unpredictable outcomes and use complex statistical analyses to generate a range of potential outputs. Stochastic models are appropriate for complex clinical trial designs that use expensive material to best understand the

cost–risk benefit of supply planning and inventory management strategy. Deterministic models are appropriate for simpler clinical trial designs that have a high degree of confidence in the demand forecast or that use less expensive material.

Notes

1 https://onlinelibrary.wiley.com/doi/10.1002/sim.8911.

2 https://www.hbs.edu/faculty/Shared%20Documents/events/163/Pooling.pdf.

14

Clinical Supply Chain Reports, Budgets, and Metrics

14.1 The Importance of Communication in Supply Chain Management

This chapter represents an inflection point in our planning journey. Through the previous six chapters, we've introduced the objectives of clinical trial supply chain management and the elements of supply chain planning from demand planning to upstream production planning to more advanced concepts in planning. Now, we turn our attention to the management and maintenance of the demand and supply plan. The first step on that path is to understand how to summarize data that's produced from within a study-specific plan to effectively communicate the supply chain strategy. The information contained in this chapter provides inputs for the business processes covered in the next chapter, Chapter 15 – Clinical Supply Chain Planning Processes. The business processes covered there provide a framework to keep the plan current and to communicate the plan and strategy to the relevant parties involved.

Effective communication is vital to clinical trial supply chain management because so many people and organizations are involved in manufacturing, releasing, and distributing the product to clinical trial sites. In earlier chapters, we've touched on the importance of aligning internal and external stakeholders to the production and distribution plans that compose the supply chain strategy. Decisions made in the supply plan impact the sponsor and external stakeholders in several ways. The first and most critical point is the ability of the supply plan to adequately supply the clinical trials and process development requirements within the program. It's vital that internal stakeholders such as clinical operations and the chemistry, manufacturing, and controls (CMC) team understand the demand and supply-related assumptions that influence the plan. Without this alignment, the inputs that drive the plan such as the enrollment rate, lead times, and production yields will not accurately reflect the reality of the supply chain, and this situation results in potential stockouts and excess waste.

Second, the supply plan drives a tremendous amount of activity within the CMC team and at the sites tasked with carrying out production and distribution activities. We've discussed the process of succinctly communicating the production and distribution summaries to manufacturing and distribution sites in earlier chapters, but those summaries also provide valuable information for hiring decisions within the sponsor. As the plan changes, so too do the human and capital resources needed to make those activities happen. For example, if the supply plan calls for a stretch of several batches, then the quality organization needs visibility to this to ensure they appropriately prioritize batch record review and release activities. Or if an increase in manufacturing scale is necessary,

Supply Chain Planning for Clinical Trials – A Practical Guide, First Edition. Ryan Mills.

regular communication of the supply plan can give the team time to prepare for and react to that change.

A third space where the supply plan has a direct and measurable impact is on the budget of the program. Each level of the supply chain has associated costs that need to be accounted for by the sponsor. While CMC activities typically make up only a fraction of a program's development budget, total spend on manufacturing and supply chain-related work may still cost tens or even hundreds of millions of dollars during clinical development. And, for some cell and gene therapy programs, CMC activities do account for a majority of the program budget. Fortunately, a well-managed supply plan is a key input to identifying CMC spend because it contains the production and distribution orders for all materials needed to support clinical trials within the program. While it doesn't include every activity, good manufacturing practice (GMP) manufacturing and distribution activities almost always make up a majority of the CMC team's budget, so it makes sense to use the supply plan as a starting point for creating a budget. To summarize, maintaining uninterrupted supply, appropriately resourcing for manufacturing and distribution-related activities, and capturing the budgetary impact of the plan all require effective communication.

Monitoring and analyzing supply chain performance is another area where communication matters. Capturing and updating key metrics give the clinical planner and the CMC team an opportunity to identify risks in the supply chain and adjust the planning assumptions and/or the inventory management strategy to mitigate those risks. Presenting the most important metrics in a digestible format such as a dashboard gives the clinical supply chain team and senior management visibility into supply chain performance and risks. No operations leader, or any organizational leader for that matter, likes to be surprised with a situation that could have been foreseen ahead of time. Dashboards give the clinical planner an opportunity to communicate risk and maintain alignment vertically with leadership and horizontally with the CMC team and external suppliers.

In this chapter, we'll focus on three tools that enable effective communication of supply chain planning outputs and supply chain performance: reports, budgets, and metrics. First, we'll introduce a handful of reports, charts, and graphs that describe and illustrate the supply chain strategy at each planning level. Each report is created from data contained within the demand and supply plan of a clinical trial. Some examples include reports on where inventory is in the supply chain, charts for how the finished product (FP) production plan stays ahead of clinical demand, how many FP weeks-of-supply (WoS) are on hand through the distribution network, the amount and timing of upstream production, and how that upstream production translates into an inventory management strategy. We'll also demonstrate how portfolio level reports give insight into resourcing requirements and activity prioritization across programs. Then, we'll show how to create program-level budgets that react dynamically to changes in the supply plan. This is a valuable exercise for understanding how much spend is required to support manufacturing and supply chain activities, and it's also useful for holding suppliers accountable to the numbers they've quoted. We'll also introduce cost of goods (COGs) analyses as an offshoot of the budgeting process. Finally, we'll cover the most relevant metrics to clinical trial supply chain management and present an example dashboard to support executive communication.

14.2 Inventory Reporting

One simple report that can be gleaned from the demand and supply plan is how much inventory is available at each stockkeeping unit (SKU)-depot location on a lot-specific basis. Recall that the WoS summary introduced in Chapter 11 – Finished Product Supply Planning: The Distribution

Table 14.1 Example Inventory Report

Current Date	**W78**						
Protocol	**Level**	**SKU**	**Depot**	**Batch No.**	**Quantity**	**UoM**	**Expiry Date**
Study 1	Finished product	100 mg active	USA	A1A	100	Kits	31-Aug-2025
				B2B	400		31-Dec-2026
		100 mg placebo		P-A1A	100		31-Aug-2025
				P-B2B	300		31-Dec-2026
		100 mg active	Brazil	B2B	30		31-Dec-2026
		100 mg placebo		P-B2B	30		31-Dec-2026
N/A	Drug product	100 mg tablet	USA	DP-C3C	1000	Bottles	31-Jul-2028
N/A	Drug substance	Molecule 1	India	DS-A1A	0.253	Kilograms	31-Dec-2027

Plan aggregates inventory to show total on-hand inventory. It's also useful to have a report that breaks down inventory on a batch-by-batch basis. The structure of an inventory report largely mirrors the WoS summary and contains the data elements shown in Table 14.1.

Inventory reports like this can be used in several ways. First, these reports can be compared with inventory reports from the randomization and trial supply management (RTSM) system to determine if depot-to-site shipments and subject dispensations are happening as predicted. If the RTSM shows a lower number than the plan, that means more inventory is moving to sites or subjects, and the clinical planner should research why that's happening. The report can also be compared with regional depot inventory reports to ensure the model accurately reflects the amount of inventory sitting at the depot of each lot. Any discrepancy should be investigated and reconciled. Second, the inventory report can be used to analyze lead times within the supply chain. That could be one of the root causes of a discrepancy between the plan and the depot if the depot hasn't received the shipment in the expected timeframe. Third, the inventory report is useful for tracing on-hand inventory of specific batches in the event of a quality event. If a quality issue is uncovered, then all inventory of the impacted batch needs to be placed in quarantine, and this report aids quick identification of where that needs to take place.

14.3 Visualizing the Production Plan

The production plan summary, distribution plan summary, and upstream production plan summary were introduced in Chapters 9, 11, and 12, respectively. These reports summarize all FP production, FP distribution, and upstream production order activity to support a clinical program. While the summaries are useful communication tools without modification, the data contained within each table can also be transformed into charts and graphs that provide a more visual representation of the plan. Bar charts are particularly useful at showing the total amount of production by batch or by total number of units across a predefined time horizon.

Figure 14.1 shows a clear increase in drug product (DP) manufacturing from early 2026 to early 2027. This type of data doesn't jump off the page when presented in a text heavy, table-based summary report format. The chart makes plain that the CMC team should consider initiating scale up

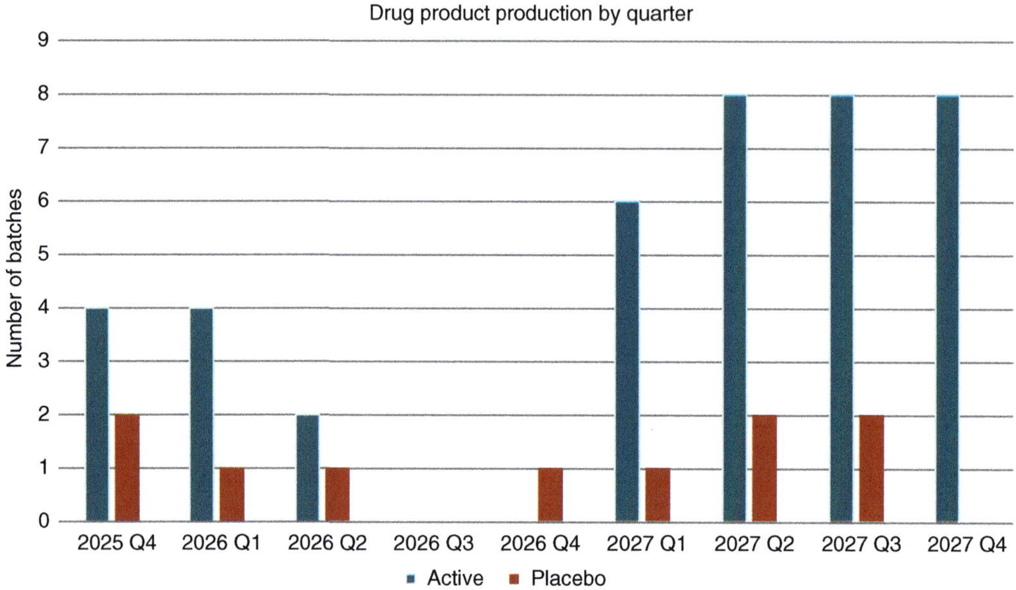

Figure 14.1 Production Report by Quarter

activities in 2026 to enable larger batch sizes in 2027 that will in turn reduce the total number of batches needed. A chart like this helps tell that story more simply and easily than the upstream production plan summary does. External suppliers also find charts like these valuable for understanding the long-term horizon of the production plan.

Another variation on this type of chart plots the previous version of the plan against a current version of the plan as in Figure 14.2. This view clearly shows how the production plan has changed

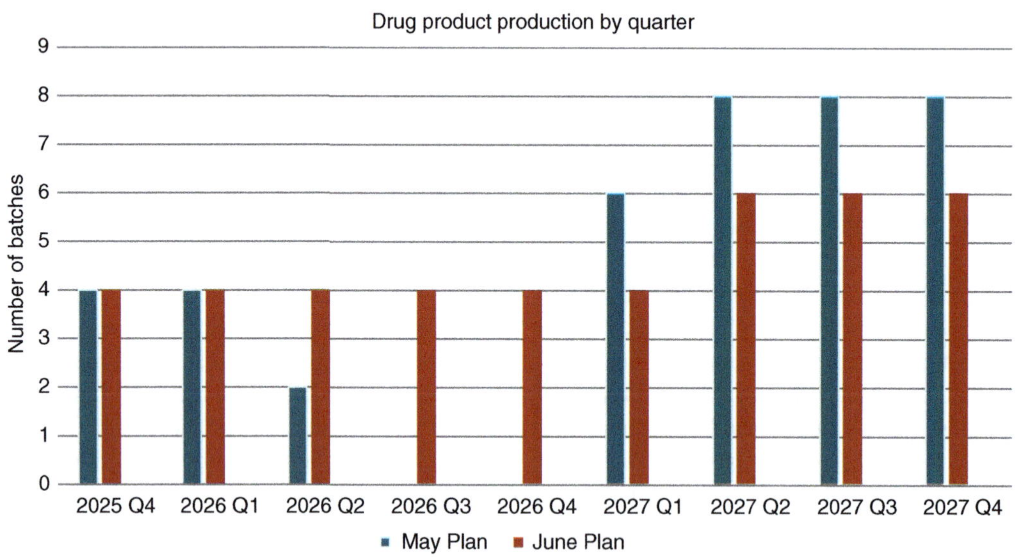

Figure 14.2 Production Report with Previous Plan versus Current Plan

from one version of the plan to another. In this example, scaling up may not have been possible or not cost effective, so instead the clinical planner smoothed the production plan by pulling in production from 2027 to 2026.

14.4 Visualizing the Supply Plan and Inventory Management Strategy

Beyond a summary of production, the production plan tables introduced in Chapter 9 – Finished Product Supply Planning: The Production Plan contain the data that makes up the FP supply plan. All that data is buried in the tables themselves though. Translating that data into charts makes understanding and analyzing the plan much easier. The most effective chart for visualizing the supply plan plots cumulative demand and cumulative supply over the entirety of the trial, as in Figure 14.3.

The beauty of this chart is its simplicity. Discrete values from the total demand and the production actual rows of the production plan table are summed into their respective cumulative totals. Making this conversion from discrete to cumulative values produces a chart that tells the story of cumulative supply (e.g. the blue line) always staying ahead of cumulative demand (e.g. the green line) for the duration of the study. There are also takeaways from this chart beyond the fact that supply meets demand. First, the slope of the enrollment curve is clearly visible. In Chapter 8 – Finished Product Demand Planning: Basic Concepts, we discussed the importance of forecasting a realistic enrollment curve to avoid excess waste. Achieving that goal is clearly visible here. Second, cumulative demand and cumulative supply should come close to touching each other at the end of the study to minimize excess waste. Third, each point where cumulative supply comes nearest to cumulative demand equals the number of weeks of safety stock at that point in time. This last point brings us to an improvement opportunity on this chart design.

While the chart in Figure 14.3 shows an effective production plan in terms of avoiding stockouts and minimizing excess waste, it doesn't tell a clear story about how optimal the production strategy is or how much risk the plan carries from an inventory perspective in accomplishing those goals.

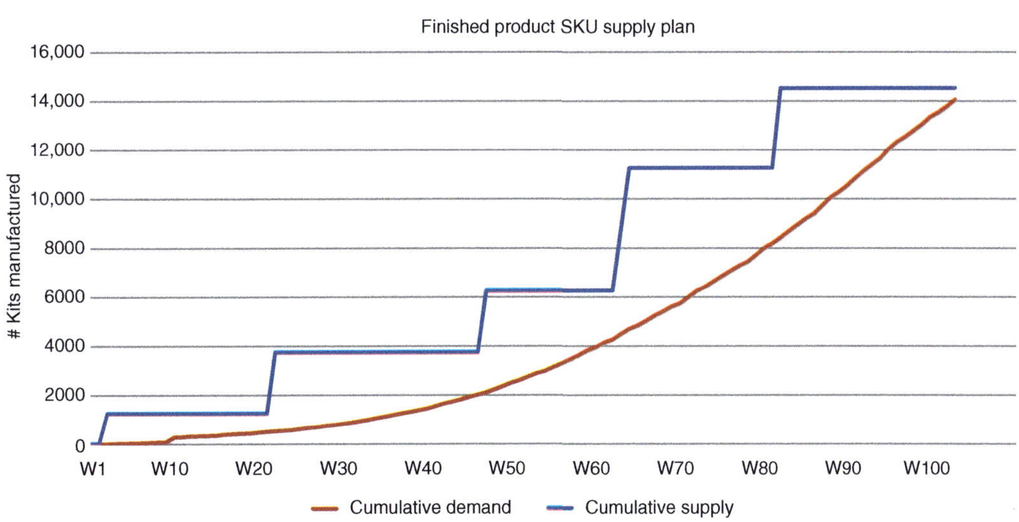

Figure 14.3 Finished Product Supply and Demand

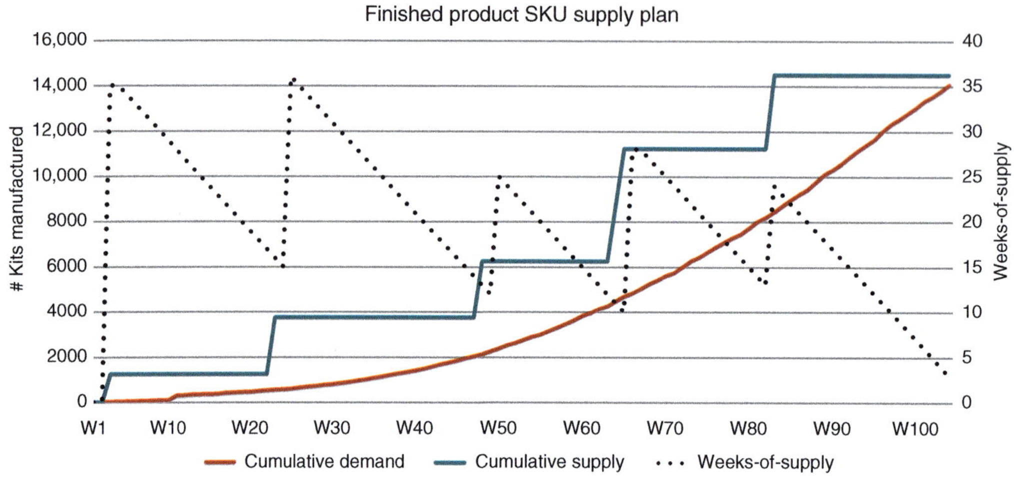

Figure 14.4 Finished Product Supply and Demand with Weeks-of-Supply

Adding aggregate FP WoS to the chart on the second *y*-axis fills that gap. And, like the total demand and production actual rows, WoS data is contained within the production plan tables.

In Figure 14.4, WoS has been added as a secondary *y*-axis to better illustrate the production strategy with respect to the desired safety stock and order target values. Let's analyze this example through the lens of WoS. The first two production orders generate 35 and 36 WoS, respectively. By the second resupply campaign in Week 49, the production strategy reverts to smaller jobs that only produce 25–30 WoS. This seems like an odd strategy because typically expiry gets better over time so it would make more sense to pack two larger jobs instead of three smaller ones to save money. Perhaps, there's an inventory constraint, but that doesn't make sense given the size of the fourth production order in Week 65. From a risk perspective, the plan seems logical, although that same resupply batch in Week 65 allows the safety stock level to drop to 10 weeks. Perhaps, this order could also be pulled in to reduce stockout risk. This type of analysis is much easier to conduct when WoS is included as a secondary data set against cumulative demand and supply.

Charts for visualizing the upstream production planning strategy can use either standard or implicit WoS levels to describe how much on-hand inventory is available. Standard WoS describes how long on-hand inventory is available until it's consumed in downstream production while implicit WoS describes the ability of upstream on-hand inventory to service future subject demand. Both data sets have value, and it's typically helpful to include at least two data points, on-hand inventory and the corresponding WoS that inventory covers. Figure 14.5 displays an example supply plan for an upstream material and its implicit WoS.

With upstream production planning, the power of including implicit WoS is even more evident. In this example, there are six manufacturing campaigns in Week 1, Week 20, Week 42, Week 60, Week 84, and Week 101. The total unit output gradually increases to keep pace with demand, but when the WoS data is added alongside production, it becomes clear that the amount of demand covered with each campaign decreases steadily. The first campaign of batches in Week 1 generates over 80 WoS, but the last campaign in Week 101 only covers 30 WoS even though it's the largest campaign yet.

Looking at this chart also tells the clinical planner if the supply plan for the upstream SKU is meeting the desired inventory management strategy for the program or if adjustments need to be made. It appears from the chart that this plan attempts to maintain at least 9–10 WoS as safety

Figure 14.5 Primary Packaging Supply Plan with Weeks-of-Supply

stock to protect against demand uncertainty at the FP level or a supply shock further upstream. This is visible from Weeks 35 to 42, Weeks 55 to 60, and Weeks 70 to 84. These are the on-hand inventory troughs that occur after FP campaigns consume the upstream SKU. Notice that WoS falls all the way to zero though in Week 100. Given this brief gap and the steadily decreasing maximum WoS filled by each campaign, the planner should consider increasing the size of the campaign in Week 84 to guarantee some safety stock is always on hand and to increase the maximum WoS total from 28 in Week 85 to a higher value.

14.5 Portfolio Reports

In addition to providing data for study level reports, the production and distribution plan summaries can also be aggregated across programs to make portfolio level reports. Portfolio level reports that capture all program manufacturing and distribution activities are particularly valuable for activity prioritization and resource planning because they show the entire scope of the sponsor's supply chain. As the demand profile of individual studies change, the importance of releasing upstream batches for further processing or FP batches for distribution to the clinic also changes. A portfolio level report helps the quality function, who is ultimately responsible for the release of batches for further processing or to the clinic, understand the highest priority batches. This is particularly important in clinical trial supply chain management where these priorities change frequently and when the team's often have limited internal resources to manage the workload. A portfolio-level report can also be organized by the responsible manufacturing or distribution site. This information helps the site determine if adequate resources such as equipment and labor are already in place or if additions are needed.

Creating a portfolio level report also enables resource planning for both clinical supply chain and quality. Internally, manufacturing and distribution events are resource drivers for both functions because every production order has a discrete amount of work associated with it to manufacture and release the batch. These activities are largely executed by clinical supply chain at the FP level and by quality at all GMP levels of the supply chain. Developing a report that projects how many batches will be made and released can be transformed into a resource planning tool to better understand hiring needs across these functions.

A portfolio level report should contain at minimum:

1) Product name
2) SKU level
3) SKU description
4) Site of manufacture or distribution
5) Batch number
6) Quantity
7) Unit of measure
8) Original target date of manufacture, release, or shipment
9) Current target date of manufacture, release, or shipment
10) Prioritization rank (for activity prioritization planning)
11) Full-Time Equivalent (FTE) estimate (for resource planning)

Table 14.2 presents an example portfolio production and batch release report with two products and nine batches scheduled to occur between Weeks 30 and 78. Using this table makes it easy to identify when a batch has fallen behind its original release date and safety stock levels are at risk. This is the case for the two FP batches at the top of the table which are 6 weeks behind their original release date. These two batches have been highlighted and prioritized accordingly for the quality team. Note that the original release date for a batch is taken from the production (or distribution) order readiness process. The current release date projection is then updated each time the production or distribution order readiness process is executed. We'll return to this process in Chapter 15 – Clinical Supply Chain Planning Processes. From a resource planning perspective, envision another column in this table with an estimate for the number of hours required to produce or release a batch. When these estimates are summed across a portfolio

Table 14.2 Portfolio Production and Batch Release Report

Current Week	**W28**

Product Name	**Level**	**SKU**	**Site/ Depot**	**Batch Number**	**Quantity**	**UoM**	**Original Release Date**	**Current Release Date**	**Prioritization Rank**
Product 1	FP	100 mg active	US	FP A1A	1000	Kits	W30	W36	1
	FP	100 mg placebo	US	FP B2B	500	Kits	W30	W36	2
	DP	100 mg active	DE	DP 1	5000	Bottles	W40	W40	5
	DP	100 mg placebo	DE	Placebo DP 1	2500	Bottles	W42	W42	6
	DS	Product 1 DS	KR	DS 1	500	Grams	W52	W52	8
Product 2	FP	1 × 50 mg/5 mL vial	CH	FP 3C3	250	Kits	W38	W34	3
	FP	1 × 50 mg/5 mL vial	CH	FP 4D4	250	Kits	W38	W34	4
	DP	50 mg/5 mL vial	CN	DP 2	1000	Vials	W46	W46	7
	DS	Product 2 DS	CN	DS 2	1000	Grams	W78	W78	9

report, they become an important input to the hiring strategy for clinical supply chain and quality leadership.

Lastly, data taken from a portfolio level report can also feed other planning inputs and operational metrics. For example, the time from date of manufacture to actual release date can be averaged across SKUs, sites, or products to inform the production lead time (PLT) inputs that are so important to the upstream production planning process. This type of analysis gives the clinical supply chain function insight into how much lead time variance there really is across the portfolio. Similar analyses can be run for distribution orders, and operational metrics such as quality batch release lead times can also be derived from these reports.

14.6 Translating the Supply Plan into a Budget Forecast

Information contained in the production and distribution summaries can also be transformed into a budget for all GMP manufacturing and distribution activities taking place within a study or program. Taking this approach has multiple benefits. First, tying the budget directly to the summaries means that when the clinical planner updates the supply plan, those changes flow directly into an updated budget projection. This is an efficient way to keep the budget forecast current with the plan. Second, when proposed changes are made to the study design, or new studies are proposed, a framework exists to estimate how much those changes will cost, and these estimates are useful for informing team decisions about whether to implement those changes. Finally, building a budget forecast that's tied to the supply plan allows the clinical supply chain team to track actual spend against the budget forecast and hold manufacturing and distribution sites responsible to the cost estimates they've provided. As actual spend accrues, the budget provides a point of comparison against the invoices received.

The clinical supply chain function typically owns FP activities, so it makes sense to begin constructing the budget at that level. To build the budget from the production and distribution summaries, take the following steps:

1) Request cost estimates from the sites responsible for FP activities within the supply chain. When secondary packaging and distribution are insourced, sites within the sponsor network should provide these figures, but when secondary packaging and distribution are outsourced, a statement of work (SOW) (e.g. a quote) should be requested that contains all cost estimates. See Chapter 20 – Contract Organization Management and Oversight for more on supplier selection and SOWs.
2) Identify and document the cost drivers contained within the estimates provided by the sites in Step 1. On the manufacturing side, cost drivers may include FP kit design, tooling for secondary packaging activities, DP or primary packaging receipt and handling, component sourcing, label design, label translations, label proofs, label printing, the secondary packaging and label application process itself, batch release activities, and storage of the FP at the production site. On the distribution side, cost drivers may include depot set-up activities, GMP and/or import license applications, receipt of the FP at each depot, storage of the FP at each depot, clinical site shipment preparation and execution, shipping container fees, temperature monitors, and return and/or destruction activities. Distribution cost drivers should be broken out at the depot level because costs can vary widely from one depot to another and because depots are active at different points in the trial. Typically, one or two depots start distributing material at the start of the study, others come online after competent health authority approval later in the study, and others may not be used at all.

3) Identify and document the costs associated with each of the cost drivers and the unit basis of the charge. For example, a packaging campaign may cost $5000 to set up the room, $10 per kit charged, $5 per kit for components, and $1000 for a quality review and release of the batch. Some cost drivers may require additional calculations to arrive at the total cost. For example, site shipment fees are usually charged on a per shipment basis but are made up of shipment preparation, shipment component costs, and shipment documents (e.g. import licenses) that need to be summed to arrive at the true total cost.
4) Categorize the cost drivers by their frequency and association to events contained within the plan. There are typically four categories: one-time fees, batch-based fees, kit-based fees, and monthly recurring fees. In the example above, the set-up fee and the quality review are batch-based fees, while the per kit and component charges are kit-based fees.
5) Map the cost drivers to activity occurrences within the production plan and distribution plan summaries on an annualized basis. Most biopharma companies prepare yearly budgets so the budget forecast should match that structure. The key to this step is to pull occurrence data directly from the production and distribution plan summaries where relevant. This is most easily done for the batch-based and kit-based fees which are linked to production and distribution order events. For the one-time events, these may require manual intervention to document which year the set-up activity takes place as these can happen months in advance of the first production or distribution occurrence. Populating monthly event occurrences can be taken from the first date a shipment arrives at a depot to the last subject dispensing visit from that depot. Figure 14.6 shows an example the output of this cost mapping exercise.

   In this example, the one-time fees occur in the first year of the study. Meanwhile, the batch-based fee and kit-based fee event occurrences are taken directly from the production plan summary. Specifically, the number of runs, number of kits per run, and the manufacturing start/release dates are used to populate the correct quantities and dates. The number of months in each year for the monthly fee cost drivers is taken from the distribution plan summary. Specifically, the date the first bulk shipment is received at the depot is used to

							SOW 1		SOW 2
Cost drivers				Event occurrences			Budget		
Category	Cost driver	Cost	Per unit	Year 1	Year 2	Year 3	Year 1	Year 2	Year 3
One-time	Study set-up	$5000	/Study	1	0	0	$5000	$0	$0
	Tooling	$10,000	/Study	1	0	0	$10,000	$0	$0
Per run	Batch set-up	$5000	/Batch	1	2	2	$10,000	$10,000	$10,000
	Quality review	$1000	/Batch	1	2	2	$1000	$2000	$2000
	Bulk shipment to reg. depot	$25,000	/Batch	1	2	2	$25,000	$50,000	$50,000
Per kit	Label printing	$10	/Kit	2000	8000	15,000	$20,000	$80,000	$150,000
	Components	$8	/Kit	2000	8000	15,000	$16,000	$64,000	$120,000
Monthly	Inventory storage depot 1	$3000	/Month	6	12	12	$18,000	$36,000	$36,000
	Inventory storage depot 2	$2000	/Month	0	9	12	$0	$18,000	$24,000
	Site shipments depot 1	$4000	/Month	6	12	12	$24,000	$48,000	$48,000
	Site shipments depot 2	$1000	/Month	0	9	12	$0	$9000	$12,000
					TOTAL		$129,000	$317,000	$532,000

Figure 14.6 Example Finished Product Supply Chain Budget

calculate how many months in the first year the depot will be active in Year 1. Depot 1 comes online 9 months earlier than depot 2 in this study.

6) Group the event occurrences and itemized budgets by the SOW issued by the manufacturing and distribution networks. As the study design and/or the trial supply chain changes over time, new cost estimates are typically needed from the sites or supplier to support those activities. This step highlights which activities are covered by specific SOWs to help manage multiple quotes in an organized fashion. In Figure 14.6, the blue-shaded cells are covered by the initial SOW for the study, while the green-shaded cells represent activities covered by a second SOW.
7) Total the amount of projected spend across each year. This total figure is then provided to the CMC team and/or the finance function as dictated by the financial planning cycle. Totals may also be subdivided across activity area. For example, subtotaling by monthly fees can provide a representative example of how much it would cost to add another depot to the supply chain for a study because monthly fees typically include a storage fee at the depot and site shipment fees out of that depot.
8) Revisit the event occurrence inputs on a regular (e.g. quarterly or semiannual) basis to ensure they still represent the study strategy. This is especially important for the monthly fees because the country and depot start up timelines often pull into an earlier start or move out to a later date so it's important to check the accuracy of the budget in that respect.

Of course, the supply chain is more than just the FP. While creating detailed budgets for DP and drug substance (DS) manufacturing and process development activities are outside the scope of this book, total costs per batch should be available from the CMC team. Using the same approach for the batch-based fees outlined in Figure 14.6, DP and DS costs can also be tabulated on an annual basis using production order data from the upstream production plan summary. The number of batches per year is readily available, and annual process or analytical development activities can be included as an additional row at the bottom of the budget. The CMC team can also indicate if raw material and excipient costs are included in the DS and DP per batch totals or not. In some cases, the sponsor may be responsible for sourcing their own raw materials, and when that's the case then the raw materials should be segregated and captured as a separate line item in the program budget. When raw materials or excipients are sourced by the manufacturing site of the material, they are typically included in the per batch fee.

The last category of spend that should be included in the budget is logistics costs for shipments of DS and DP to the next downstream node of the supply chain. International logistics is a complex enterprise, and the per shipment cost structure reflects this. There are shipping costs such as service fees, customs broker fees, container fees, and fuel charges. Customs and duties also need to be tracked because some of these may be recoverable, such as value-added tax (VAT) charges that most countries outside of the United States exercise on imports. Strategies for managing VAT are covered more extensively in Chapter 21 – International Logistics for Clinical Biopharmaceutical Products. Once the total cost per DS or DP shipment is tallied, these can be mapped to the DS and DP campaigns listed in the upstream production plan summary and added into the budget for the program.

Once a complete budget forecast is in place that includes FP-, DP-, DS-, and logistics-related costs, the organization has a full picture of GMP manufacturing and distribution activities. Then, process development, analytical development, and raw material costs (if applicable) can be added to complete a full list of activities and spend for the program. Clinical supply chain may not always be responsible for budget development, but the supply plan unlocks the ability to both create the budget and maintain an accurate view of spend over time.

14.7 Cost of Goods Analyses

Preparing a program-level budget enables another type of financial analysis, the COGs to manufacture a product. COGs measure how expensive a product is to make and are used to evaluate the financial viability of a drug and in long-term budget forecasting after a drug is approved for commercial sale. Typically, COGs are composed of the direct manufacturing costs of the drug and are estimated on an annualized basis. COGs are also measured over time during clinical development to determine the effectiveness of manufacturing process development efforts.

To calculate the COGs, each level of the supply chain must be considered including raw materials, DS, DP, primary packaging, and FP manufacturing. One method for calculating small-molecule COGs involves estimating the per unit cost at each level of the supply and then adding the total cost. Many of these costs can be inferred from the cost per batch estimates used in the budgeting process. For example, if a batch of DS costs $250,000 and yields 100 kg, then the cost per kg of the DS is $2500 ($250,000/100 kg). To illustrate how to calculate COGs, consider the following inputs:

- Raw material 1 = $1000/kg × 1 kg required per kg of DS = $1000 per kg of DS
- Raw material 2 = $5000/kg × 1 kg required per kg of DS = $5000 per kg of DS
- DS = $500,000 to manufacture a 100 kg batch = $5000 per kg of DS
- Raw materials + DS total = $11,000 per kg of DS
- 100 mg tablet manufacturing cost = $0.25 per tablet with an 80% manufacturing yield
- 30-count bottle packaging cost = $2 per unit
- Once daily dosing frequency

In this example, the total cost to produce one kilogram of DS is $11,000. The DP is a 100 mg tablet with an 80% yield, which is equivalent to 0.000125 kg or $1.38 worth of DS. The cost to manufacture a tablet is $0.25 making the total tablet COGs equal to $1.63. A bottle containing 30 tablets therefore has $48.90 worth of tablets. Add the $2 packaging cost per unit, and this brings the COGs for a single bottle to $50.90. Approximately, 12 bottles will be taken by a patient each year, for a total annualized COGs figure of $610.80 per subject.

A different, more simplistic approach for large-molecule COGs involves summing the total estimated cost of manufacturing DS and DP and dividing that into the estimated yield from a batch of material, and then adding the packaging cost. Like the small-molecule example, all cost per batch and cost per kit estimates can be taken from the budget outputs. Consider another set of inputs:

- DS = $3,000,000 per batch
- DP = $500,000 per batch with two DS batches required to produce one DP batch
- Number of vials yielded per DP batch = 20,000
- Secondary packaging cost = $5 per kit in a single vial configuration
- Every other week dosing frequency

Here, the total DS and DP manufacturing cost is $6,500,000 which is the sum of two DS batches and one DP batch to yield 20,000 vials. The per vial cost is then equal to $325/vial. Adding the secondary packaging cost brings the total to $330/kit. The drug will be administered 26 times per year, bringing the annualized COGs to $8580 per subject.

14.8 Metrics

All supply chains run on data, and metrics are pieces of data that summarize how well the supply chain is performing. In Chapter 1 – Supply Chain Management, nine commonly used supply chain metrics covering forecast accuracy, order fulfillment, inventory management, and supplier performance were introduced as relevant to measuring clinical trial supply chain management. Not all metrics are created equally, though. The most valuable metrics for supply chain planning are the inventory management metrics – WoS, stockouts, and waste. WoS measures how much inventory is on hand relative to future demand, and it can be expressed in numerous ways. One expression is the target WoS as determined by the safety stock and order target supply planning parameters. Another expression is the current number of WoS on hand at a specific point in time. The value of WoS has been covered extensively in previous chapters. It provides the best indicator of resiliency at each level of the supply chain and is the simplest metric for communicating inventory health and potential stockout risk.

The second inventory management metric, stockout rate, measures the rate at which doses and/or dispensing visits are missed due to unavailable drug at the clinical site or depot. Stockout rate is measured using the formula outlined in Chapter 1, but with slightly different nomenclature for clinical trial supply chain management:

$$\text{Stockout rate} = \frac{\text{Number of missed doses or visits}}{\text{Total number of doses or visits}}$$

A corollary to the stockout rate is the service level, which is defined as the percentage of subject visits successfully serviced by on-hand inventory. The service level is calculated using the following formula:

$$\text{Service level} = 1 - \left(\frac{\text{Number of missed doses or visits}}{\text{Total number of doses or visits}} \right)$$

Ideally, the stockout rate is 0% (e.g. a 100% service level), but that's not always realistic given the scale and complexity of global clinical trials. A better goal is to limit the impact of a stockout event to a single dose or visit. Those are typically the result of problems within the physical supply chain as opposed to strategic problems within the supply plan. When a stockout impacts multiple doses and/or multiple clinical sites simultaneously, that's usually because inventory has not been managed well at the depot level or because of multiple catastrophic events happening simultaneously such as a spike in demand coinciding with a lost batch of DS or DP. When capturing stockout rate, it's also a best practice to capture the root cause of the stockout and to identify corrective actions so the stockout does not occur again. Preventive measures can then be shared and implemented across clinical supply chain and other relevant functions.

Waste is another valuable metric in clinical trial supply chain management that measures the amount of unused investigational medicinal product (IMP) against the total amount of IMP produced to support that trial. The first point to consider is that there are two types of waste, inherent waste and excess waste. Inherent waste is structurally imbedded in the clinical protocol design elements and the physical supply chain of the study. Two examples of inherent waste are the RTSM minimum values that ensure some amount of drug will be leftover at the clinical sites and the capacity constraints in upstream manufacturing that force a minimum amount of DP or DS to be produced with each batch. The other much more prevalent type of waste is excess waste, which is waste that occurs when material expires or simply goes unused by the end of the study because supply has exceeded demand.

Waste is a complex metric not only because there are two types of waste but also because waste exists at every level of the supply chain. These complexities are captured in the two formulas below:

$$\text{Total waste} = \Sigma\left(\frac{\text{Number of unused FP kits}}{\text{Number of FP kits produced}}\right)+\left(\frac{\text{Number of unused DP units}}{\text{Number of DP units produced}}\right) + \left(\frac{\text{amount of DS remaining}}{\text{amount of DS produced}}\right)$$

$$\text{Excess waste} = \text{total waste} - \text{inherent waste}$$

The total waste formula can be further expanded for small-molecule programs to the unconsumed capsule or tablet level, although this data is more challenging to retrieve because it's not always clear how many pills were consumed by subjects as opposed to returned to the clinical sites unused. This data is available in accountability logs, but collecting this information is difficult and time consuming. Large-molecule programs that package multiple vials into a kit may also face this issue, which is another reason to pack single vial kits wherever possible.

The best time to measure waste is after a clinical trial has concluded. At that point, it's a relatively straightforward exercise to calculate how many dispensing visits occurred and how many kits either expired or were left unconsumed because this data is all readily available in the RTSM system. The amount of leftover DP and DS that was allocated to the study can also be tallied. Attempting to measure waste during the trial is challenging because the numerator (e.g. the number of unused kits or units) and the denominator (e.g. the total number of kits produced) is a moving target, so measuring waste in the middle of a trial is not indicative of where the final waste tally will land.

Measuring waste informs how well the planning process aligned supply to demand, but due to the variety and complexity of the clinical trial landscape, there is no defined standard for how much waste is acceptable. One study[1] found that on average, only 45% of the total amount of drug manufactured for a study was eventually dispensed to a subject. In other words, more than half went to waste. More recent studies have found that on average up to 70%[2] of the DS manufactured for clinical trials goes to waste. Based on these figures, achieving a target of less than 20–30% FP excess waste seems reasonable. No matter where the metric lands at the end of the trial, analyzing what happened and searching for improvement opportunities within the planning process is a valuable exercise.

The fourth inventory management metric is working capital in inventory. This is the total amount of cash tied up in inventory, and it's particularly useful in clinical trial supply chain management for measuring how many dollars are invested in sustaining DP and DS inventory levels. Working capital in inventory translates the upstream inventory management strategy into a cash figure that can be used to understand trade-offs between increasing supply chain resiliency and decreasing the amount of cash available to the company. This analysis is applicable at both the product and the portfolio levels. We'll expand on how to incorporate working capital in inventory into portfolio level analyses in Chapter 15 – Clinical Supply Chain Planning Processes.

Order fulfillment and supplier metrics are also important tools for understanding how manufacturing and distribution sites are performing with respect to executing the plan. On time, in full (OTIF) is a challenging metric for DS and DP manufacturers to achieve in early clinical development when process and analytical development are ongoing. OTIF is more valuable for FP manufacturing and distribution though where a high percentage rate is expected even for early phase studies and supplies. PLT was covered extensively in Chapter 12 – Upstream Supply Planning: Primary Packaging, Drug Product, and Drug Substance. PLT is a key input to the supply planning process, and it can be measured not just in absolute terms, but also with respect to variance via PLT standard deviations. As PLT standard deviations increase, supply chain reliability decreases. Defect rates offer diagnostic value and should be tied to OTIF rates. Finally, customer service level is an excellent qualitative metric for both internal and external suppliers. To get maximal value out of customer service levels, a justification for the ranking is required so that improvement opportunities are clear and actionable. Each of these metrics takes on added importance and value when manufacturing and distribution activities are outsourced to third party suppliers. We'll return to this topic in Chapter 20 – Contract Organization Management and Oversight.

That leaves the final metric discussed in Chapter 1, forecast accuracy. Measuring forecast accuracy in the clinical trial space is controversial because it's a less useful metric here than it is in consumer good planning. Unlike consumer supply chains, or even commercial biopharmaceutical supply chains for that matter, clinical supply chain demand forecasts start with a fixed view of demand based on the clinical protocol and its associated design elements (e.g. sample size and duration of treatment). Often, changes to the protocol are made, and this alters the context against which the original demand forecast was created. In effect, these changes reset the forecast accuracy baseline, so what was accurate last month against an original set of assumptions will no longer be accurate against the new set of assumptions. Chasing forecast accuracy in clinical trial supply chain management only works if the forecast parameters are static, which they almost never are.

Think about what measuring forecast accuracy seeks to accomplish, too. Commercial supply chains track forecast accuracy because it translates to higher customer service levels and lower inventory carrying costs. These are the exact goals of the clinical planner but utilizing the demand and supply planning strategies laid out in earlier chapters combined with utilizing other metrics such as WoS, stockout rates, and waste offer far more value in accomplishing this goal for clinical trial supply chains. Keep in mind though that not relying on forecast accuracy to improve supply chain performance doesn't mean the clinical planner shouldn't try to be accurate! Chapter 13 – Advanced Concepts in Demand and Supply Planning explains why attempting to build an accurate forecast is the right approach, but it's important not to confuse attempting to forecast accurately with measuring forecast accuracy.

14.9 Dashboards

The most efficient way to communicate the supply chain metrics for a program is through development and implementation of a dashboard. A dashboard is a reporting tool that displays a combination of metrics for review and analysis of where supply chain performance is lagging. An example dashboard is presented in Figure 14.7.

From a structural perspective, the dashboard identifies the respective molecule or study in each row and the metric in each column. Each metric has a unit of measure and a target. The DS, DP, and FP inventory on hand and working capital in inventory targets are all program or study dependent so the targets are captured below each metric value. Meanwhile, the service level and waste metrics are set globally so the targets are identified in the column header. Note that in this example, the service level target is set to 99.7% which is equal to six standard deviations from the mean, but this may not be high enough for mature clinical supply chain organizations. The waste metric is limited to the FP level for relative ease of capture, and the target is set to 35%, an improvement upon historical clinical trial supply chain performance.

With the dashboard setup, it's easy to see the rich classification of information through the metric color coding. In this example, green represents "On track or low risk," yellow represents

Program	DS inventory on hand (*Current weeks-of-supply*)	DP inventory on hand (*Current weeks-of-supply*)	Working capital in inventory (*Dollars*)
Molecule 1 (*Large molecule*)	52 Weeks-of-supply (*Target 52 WoS*)	0 Weeks-of-supply (*Target 0 WoS*)	$3,000,000 (DS) (*Target $3M*)
Molecule 2 (*Small molecule*)	26 Weeks-of-supply (*Target 52 WoS*)	26 Weeks-of-supply (*Target 0 WoS*)	$1,000,000 (DS) $250,000 (DP) (*Target $2M*)
Molecule 3 (*Large molecule*)	0 Weeks-of-supply (*Target 0 WoS*)	0 Weeks-of-supply (*Target 52 WoS*)	$0 (*Target $4M*)

Study	Status	FP inventory on hand (*Current weeks-of-supply*)	Service level (*SL target = 99.7%*)	Finished product waste (*Waste target = 35%*)
Study 1	Complete	N/A	1 Stock out against 1,478 doses SL = 99.932%	40.99% = (7648 kits unused/18,660 packed out)
Study 2	Complete	N/A	6 Stock outs against 3,540 doses SL = 99.83%	56.83% = (4660 kits unused/8200 packed out)
Study 3	Enrolling	26 Weeks-of-supply (*Target 26 WoS*)	2 Stock outs against 820 doses SL = 99.75%	N/A
Study 4	Enrolling	13 Weeks-of-supply (*Target 39 WoS*)	5 Stock outs against 437 doses SL = 98.86%	N/A
Study 5	Planned	52 Weeks-of-supply (*Target 39 WoS*)	N/A	N/A

Figure 14.7 Clinical Program and Trial Supply Chain Dashboards

"Off target or medium risk," and red represents "Off target and/or high risk." From the program dashboard, it's clear that Molecule 1, shaded all in green, is in excellent shape because all three metrics are on target with plan. Molecule 2 has some risk identified in the DS and DP inventory on-hand metric because DS WoS are running 26 weeks below target, and it seems like this is due to DS inventory getting converted to DP but then getting held at that level of the supply chain. This could be undesirable because the DP has a relatively short shelf life (e.g. <24 months). Meanwhile, Molecule 3 is in the worst shape with no DS or DP inventory on hand. This program is likely enduring a supply shock where a batch or batches of DS have been lost, and the supply chain is struggling to get back on track.

At the study level, Studies 1 and 2 have both completed and performed relatively well with respect to service level but each struggled with waste to varying degrees. From a stockout perspective, it would be interesting to further understand what led to the six stockouts in Study 2. Was it an isolated event that caused multiple missed doses or where there multiple events that led to individual missed doses? To answer these questions, implementing a stockout dashboard is a helpful complement to the program and study dashboards. With respect to waste, Study 1 performed relatively well compared to Study 2, so it's natural to wonder what happened in the planning process for Study 2 that led to so many kits going unused. Did this occur due to a major delay in enrollment, poor expiry, poor supply planning, or a combination of these events and more? Study 3 is performing well overall, but Study 4 has already experienced a relatively high number of stockouts against the number of administered doses. Follow up there to understand why those stockouts are happening would be a valuable exercise. Finally, Study 5 is in planning and seems to be on track with the initial amount of FP inventory on hand to support study start-up.

This is a relatively simple example of a dashboard, but it shows the power of summarizing and highlighting important data points within the supply chain to prompt action where it's needed. Careful thought needs to be put into the metrics and appropriate targets for each program and study and into how often the dashboards are refreshed for analysis. In an ideal state, dashboards are automated through technology and show real-time information, but this may not be possible in younger organizations. If dashboards require a large resource investment to maintain, then updating them on a monthly or even quarterly basis may be the right cadence.

14.10 Key Takeaways

1) Effectively communicating the supply plan and inventory management strategy for a program is necessary for stakeholder alignment across the supply chain. This alignment is important because the supply plan meets the demands of the clinical program, drives resource needs within the sponsor and at sites within the supply chain, and has a significant impact on the program budget. Data, metrics, and dashboards are tools that communicate supply chain performance and risks to the clinical supply chain function and executive leadership.
2) Charts that provide a visual representation of the FP and upstream supply plans are especially useful communication tools. They not only illustrate the plan's effectiveness at avoiding FP stockouts but also support optimization analyses and show how well the supply plan meets the desired supply planning parameters (e.g. safety stock and order target values) of the program. These analyses are enabled by adding WoS data to complement demand and supply.
3) Reports that capture GMP production and distribution activity across the portfolio also provide important insights for activity prioritization and human resource planning. These reports aggregate production and/or distribution event data from production and/or distribution summaries and enable the clinical supply chain and quality functions to identify the highest priority activities. Hourly work estimates can be added to these reports and then summed across the portfolio to understand hiring needs.
4) Information from the production plan and distribution plan summaries can also be transformed into a budget for the program. Creating a direct connection between the supply plan and the budget has multiple benefits including making the budgeting process more efficient and enabling quick cost estimates for proposed changes to the study design that impact the supply chain.
5) At the FP level, creating the budget is a multi-step process that involves identifying cost drivers within the FP supply chain. After cost drives and specific costs associated with those drivers are categorized, they should then be mapped to activity occurrences within the production plan and distribution plan summaries. Inputs to the budget should be reviewed on a quarterly or semiannual basis for accuracy.
6) A complete budget for GMP activities can be built when DP and DS costs and logistics-related costs are included. DP and DS manufacturing costs can be pulled directly from the upstream production summary using the same methodology used at the FP level. COGs analyses can also be built from the same inputs that make up the overall program budget.
7) Metrics capture how well the supply chain is performing, but not all metrics are created equally. In clinical trial supply chain management, the most important are inventory management metrics such as WoS, stockout rate, service level, and excess waste. Excess waste occurs when supply expires or goes unused by the end of a clinical trial. Measuring waste is complex because it exists at every level of the supply chain, and getting an accurate measurement of waste is only possible at the end of the trial.
8) The most efficient way to communicate metrics for a program is through the development and implementation of a dashboard, which is a reporting tool that displays a combination of metrics for review. Dashboards enable quick analysis by summarizing the most important metrics and creating a visual representation of poor performance and risks within the supply chain via a color-coding system.

Notes

1 Tufts Applied Clinical Trials article (downloaded).

2 https://www.clinicaltrialsarena.com/sponsored/how-to-cut-drug-supply-costs-with-seamless-forecasting-and-irt-systems/.

15

Clinical Supply Chain Planning Processes

15.1 The Clinical Supply Chain Planning Processes

In this chapter, we review the business processes that the clinical supply chain function uses for supply chain planning and execution. Business processes are a cornerstone of supply chain management because they provide a structured approach to performing a sequence of interconnected tasks. The business processes we'll cover in this chapter are designed to accomplish three things. First, they drive consistency across the creation and maintenance of demand and supply plans. Second, they ensure that the most current information and assumptions needed to revise the plan are made available to the clinical planner in a timely fashion. Third, they facilitate effective communication to the stakeholders who act on the supply plan outputs. While this isn't an exhaustive list of business processes, these are the six most impactful:

1) *Demand and Supply Plan Creation*: The process of building the initial demand and supply plan for a new clinical trial or program.
2) *Demand and Supply Planning*: The process of updating the demand and supply plan for each study and program. This process is composed of four steps: actualization, reconciliation, requirements review, and allocation.
3) *Clinical Demand and Operations Planning*: The process of obtaining the inputs and assumptions required to update the demand and supply plan. This process also creates a formalized pathway for review and communication of the final supply plan to relevant stakeholders.
4) *Strategic Demand and Operations Planning*: The process of obtaining stakeholder alignment on the upstream manufacturing strategy of each program and the associated cost–risk trade-offs of production decisions.
5) *Supply Plan Handover and Order Readiness*: The process of providing the supply plan to the manufacturing and/or distribution sites responsible for its execution. This process includes tracking order readiness activities and feeding order status back to the clinical planning function.
6) *Global Supply Execution Readiness*: The process of planning and executing all activities required to enable the first shipments of investigational medicinal product (IMP) into a specific country or region.

Executing these processes takes on added importance in clinical trial supply chain management because of the large number of inputs required to create and maintain the demand and supply plan and because of the many stakeholders who are impacted by the supply plan. There are dozens of variables the clinical planner needs to collect and synthesize in their analyses, and the best way to organize that information and keep it current is through these processes. Without the structure these processes provide, each clinical planner is left to find their own way when creating and

Supply Chain Planning for Clinical Trials – A Practical Guide, First Edition. Ryan Mills.

maintaining their plans and ensuring they are executed properly. The result is inefficiency and an increased risk of stockouts and excess waste.

Well-designed business processes share a set of common characteristics that define the purpose of the process, the activities required to execute the process, and those responsible for process execution. Each of these characteristics matters because without them the business process becomes inefficient or worse, ineffective. These characteristics are listed below:

- *Process Objective*: The goal, objective, and/or outputs of the business process. Defining an objective gives the process direction and helps define its scope.
- *Process Scope*: The boundaries of the business process and what activities and/or tasks it will and will not cover. Defining a scope gives the process focus while limiting the scope prevents distraction and inefficiency.
- *Process Owner*: The business function or role that is accountable for the business process to be executed. The owner is not necessarily responsible for each step within the process. Having a single process owner means that one person or function takes accountability for process adherence and maintenance as the organization grows and changes over time.
- *Process Trigger or Cadence*: An event that triggers the process to begin or a prespecified cadence within which the process should be completed. Identifying a clearly defined trigger or cadence ensures the process is executed in a timely fashion to meet its objective.
- *Process Map*: A diagram containing the sequence of steps or tasks required to complete the process goal or objective. Each step has a description of what to do and who is responsible for executing that activity or task. The process map also includes clear linkages and interdependencies between process steps. Building a process map creates a visual illustration of the process and provides a training document for new team members.
- *Process Responsibilities*: An assigned role that is responsible for completion of each step. Process maps typically include both the responsible party for each step and the functional area or role responsible for providing inputs to that process. Documenting responsibilities and functional areas that provide inputs to the process eliminates confusion about who is supposed to complete a task and where inputs for that task should come from. The Responsible-Accountable-Consult-Inform (RACI) framework is commonly used in business process management to identify who is responsible for a task, who is accountable for the task (e.g. the process owner), who needs to be consulted to execute a task, and who needs to be informed of the task.

Over the remainder of this chapter, we'll cover the six business processes through the framework of theses process characteristics. Most of our time will be spent on the process map. We'll walk through each step and then illustrate how the processes relate to and reinforce each other. We'll also touch on supply network planning (SNP), a process that's typically reserved for commercial-stage biopharmaceutical companies and three keys to business process success. At the conclusion of this chapter, we'll understand how the demand and supply planning elements we've covered come together to enable clinical trial supply chain planning and execution. We'll also have a clearer picture of how the clinical planner and clinical supply chain function work on a day-to-day basis to deliver the supply chain.

15.2 Demand and Supply Plan Creation

Demand and supply plan creation is the logical place to begin our review. Creating a holistic demand and supply plan for a study and program is the objective of this process, and the scope includes both the finished product and the upstream drug product (DP) and drug substance (DS)

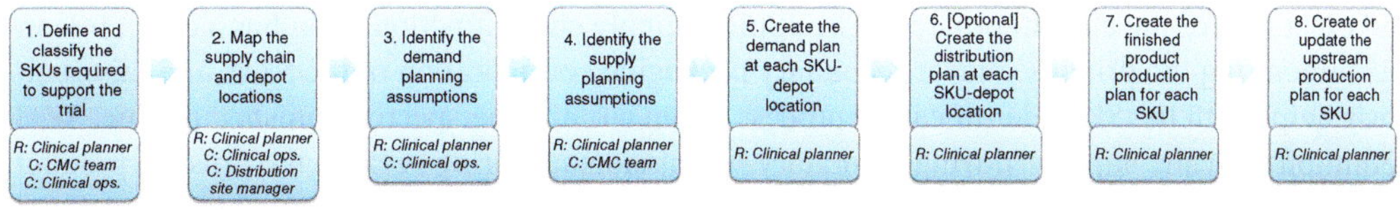

Figure 15.1 Demand and Supply Plan Creation Process Map

materials required to support a trial. Raw materials may or may not be in scope depending on whether or not the clinical supply chain function has the responsibility to plan for them. Additionally, not all clinical trials may require an end-to-end demand and supply plan, particularly Phase I studies that use a single site, have a single dose regimen, and/or use a single stockkeeping unit (SKU). However, as trial complexity increases beyond those parameters, a demand and supply plan should be created even if the trial is still in Phase I.

Process ownership rests with the head of the clinical supply chain function who determines which trials require creation of a demand and supply plan and which clinical planners will be assigned to a given program and/or trial. The trigger for creating the first demand and supply plan is the availability of the clinical protocol which contains the intrinsic design elements required to create the demand plan. Once the protocol is available, the clinical planner should take the steps outlined in Figure 15.1 to create the plan.

Collaboration is fundamental to demand and supply plan creation. The clinical planner is responsible for each step, but there are many other stakeholders they need to consult with along the way. In Step 1, the chemistry, manufacturing, and controls (CMC) team provides the formulations and/or doses that will be manufactured, while the clinical operations team provides input on how those doses should be packaged or bundled in the finished product. In many cases, there is a core team for the product with members from the CMC team and clinical development who agree on the initial formulation and dose that will be manufactured to support the first study in advance of protocol availability. In Step 2, clinical operations provides the country list, and the distribution site manager informs the clinical planner of which depots supply the countries within that list. Clinical operations also provides many of the demand planning assumptions in Step 3 via the clinical development plan and the study protocol. Interpreting the intrinsic design elements of a protocol is not always straightforward, especially when the study design is complex and involves multiple arms and/or cohorts so reviewing the assumptions with clinical operations is a useful approach. Similarly, the clinical planner needs to collaborate with the CMC team on identifying the supply planning assumptions in Step 4. The DS and DP technical leads provide estimated bill of material (BOM) quantities and yields.

Also note that the process map is silent on the technology solution used to create the plan. That's because these process steps are applicable no matter which system is used for planning. At companies where a dedicated planning technology has been implemented, there are usually detailed instructions for how to set up the initial plan in the system. More information on technology is provided in Chapter 22 – Clinical Supply Chain Planning Tools and Technology.

Demand and supply plan creation begins with demand planning assumption identification and documentation. From Chapter 8 – Finished Product Demand Planning Basic Concepts, the first step in the demand planning process is to define the SKUs that will be supplied to the trial and identify the manufacturing site(s) where secondary packaging will take place for the study. SKU definition happens through collaboration between the CMC team responsible for the manufacture and release of the drug and the clinical development team responsible for authoring, publication, and execution of the clinical protocol. The clinical supply chain function often has a direct role in recommending the primary and/or secondary packaging configuration of the SKU, and this obviously helps with the SKU identification process. From a timing perspective, finished product SKU definition typically happens during regulatory submission authoring because the finished product configuration must be listed in the investigational new drug (IND) application or Investigational Medicinal Product Dossier (IMPD) so this information needs to be available by the time IND or clinical trial application (CTA) authoring begins.

Mapping the supply chain and the distribution network is the second step, and the process for this was covered in Chapter 11 – Finished Product Supply Planning: The Distribution Plan. The clinical supply chain function typically owns the secondary packaging supplier and location selection process. This step is necessary to create the appropriate SKU–depot locations within the plan and the connections between the supplying central depot(s) and the receiving regional depots. Once the packaging design and the depot locations are selected, the remaining demand planning assumptions need to be identified and documented in the third step. These assumptions include:

- Selecting a weekly or monthly planning interval.
- Determining the individual subject demand curve for each arm of the study (see Chapter 8 for how to create this).
- Determining projected screen failure and enrollment rates.
- Selecting a demand overage multiplier.
- Documenting the site activation schedule and calculating site seeding quantities.

The fourth step is to identify and document the supply planning assumptions. These include:

- Selecting the finished product safety stock and order target values at the manufacturing and distribution depots.
- Identifying the finished product production lead time.
- Identifying the central and regional depot distribution lead times.
- Identifying the randomization and trial supply management (RTSM) expiry parameters (e.g. do not dispense, do not ship, and do not count).
- Identifying the DP and DS expiry period.
- Determining the upstream safety stock and order target values for each applicable SKU.
- Identifying the upstream production planning assumptions including the production lead time, BOM quantities, and production yields.

Once all the demand and supply planning assumptions have been gathered and the SKU–depot locations have been created, the fifth step is to create a demand forecast at each applicable SKU–depot location. Chapter 8 walks through a step-by-step process for building the demand plan. Recall from Chapter 13 – Advanced Concepts in Demand and Supply Planning that when the initial demand forecast is built it's strongly recommended to use the aggregate forecasting technique to place all individual subject demand at the depot location with the highest expected enrollment. If aggregate forecasting is used, the only source of demand at regional depots should be site seeding demand. If aggregate forecasting is not used, then regional depots should have subject and site seeding demand. Central depots should have subject demand if they supply clinical sites, site seeding demand, and distribution demand from all applicable regional depots.

The sixth step is optional and involves building the distribution plan at each applicable SKU–regional depot location. Some studies may only utilize one depot, and, in that case, this step may be skipped. But, if even one regional depot is utilized, this step needs to be taken before going to Step 7. Assuming the aggregate forecasting step has been implemented in Step 5, the relevant site activation schedule at each regional depot should be populated, and the initial site stocking quantities should be calculated and entered in the plan. The site stocking quantities create demand at each regional depot. The clinical planner then enters distribution orders to provide sufficient supply at each regional depot to support site activations. These distribution orders then become distribution demand at the central depot location. Chapter 11 covers this entire process. All sources of demand are now accounted for and should be translated into production or distribution plans at each depot location.

Creating the finished product production plan for each applicable SKU at the central depot is the next step. Do not take this step before Step 6. If distribution demand is added to regional depot locations after the finished product production plan has been finalized, then that impacts the finished product production plan by increasing the requirement for more finished product supply to service that demand. If this happens, the plan must be reassessed. Expiry considerations also need to be layered into the finished product production plan during this step. Strategies for finished product production and expiry planning are laid out in Chapter 9 – Finished Product Supply Planning: The Production Plan and Chapter 10 – Finished Product Expiry Planning, respectively.

The final step in the process is to create the production plan for each upstream SKU. If the clinical trial will be supplied with a new primary packaging or DP SKU and/or if the DP will be manufactured from a new DS SKU, then a new upstream production plan needs to be created. On the other hand, if the upstream primary packaging or DP SKUs exist and are actively used in the manufacture of other finished product SKUs, the production plan for the primary packaging or DP SKU needs to be updated with demand from the new finished product SKU placed against it. The production plan for each upstream SKU should service downstream demand and adhere to the desired inventory management strategy of the company. Upstream production planning strategies are laid out in Chapter 12 – Upstream Supply Planning: Primary Packaging, Drug Product, and Drug Substance.

Building the initial demand and supply plan brings together many of the concepts we've covered so far from basic demand planning concepts through to advanced planning techniques. Demand and supply plan creation is the first step on the supply chain management journey of the trial. Almost as soon as the first plan is built though, it becomes outdated as the assumptions it rests on begin to evolve. Changes to demand assumptions, supply assumptions, and the expiry period all impact the plan, and we'll learn how to address these changes via the demand and supply planning process.

15.3 Demand and Supply Planning

Demand and supply planning is the process of updating the demand and supply plan when the inputs and assumptions that underpin the plan change. Chapter 7 – Supply Chain Planning for Clinical Trials summarized the objective of demand and supply planning which is to ensure that investigational medicinal products (IMPs) are delivered to all clinical trial sites and depots in sufficient quantities to support all IMP dispensing visits when those visits happen, and to accomplish this for the least amount of monetary investment possible. Like demand and supply plan creation, process ownership resides with the head of clinical planning because they are accountable for the supply plan outputs of the clinical supply chain team.

The scope of the process includes every active clinical trial demand and supply plan that generates a supply plan for finished product, primary packaging, DP, and DS SKUs. Raw materials may also be in scope if the clinical supply chain function owns raw material planning. Bear in mind that an early phase program with only a single Phase I trial may be simple enough to calculate demand and supply with just a few inputs. In that case, a finished product demand and supply plan may not be necessary, and only the DP and DS supply plans are necessary. The head of clinical planning should judge which programs and studies fall under the scope of the process.

The trigger for demand and supply planning is typically based on a predefined cadence. At minimum, the supply plan for an actively enrolling trial should be reviewed monthly, but if there is significant site activation, screening, or enrollment activity, then the plan may need to be updated more frequently. Similarly, if there are high impact changes on the supply side such as a failed batch or a major shift in the upstream production schedule, the plan needs to be updated to understand the impact of these changes. As the business process owner, the head of clinical planning sets the cadence. The clinical planner should also be empowered to update the plan on an as-needed basis when circumstances dictate refreshing the plan.

The business process map for demand and supply planning is summarized in Figure 15.2 where there are three primary steps: (1) actualization; (2) requirements review and allocation; (3) reconciliation and allocation. Each step contains several subtasks, and these should be followed in the prescribed order to make the process as efficient as possible.

The first step in the process is actualization, which is defined as the conversion of forecasted values into actual values. Actualization takes place at three points in the demand and supply plan: within the demand forecast, within the production plan, and when expiry period extensions occur that impact finished product, DP, and/or DS batches. Actual subject enrollment and demand data is warehoused in the Randomization and Trial Supply Management (RTSM) system and can be accessed via the subject dispensing report (SDR). The SDR contains:

- The SKU description.
- The number of kits dispensed at each visit.
- The date the kits were dispensed.
- The country of the site where the subject is enrolled.
- The supplying depot for each site.

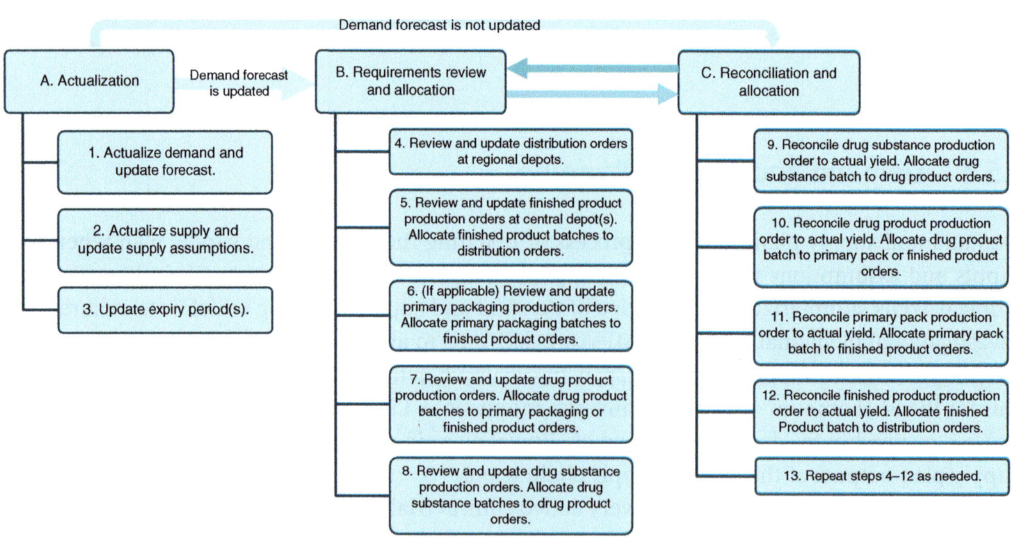

Figure 15.2 Demand and Supply Planning Process Map

Table 15.1 Actualized Subject Demand Table

Planning Interval: Week 10		
Kit Type	**Depot Location**	**Quantity of Kits Dispensed**
Active	United States	10
Active	Germany	8
Placebo	United States	12
Placebo	Germany	2

To actualize subject demand within a given planning interval, SDR data can be summarized in a table that totals the number of kits dispensed by kit type and depot location as in Table 15.1.

Most RTSM systems can generate the SDR as a spreadsheet where data can be further summarized into tables. The number of kits dispensed can then be manually transferred into the planning system directly by the clinical planner. Manually actualizing demand is a resource-intensive process, but it yields one notable advantage. The clinical planner sees exactly how actual demand varies from the forecast. Many global biopharmaceutical companies integrate RTSM systems directly into their planning solution, so demand is actualized in real time and the forecast is automatically adjusted. This approach results in a massive time savings for the organization, but it also removes an opportunity for the clinical planner to analyze how demand is changing if the solution does not show the delta between the original forecast and the actualized demand values. The best planning tools highlight those changes so the clinical planner has visibility to demand trends and can act accordingly.

Actualizing demand is the first half of this subtask. The clinical planner also needs to assess if adjustments to future demand are necessary and update the forecast accordingly at each SKU–depot location. For example, when actual demand is lower than the forecast, subject enrollment needs to be pushed out to a future date when it hasn't happened. If this trend continues over several months, then the enrollment rate projection may need to be revisited and the enrollment curve for the study amended. Conversely, if subject demand is higher than projected, that means enrollment is going faster than originally forecast and future subject demand needs to be pulled in. Additionally, if aggregate forecasting has been utilized, then subject enrollment needs to be actualized and entered at the correct SKU–depot location while removing subject demand from the depot location where the aggregated forecast was being held.

The next step in the actualization process is to convert planned production and distribution orders into actual receipts. There are two elements of each order that need to be actualized within the plan, the quantity received that's available for clinical use and the date the batch is available for use. On the production side, the quality organization is responsible for the batch release process which is a prerequisite for shipment of a batch to a clinical site. Quality needs to communicate four items to the clinical planner as an output of their batch release process: confirmation of the release, the date of release, the quantity of units available for use, and the countries in which the batch is released for use. The confirmation and date of release are straightforward, but as we've learned, the quantity available for use must account for routine and nonroutine sampling requirements such as in-process testing requirements and stability retain samples. The country list matters as well so the clinical planner is able to ensure regulatory compliance with the approved application in each country.

In clinical-stage biopharma companies, batch release is typically a paper-based process, so actualization of supply also needs to be performed manually by the clinical planner. In commercial-stage organizations, there is usually an enterprise resource planning (ERP) technology that contains an electronic record of all good manufacturing practice (GMP) inventory and serves as the mechanism to manage batch disposition status. The ERP system may or may not be directly integrated with the demand and supply planning technology solution, but when it is the same point around visibility into how actuals differ from the forecast also apply. The clinical planner must see the variance between planned order size and actual order size. These variances can then be analyzed to determine if updates need to be made to supply planning inputs like the BOM quantities, lead times, and yields. This step is even more important for supply planning than demand planning because there are many fewer production-related events than demand-related events. Getting the supply assumptions right also matters more because making a mistake here has a greater impact on the plan's ability to supply the study. Given the importance of getting these assumptions right, there is a separate business process dedicated to managing demand and supply plan inputs which we'll cover in the next section.

The third sub-step in the actualization process is to update the expiry period for DS and DP batches when extensions are issued. Finished product expiry dates should only be extended if the batch has a US-only label with no expiry date printed on it or when a shelf-life extension over-labeling operation has been completed. Otherwise, expiry extensions at the DP level do not impact the expiry dates of finished product batches because the expiry date is printed on the clinical label, and this is not automatically updated when the DP expiry period is extended.

Each actualization step impacts the demand and supply plan in different ways. Actualizing demand decreases or increases the requirement for supply at that SKU–depot location. Actualizing supply decreases or increases the amount of finished product, primary packaging, DP, or DS material available to allocate to downstream distribution or production orders. Actualizing expiry increases the amount of time an SKU is available for consumption, potentially decreasing future requirements for resupply material. Fortunately, there is a logical flow to analyzing these changes, and this begins with requirements review, the second step in the demand and supply planning process.

Requirements review is a sequential assessment of how changes at a downstream level of the supply chain impact requirements for supply at the upstream level. Requirements review begins at the most downstream level of the supply chain, the regional depots responsible for distribution of IMP from the depot to clinical sites. When the demand forecast is updated at a regional depot the requirement for the finished product SKU from the supplying depot changes in the following ways:

- When total demand is pushed out, then a resupply batch is needed later than originally forecast.
- When total demand is pulled in, then a resupply batch is needed earlier than originally forecast.
- When total demand is added to an SKU–regional depot location, this increases the quantity of the resupply batch over what was originally forecast.
- When total demand is removed from an SKU–regional depot location, this reduces the quantity of the resupply batch below what was originally forecast.

After changes to demand occur at a regional depot, the clinical planner needs to assess if the distribution orders from the previous version of the plan are still acceptable or if they need to be updated to meet the revised demand forecast. Early in the study, site activation dates may move around, or sites may be added or removed entirely, changing the timing and amount of supply required to stock up those sites. Later, in the study, the weeks-of-supply summary introduced in Chapter 11 – Finished Product Supply Planning: The Distribution Plan helps to identify which depots have lower or higher-than-expected inventory levels after demand actualization. The weeks-of-supply summary shows how many weeks-of-supply of each SKU are available at each depot, so it's easy to see when imbalances occur between SKU–depot combinations. If one SKU–depot combination is lower than the others, this is a signal that more of that SKU needs to be supplied to that depot sooner. If one SKU–depot combination is higher than the others, this is a signal that less of that SKU needs to be supplied to that depot or that distribution orders should be pushed out.

Recall that a distribution order at the regional depot level creates demand at the upstream supplying depot. Thus, after the distribution plan has been updated at each regional depot, the next step is to analyze how those changes have impacted the supply plan at the upstream supplying depot. In most supply chains, the upstream supplying depot is the central depot (e.g. the manufacturing site of the finished product), but in some cases, the upstream supplying depot may be another regional depot. In the latter case, the distribution plan of the downstream regional depot needs to be updated before the upstream regional depot plan is reviewed.

Changes to distribution orders impact the upstream supplying depot in two ways. First, the clinical planner needs to analyze the allocation strategy at the upstream supplying depot. If distribution orders have pulled in or increased, the clinical planner must assess if sufficient inventory is available at the supplying depot to meet this increase in demand. If distribution orders have pushed out or decreased, the clinical planner must assess if it still makes sense to allocate the same batch as in the previous plan. Remember that distribution planning uses a combination of the first in first out (FIFO) and last in first out (LIFO) strategies, so there might be a newer batch with better expiry dating that makes sense to send to a regional depot when demand has pushed out. Second, the clinical planner must assess the impact of these demand and allocation changes on the finished product production strategy. The logic here mirrors that in the bullets listed above. When distribution orders push out or are reduced, then finished product production orders may need to be delayed or reduced, and when distribution orders pull in or increase in size, then finished product production orders may need to be accelerated or increased.

This logic continues from downstream sites to upstream sites working through each level of the supply chain. If finished product production orders change, this impacts the requirement for primary packaging or DP supply. The clinical planner needs to assess the primary packaging or DP allocation strategy to support finished product production and then determine if changes to the upstream production plan are necessary. If changes to the DP production plan are made, this impacts requirements for DS, so the allocation strategy and production plan may also need to be changed.

Expiry period updates at the finished product level that lead to shelf-life extensions of finished product batches also trigger the requirements review step. That's because shelf-life extensions increase on-hand inventory beyond the original expiry date of the batch, and this both delays the need for a resupply production order and reduces the quantity of the next resupply order. These events impact the finished product production plan which then impacts the upstream allocation and production strategy. Similarly, expiry period extensions at the DP and/or DS can change the allocation strategy for downstream production orders because batches are good for use for longer periods of time.

As the requirements review process step completes, it's important to remember that the production plan cannot always be changed to meet downstream requirements, especially at the DP and DS levels. When upstream production orders are locked, the clinical planner needs to ensure that downstream orders don't exceed the projected available quantities to allocate to the downstream order. This assessment kicks off the reconciliation process step.

A final comment on the requirements review step before moving to reconciliation. If there have been no changes to the demand forecast or to the expiry dates of finished product batches, the requirements review step may be skipped entirely. Once subjects begin enrolling, this situation is relatively rare as forecasted demand actualizes, but before enrollment starts, it's not uncommon for production events to occur without changes to the demand plan. When this happens, the clinical planner can proceed directly from actualization to reconciliation because the downstream requirements for upstream material have not changed.

Reconciliation is the process of aligning downstream production and distribution orders to the actual quantities and release dates of the allocated upstream order. Reconciliation occurs in two situations, when the production plan is locked and DS and/or DP delivery quantities can't be changed or when the supply plan is actualized. In either event, the downstream plan needs to be reconciled with the quantities and timing of upstream orders. The reconciliation process flows upstream to downstream because as the quantity or timing of a DS production order changes, then the DP production order that consumes that DS batch changes. If the DP production order is impacted, then the finished product batch(es) consuming that DP batch are impacted. And, if the finished product batch(es) are impacted, then the finished product distribution plan must be updated.

After the reconciliation step takes place and orders have been adjusted, the clinical planner may need to go through the requirements review process again. This is particularly true when the supply plan delivers significantly more or less than expected. As finished product orders are adjusted to match what's available from recently released DS or DP batches, those adjustments change the next resupply order requirement in the following ways:

- When more DS or DP than expected is made, this pushes out and reduces the quantity of future finished product resupply orders. This change in future finished product supply pushes out and reduces the amount of future DP and/or DS required to support the trial.
- When less DS or DP than expected is made, this pulls in and increases the quantity of future finished product resupply orders. This change in future finished product supply pulls in and increases the amount of future DP and/or DS required to support the trial.

Figure 15.3 captures how the reconciliation process flows naturally back into the requirements review process. As DS or DP production actualizes, the clinical planner needs to assess the impact of those yields on the entire plan. But, this can only be done accurately and efficiently by taking the methodical, step-by-step approach outlined in the demand and supply planning process. If the steps are taken out of order, the result is time-consuming rework. An even worse mistake though is to simply not complete all the steps in the process. Many organizations separate the finished product supply planning role from DP and DS supply planning. This can work, but only when

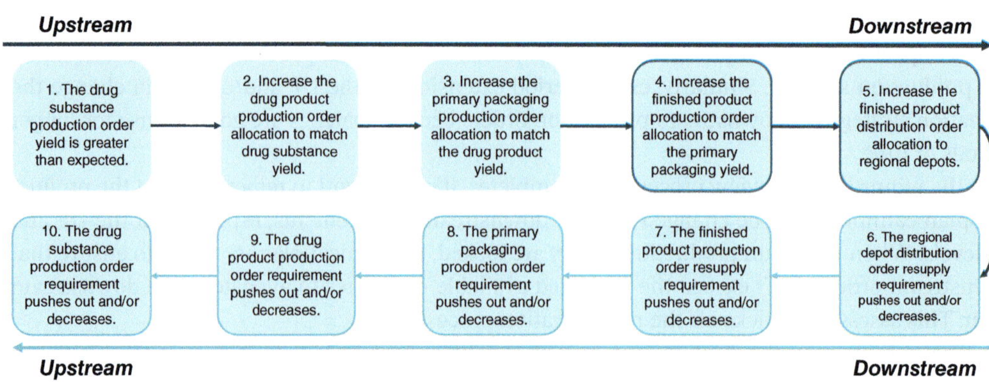

Figure 15.3 The Reconciliation and Requirement Review Cycle

there is a clear signal from the finished product clinical planner to the upstream clinical planner that requirements for DP and DS have changed. The upstream clinical planner must also communicate back to the finished product planner when reconciliation occurs so those changes are translated back into the finished product supply plan.

While we've focused on how actualization drives requirements review and reconciliation, a change to any aspect of the demand and supply plan should trigger this process. For example, making updates to safety stock and order target supply planning parameters or changing the expiry extension strategy should also cause the clinical planner to work through the requirements review step because the finished product production plan is impacted by those decisions.

At the completion of the demand and supply planning process, the clinical planner should always review the supply plan outputs contained within the distribution plan summary, production plan summary, and upstream production plan summary reports. When reviewing the distribution plan summary, the planner should check their allocation strategy, the timing and quantities of shipments, and that resupply shipments containing multiple SKUs are aligned with each other when intended. When reviewing the production plan summaries, the planner should check that the allocation strategy from upstream to downstream batches is sufficient from a timing, quantity, and expiry perspective.

15.4 Operational Clinical Demand and Operations Planning

The demand and supply plan leans on several inputs and assumptions that are not owned by the clinical supply chain function. On the demand side, the study design, enrollment rate, and country list are necessary to create and maintain a forecast. On the supply side, yields, manufacturing lead times, and batch release lead times are typically not consistent through the duration of the trial. When changes happen, it's often not the first instinct of the function who owns those inputs to notify the clinical planner so they can be incorporated into the demand and supply plan. The clinical planner benefits greatly from a structured approach to collecting and reviewing these inputs on a regular basis from their owners. We've also touched on the importance of effectively communicating supply plan outputs to impacted stakeholders, and the same process that facilitates review of planning inputs also supports communicating those outputs.

Clinical demand and operations planning (CD&OP) is the process of aligning on plan inputs and communicating the output of the demand and supply planning process. CD&OP borrows heavily from sales and operations planning (S&OP). S&OP was introduced in Chapter 1 – Supply Chain Management as the process that drives alignment between the sales (e.g. demand-driven) function and the operations (e.g. supply-driven) function. In a standard S&OP process, sales and operations meet with finance to review the sales forecast, the production plan, and the financial implications of the strategy. Effective clinical trial supply chain management also requires bringing the demand-driven functions (e.g. clinical operations) and the supply-driven functions (e.g. the CMC team) together with clinical supply chain.

The head of clinical planning is also the business process owner for CD&OP, but this is the first process that has a sustained collaboration between other functions like clinical operations, quality, and the DS and DP technical leads. The leadership of those functional areas also needs to be aware of the CD&OP process, and they need to champion this work within their own functions. It's the responsibility of the CD&OP business process owner to build relationships with these functional leaders, to communicate the value of CD&OP to their organizations, and to measure process adherence over time.

The scope of CD&OP covers all programs and trials that have an end-to-end demand and supply plan. If a program and/or trial does not have an end-to-end plan, it's out of scope. CD&OP has two distinct subprocesses, operational CD&OP, which we'll focus on in this section, and strategic CD&OP, the focus of the next section. The scope of operational CD&OP focuses on trial and product-specific attributes that drive the demand and supply plan. This includes discussion and alignment on the intrinsic and extrinsic trial design elements that influence the demand forecast and analysis of manufacturing and quality activities that influence the supply plan. Strategic CD&OP, on the other hand, focuses more on the upstream production and inventory management strategies that dictate how much finished product, DP, and DS is made and held in the supply chain.

Like the demand and supply planning process, operational CD&OP typically runs on a monthly cadence and consists of a cyclical sequence of four process steps which are illustrated in Figure 15.4. The first three steps are meetings between clinical supply chain and stakeholder functions, while the last step is the formal publication of the plan.

The process steps are organized sequentially with each taking place during a predefined week of the month. For example, demand review could happen in the first week of the month, supply review in the second week of the month, plan review in the third week of the month, and plan publication in the fourth week of the month. That said, rigidly adhering to these predefined dates does not always make sense in the clinical space. Sometimes changes to the plan happen quickly and the cycle needs to accelerate. At other times, there aren't updates to share and meetings can be cancelled. The clinical planner should have discretion to decide when meetings are unnecessary while the business process owner holds the team accountable to overall process adherence.

The operational CD&OP cycle begins with a demand review meeting between the clinical planner and clinical operations to align on demand forecast inputs. Before the meeting, the planner documents the set of intrinsic and extrinsic assumptions they're using in the plan. Two items that

Figure 15.4 The Clinical Demand and Operations Process

are especially useful to discuss at the meeting are the study design and the projected enrollment curve provided by the contract research organization (CRO) or clinical operations. Reviewing the study design serves as a reminder of potential amendments to the protocol that impact supply requirements. Meanwhile, it's typical for the CRO or clinical operations to make updates to the enrollment curve of the study on a regular basis, and it's critical for the clinical planner to understand the assumptions the clinical operations team is planning toward. That doesn't mean the enrollment curves need to match exactly, but they also shouldn't be far apart from each other. When enrollment rate projections aren't aligned, that usually means clinical operations is projecting a more aggressive enrollment forecast than the clinical planner. When this happens the clinical planner should escalate to the head of clinical planning to determine whether this warrants inclusion in the next strategic CD&OP process cycle.

A final topic to cover at demand review is the timing of key activities. Clinical operations share updates on the first site activation (FSA) dates and when subjects are expected to begin screening in specific countries while the clinical planner confirms when supply will be available for shipment to sites out of each regional depot. Demand review should take place at least monthly, and more often if needed. At each demand review meeting, the clinical planner brings forward the set of assumptions from the previous meeting and updates can be documented which then become the baseline for the next meeting.

Supply review is the second meeting in the cycle where the clinical planner meets with relevant members of the CMC team to align on supply-plan-related assumptions. Specifically, these are the BOM quantities and conversion ratios, estimated yields, and the lead times associated with manufacturing and release of each batch of IMP. Expiry period extension events are also in scope at supply review. For the BOM quantity and conversion ratios, the respective technical leads for DS and DP are the best sources of information. For the manufacturing lead times, the DS or DP lead may have this information, or it might come from the CMC project manager if they're responsible for the overall project plan at the manufacturing site. The batch release lead times should be reviewed with the quality lead. These supply assumptions should also be reviewed at the completion of each manufacturing campaign following final quality release of the batch(es). That means the supply review meeting may not need to happen every month, but it's vital that it does take place after manufacturing or expiry extension events so the clinical planner can get the most current and accurate assumptions into the plan. Once agreed upon, the assumptions should be documented and made available to the entire CMC team.

Plan review is the third meeting in the sequence where the clinical planner shares the supply plan output for endorsement by the CMC team. At minimum, the CMC team needs to be aware of the supply plan outputs for two reasons. First, the CMC lead needs to ensure the production plan fits with the overall CMC strategy and that manufacturing events support both clinical demand and process development activities for the program. Second, the wider CMC team needs to understand when manufacturing campaigns are planned and what quantities are required out of those campaigns so they can plan their activities in alignment with this plan. At this time, adjustments are made to the DP or DS plan as needed to fit the team's needs.

Prior to taking the supply plan to a formal plan review meeting with the CMC team, there is an optional step to conduct an informal plan review with the head of clinical planning, and/or another clinical planner. Demand and supply planning is a complex process so it's always helpful to talk through the assumptions and the thought process behind the plan with the head of the function or another teammate. Conducting an informal plan review with other clinical planners creates that opportunity and helps everyone on the team develop their planning skills in a less pressure-filled environment than more formal plan reviews with the CMC team.

The final step of the CD&OP process is plan publication where the supply plan is made public to key stakeholders including the manufacturing and distribution sites responsible for its execution and other relevant functions such as finance. Some sites may prefer frequent updates even when there are no changes to the plan, while others may want to see the plan only when changes have been made. Supply Plan Handover and Order Readiness is the process of transmitting the supply plan to manufacturing sites and following through on order readiness activities, which is covered later in this chapter.

Operational CD&OP is distinct from yet closely linked to the demand and supply planning process, and Figure 15.5 shows how the two processes merge together. The clinical planner performs actualization before the demand review and supply review meetings. Actualizing demand prior to demand review gives the clinical planner the latest site activation and enrollment trends, and this is useful information to review with clinical operations. One warning, it's absolutely forbidden to discuss or share any potentially unblinding information at a demand review meeting, and for blinded trials documenting a record of the meeting to demonstrate no such information was shared is a best practice. Meanwhile, actualizing supply forms the basis for the supply review meeting because it's not possible to create or update the supply planning assumptions without the yield information from the latest DS, DP, or primary packaging manufacturing event.

After actualization, demand review, and/or supply review complete, the clinical planner has the most current inputs and assumptions to plug into their plan. The planner then makes final adjustments to the demand forecast where necessary and begins the requirements review and reconciliation process steps. Once the demand and supply planning process finishes, the supply plan is ready for plan review with clinical supply chain leadership and/or the CMC team. At the plan review meeting, feedback from functional leadership should be incorporated into the plan via the requirements review and reconciliation steps.

Figure 15.5 Demand and Supply Planning Integrated with Operational CD&OP

Stated another way, the demand and supply planning process always takes place within the context of the operational CD&OP process, but operational CD&OP isn't a prerequisite for demand and supply planning. There will be times when demand review doesn't happen for a lack of topics or supply review isn't necessary because there hasn't been a recent manufacturing campaign. In either situation, the clinical planner must be empowered to update the plan at their discretion. In Chapter 16 – A Week in the Life of a Clinical Planner, we'll encounter situations where updating the plan outside a standard operational CD&OP framework is necessary to react to unexpected events.

15.5 Strategic Clinical Demand and Operations Planning

Strategic clinical demand and operations planning (SCD&OP) steps back from the operational inputs of the demand and supply plan to assess the strategy behind the DS and DP supply plans. Strategic CD&OP is the process of aligning on the upstream production and inventory management strategy of each program by analyzing demand and supply risk against the cost investment to hedge against those risks. This review is completed at an executive level so leadership from clinical supply chain, clinical development, manufacturing, and finance are able to assess the total cost of maintaining the supply chain for a particular program. The head of clinical supply chain is the process owner of SCD&OP. While clinical planners own individual programs, and the head of clinical planning owns the demand and supply planning and operational CD&OP processes, the head of clinical supply chain owns the end-to-end clinical supply chain strategy for the company.

Unlike operational CD&OP, SCD&OP's scope doesn't cover every clinical-stage program. Most early phase programs may require only one or two batches of DS and/or DP to facilitate a Phase I clinical trial, limiting the strategic considerations of these programs. Instead, SCD&OP focuses more on late-phase programs that are building inventory to support Phase II and Phase III clinical demand and to ultimately prepare for an eventual commercial launch. These programs often spend millions of dollars on multiple batches of DS and DP, and, in some cases, millions of dollars on comparator products as well.

Strategic CD&OP does not follow a structured cadence the way that operational CD&OP does. Instead, the trigger for a strategic review of a program is the point when a company must commit to a DS or DP campaign. When upstream production is outsourced to a contract manufacturing organization (CMO), there is a time point (e.g. a commitment window) when production can no longer be changed or cancelled without incurring a large cancellation fee. Similarly, when upstream production is insourced, there is a commitment window when labor, materials, and capacity can't be reorganized without incurring significant costs. For example, let's assume that a sponsor works with a CMO who has a 12-month commitment window. If a campaign is cancelled less than 12 months from the agreed upon start date of production, a large cancellation penalty is applied. Therefore, it makes sense to trigger the SCD&OP process several weeks before the 12-months commitment window is reached to assess whether moving forward with the proposed campaign is still desirable.

The process steps for SCD&OP are straightforward and presented in Figure 15.6.

The first step of SCD&OP is to prepare a program update which should contain the following sections:

- *Demand Section Including Planning Assumptions and an Upside/Downside Demand Risk Assessment*: The demand section documents the site activation schedule, enrollment rate, and

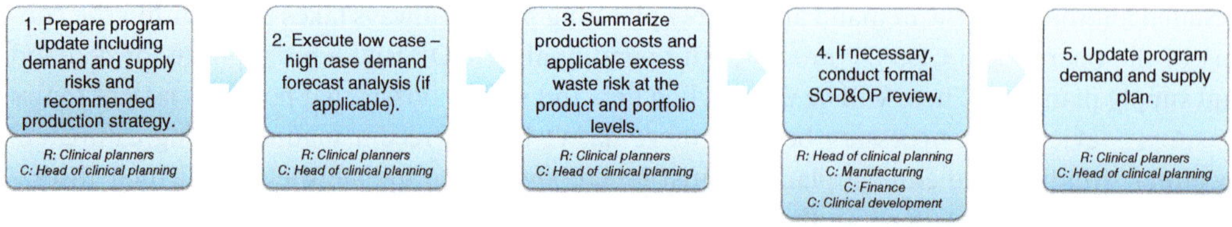

Figure 15.6 Strategic Clinical Demand and Operations Planning Process

enrollment curve for each planned late phase trial. The team then needs to assess the potential for upside/downside risk to the enrollment curve of the study. Both the type of risk and the degree of potential variance should be noted in this assessment. Studies with a high potential variance (e.g. demand could be much faster or slower than expected) are graded with a higher risk score while studies with low potential variance have a lower risk score.

- *Supply Section Including Planning Assumptions and a Supply-side Risk Assessment*: The supply section documents the expiry period of the DS and DP, and the safety stock and order target parameters at each level of the supply chain. The team should also perform a supply-side risk assessment of the supply chain's ability to deliver material consistently to the clinic. Including metrics such as the manufacturing and release lead time variance and the number of deviations associated with each campaign quantify the reliability of DS and DP manufacturing sites. External risks such as geopolitical considerations are also worth highlighting in this analysis.
- *Recommended Production Strategy*: This section contains a recommendation of whether to commit to the upcoming campaign and any additional recommended changes to the production strategy based on the demand and supply analyses.

In the second step, an optional low-case–high-case planning analysis is performed that quantifies the stockout risk and excess waste risk associated with the recommended supply plan. This is especially useful when there is a large amount of demand uncertainty in the clinical development plan, and the trade-offs between stockout risk and excess waste risk are clear at the program level. Performing an entire low-case–high-case analysis may not be necessary though if there is a low degree of demand uncertainty. Alternative production strategies may also be presented here that show increased production versus the recommendation or decreased production versus the recommendation. In the case of increased production, up-front investment and excess waste risk increase, but stockout risk decreases. Conversely, when decreased production is shown, up-front investment and excess waste risk decrease, but stockout risk increases. The clinical planner has an opportunity to quantify and showcase those trade-offs here. Then, in the third step, the portfolio level view is layered into the analysis. Data showing production costs across all products in the portfolio and whether previous decisions erred on the conservative side (e.g. increasing production) or the aggressive side (e.g. reducing production) provides important context for each SCD&OP review.

The fourth step of the SCD&OP process is to determine whether a formal review with all stakeholders is necessary. In some situations, particularly those where there is low-demand uncertainty, it's obvious that upstream production is needed to maintain uninterrupted supply. When the decision isn't obvious though, and there's a real choice to make between balancing up-front investment versus stockout risk, then clinical supply chain, clinical development, manufacturing, and finance need to come together and align on a path forward. At the review, the functional representatives provide feedback on the recommended supply plans for each program and have an opportunity to suggest changes or improvements to the plan.

Strategic CD&OP completes with the clinical supply chain function taking feedback from the meeting and incorporating that into the demand and supply plans where applicable. Here, the process links back to the requirements review and reconciliation steps. The head of clinical planning should review all program plans coming out of SCD&OP to ensure they fit the recommended strategy and guidance from the SCD&OP review meeting before moving on to plan publication.

15.6 Supply Plan Handover and Order Readiness

The supply plan handover and order readiness process transmits the supply plan from the plan publication step of operational CD&OP to the sites responsible for its execution. It also formalizes many of the activities described in the production and distribution order readiness sections within Chapters 9 and 11, respectively. The purpose of this process is to ensure that new orders are handed over with the appropriate time horizons, order readiness activities are actively managed, and feedback about order completion dates and quantities are fed back to the clinical planner responsible for that program. The scope of this process covers at minimum all finished product production and distribution orders and may also be expanded to cover DP and DS orders. Like the operational CD&OP process, the supply plan handover and order readiness process follows a routine cadence, but these check-ins should happen at least weekly for finished product orders and biweekly for upstream production orders.

Typically, the head of clinical planning is the business process owner for supply plan handover, and they need to identify who is responsible for providing oversight to and management of the manufacturing and distribution sites within the supply chain. It's not uncommon for different functional areas to manage DS and DP sites versus finished product sites. Similarly, different functions may manage internal manufacturing and distribution sites versus contract manufacturing sites. For simplicity's sake, we'll refer to the group of functions who oversee manufacturing and distribution sites as site managers. The supply plan handover and order readiness process takes the steps shown in Figure 15.7.

The process begins with transferring the supply plan to the respective site managers at a regularly scheduled handover meeting. As a reminder, the planning horizon is critical to define, particularly for DS and DP materials where significant lead times (e.g. >12 months) may be required to book capacity and source raw materials for the planned batch(es). Finished product production orders should be handed over approximately 6 months from the desired release date, and distribution orders should be handed over at least 1 month before order readiness activities such as import licenses applications need to begin. The production plan summaries and distribution plan summary contain all the necessary order information to begin order readiness so these can be used as a template to communicate the handover.

Once the orders are handed off to the site managers, they communicate the requested order dates and quantities to the responsible sites. In some cases, the site manager's role is to act as a

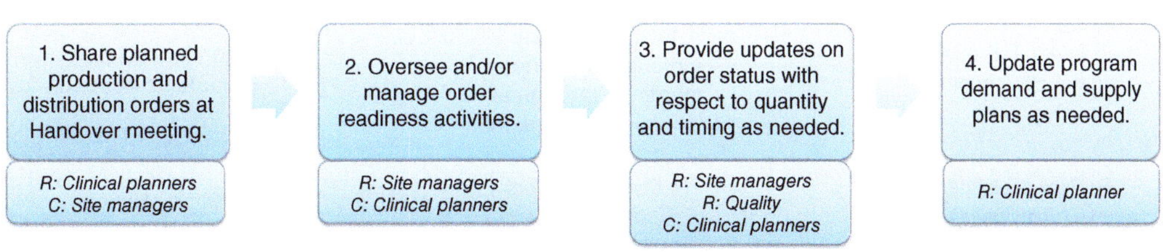

Figure 15.7 Supply Plan Handover and Order Readiness Process

point of contact with the site for updates on order readiness while in other cases, the site manager may directly oversee and execute order readiness. In either situation though, the next process step instructs the site manager to feed information back to the clinical planner as needed. As needed means that order status should flow back through the handover meeting at minimum, but if there is a production delay or if the yield is impacted during the production process, the clinical planner should be notified as soon as that event occurs.

The final step in the process connects back to the demand and supply planning process and specifically the reconciliation and requirements loop outlined in Figure 15.3. We know that when order quantities and timing change through production and release, the clinical planner needs to understand and quantify that impact. Feeding these changes back into the plan gives the planner an opportunity to prevent stockouts and/or reduce excess waste by reacting quickly.

15.7 Global Supply Execution Readiness

Global supply execution readiness (GSER) is the final core business process that we'll cover, and it's also the only process not directly linked to the demand and supply planning process. The purpose of GSER is to ensure finished product availability within a newly activated country in a clinical trial. The business process owner for GSER is the head of clinical planning who communicates the value of GSER to leadership within clinical operations, quality, and regulatory functions. The scope of GSER covers clinical trials with a global scope. Putting smaller trials (e.g. <5 total countries) through GSER doesn't always make sense, but as trials expand globally into Latin America, Asia–Pacific, or Eastern Europe, GSER becomes more valuable. That's because these regions often have lengthy regulatory approval timelines, regulatory restrictions that need to be managed, and challenging import/export requirements.

There is a two-part trigger for GSER. The first part is identification of a clinical trial that fits within the scope, and the second part is the timing of the FSA within the first country. The process formally begins about 4–6 months before the FSA date in each region. At that point, the process takes the steps shown in Figure 15.8.

The first step in the process is for the clinical planner to identify the relevant stakeholders and tasks associated with enabling the initial shipment of material to a clinical site. The example presented in Table 15.2 identifies seven different functional roles who are responsible for activities required to ship product to a clinical site. In early-stage biopharmaceutical companies, these roles may be combined under one person, while in mature, commercial-stage companies, the roles, and responsibilities may be further broken down into additional roles. The responsibility of bringing this team together rests with the clinical planner so once the team is identified, the next process step is to schedule a reoccurring supply execution readiness meeting.

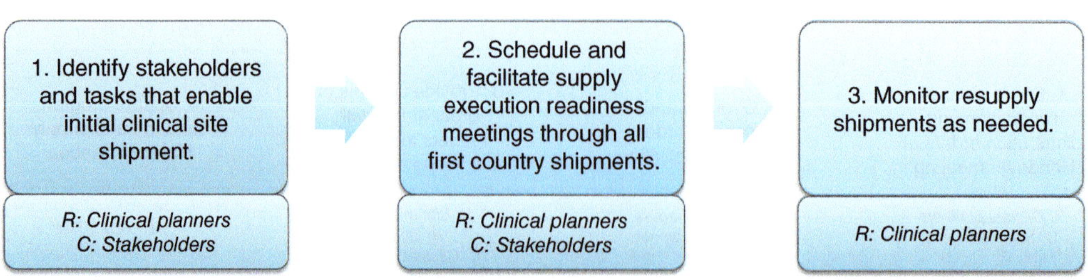

Figure 15.8 Global Supply Execution Readiness Process

Table 15.2 Global Supply Execution Readiness Roles and Responsibilities

Function	Role	Responsibilities
Clinical operations	Study lead	Provide estimated first site activation (FSA) date and estimated first subject screening date by country.
Clinical operations (or RTSM supplier)	RTSM lead	Provide estimated RTSM activation date and ensure RTSM settings are correct to enable first site shipment.
Regulatory affairs	Study or program lead	Provide estimated competent health authority and ethics committee approval dates.
Quality	Product lead	Provide status of finished product batch release and underlying drug product and drug substance batches, if applicable.
Quality	QP coordinator	Provide QP certification issuance status and remaining requirements.
Clinical supply chain	Regional depot site manager	Execute distribution order readiness activities for import to regional depots including import license application and approval.
Clinical supply chain	Clinical planner	Monitor resupply shipments to ensure RTSM predictive resupply algorithm is functioning appropriately.

The GSER meeting focuses on the checklist of activities required to facilitate initial country shipments. At the meeting, clinical operations begins with an update on the estimated FSA and first subject screening dates. These target dates set a range within which supply should be available to ship. While it's ideal to have the finished product material ready by FSA, this may not always be possible, so the first subject screened date provides the latest need by date. If supply won't be ready by the first subject screen date, then clinical operations gains visibility to this and can work with the CRO or with the site directly to reschedule subject screening to the estimated supply availability date. The other functions provide input on their enabling tasks. If delays occur with respect to regulatory submissions and approval, batch releases, or system readiness, these issues are surfaced at the meeting, and the team has an opportunity to align on new readiness dates. Note that any revised site activation dates can be updated in the demand forecast of the plan, but the expectation is for clinical operations to confirm this internally and then provide firm activation dates during the demand review meeting of operational CD&OP.

After a site has been activated and the initial shipment has been received, the work is not quite done. In the last process step, the clinical planner needs to monitor resupply shipments to ensure the distribution supply chain is functioning smoothly and the RTSM parameters are appropriately set. It's not uncommon for issues to crop up when resupply orders are triggered by the RTSM and then translated to the regional depot responsible for executing the resupply shipment. Both the timing and quantity of resupply orders should be checked on to make sure the supply is getting an appropriate resupply of material when they need it.

15.8 Supply Network Planning

For global biopharmaceutical companies with large clinical and commercial portfolios, there is often a qualified network of internal and manufacturing sites within their supply chain. When a new product enters the portfolio, or an existing product needs to transfer to a new site, then a decision must be taken regarding site selection. If the current network isn't suitable, then further strategic changes to the supply chain may be necessary like choosing to acquire a site from a third party, build a new site, or divest an existing site. SNP has two objectives. The first is to optimally allocate manufacturing campaigns across a network of available sites, and the second is to assess the entire supply chain network and recommend changes based on the future needs of the portfolio.

SNP is not typically applicable to precommercial sponsors for a few reasons. First, their portfolios are small enough that supply chains are built for each product on an as-needed basis, and a single-sourcing strategy is typically preferred to save money. Second, these companies tend to outsource the majority of their GMP manufacturing and distribution work to contract manufacturers. Due to this, there are far fewer decisions to make regarding which site a product should be placed at or whether the existing network of sites is sufficient to handle a broad portfolio. In large commercial biopharmaceutical companies, it's not unusual to have multiple internal manufacturing sites and external contract manufacturing sites that are qualified to perform the work. As a result, these organizations typically establish a function dedicated to global supply chain strategy that owns the SNP process and is responsible for connecting it to the other clinical and commercial planning processes in the company.

From a process perspective, SNP is divided between individual product placement and network strategy. In larger organizations, production planning takes place on a monthly cadence with a 3-year (or longer) planning horizon which creates visibility into medium- and long-term product supply requirements. Detailed manufacturing process and technical requirements about each product are also shared with the supply network planner to provide the information they need to make placement and network strategy decisions. One key ingredient to effective SNP is the inclusion of both the clinical and commercial supply chain functions in the process. The entire portfolio of products and each product's specific needs must be addressed by the supply chain. Performing SNP through one lens or the other often leads to suboptimal outcomes because many sites are capable of supporting both types of products.

With respect to product placement, there are two triggers, new clinical or commercial products and products that require a technical transfer to a new site due to process or scale changes. There are many inputs to consider in this decision, some of which are outlined below:

- *Qualification*: Is the site currently qualified to support the work?
- *Capability*: Does the site have the technical ability to execute the work?
- *Agility*: Can the site meet the lead-time requirements of the product and the clinical development plan?
- *Capacity*: Does the site have the available capacity necessary to meet the product's needs?
- *Control*: Is the site under internal control or the control of the third party?
- *Cost*: How much will the product cost at the site?
- *Other*: Are there other factors (e.g. past experience, proximity, and tax considerations) to consider when evaluating this site?

From a network strategy perspective, the trigger is less event driven (e.g. a new product or a scale change) and more portfolio driven. The supply network planner needs to consider at least a 5-year horizon because building or purchasing a new site is a multiyear process, and lead time needs to be factored into the strategy. Over that period, the supply network planner also needs to assess the total number of products entering both the clinical and commercial portfolios, the modality of those products, the manufacturing capabilities required to make those products, and the total demand profile of each product. Each of these inputs is weighed against the supply network's ability to support the portfolio. The same considerations used for individual product placement are equally applicable when assessing the portfolio as a whole. Capability, agility, capacity, control, and cost all need to be incorporated into an analysis of whether or not to expand or shrink the network.

There are specific examples of SNP that we'll touch on in future chapters. In Chapter 18 – Make-to-Order and Just-in-Time Supply Chains in Clinical Trials, the topic of cell therapy supply chain strategy is covered including the decision of whether or not to build internal capabilities for raw material and DS manufacturing. Performing a make-versus-buy analysis is also relevant for companies with traditional modalities in their portfolios when there is sufficient demand. This topic is addressed in Chapter 20 – Contract Organization Management and Oversight.

15.9 Three Keys to Business Process Success

The business processes outlined in this chapter share two characteristics that help them meet their desired objectives. First, the head of clinical planning and head of clinical supply chain are the business process owners. For some processes such as demand and supply planning, this is a no-brainer. But, others, like operational CD&OP and GSER, aren't effective if clinical supply chain doesn't communicate with and gain buy-in from other functional leaders on their importance. A case needs to be made that these processes create value for the organization because they facilitate cross-functional collaboration that's necessary to optimize the supply chain, prevent stockouts, and save millions of dollars. Clinical supply chain leadership also plays a crucial role of holding the function accountable for process adherence. The processes don't work if the clinical planners aren't following through on updating the plans, scheduling the meetings, and driving for process results. Finally, business process management takes incredible time, energy, and knowledge on the part of the team. The clinical supply chain function needs support from company leadership to adequately staff, train, and build the organization as the portfolio grows. It's the responsibility of clinical supply chain functional leadership to build the business case for the people and roles needed to execute these processes.

Second, the processes are designed to integrate with and reinforce each other. Figure 15.9 illustrates how demand and supply plan creation, demand and supply planning, operational CD&OP, strategic CD&OP, and supply plan handover connect. Notice how the plan publication step triggers the strategic CD&OP process and the supply plan handover process. Then, the supply plan handover and SCD&OP processes each generate feedback that trigger a cycle of reconciliation and requirements review. And, finally, notice how all roads eventually lead back to the start of the demand and supply planning process, actualization. This integration is important because the processes reinforce each other by supplying each other with the inputs and outputs needed to create, maintain, and communicate the plan for each program. Using this framework, the clinical planner also sees where they are in the planning cycle and is able to identify which process should be initiated or which process step should come next. We'll show examples of this in Chapter 16 – A Week in the Life of a Clinical Planner.

A third key to process success is to rightsize them for the organization's maturity level. In a precommercial biopharmaceutical company with only one or two products in the clinic, the process steps may be executed with a relatively small number of team members and through a series of relatively informal meetings because planning inputs and outputs are housed in a few places and managed by a handful of people. In commercial organizations with a dozen or more clinical programs, the level of process ownership, formality, and training must increase. Planning inputs

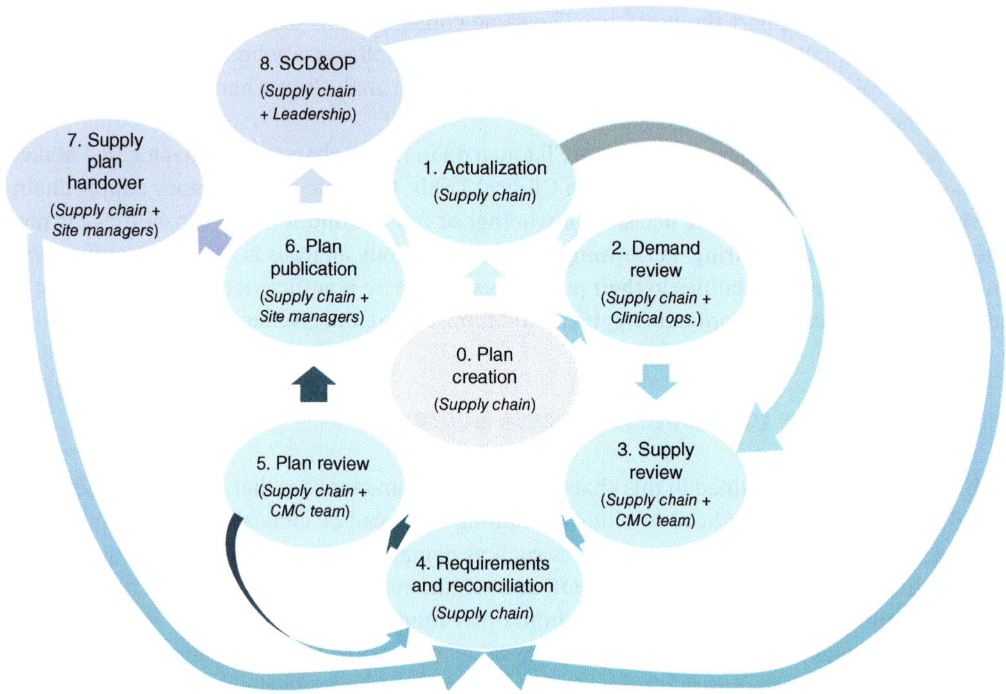

Figure 15.9 Supply Chain Planning Process Integration

and outputs may be housed in multiple technology solutions, and they may be managed by a highly diverse set of functions and stakeholders. As an organization grows and matures, the processes need to be revisited by the business process owners so they reflect the organizational structure, clinical program dynamics, and any new tools and technology that help facilitate their execution. This is another reason why placing business process ownership with functional leadership is critical.

15.10 Key Takeaways

1) Business processes are important in clinical trial supply chain management because they drive consistency, enable information sharing, and facilitate effective communication. There are six core business processes in clinical trial supply chain management: demand and supply plan creation, demand and supply planning, operational clinical demand and operations planning, SCD&OP, supply plan handover and order readiness, and GSER.
2) There are also six common characteristics to each business process: objective, scope, owner, trigger, map, and responsibilities. Each characteristic gives the process structure and clarity to those who are responsible for executing it.
3) Demand and supply plan creation is the process of creating the first demand and supply plan for a clinical trial. The process follows eight steps including SKU definition, supply chain mapping, assumption identification, and building the plan. This process brings together many concepts from demand planning, finished product supply planning, expiry planning, distribution planning, upstream planning, and advanced planning techniques.

4) Demand and supply planning is the process of updating the demand and supply plan when the inputs and assumptions that underpin the plan change. Demand and supply planning follows a three-step process: actualization, requirements review, and reconciliation. Actualization is the process of converting forecasted values to actual values and updating the demand forecast or supply plan inputs as needed. Requirements review is a sequential assessment of how changes at a downstream level of the supply chain impact requirements for supply at the upstream level, and it should always be executed from most downstream to most upstream level in the supply chain. Reconciliation is the process of aligning downstream production and distribution orders to the actual quantities and release dates of the allocated upstream order, and it should always be executed from most upstream to most downstream level in the supply chain.
5) CD&OP is the process of aligning on plan inputs and communicating the output of the demand and supply planning process. CD&OP comprises two subprocesses, operational CD&OP and strategic CD&OP. Operational CD&OP has four steps, demand review, supply review, plan review, and plan publication. These steps are closely interlinked with the demand and supply planning process but shouldn't be viewed as a prerequisite to refreshing the plan when needed.
6) SCD&OP is the process of aligning on the production and inventory management strategy of each program by analyzing demand and supply risk against the required investment to hedge against those risks. SCD&OP comprises five steps including preparation of program updates, executing a low-case–high-case analysis, summarizing costs at the product and portfolio level, conducting the SCD&OP review, and updating the plan.
7) Supply plan handover and order readiness is the process of transmitting the supply plan to the sites responsible for its execution, ensuring order readiness activities are actively managed, and communicating feedback about order status back to the clinical planner. Supply plan handover has four process steps: sharing the planned orders at a predefined handover meeting, overseeing order readiness activities at the sites, providing updates on order status, and then updating the plan as needed.
8) GSER is the process of aligning on target FSA and first subject screening dates in a particular country and ensuring finished product availability to meet those dates. GSER has three process steps: stakeholder and task identification, facilitation of supply execution readiness meetings, and monitoring of resupply shipments as needed.
9) SNP is the process of optimally allocating products across a network of available sites and of making recommended changes to the supply chain (e.g. site additions or divestitures) based on the needs of the portfolio. SNP decision-making must include site qualification status, capability, agility, capacity, control, and cost as well as other relevant inputs.
10) There are three keys to process success: placing business process ownership with the heads of clinical supply chain and clinical planning; designing the processes so they integrate with and reinforce each other; and rightsizing the business processes to the maturity level of the company.

16

A Week in the Life of a Clinical Planner

16.1 Before the Week Begins

Welcome to the life of a clinical planner, or CP for short. In this chapter, you'll see through the eyes of CP, a planner working at a mid-sized biotechnology company as they work through a handful of situations and scenarios. CP's time is filled with information coming at them from a variety of sources. The clinical study team, the chemistry, manufacturing, and controls (CMC) team, the clinical supply chain team, the network of contract manufacturing and/or distribution organizations, and other functional stakeholders like finance or project team leads are either providing updates or asking questions that only CP can answer.

Before jumping in and immediately reacting to new information or a request to run a scenario though, the best action to take is to not act at all. Instead, pause, absorb, and think through the situation by asking the following questions:

- What is the specific question or problem(s) that needs to be solved?
- Can a response be provided to that question or problem(s) without requiring an update to the demand and supply plan for that trial or program?
- If an update to the plan is required, what steps need to be taken within the plan to provide that answer?
- If an update to the plan is required, where does this analysis fit within the current planning cycle?
- If an update to the plan is required, does it require accelerating the planning cycle to communicate the outputs?
- How should the response or solution be communicated to the appropriate stakeholders?

These questions offer a framework to manage an oft-overwhelming amount of information and data. The first question is the most important because it's not always obvious what question or problem needs to be accounted for when a situation presents itself. Without defining the question or problem, it's not clear what next step to take. Once the problem statement is clear, the next question to ask is whether a response can be provided without making a complete update to the demand and supply plan because responding to some questions doesn't require updating the plan at all. The answer may already exist in the current plan, in an external data set, or it can be provided through a simpler analysis. Taking a moment to properly diagnose the question and identify the quickest path to an answer saves a tremendous amount of time for the planner.

If an update to the plan is required, we now have the concepts, tools, and processes in place to move through the remaining questions in the framework. Understanding what steps should be taken in the plan, where the analysis fits in the planning cycle, how to accelerate that cycle, and

Supply Chain Planning for Clinical Trials – A Practical Guide, First Edition. Ryan Mills.

how best to communicate the plan have all been covered extensively in previous chapters. Now, we'll bring these concepts together in the remainder of this chapter as we step into CP's world. Over the course of their week, CP encounters demand-based scenarios, supply-based scenarios, and a unique trial design. All of these require CP to stop, analyze the actions they need to take, and communicate effectively about those actions with their fellow colleagues. At the end of this chapter, we'll summarize the characteristics and attributes CP displays to navigate these challenges and ultimately earn a just reward.

16.2 Monday: Scenarios Incoming

Ding. It's Monday morning and your inbox is alive. A notification lights up on your monitor with an email from the clinical operations lead (COL) of your small-molecule program. The COL heads a Phase III study that's three months away from activating its first clinical site. "Regulatory Feedback and Impact to Study Design" reads the subject headline. "Impact to Study Design" you think to yourself as you start reading. "That could mean any of a hundred possible changes to the demand and supply plan for that study." As you digest the email, your brain starts working through the follow-up questions you need to ask clinical operations and the potential updates you'll need to make to the plan.

You leave your desk and head to the chemistry, manufacturing, and controls (CMC) team meeting of your second program, a large-molecule program that's shown promising data and is now advancing to a late phase trial. When you walk in the door, you can already tell the mood is off from what you're used to, and the quality lead for the program looks particularly unhappy. You take your seat, and the CMC project manager immediately turns to the quality lead and asks "Again? Really? I thought we had a commitment from our contract manufacturer that they had solved the issue on the last batch?" The quality lead turns and looks straight at you. "I'm sorry CP, but it looks like the same issue has reared its head again, and we're going to lose about 50% of our latest Drug Substance campaign." With a slight outward groan, your mind races ahead to the implications on the supply plan. "Can you let us know the impact by the end of this week?" You know that question is coming, so you tell the team you'll do your best to work through the analysis as soon as you're able.

After the meeting you trudge back to your desk and open your laptop. Another email waits for you, this time with the headline "Potential Specification Modifications." The CMC regulatory lead of the second molecule has received feedback from multiple competent health authorities (CHAs) on an amendment to the investigational medicinal product dossier (IMPD) for the ongoing Phase I/II study. Two agencies have requested tighter concentration limits before they will approve the latest version of the IMPD. The CMC regulatory lead wants to pull the CMC team back together on Thursday morning to discuss and align on a strategy to manage this. "Wow, it's not even 11:00 yet and I have three major issues to address, one demand-related and two more supply-related." This is turning into a very busy week already.

That afternoon, your boss, the head of clinical planning, HP, calls you into her office. "CP, you've done such a good job covering your molecules that we'd like to move you onto one of our fast-moving new programs. How does that sound?" You pause, smile, say "Thank you! That sounds like an excellent opportunity" and wait for the other shoe to drop. "Excellent! Nothing has been set up yet on this program. We need you to join the team immediately, create the demand and supply plan, and develop a recommended production and inventory management strategy." Ah. There's the other shoe.

"There's just one catch," the head of planning goes on. "This team is planning to execute an Adaptive Trial Design. That means the trial begins as a Phase I study, but the design can be modified so that it eventually becomes registrational. Those changes may include adding subjects, arms, or doses, so it's not a straightforward design at all. You'll need to take all that into account when you present the production and inventory management strategy at our Plan Review meeting later this month. Does that sound reasonable?"

"Yes, I think so," you respond.

As your day ends, you take a few minutes to reflect on the issues demanding your attention. Your mind first goes to the "Regulatory Feedback and Impact to Study Design" email from the COL of this late-phase study. Fortunately, you scan your calendar and realize that you have a demand review meeting first thing in the morning to discuss those study design implications. One day down, four to go.

16.3 Tuesday: Study Design Scenario Planning

It's Tuesday morning, and you're sitting across from the COL. They tell you the team is weighing a study protocol amendment with two significant changes to the design including:

- Switching from a 2-arm study design with 500 mg and placebo doses to a 3-arm study design with 500 mg, 250 mg, and placebo doses.
- Increasing the sample size from 500 subjects to 750 subjects.

Given the sample size increase from 500 to 750 subjects, the team is also considering adding up to five new countries to maintain the enrollment timeline. The countries under consideration are Colombia, Mexico, South Korea, Taiwan, Australia, Turkey, and Israel. "So," the COL says to you. "What's your assessment of these proposed changes and the potential impact they'll have on the supply chain for the study?"

"These are fundamental changes to both intrinsic and extrinsic design elements," you respond. "Let me attempt to clarify the impacts you're asking to me to assess. The first, and perhaps most important from a Clinical Operations perspective, is a potential delay to the First Site Activation (FSA) date. The second, and perhaps most important from a CMC perspective, is an increase in the amount of drug we'll need to support a second active dose. And the third, and perhaps most important from a Finance perspective, is the increased spend required to support this change. Does that summarize the assessment that you're looking for from me?"

"Yes, although I'm primarily interested in the first point, I can see how they're all relevant to this decision. We'd also like you to provide supply chain-related feedback on potential country additions. Let us know how quickly IMP could be ready for them, potential challenges associated with importation, and a cost breakdown for each country. Those details will flow into a larger assessment of whether we move forward with these potential changes."

"That sounds doable. When do you need this analysis by?"

"How does Friday sound?"

You hesitate, not wanting to overcommit to so many deliverables. "Would next Wednesday work?"

"That's manageable from our side," the COL replies. "But let us know if you get it done more quickly," they say with a smile.

You walk away from the conversation with a clear set of questions that need a response:

- What is the potential impact to the FSA date?
- How much more drug will the CMC team need to produce to meet the increase in demand?
- What is the financial implication of the increased supply plan?

- For each potential country, assess how quickly IMP can be ready, potential importation risks, and how much it will cost to supply each country over the duration of the study.

As you get back to your desk, you open the scenario analysis framework and begin working through each question. On the potential impact to the FSA date, you determine that this does not require updating the model. Instead, you realize that a new finished product stockkeeping unit (SKU) will need to be manufactured to support the new 250 mg dose. Your first instinct is to check the finished product manufacturing and release lead time assumptions that go into your model and see that they are 12 weeks. "Whew!" you think to yourself. "This may not impact supply availability after all because that's before our currently scheduled FSA date." Before you get too excited though, you decide to give your packaging site manager a call to discuss the situation and what activities are contained within the 12-week lead time quoted in your planning assumptions.

After describing the situation, your packaging site manager says, "I'm glad you came to me first, CP. Have a seat. The first point to understand is that the 12-week lead time only covers pre-production, manufacturing, and batch release activities. If a new label or a new packaging design needs to be developed, that adds lead time to the 12 weeks. The second point is that the 12-week lead time assumes the order has been given to our packaging site at least 6 months before the order start date to allow for component ordering and availability. Now, you're in luck because we've already designed this as a wallet containing blister strips, so we don't need a new packaging design to accommodate the dose. That said, I do need to check with the site if we have enough components available to meet the revised order quantities. Do you have those ready?"

"Not yet," you reply. "There are a few other parts of this analysis that will require me to update the planning model to assess. If I get those numbers over to you in the next 24 hours, when do you think you could have an answer from the packaging site?"

"Within 2 days of receiving your revised order quantity."

"Great. I'll get those numbers over to you soon and then check back on Thursday for your assessment."

You set the first question aside in your mind and move on the implications of increased demand on the supply plan and the overall budget for the program. Understanding this requires an end-to-end update of the plan. Before you begin that process, you go to the next step of the framework and start writing down a step-by-step approach:

1) Create new SKU–depot locations for the 250 mg dose at each applicable central and regional depot location.
2) Add the clinical site activation schedule to each new tab. The site activation schedule for each depot can be taken from the respective 500 mg SKU–depot or Placebo-To-Match (PTM) SKU–depot forecasts because each site needs to be supplied with material from each SKU.
3) Review the randomization and trial supply management (RTSM) site supply strategies for potential changes given the introduction of the new SKU. In this case, since the number of subjects increases in proportion to the number of arms, it's unlikely that an update is required. The initial site supply values from the 500 mg and placebo SKUs can be applied to the 250 mg SKU.
4) Plan initial depot stocking quantities at each 250 mg SKU–depot location. These should mirror the initial depot stocking quantities for the 500 mg and placebo SKUs because the initial site supply values should be the same across SKUs.
5) Link the 250 mg SKU–regional depot locations back to the appropriate central depot location and ensure distribution demand from the site stocking quantities is visible and accurate.

6) Aggregate subject demand for the 250 mg SKU at the depot with the highest ratio of planned enrollment, which is the central depot for the 500 mg and placebo SKUs. Forecast total subject demand at the central depot.
7) Apply the desired overage to total subject demand at the central depot and then add site demand and distribution demand from the regional depots to arrive at a total demand forecast for the 250 mg SKU.
8) Translate the demand forecast to a production plan using the same supply planning parameters used for the 500 mg and placebo SKUs. This scenario has a 1:1:1 randomization ratio, which means the finished product production plans for each SKU should be identical. If that ratio was not proportional, then the finished product production plans for the SKUs would also not be proportional.
9) Link the additional finished product production back to the underlying primary packaging SKU, in this case, a blister pack containing 250 mg tablets. Update and increase the production plan for the blister pack to accommodate the additional demand from the 250 mg finished product SKU.
10) Update and increase the drug product (DP) production plan to accommodate the additional demand for more blister packs.
11) Update and increase the drug substance (DS) production plan to accommodate the additional demand for more DP.
12) Work through the reconciliation and requirements review cycle until DS, DP, and finished product orders are fully allocated throughout the plan.
13) Analyze the difference in the plans between the original version of the model and this new scenario. Use the production and distribution summaries to quantify the increased number of distribution orders and finished product production orders; the increased number of blister packs, tablets, and kilograms of DS that will need to be manufacture; and the increased number of batches that will need to be manufactured through the supply chain.
14) Collect the cost per batch figures for DS, DP, and finished product from the respective site managers. Compare the budget for the original plan against the budget for the new plan inclusive of both production and distribution activity.
15) Use data visualization techniques such as charts and graphs to summarize the analysis.

Before you begin working up the scenario you call HP to discuss and review your approach. “Hi HP. I’ve been given a complex scenario to work through by the Clinical Team that potentially impacts our study timeline, demand plan, supply plan, and budgets. I’ve been working with our Packaging Site Manager on the study timeline, but I wanted to review the scenario and my approach with you,”

“Sounds great, CP. Thanks for coming to me, and take me through what you’ve got so far,” HP says. And, so you do.

“Nice work pulling this together, CP, and great recognition that the timeline response needs further analysis by the Packaging Site Manager. On your proposed updates to the plan, let’s keep in mind this is still just a scenario for now. Let’s move forward with a separate version of the plan under these conditions to share with the team. Have you thought about the stakeholders who need to review the outputs of this scenario?”

“Yes I have. First, I need to get the Finished Product quantities back to the Packaging Site Manager so he can complete the lead time analysis. Second, we have our weekly CMC team meeting on Thursday afternoon this week, so I’m planning to get this on the agenda to review the outputs and feasibility of the plan. Fortunately, we’ve been carrying a minimum 6 months of Drug Substance inventory, so I’m hopeful we’ll be able to absorb this additional demand without having to press our manufacturing sites too hard, but the analysis will tell. I'm also planning to set up a separate meeting with the CMC Team Leader and the Finance Lead for the project to review the potential CMC budget impact. Then I'm planning to take feedback from CMC and Finance and wrap that into my analysis for inclusion in the Clinical Team's broader case for making this planned change."

"Excellent, CP. You have a solid plan for pulling this analysis together and getting it in front of the right people before it goes to the full team for a thumbs up or down. If this scenario is approved, let's regroup to discuss how it fits into the planning cycle," HP tells you. "One more item before I go. Have you worked up the country-level analysis that Clinical Operations has requested from you?"

"Not yet because I've been focused on addressing the first three questions. Do you have some advice for me on that one?"

"Sure. First suggestion I have for you is to loop back with your Packaging Site Manager. Adding countries will impact the label, and this can be a long lead time activity, up to 4–6 months in some cases. Next, I'd check in with our resident Distribution Team Lead (DTL) who is an expert on Finished Product logistics and they can walk you through the distribution lead times and the activities associated with those specific countries. DTL can also help you work up a risk assessment for Colombia, Mexico, South Korea, Taiwan, Australia, Turkey, and Israel based on other team's recent experience there."

"Fantastic, HP! The one piece that should be straightforward to document is the budget for each country. Fortunately, I know our team has pulled together detailed budgets for other programs that have operated sites in those countries, and I can add those figures to the feedback from DTL to complete the assessment. Thank you for the help and support!"

With your plan in place for all four questions, you get to work on building out the scenario. You also schedule follow up meetings with the CMC team lead and finance lead to cover the budget, the packaging site lead to discuss label preparation, and with the distribution team lead (DTL) to work up the country list assessment. These are all set for Friday, so you also write back to the COL and let them know the analysis is on track for completion by your mid-next week target. The COL pings you back with thanks for the update.

It's nearing the end of the day, and you're just finishing up the scenario in your planning tool. You're feeling pretty good about where you are until you look up and it's 5:00 p.m. and you realize you haven't started working on the news that dropped at your other program's CMC meeting earlier that day. "Oh no. I didn't have time to start on the reduced batch issue on my other program," you think to yourself. "And we still have the expiry restriction for our Phase II study to deal with, too." You fire off the revised order quantities to your packaging site manager and head out. Fortunately, it's still only Tuesday.

16.4 Wednesday: A Lower-than-Planned Yield

Walking into your office on Wednesday morning, you're feeling determined to work through the two supply-related issues before the end of the week. You sit down and apply the scenario management framework to the reduced batch yield issue first. This is now the second DS batch in a row to yield less than half of what was projected, so you want to address the following questions:

- How does the shortage impact finished product supply to ongoing clinical studies? Is there a potential stockout risk?
- How does the shortage impact the near-term supply plan for DS and DP? How quickly will additional campaigns need to be rescheduled? How many campaigns will need to be scheduled?

- Do the underlying DS planning assumptions need to change given this is the second batch in a row that's yielded far less than expected?

As you're writing these questions down, you're also thinking about the clinical development plan of the program. There is an ongoing open label Phase I/II study with about 100 subjects enrolled across 10 countries supplied out of four depots. Maintaining supply for that trial is a top priority, but there's also a planned Phase III study about a year away that this DS batch was also intended to support. With this reduced yield, you'll have to assess the impact to both studies. A realization dawns on you that you'll need to answer the third question about updating the planning assumptions before you can deal with the second question on potential future shortages. If the yields need to be reduced going forward, those assumptions need to be updated before going through the demand and supply planning process.

It's also clear that you'll need to update the plan to address these questions. Unlike the demand-based scenario you worked through yesterday, you know that updating the plan directly is required because this is an actual batch quantity as opposed to a hypothetical scenario. You start writing down the updates you'll need to make in the plan:

1) Actualize the DS batch quantity and reconcile the downstream DP batch quantity to the actualized DS batch quantity.
2) Reconcile the downstream finished product batches of the Phase I/II to the updated DP batch quantity.
3) Reconcile the finished product distribution orders to the updated finished product batch size. Adjust the distribution order quantities so weeks-of-supply across the depots is equalized as much as possible.
4) Review the timing and quantity of the next finished product resupply order and the requirement for supply this is placing on the next DP and DS batches.
5) Determine if there is sufficient on-hand DP and/or DS to meet the new finished product supply plan or if the upstream production schedule needs to be altered.

Before you begin, you pause to double check where in the operational clinical demand and operations planning (CD&OP) planning cycle you are. "Nice," you think to yourself. "I see that we're still in the first week of the month, which is typically Demand Review. That makes sense because I just had Demand Review on my other program yesterday!" Yikes. There's so much going on you're forgetting what happened 24 hours ago. As you scan the agenda for your CMC meeting next week, you see the placeholder for supply review, which will be your opportunity to present the latest version of the plan to the team for endorsement. If this update had been needed off-cycle, you're confident the CMC team would have requested an ad hoc supply review to assess the changes to the plan and develop a communication strategy for the DS manufacturing site.

After working through the steps to update the plan, you breathe a sigh of relief. The inventory management strategy for this program dictated carrying a minimum of 26 weeks of on-hand DS because this program is moving into late phase trials. The reduced yield from the batch only accounted for 13 weeks of finished product supply so the team should be able to move forward with pooling the remainder of this batch with some of that DS safety stock to avoid any major supply interruption. "That's the good news," you think to yourself. "Now let's get to the longer-term implications of these yield misses."

Opening your calendar, you check when the DS and quality leads are both available to talk through what's happened on the last two campaigns and whether the underlying supply planning assumptions need to be corrected. Fortunately, both are free at lunch, so you send them a message asking to discuss next steps. As the three of you grab your food and sit down, you start the conversation, "DS, help me understand what's been happening with these campaigns and how persistent this issue might be going forward?"

"Thanks for pulling us together, CP. At this point, we don't know yet exactly what's happening within the manufacturing process. This issue started when we scaled up earlier this year to meet increased demand for the planned Phase III study. It could either be a problem somewhere in the cell culture media, or it could be due to an issue further downstream in the purification process, or some combination of the two. We're just not sure at this point. It's going to take us some time to find the root cause, and we don't know when, or if, we'll be able to get future yields back to their original projections."

"Ok, thanks for your assessment, DS. It sounds like we'll have to assume the lower yield for a while. I worked through the plan earlier this morning, and we have three more Drug Substance batches scheduled this year assuming the original yield projections. Do you think it's safe to assume that we'll have the issue solved by the end of the year?"

"Unfortunately not, CP. There's a real possibility that while we're able to make some process improvements, the yield might never reach the original projections. My advice would be to assume that we'll stay at this lower yield for the next several batches and that we'll gradually make improvements from there."

You turn to the quality lead and ask, "Do you think this yield issue will have an impact on our ability to eventually release the batch?"

"I think it's low risk, CP," the quality lead replies. "While we don't have a root cause for the lower yield, we don't think it's a product quality issue or that the analytical results will miss the specification. This happened on the first batch, too, and all the process parameters from concentration to potency were acceptable. We simply got less than we were expecting. I don't see any reason to think this time will be different."

"Thank you both," you respond. "I have what I need for now to update the plan and bring the revised projections to the CMC team next Monday. Appreciate you both getting together on such short notice and over your lunch."

With these responses in mind, you head back to your desk to document the updates you'll need to make in the plan:

1) Update the DS yield assumptions to 50% of the original yield assumption for batches that fall within the next 12-months including the three batches already on the schedule.
2) Update the DS yield assumptions to 75% of the original yield assumption for batches that fall after the 12-month period.
3) Plan additional DS and DP campaigns as needed to meet finished product demand for the ongoing Phase I/II and the planned Phase III studies by moving through the reconciliation and requirement review cycle.
4) Identify how many batches fall within the 12-month planning horizon the DS manufacturing site typically maintains for new batches.
5) Prepare side-by-side comparison charts of the total number of DS and DP batches from the previous plan to this new version of the plan for communication to the CMC team.

You spend most of your afternoon making these changes and then creating charts that visualize the updates to the DS and DP plans. When you're done, you ping HP and ask if they have a few minutes to look over the supply plan outputs and discuss next steps.

"CP, this is another nice piece of work. The plan looks fully reconciled from Drug Substance all the way downstream to Finished Product production and distribution. It's also important that you were able to collect feedback from your Drug Substance and Quality Leads on the revised assumptions before you made the updates. I have two follow up questions for you. First, have you considered that one of your additional batches falls within the planning horizon of the Drug Substance manufacturing site? And second, have you checked in with your CMC Lead or Finance Lead on the potential budget impacts of these additional batches?"

"Thanks for the positive feedback, HP. No, we haven't checked with the DS manufacturing site on the additional batch, but this site has shown a reasonable degree of agility in the past when we had to pull in this same batch by 2 months to maintain our desired safety stock levels. I'm confident they can do it again, but if they can't we still have enough DS on hand to tolerate keeping manufacturing start outside the horizon. On your second question, I checked the planning cycle and saw we had Supply Review at this CMC meeting next week. I wanted to get the team's alignment on the new plan before sharing with Finance."

"Ok great. That strategy makes sense, but you might want to give Finance a heads up that these changes to the plan are coming. They'll appreciate some advance warning, and if they need the numbers sooner, they'll let you know, especially because this will represent a big hit to our budget adding so many additional batches. We'll likely need to run this through a Strategic CD&OP cycle as well because of that."

"Sounds good, HP," you respond with a smile. You haven't participated in a formal strategic clinical demand and operations planning (SCD&OP) review yet so this will be a nice opportunity to get visibility into executive level thinking. Oh, and two analyses down, two to go. The week is brightening!

16.5 Thursday Morning: Regulatory Restrictions

Fortunately, there's still an hour left on Wednesday for you to prepare for the meeting tomorrow on the CHA feedback to tighten the limits of your DS specification. You open the batch genealogy table to refresh your memory on the batch history of the program and the regulatory relevant attributes. You see there have been four DS batches manufactured to date using two different specification versions, but all four batches are categorized as Process Version 1.0. "This should be an interesting conversation tomorrow," you think. If some of the previously manufactured batches don't meet the new limit, this has the potential to change the entire distribution strategy for the Phase I/II.

Walking into the meeting the next day, you see the entire CMC team sitting around the table. The CMC team lead, CMC project manager, DS technical lead, DS analytical lead, the respective DP leads, quality lead, and regulatory CMC lead are all there. As the regulatory CMC lead begins talking, you lean in. "Team, this is challenging feedback that we've received from these health authorities. They've both requested us to tighten up one of our key specification criteria to the point that only one of our previously manufactured Drug Substance batches is acceptable for use. I've already discussed this with my Clinical Regulatory counterpart, and both countries are very promising from a recruitment perspective so there is pressure on us to find a way to accept these changes and make this work."

The CMC project lead speaks up. "When we get feedback like this, we generally have two choices. We can either move forward with a single, tighter specification that applies to all batches, or we can create separate versions of our specifications. In the latter case, we will retain our original specification for those countries that have already approved it, and we'll create a second, tighter version of the specification for these countries that have requested the updated limit."

"I don't think we have a choice here," the DS analytical lead chimes in. "We were all at the meeting on Monday morning where we discussed the yield issues with this second batch. We don't know what's causing the issues there, and while that's likely not directly related to product concentration, we simply can't guarantee that we'll be able to hit that tighter limit going forward. I don't see a path for a single global specification."

"When will we have the analytical test results from this most recent batch?" the quality lead asks.

"Typically those come in about 8 weeks from completion of the batch," the DS analytical lead responds.

"Ok. CP, we're going to need your help here," the CMC project lead says. "We need to understand if there's enough Drug Substance available from the sole batch we have that meets this tighter limit to supply these two countries. If not, we'll have to wait on the results of this most recent batch. If that batch doesn't meet the tighter limit, we might be out of luck. If that batch does meet the limit, then would those two batches be sufficient to supply these two countries? Do you think you can have a preliminary look at that assessment today so we can get back to Clinical Operations?"

You think quickly because you know you'll need to analyze the distribution plan and potentially make changes to properly answer the question. But, it's also tricky because you know the country-level demand forecasts in the plan are probably wrong, and even if the plan currently had enough supply from the first batch, over-enrollment at one or both countries could quickly deplete whatever is available from the first batch. "Here's what I can do," you say. "I can tell you how many subjects across both countries could be supplied from the first batch and how many subjects we could support with a second batch. I don't know if I can get to that answer by today, but I'll do my best. I'll send an update by this afternoon if that works for the team."

The CMC project lead nods, and the team moves on to other technical aspects of the discussion.

You get back to your desk and begin writing up the analysis:

1) Check inventory levels at the finished product central depot location to determine how much inventory of the batch with the acceptable limit is available for distribution.
2) Analyze the distribution plan to determine how much of the acceptable batch can be reallocated from countries that have approved the wider limit to countries that will only approve the narrower limit.
3) Review the site stocking and depot stocking quantities for each of the respective depots, sum the total amount, and then subtract that total from the amount of inventory available in Step 2.
4) Calculate the total number of kits required to support an entire course of treatment for one subject. Account for the weight-based dosing and round up the number of kits dispensed at each visit to the top end of the range.
5) Take the total amount of inventory from Step 3 and divide that by the number of kits required to support a single subject from Step 4. Round that number down to the nearest integer to arrive at the number of subjects the batch can support.
6) To calculate how many subjects the second batch can support take the total estimated yield from that batch, subtract the site and depot stocking quantities from Step 3, then divide that total from the number of kits required to support a single subject from Step 4.

It takes you about an hour to work through the analysis, and the numbers are grim. There's only enough material left from the batch to support one total subject across the two countries. If this is going to work, the team needs the second batch of material to meet the tighter specification limit. Assuming that happens, there will be enough material to support up to 10 total subjects, a more palatable figure given the high potential enrollment from the two countries. Before you start writing up your response, you open the plan and review the distribution plan summary to note the different process versions of the batches. You'll need that information to explain the strategy at your next plan review in 2 weeks. Then, you write up your note to the CMC team detailing the findings of your analysis and reiterating the importance of delivering that second batch to support opening recruitment activities in the impacted countries. You close the note asking the CMC team lead who will communicate this information back to the study team because you know you'll get questions at the next study execution team meeting and want to be sure you're aligned with the messaging before then. Feeling satisfied with how the day has begun, you close your computer, and head for a quick lunch break.

16.6 Thursday Afternoon and Friday: The Adaptive Trial Design

Now, for the task you've been most excited about since the start of the week, creating the initial demand and supply plan for a new program. But, this isn't a typical Phase I program. Instead, it's an adaptive trial design, meaning the study includes a set of predefined changes based on emerging data which enable the trial to potentially become registrational without a second supportive study. That context brings added pressure to ensure uninterrupted supply while also being mindful of the high cost to manufacture the DS for this new molecule. Earlier in the week, you had reached out to key members of the CMC team the COLs for the program. From the CMC team, you've requested a host of information and planning inputs including the intended formulation and container closure system of the DP. From the COLs, you've requested a copy of the clinical development plan for the program, the protocol summary, and the initial country and site list for the study. Both groups have responded to you, and you decide to begin with the CMC feedback which has been summarized in an email. You read:

"Hello CP,

Here are the CMC assumptions you requested. Please give me a call if you have questions or want to discuss these inputs:

- *Active Formulation*: 40 mg/mL in a 10 mL vial = 400 mg total dose available per vial
- *Placebo Formulation*: 0 mg/mL in a 10 mL vial
- *DS Assumptions*:
  - *Manufacturing Site*: Drug Substance Kings (China)
  - *Reactor Size*: 1000 L
  - *Titer*: 2.0 g/L
  - *Harvest Yield (%)*: 80%
  - *Purification Yield (%)*: 30%
  - *Total DS Yield (g)*: 480 g
  - *Manufacturing Lead Time*: 12 weeks
  - *Quality Lead Time*: 8 weeks
  - *Total Lead Time*: 20 weeks
  - Team intends to scale up to 2000 L for GMP production in Year +2
- *DP Assumptions*:
  - *Manufacturing Site*: Drug Product Aces (China)
  - *DP Yield (%)*: 90%
  - *DP BOM Quantity (mg)*: 444.4 mg
  - *DP Yield (vials)*: ~1080 vials
  - *Routine Sampling*: 250 vials/campaign

- *Nonroutine Sampling*: 300 vials/study
- *Expected Yield After Sampling*: ~530 vials
- *Manufacturing Lead Time*: 6 weeks
- *Quality Lead Time*: 8 weeks
- *Total Lead Time*: 14 weeks
- *Current DP Expiry Period*: 12 months
- *Planned DP Expiry Period*: 24 months

We're looking forward to bringing you onto the team for this important and exciting program! – CMC-L"

As you read through the assumptions your first thought is that a minimum of two SKUs will be necessary for the study, a single-vial configuration for each of the active and placebo formulations. The only question left on the SKU strategy is whether a US-only label makes sense to start the trial or if going straight to a global label is better. Next, you see that only 530 vials will be available from each batch after routine and non-routine sampling. "We may have to look at pooling Drug Substance batches early on in the study," you think to yourself. You wonder if the CMC team has had conversations about potentially scaling up the reactor size, but then you catch yourself and realize you need to see the clinical development plan (CDP) before jumping to conclusions. You draft up a short response to the CMC Lead asking for the CMC strategy, and, fortunately, they're at their computer because a note comes back to you almost immediately.

"Hi CP – Attached is the high-level CMC strategy – shown in Figure 16.1. Our first DS campaign just started this quarter, and we're running two batches. Our strategy is to pool these batches into a single Drug Product batch for initial trial supply later this year. We'll run our second DS campaign at the end of this year and utilize the same two-for-one pooling strategy. Concurrently, we're planning to initiate scale-up work at our Drug Substance manufacturing site later this year so we're ready to support clinical supply and reach commercial scale in time for the qualification batches scheduled in Year +3. You'll see placeholder campaigns in subsequent years for clinical resupply and qualification. Feel free to move the clinical resupplies around after the Current Year batches, but we'll need all three Process Performance Qualification (PPQ) batches in Year +3, so those are locked in from a timing perspective. Theoretically those batches could be used for commercial supply, but if we need them for clinical supply, we should allocate them to the study."

"I'm on it, CMC-L. Thanks for the vote of confidence."

You press send and turn your attention to the response from the COLs. They've also sent a reply and have attached the requested documents including the CDP, the protocol summary, and a table containing the country list, number of sites, and desired site activation target dates.

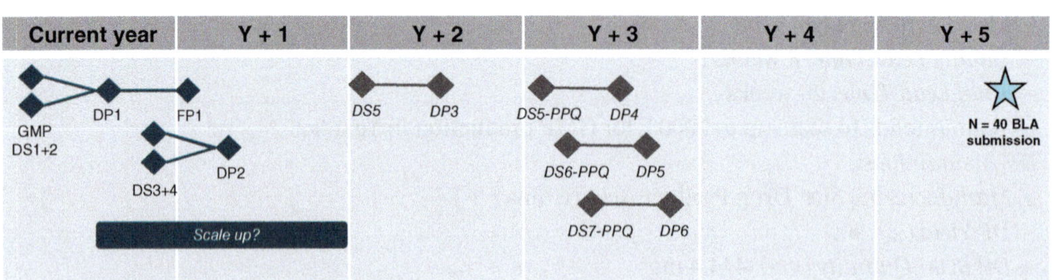

Figure 16.1 Adaptive Trial Design CMC Strategy

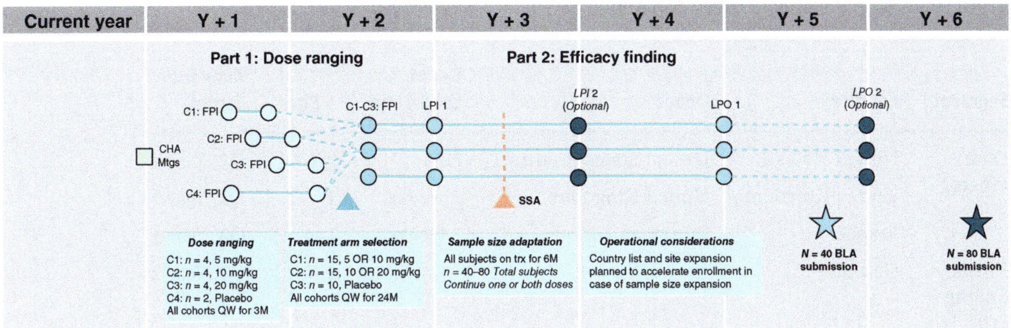

Figure 16.2 Adaptive Trial Design CDP

You skim through the CDP shown in Figure 16.2 and note the study has two segments, dose ranging and efficacy finding. In the dose-ranging segment, there are four cohorts testing three different doses of active IMP and a single placebo control. All four cohorts dose every week for 13 weeks total. After completion of the dose-ranging cohort, treatment arm selection takes place where one or two of the active doses are taken forward. Then, the efficacy-finding segment of the study begins with two active cohorts and one placebo cohort. All subjects within the efficacy-finding segment take drug once a week for 104 weeks total. Once all subjects have reached a minimum of 6 months of treatment, a sample size adaptation analysis takes place which has the potential to increase the total sample size from 40 subjects to 80 subjects. If the sample size remains at 40 subjects, last patient in (LPI 1) occurs at the end of Year 2. If the sample size is adjusted upward, recruitment and enrollment continue until last patient in (LPI 2) occurs at the end of Year 3. The CDP notes that the number of countries and sites increases if the sample size is estimated upward to retain the last patient in timeline.

Reading through the protocol summary, you also notice that subjects enrolled in the dose-ranging segment of the study may continue treatment on their respective dose through the treatment arm selection milestone and into the efficacy-finding segment. Meanwhile, subjects assigned to the placebo cohort in the dose-ranging segment are rerandomized to one of the active arms. That also explains the relatively short enrollment timeline from first patient in (FPI) of the efficacy-finding segment to the LPI 1 date. If all 14 subjects carry over, then only 26 more subjects are needed to fully enroll the study before the sample size adaptation analysis.

Another thought hits you as you look over the CDP one more time. This is a large molecule-program with a mg/kg dose and that means weight-based dosing. You go back to the protocol summary again but don't see an average patient weight listed. Before panicking you scroll to the inclusion/exclusion criteria and see that this is a rare disease, and the acceptable age range is 20–60. You write down 80 kg as a first pass, knowing this is likely high and send a brief email to the COL asking them to confirm.

Next, you look at the country and site list in Table 16.1 and see there are four sites planned across three countries for the dose-ranging study, but that quickly expands to 20 sites across 10 countries by the start of the efficacy-finding segment. Countries within the efficacy-finding segment are broken into two tiers. You look back at the CDP and realize the Tier 1 countries and sites are intended for the start of efficacy finding while the Tier 2 countries are back up countries and sites to support additional recruitment efforts in the event of a sample size adaptation. You ping the distribution site manager who's been assigned to the program and ask if they're available for a quick chat.

"Hi DSM, it's CP. I have the country list for our new program, the one with the Adaptive Design, and I have two questions for you. Can you please let me know the distribution network we're

Table 16.1 Adaptive Trial Country and Site List

Segment	Country	Depot	Depot Lead Time	# Sites	Activation Target
Dose ranging	United States	United States (Central)	N/A	2	Q2, Year 1
	United Kingdom	United Kingdom	3 weeks	1	Q2, Year 1
	France	European Union	3 weeks	1	Q3, Year 1
Efficacy finding (Tier 1)	Germany			2	Q2, Year 2
	Italy			2	
	Spain			3	
	Israel	Israel	4 weeks	2	
	Turkey	Turkey	4 weeks	3	
	Japan	Japan	3 weeks	2	
	China	China	8 weeks	2	
Efficacy finding (Tier 2)	Brazil	Brazil	7 weeks	2	Q2, Year 3
	Mexico	Mexico	8 weeks	3	
	South Korea	Asia–Pacific	3 weeks	2	
	Taiwan			2	
	Vietnam			1	

planning to use for this program, and can you please email over the corresponding depots for each of the countries in the list?"

"Yes I can get that over to you right now, CP," the Distribution Lead responds. "Keep in mind that the United States depot will also be the Central Depot as it's the only site in this network with secondary packaging and labeling capabilities. I'll also include the depot-to-depot lead times for inclusion in your plan."

Five minutes later, the depot list pops into your inbox, and you update the table that clinical operations has sent over to you.

You pull up the business process map for demand and supply plan creation and walk through the steps:

1) *Define and Classify the SKUs*: Now that you have the country and depot list in front of you, you decide to start the study with a global label. The United Kingdom and France join the United States in the dose-finding segment so there's no advantage to creating a US-only SKU because then a separate label for the United Kingdom and France would be needed and this would add unnecessary cost and SKU complexity. Instead, you decide to create a label with all dose-ranging and efficacy-finding countries given how quickly the study could transition to not only the Tier 1 countries but also the Tier 2 countries. This strategy will be slightly more expensive than creating a label with just the United States, the United Kingdom, and France, but it increases flexibility because product labeled for dose ranging could also support efficacy-finding countries when the transition happens.
2) *Map the Supply Chain and Depots*: Table 16.1 captures the countries, depots, and lead times.
3) *Identify the Demand Planning Assumptions*: You have the intrinsic design elements from the CDP and protocol summary that are necessary to create individual subject demand curves. You also have the extrinsic design elements from the country list and site activation schedule. The only missing variable is the enrollment rate, but you can extrapolate from the CDP how many

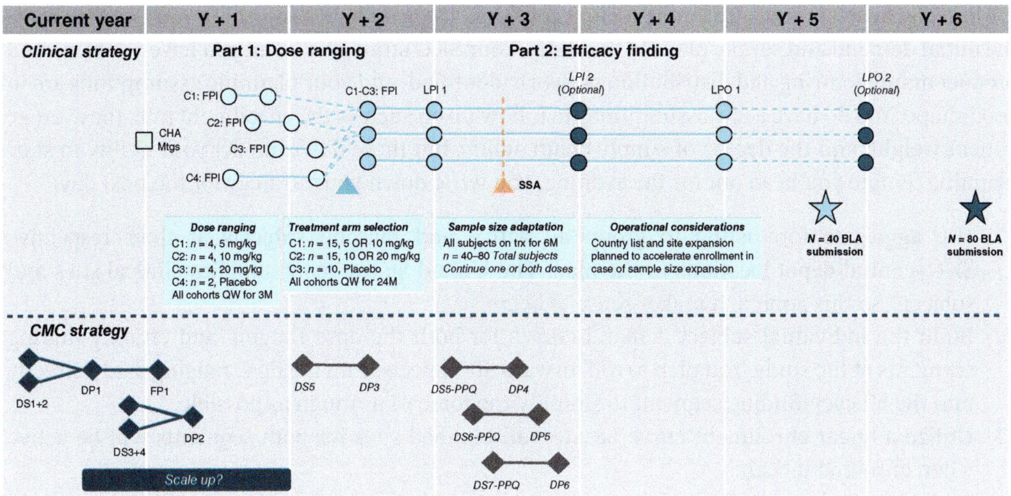

Figure 16.3 Adaptive Trial Design CDP and CMC Strategy

subjects will enroll in both segments and the amount of time from FPI to LPI (LPI 1 and LPI 2) with and without sample size adaptation. Given the aggressive site activation schedule, you're comfortable assuming a linear enrollment rate. This isn't typical for other programs, so you also make a note to revisit the enrollment rate assumptions with clinical operations at the first demand review meeting for the program.

4) *Identify the Supply Planning Assumptions*: You have the main supply planning assumptions from the CMC lead including the initial GMP production strategy from the CMC strategy. You take a few minutes to map the CMC strategy against the CDP and come up with the image in Figure 16.3. Based on the 24-month expiry, you can see that additional DP campaigns may be needed beyond the PPQ lots because the study could extend into Year +6. You make another note to ask the CMC lead how much flexibility the DS CMO has with respect to scheduling and cancelling DS and DP orders. The sample size adaptation milestone occurs halfway through Year +3 and that could lead to disruptive changes in the DS and DP supply plan. Understanding those limitations could help the team avoid unnecessary cancellation fees in the event the sample size stays at $n = 40$ and runs need to be cancelled or deferred.

The only remaining assumptions to document are the supply planning parameters at each level of the supply chain. Your team has a standard set of parameters for new large-molecule programs that you plan to use which are shown in Table 16.2.

Table 16.2 Supply Planning Parameters by Planning Level

Planning Level	Safety Stock (weeks)	Order Target (weeks)
Finished product regional depot	9	26
Finished product central depot	13	52
Drug product	26	52
Drug substance	0	52

Thursday nears its end, and you're feeling confident that you have everything you need to create the initial demand and supply plan for the study. Your SKU strategy is clear, you have your finished product manufacturing and distribution network identified, and your planning assumptions are in good shape. You do have a few assumptions to follow up on such as the enrollment rate, the average patient weight, and the degree of supply chain agility, but these don't impact your ability to start planning. Before you head out for the evening, you write down your strategy for the next day:

1) Use aggregate forecasting to group all active and placebo subjects at their respective SKU–central depot locations in the plan. The United States depot supplies clinical sites and subjects, so this approach makes sense to begin.
2) Build the individual subject demand curves for both the dose-ranging and efficacy-finding segments of the study. You plan to roll forward all subjects from the dose ranging study directly into the efficacy-finding segment to simplify the forecast as much as possible.
3) Utilize a linear enrollment curve because all planned sites for both segments will be active when enrollment begins.
4) Add a 10% overage multiplier to account for weight-based dosing, but keep this low to begin until clinical operations confirms 80 kg or provides a more accurate figure.
5) Estimate the RTSM initial site supply values and determine an initial depot stocking strategy at each regional depot with respect to both timing and quantity. Check that the distribution orders flow back to the central depot as distribution demand.
6) Total all subject, site, and distribution demand at the United States depot and begin the finished product production planning process. Based on the relatively limited shelf life, consider splitting the first DP batch into multiple finished product batches to optimize for expiry.
7) Check the allocation strategy for the initial distribution orders to all regional depots for sufficient shelf life. Plan resupply distribution and production orders as needed to complete the dose-ranging segment of the study.
8) Link the finished product production plan to the DP production plan and the DP production plan to the DS production plan. Enter the supply-related assumptions for DP and DS and the initial upstream supply planning parameters.
9) Execute the reconciliation and requirements review cycle through the duration of the clinical program.
    a) Align the initial DS and DP batches (DS1+2 and DP1) to the CMC strategy because these batches are likely locked on the production schedule.
    b) Assume 24 months of shelf life for all finished product batches made from Year +2 onward.
10) Highlight when more than four DS batches will be needed in a single year as a potential target date for implementation of a larger DS batch size.

You scan through the list one more time before shutting your computer down, feeling like you're missing something. Then, it hits you that the sample size adaptation creates two scenarios within the CDP, a low-side demand case where only 40 subjects enroll and a high-side demand case where up to 80 subjects enroll. You think hard about which scenario to use in your forecast and add a few more steps to your list:

11) Conduct a low-case–high-case analysis:
    a) Create the original demand forecast using the upside efficacy-finding sample size of $n = 80$ as the high case and run Steps 7–10 as planned.
    b) Create a low-case scenario with the demand forecast based on an efficacy-finding sample size of $n = 40$.

c) Take the high-case finished product, DP, and DS production plans copy them into the low-case scenario forecast. Analyze the plan to determine the amount of excess waste risk. Reduce the finished product production plan as needed to eliminate waste. Update the DP and DS production plans accordingly.
d) Take the updated DP and DS production plans and copy them back into the upside demand forecast. Analyze the plan to quantify the stockout risk of producing fewer kits than the standard supply planning parameters called for in Step 11a.
e) Iterate through Steps 11c–d to determine if an optimal production and inventory management strategy exists that avoids both excess waste and stockout risk in either the high-case or low-case demand scenarios.

You send a final email to HP asking if she has time to chat first thing in the morning and shut down for the night. Friday is on the horizon.

"Hello CP. Good to see you again. Third time this week is it?"

"Actually, it's the fourth time, but who's counting right?" You laugh. "I've been working on my approach to the initial Adaptive Trial Design plan and wanted to run through the key points for feedback," you say as you pull open the combined CDP and CMC strategies and your notes from the previous afternoon.

HP spends a few minutes scanning through them while tapping a finger to her lips. "Good recognition that a low case – high case analysis is appropriate for this scenario. I know you've run through these for your other programs but that was in the context of our Strategic CD&OP process for late-phase programs. Have you thought about how you're going to present the output for Plan Review and ultimately to the CMC team?"

"Not yet," you respond. "Do you have guidance for me?"

"First, remember that they're different audiences. At Plan Review, we'll be interested in reviewing the assumptions of your plan and how you arrived at your recommended strategy. The CMC team cares more about the output, and how that recommendation impacts the overall CMC strategy. I've been so impressed with your work this week; I think that's all the feedback you need for now to get started. Let's plan to review your communication strategy first thing on Monday. How does that sound?"

"That works, HP. Time to get to planning!"

"One more item before you go, CP. I need you to stop by Conference Room One before we head out for the day. See you at 4:00 p.m., and good luck with the analysis."

16.7 Friday Afternoon: Excellence in Planning

You're tired but satisfied. It's almost 4:00, and you've had a week. But you're happy because you've finished up both of your major analyses. On the study design scenario, you connected with the CMC team lead and finance lead to review the budget impact; met with the packaging site lead to align on label strategy; and saw the distribution team lead to finalize the country list assessment. While you still need to organize the final analysis for the COL, you're confident you can complete this on Monday. Over the rest of the day you worked up both the low-case and high-case plans, and you're almost done putting the finishing touches on the recommended production and inventory management strategy for the adaptive trial design program. While you wanted to finish it today, the plan review meeting isn't until the end of the month, so you still have time to polish both the analysis and the communication plan over the coming two weeks.

Walking over to the conference room, your curiosity perks up. It's unlike HP to ask for a spontaneous meeting this late on a Friday afternoon. She's usually focused on starting her weekend like most of your colleagues. You open the door and are surprised to see that HP isn't alone, the head of clinical supply chain and global head of supply chain are sitting with her at one side of the table. Your stomach drops a bit because you don't usually take meetings with the global head. This is going to be either very good or very bad.

"Have a seat CP," HP instructs. "You're probably thinking this is either going to be very good or very bad. I'll kill the suspense now and tell you that you're not getting fired," she says with a laugh. "Head of Clinical Supply, why don't you tell CP why they're here?"

"Well. We're here this afternoon to congratulate you on your promotion to Senior Planner, CP. Please, sit down. We'd like to share some of the thought process and reasoning that went into this decision."

You break into a huge smile. Yes, this is very, very good after all. "Wow. Thank you all very much for the support and guidance over the last two years that I've been here. I've learned so much, and I'm excited to continue our mission of helping patients."

The global head of supply chain speaks first. "I'd like to start by saying that your hard work, positive attitude, and hunger to learn hasn't gone unnoticed. These traits are foundational to advancement in our organization, and I want to personally give you that feedback. Thank you for your contributions to our programs and to our team. We're looking forward to seeing what you can do as a Senior Planner over the next several years. Congratulations, again."

Your boss picks up the from there. "Beyond those traits, you've also demonstrated mastery of the skills required to be an excellent Clinical Planner. You've become an expert in the concepts of planning including demand forecasting, supply planning, distribution planning, and upstream planning. You're competent at using our planning solution for both demand and supply planning and for scenario planning. The CMC teams you work with have provided feedback that you understand your role and how you can add value by providing analytical and strategic input to the team. I've also noticed that you have a strong understanding of the business processes we use to manage the clinical supply chain, and you're able to recognize when processes should be adhered to within our planning cadence and when ad hoc work needs to get done to keep the supply chain moving.

Lastly, and this was a deciding factor in your promotion, you've become an excellent communicator within our team and to the external stakeholders that rely on our work. You're responsive, concise, and you frame complex ideas and analyses in a digestible manner for team members who aren't supply chain professionals. I also appreciate the relationship that we've built over the last 18 months and how you always bring proposed approaches to your planning work for my input rather than simply asking me to give you the answer."

"You embody the values we look for in our Planning team, CP," the head of clinical supply chain says. "Thoughtful. Analytical. Strategic. Agile. Responsive. Communicative. You have a very bright future in our organization."

"Thank you all," you say. "This has been such a rewarding journey, to be a part of a great team that's dedicated to serving patients and pushing our medicines forward. I can't wait to see what the future holds either."

Happy Friday, indeed, you think as you leave the conference room. Time to celebrate over the weekend before getting back to it on Monday.

16.8 Key Takeaways

1) While CP is a fictional character, the week's events capture what CPs are regularly asked to confront – unexpected situations, complex scenarios, and a wealth of ambiguity. Fortunately, CP had the knowledge, skills, processes, tools, and management support to successfully navigate these situations. Effective clinical supply chain management and demand and supply planning rely on all of these factors.
2) When new information that affects the plan becomes available or questions come in from team members, the best strategy is to pause, absorb, and think through the scenario or problem. Defining what needs to be solved, determining whether the plan needs to be updated, and writing down a step-by-step approach to getting the answer enables proper diagnosis of the issue and saves time in execution.
3) Documenting the impacted stakeholders and a communication plan is another key to success when making major changes to the plan. Business processes enable information flows, but it's easy to overlook a team or function when events are moving quickly, and scenario outcomes need to be distributed broadly.
4) Two effective tactics when learning about a new scenario are to clarify what result or outcome is needed and to set appropriate expectations for when a response will be ready. Clarifying the scenario aids the CP in delivering a concise and effective response. Setting appropriate expectations on when the analysis will be complete gives the CP time to fully assess the problem and space to prioritize the work as needed.
5) Use subject matter experts in the organization to reconfirm assumptions before diving into the analysis because the clinical supply chain is constantly evolving. Taking a few minutes to have a conversation and get the latest data and inputs into the plan also saves time. The most important subject matter experts to rely on when making an update to the plan are fellow CPs and/or the head of the planning function.
6) Great planning relies on a combination of many factors. Individual knowledge, skill, and tenacity is one part of the recipe. Well-designed business processes and a solid planning tool or model is another part. A supportive management team and organizational culture is yet another part. When those elements come together, optimal plans are the result.

17

Comparator and Ancillary Planning

17.1 Comparator and Ancillary Definitions

This chapter marks the transition from the techniques and approaches used in supply chain planning for clinical trials to alternative clinical supply chains that require planning. Interventional clinical trials are designed to test novel investigational medicinal products (IMPs), but in most cases, these aren't the only materials needed to conduct the study. A variety of other therapeutics, supplies, and equipment may also be provided to sites and participating subjects during the trial. In this chapter, we'll explore the world of comparator drugs that are classified as investigational products, comparator drugs that are classified as non-investigational (e.g. non-IMP) products, and other ancillary supplies. Let's begin by defining and categorizing IMP comparators, non-IMPs comparators, and ancillaries because they're often used interchangeably when in fact they have different meanings and uses in a clinical study.

There are two categories of products supplied to a clinical trial, those that are classified as IMPs and those that are classified as non-IMPs. The novel therapeutic is always categorized as an IMP, but so are many comparator products such as placebos and commercially available medications used in a trial. A comparator is defined as a medicinal product used in a clinical trial that is not the novel IMP. Comparators classified as IMP are medicinal products that are used in the analysis of whether or not the trial met its endpoints. Conversely, non-IMP comparators are not used in the analysis of the clinical trial's endpoints.

Within, the IMP comparator bucket is placebo-control comparators and active-control comparators. A placebo comparator is simply "an arm type in which a group of participants receives a placebo during a clinical trial.[1]" On the other hand, an active comparator is defined as a commercially available "treatment that is considered to be effective (or active) by health care providers.[1]" Active comparators are commercially registered products and are available for on the open market for use in a clinical trial. Active IMP comparators are typically standard of care (SOC) products that are used as a control arm in the study. An SOC is defined as "a treatment that is accepted by medical experts as a proper treatment for a certain type of disease and that is widely used by healthcare professionals.[2]" When used as a control arm, the active comparator may be administered open label, or it may be blinded against the novel IMP.

Meanwhile, a non-IMP product is "a medicinal product used for the needs of a clinical trial as described in the protocol, but not as an investigational medicinal product.[3]" In the European Union, non-IMPs are also known as auxiliary medicinal products (AMPs). Examples of AMPs include rescue medications, challenge agents that are co-administered with the test drug in drug–drug

Supply Chain Planning for Clinical Trials – A Practical Guide, First Edition. Ryan Mills.

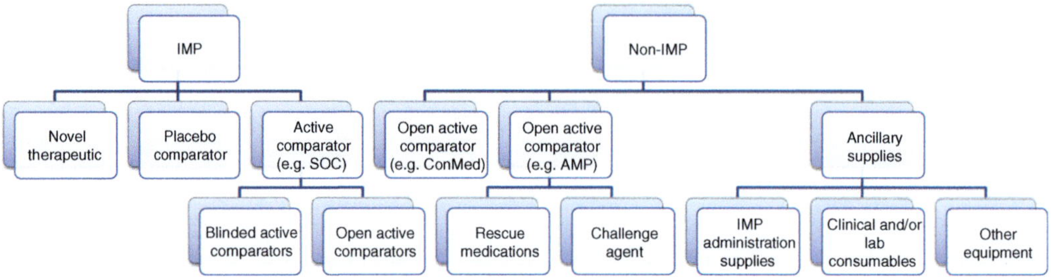

Figure 17.1 Investigational Product, Comparator, and Ancillary Categorization Framework

interaction studies, and background medications. Another example of a non-IMP product is an SOC treatment administered as a concomitant medication (ConMed) in the trial. A ConMed is defined as "a medication a person is taking that is not being studied in the clinical trial he or she is taking part in.[4]" Within the European Union, ConMeds are distinguished from AMPs because AMPs are directly relevant to the conduct of the trial while ConMeds are not.

The final category of products used in clinical trials are ancillary supplies. Ancillaries are defined as any material required to conduct a clinical trial beyond the novel therapeutic and any comparator products used in the trial. All ancillaries are non-IMP, and there are three subcategories within this group. The first are materials required to facilitate administration of a drug such as infusion pumps, saline, and sterile water for large molecules. The second category is lab kits and/or supplies required to collect and process samples taken within the study. And, the third category captures all other supplies needed to conduct the trial such as specific equipment needed to run specific assessments. Typically, the clinical supply chain function is solely responsible for ensuring availability of IMP materials, but it's not uncommon for clinical supply chain to also support the non-IMP, AMP, ConMed, and ancillary supply chains.

Figure 17.1 summarizes how IMP and non-IMP products are categorized between each other. With the definitions and framework above in mind, we'll review how good manufacturing practice (GMP) regulations apply to IMPs and non-IMPs. These regulations dictate the degree to which GMP activities and requirements must be applied to each category, and this has implications for how the supply chain is managed. Understanding the GMP landscape for each material type also provides context for a review of supply chain-specific considerations that are relevant for IMP comparators, blinded IMP comparators, non-IMP comparators, and ancillary supplies. For ancillaries, we'll also introduce a new business process for identifying, procuring, and supplying these items. There are several activities that require a significant amount of cross-functional activity to coordinate for ancillary sourcing, so this is an area where implementing a clear business process supports efficient supply chain planning and execution.

17.2 Regulatory Implications for IMP and Non-IMP Comparators

From a regulatory perspective, placebo-control and active-control comparators are categorized as IMP and are therefore treated similarly to the investigational product of the study. This categorization has three significant implications on labeling requirements, potential modifications to the active comparator for blinding purposes, and on regulatory application requirements.

First, all IMP comparators must have a clinical label applied to them for use in a trial. Both American and European Union regulations are clear on this point. According to recent Food and Drug

Administration (FDA) guidance, novel therapeutics, placebo-control comparators, and active-control comparators are all covered under the purview of the investigational new drug (IND) application.[5] Therefore, all IMP comparator products are required to comply with the FDA's IND regulations, and this also means that all placebo and active comparator products must be labeled in accordance with 21 Code of Federal Regulation (CFR) Part 312. The label indicates that the comparator product is limited to investigational use in the context of the clinical trial for which it's being provided. The scope of CFR Part 211 encompasses packaging and label control, so by definition, the labeling of comparator IMPs falls under GMP in the United States.[6] In the European Union, the Clinical Trial Regulation, Annex VI requires that all investigational therapeutics and unauthorized AMPs undergo a labeling operation to enable distribution of the product through the EU.[7] A clinical label is not required for commercially approved AMPs, "unless specific circumstances are required in order to ensure subject safety and the reliability and robustness of data generated clinical trials.[8]" To summarize this, comparator products used as IMP need to be labeled because appropriate administration of the comparator is required to ensure the reliability and robustness of data generated in the clinical trial.

The second implication of the investigational product classification is that any modification to an IMP comparator falls under the scope of GMP. Modifications to the product include labeling as well as any manufacturing activities required to blind the product against the novel IMP. Blinding a large molecule product may involve changing the primary or secondary packaging design to conceal the contents of the comparator and the novel therapeutic. Blinding a small molecule may also require over-encapsulating the drug product form (e.g. a capsule or tablet) so that both compounds appear identical to each other when the primary packaging is opened. When these modifications are executed outside the scope of the commercial marketing authorization of the comparator, they must be executed under GMP. In the European Union, a set of recommendations contained in EudraLex Volume 10, Chapter III states that "where an authorised AMP is modified while such modification is not covered by a marketing authorization, it shall be manufactured according to Good Manufacturing Practice.[7]"

A third implication of the IMP categorization and investigational product classification of active-control comparators is the regulatory application content requirements. All manufacturing steps performed prior to acquisition of the active-control comparator product do not need to be listed in the IND or CTA. Instead, reference is made to the United States Prescribing Information (USPI) in the IND or to the Summary of Product Characteristics (SmPC) in the CTA. The USPI "contains a summary of the essential scientific information needed for the safe and effective use of the ... drug.[9]" In the European Union, the SmPC is a "legal document approved as part of the marketing authorization of each medicine," and it contains information on how the commercial product was manufactured, packaged, and how it should be used as a medication.[10] However, any GMP manufacturing step performed after the active-control comparator has been procured must be filed in the clinical regulatory application with all the same requirements as a novel therapeutic or a placebo control. At minimum, the label text and labeling site must be included in the application. If additional GMP steps are performed, such as over-encapsulation of a small molecule drug product, then content commensurate to that step must also be filed as part of the application. Inclusion of this content means that changes to the comparator product are subject to the IMP lifecycle management and process version considerations outlined in Chapter 5 – Regulatory Considerations.

Non-IMPs are not subject to the same regulatory and GMP requirements as IMP products precisely because they are not modified in any way from their commercially available state and are only intended for use under the commercial label. Thus, concomitant medications, rescue medications, challenge agents, and ancillary supplies do not require additional labeling activities for use in a trial and no information regarding the use of these products needs to be included in the regulatory application. However, non-IMPs have a different set of supply chain complexities, which we'll return to later in this chapter.

17.3 Active IMP Comparator Considerations

As we've just seen, active IMP comparators share some common traits with novel investigational drugs and placebo comparators – their classification as an investigational drug when used as a control agent in a clinical trial, the requirement for a clinical label to be applied to the product, and their inclusion in regulatory applications. Active IMP comparators are also defined by the fact that they're both manufactured and commercialized by a third party, and this means they have a variety of unique supply chain considerations that don't apply to traditional IMPs and placebos. These include developing a sourcing and procurement strategy for the comparator, incorporating the comparator into the demand and supply planning process, and managing supply availability challenges.

Unlike novel therapeutics where the entire supply chain is entirely under the control of the sponsor company, active IMP comparators are manufactured and released by another biopharmaceutical company. To use an active IMP comparator in a clinical trial, it must be purchased on the open market and that typically requires working with a sourcing agent who negotiates with the manufacturer directly or with a wholesaler who acquires the product directly. Identifying a source is only the beginning though. Developing the sourcing and procurement strategy ultimately depends on the country list of the trial.

The country list dictates the sourcing strategy for comparator products for two reasons, one regulatory related and the other financial related. From a regulatory perspective, the active IMP comparator product quality information that's filed in one country or region may differ from another country or region. These differences then limit which countries may export or import the product if it was manufactured in a different country or region. For this reason, it's vital to work with the sourcing agent or to liaise directly with the manufacturer to understand the regulatory landscape of the product and map that to the country list of the trial. One mechanism to enable use of comparators that have been approved in different regions is through an equivalency statement. An equivalency statement declares that the product quality attributes of a comparator manufactured in one country or region meet the product quality attributes that have been filed and approved in another country or region. Equivalency statements may be provided with a clinical trial application to secure approval for use of a comparator in that country.

A second influence on the comparator sourcing strategy is whether the trial will be conducted in the United States. When this is the case, comparator manufacturers often require that a drug intended for use in the United States be purchased within the United States. This is because branded commercial drugs in the United States are often much more expensive to purchase than they are in other countries. It's therefore more profitable for the manufacturer to sell the comparator at the higher American price for use in the United States than it is to sell product made outside the United States for a cheaper price. In global clinical trials that operate inside the United States, it's typically the case that the comparator must be sourced from the United States for use in the United States and sourced outside the United States for use in other countries. This strategy ultimately saves money when there is a large cost difference between US and ex-US supply, but it adds stockkeeping unit (SKU) complexity to the clinical trial supply chain. For generic commercial drugs, these cost differences may be much smaller, so it's easier to source the comparator from a single country and logical to do so from a financial perspective.

From a demand and supply planning perspective, managing comparator products takes the same approach as managing novel therapeutics. The first step is demand and supply plan creation which involves defining the SKU strategy, mapping the supply chain and depots, collecting the demand and supply planning assumptions, and creating the initial demand and supply plan. There are special considerations for comparators at each point in the process.

First, the comparator sourcing strategy determines the SKUs that need to be created for the demand and supply plan. A unique SKU needs to be created for each source location from which the comparator is procured. Unique SKUs also need to be created for each finished product comparator SKU. These SKUs then need to be mapped to the respective central and regional depots. To illustrate this, let's assume a study is testing a novel therapeutic against an active-control comparator in eight countries across the United States, European Union and Latin America. Let's also assume the active-control comparator needs to be sourced from three regions: the United States, the European Union, and Latin America. The supply chain map for this program might look like Figure 17.2.

In this example, there are six total SKU–depot combinations for the novel therapeutic: the drug substance SKU, drug product SKU, finished product-central depot SKU, and three finished product-regional depot SKUs. On the comparator side, there are eight total SKU–depot combinations: one SKU each for the wholesale comparator that's approved for use in the respective region, one SKU each for the respective finished product comparator SKUs, and two additional Latin America regional depot SKUs. The reason for these two additional SKUs is because the Latin America comparator is labeled out of the EU-based central depot, and then it needs to be shipped to the local Latin American depots. It may seem counterintuitive for the Latin America supply to be labeled in the EU, but a scenario like this isn't entirely farfetched. Global commercial supply chains have their own nuances and complexities, which makes this kind of supply chain mapping exercise especially important for comparators. Also note that finished product comparator labels should only include the countries for which those comparators are approved for use to prevent accidental importation to the wrong country.

With respect to demand planning assumptions, the overage multiplier deserves special attention. Utilizing a relatively low overage multiplier early in the planning process helps create a more accurate demand forecast. Forecasting accurately for active IMP comparators is especially important for two reasons. First, comparators are often expensive to procure, and if the demand forecast

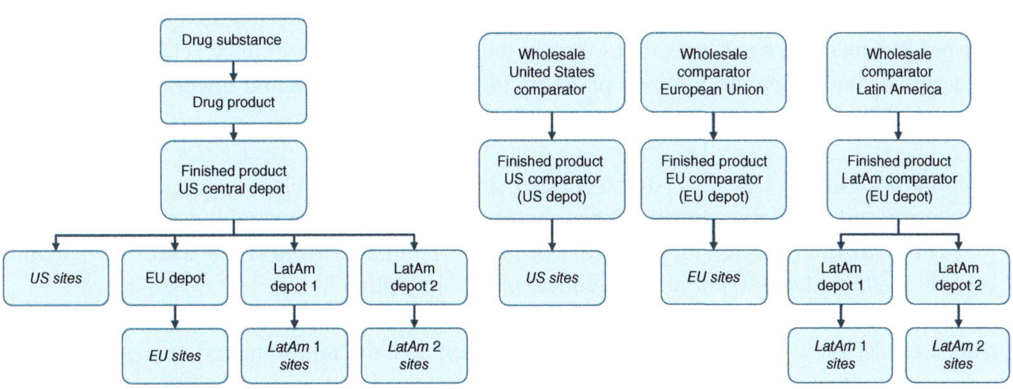

Figure 17.2 Comparator SKU–Depot Supply Chain Map

has excess overages built into it this leads to not only excess waste physical waste but also significant financial waste. Second, comparators are sometimes difficult to procure on a consistent basis. Market shortages occasionally happen, and over-forecasting demand takes supply away from future purchases.

With respect to supply planning assumptions, there are three inputs to focus on: the cost, the expiry period, and the relative supply availability of the product in the open market. On cost, when the product is relatively cheap, more liberal decisions can be taken with increasing the purchase quantities of each order, but as the product gets more expensive, care must be taken to avoid overspending and excess waste.

Second, the comparator product's expiry period is also a notable input. Unlike investigational products, the expiry period of a commercial product is highly unlikely to change over the duration of the clinical trial. If the comparator product has a relatively short expiry period (e.g. 18–24 months), this becomes a challenge to plan for because of the time needed to manufacture and release the comparator, procure the comparator, and then label the comparator for use in the clinical trial. The comparator manufacturer usually times production campaigns to fit that product's commercial demand profile, not to fit the needs of the sponsor running a clinical trial. This creates a situation where expiry becomes constrained by the time the product is ready for use in the trial, and this leads to a purchasing plan with frequent resupply orders to keep non-expired product available.

Third, the relative supply availability of the product also needs to be assessed. This dictates when and how much product is available for purchase at a given time. Supply availability should always be balanced against cost and expiry considerations though. Purchasing excessive quantities to hedge against future shortages doesn't make sense when that material will expire before it's consumed by demand. Supply availability is also a subjective assessment because market conditions change, so identifying a sourcing agent with experience is recommended because they can share historical insight on a particular product.

Once all the demand- and supply-related assumptions are in place, the planning process for comparators mimics that used for the investigational drug and for placebos. The same principles that apply to finished product demand forecasting, distribution planning, expiry planning, and finished product production planning also apply to the comparator finished product SKUs. The wholesale comparator SKU is treated as the upstream material for the finished product comparator SKU, and the other supply planning assumptions such as bill of material (BOM) quantity and yield also need to be reflected accurately in the plan.

The planning process for comparators includes both the finished product packaging strategy and the purchasing strategy of the wholesale product. The packaging strategy is largely driven by the same demand (e.g. subject and distribution) and expiry considerations as planning for novel investigational therapeutics, but it's also intertwined with the purchasing strategy, which has unique considerations for comparators. When planning for comparators, demand uncertainty, cost, and expiry should be considered in combination:

- *Demand Uncertainty*: Like novel therapeutics, demand uncertainty plays a significant role in the purchasing and packaging strategy. When the demand forecast is uncertain (e.g. few or no subjects enrolled), than smaller, more frequent orders are optimal to avoid excess waste. As demand becomes certain, then larger, more infrequent orders are optimal.
- *Cost*: When the comparator cost is high, smaller, more frequent purchases and packaging campaigns are desirable to avoid the risk of excess waste if demand fails to materialize. As the comparator cost decreases, larger batch sizes should be considered, but never to the extent that supply exceeds demand over the assigned expiry period.

- *Expiry*: When the expiry period is relatively short, smaller, more frequent purchases are necessary to avoid excess waste. As the expiry period increases, larger, more infrequent purchases should be considered. Proceed with caution though. The cost of the comparator is a more important factor to consider than expiry, especially when the demand forecast is highly uncertain.

Supply availability challenges are another consideration to manage when planning comparators. Market shortages are not under the control of the organization purchasing the comparator product, but there are some steps the clinical planner should take to prepare for this scenario. The best tactic is to share an end-to-end demand forecast for the comparator product with the sourcing agent as early as possible in the process. This gives the comparator manufacturer visibility to the additional demand for the product which they can add to their manufacturing plan. It's also important to update the forecast when demand for the product changes. The sourcing agent is a key stakeholder to include in the plan publication step of the operational clinical demand and operations planning (CD&OP) business process and in the supply plan handover process. Whoever the internal site manager is that liaises with the sourcing agent needs to receive and communicate updated purchasing requirements for comparator products as they're published.

Finally, an excellent technique to use when developing the purchasing strategy for expensive comparators is the low-case–high-case analysis. Working through a low-case–high-case analysis generates an optimal purchasing strategy that provides enough comparator to support subject demand in the high case while also avoiding excess waste (if possible) in the low case. If an optimal plan that meets both goals isn't possible, low case–high case identifies the potential cost of excess waste if the low-case demand forecast is realized and the potential stockout risk if the high-case demand forecast is realized.

17.4 Additional Blinded IMP Comparator Considerations

Blinding active-control comparators adds another layer of complexity to the clinical supply chain because GMP manufacturing and/or repackaging steps are also necessary to make the finished product. If the clinical trial is registrational, this adds even more pressure to design a supply chain and purchasing strategy for the comparator that performs and delivers supply to the trial consistently. Each GMP step is a unique upstream manufacturing event and needs to be treated as such in the demand and supply plan. Supply chain implications from blinding comparators include:

- Developing a manufacturing process and packaging design that ensure the blind between the investigational product and the blinded active comparator.
- Identifying a manufacturing site (or sites) capable of executing the manufacturing process and label design and qualifying those sites for use if they are not already approved.
- Adding sections to the regulatory application for each GMP step in the process and effectively managing changes that impact the content filed in those applications.
- Creating additional upstream SKUs as needed in the demand and supply plan. For example, if over-encapsulation of a small molecule is required, then an SKU for the over-encapsulated product, an SKU for the packaged over-encapsulated product, and an SKU for the labeled finished product may all be required.
- Collection of all supply-related planning assumptions for each upstream manufacturing step. Lead times are particularly important because the amount of time to execute comparator modification eats into the amount of shelf life available on each batch of product. In cases where the

active comparator has a relatively limited expiry period, modification may not be feasible because there's not enough time to modify, package, label, and distribute the product before it expires.

In some cases, blinding active comparators is prohibitively costly and complex to execute. For this reason, the clinical supply chain and other impacted functions should perform an assessment of the risks and challenges associated with the considerations listed above. The output of this should be brought forward to the project team for alignment and endorsement before moving forward.

17.5 Non-IMP Comparator Medication Considerations

Non-IMP comparator medications like concomitant medications, rescue medications, and challenge agents present an entirely different set of supply chain considerations than their IMP comparator counterparts. On the bright side, non-IMPs are always commercially sourced and used on label, and therefore do not require labeling or any other GMP manufacturing step to be performed on them. That's why it's rare for the clinical supply chain function to get actively involved in planning non-IMP supply, with one important exception that we'll return to shortly.

Before that, it's important to understand how non-IMPs are typically sourced and supplied to sites and subjects. Non-IMPs are also usually available to clinical sites within their own countries so they're able to source these items themselves. This type of sourcing is known as local sourcing. However, there are situations where a non-IMP can't be sourced in-country. In this scenario, an alternate source of supply needs to be identified from outside of that country. This type of sourcing is known as global sourcing.

When global sourcing is required, the process becomes more complex. Like IMP comparators, a sourcing agent is typically required who can find the desired non-IMP comparator on the open market, ensure that it's approved for use in the country that it needs to be supplied to, and arrange for the purchase and delivery of the material. Traditionally, these activities are managed by clinical operations and/or the contract research organization (CRO) responsible for overseeing study conduct. It's not uncommon though for the clinical supply chain function to also get involved with non-IMP medication sourcing when IMP active comparators are also used in a study. That's because it usually makes the most sense to use the same sourcing agent for both IMP active comparators and non-IMP medications.

In a situation where non-IMP medications are globally sourced, centrally warehoused, and distributed from that warehouse to sites and subjects over the course of an entire study, then additional demand and supply planning activities may be necessary. That's because non-IMP medications have expiry periods and shelf-life assignments, and if these aren't actively managed, expiry-related stockouts become a real risk.

17.6 Ancillary Supply Considerations

Ancillaries encompass all non-medication, non-IMP supplies that need to be provided to clinical sites during the course of a clinical trial. These include supplies required to enable administration of the investigational product or active comparator product, lab-related supplies and consumables, and any equipment required to conduct assessments or other aspects of the clinical trial. On the surface, the fact that these aren't medications and that no GMP operations are performed theoretically simplifies the supply chain, but that's not necessarily true. Ancillaries are often a deceptively complex supply chain for a few reasons.

First of these is the sheer diversity of clinical ancillaries. Some, like those that facilitate administration of an IMP, have expiration dates and that means inventory levels need to be actively managed. Others, like lab supplies, need to be identical across all sites so they need to be centrally sourced for consistency. This diversity also increases the number of stakeholders involved in the ancillary sourcing and supply management process. Clinical operations is typically accountable for ensuring that clinical sites have all the materials and equipment they need to run the study, but they're not always interacting directly with the clinical sites. That responsibility falls on the CRO, who typically has local personnel in the countries where the clinical trial takes place so they can liaise directly with the sites. Internally, there are also separate functions that define which ancillaries are needed to support administration of the drug, which lab kits are required for collecting samples, and which equipment is needed to conduct specific assessments. Input from these other functional areas is needed to make sure the correct ancillaries are identified and sourced.

Another area of complexity is that unlike IMP comparators and non-IMP medications, ancillary supply requirements usually differ from site to site within the same study. One site in the United States may have different supply needs than a site in the United Kingdom than a site in France. This variation means that identifying individual site requirements early in study start-up is essential so there's enough time to source everything that's required. Varying site requirements also means the sourcing strategy of a particular item may vary from site to site. At the one site, an ancillary may be found locally while at another site that same ancillary may need to be sourced globally because it can't be found within the country.

Effectively coordinating the identification of which ancillaries are required, how those ancillaries will be sourced, and who will ensure these activities are driven to completion requires an efficient, repeatable business process. The objective of the ancillary supply business process is to ensure that all ancillary supplies required to conduct the trial are identified and delivered in a timely fashion to enable study execution. All non-IMP products including non-IMP medications are included in the scope of the process. Clinical operations is the function typically responsible for ancillary supply and is therefore the logical choice as business process owner. The ancillary supply process begins as soon as the responsible functional areas have identified supplies that will be needed for the study, although the process formally begins with completion of the full clinical protocol. The business process steps and responsibilities are outlined in Figure 17.3.

Identification of the ancillary supplies needed for the trial depends on the type of supplies provided. In the case of ancillaries required to support administration of a biological investigational

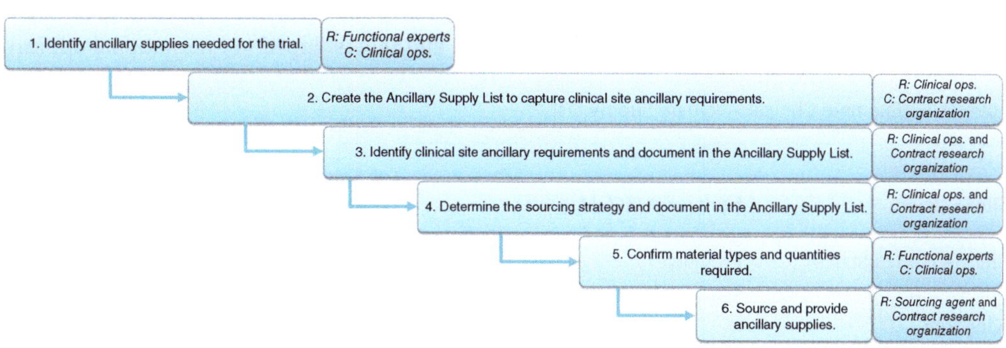

Figure 17.3 Ancillary Supply Chain Management Business Process Map

product, this work begins during formulation development. The chemistry, manufacturing, and controls (CMC) team and the drug product lead(s) in particular are responsible for developing the drug's route of administration. When a drug needs to be given intravenously or intrathecally, for example, the CMC team also develops the process and materials to support preparation and administration. This information is then written into the pharmacy manual (e.g. manual of procedures). The pharmacy manual provides detailed instructions on how IMPs are to be handled, prepared, administered, and returned or destroyed over the life of the trial. They contain detailed instructions on the types of ancillary materials that are both necessary and acceptable when giving the drug to a trial participant.

Identification of lab kits and lab-related consumables begins with the schedule of assessments to understand which tests need to be conducted and which supplies are necessary to capture samples from those tests. The labs responsible for executing sample analyses also need to provide input on specific material needs. Finally, identification of any remaining equipment needed to execute study assessments should be provided by medical experts within the sponsor or the CRO responsible for overseeing the study. At each step, clinical operations should be informed of the requirements, which leads to the second step of the process.

In the second step of the process map, clinical operations starts to compile the ancillary supply list which serves as a single source of truth for ancillary supply requirements and status across the study. The ancillary supply list also captures several pieces of information that help facilitate the sourcing process. Along with listing all ancillary supply requirements for each site, it also documents the sourcing strategy and sourcing status of each item as well as any country-specific importation requirements for each item. Creation of the ancillary supply list enables the third step in the process, identification, and documentation of each site's supply requirements.

During study start-up, clinical operations and/or the CRO run a site feasibility process where they work with potential clinical sites to understand the site's ability to conduct the trial. As part of that site feasibility process, the list of ancillary supply requirements should be shared with the site to collect their feedback. Following site feasibility, clinical sites are selected for participation in the study, and the site activation process begins. At this point, if there have been changes or additions to the ancillary supply list, it should be shared with the site again for review and feedback on whether the site has the supplies or whether they will need to be sourced for them. As that feedback comes in, clinical operations and/or the CRO updates the ancillary supply list with the information provided as in Table 17.1.

The fourth step in the process is to determine and document the sourcing strategy for each ancillary in the ancillary supply list. Feedback from the site and the sourcing agent is needed to determine if an item can be sourced locally or if it needs to be sourced globally. Depending on the item type there may be special importation requirements if a product needs to be provided from another country. When ancillaries need to be centrally sourced and then shipped to international clinical

Table 17.1 Example Clinical Ancillary Supply List

Site Location	Equipment (ECG Machine)	Lab Supplies (Kits, Manuals)	Other Materials (Pharmacy Manual)
Site 1 – United States	Provided	Sourcing globally	Provided
Site 2 – United Kingdom	Not provided – sourcing locally		Provided
Site 3 – France	Not provided – sourcing globally		Requested

sites, understanding international logistics requirements is important to avoid shipment delays that could impact study start up timelines. It's also useful to document special importation requirements in the ancillary supply list.

Ancillary supplies also vary with respect to the number of items that need to be provided over the duration of the study, so the fifth step in the process calls for a confirmation of order quantities. Some are single-item pieces of equipment like the machines needed to run assessments. Lab supplies need to be provided at each visit where samples are taken. Items required to support administration of the drug may need to be provided at each dispensing visit or with each kit of material supplied to a subject. Therefore, determining the appropriate quantity of ancillaries to source may require input from subject matter experts like clinical supply chain. After the quantities are confirmed, the sourcing and delivery process is triggered, and the supplies are shipped to the clinical site for use in the study.

Finally, notice how the business process for ancillary supply management has concurrent, overlapping activities from Steps 2 to 6. That's because the clinical site feasibility and activation processes can take place over a lengthy period of time. Early activating sites may go through Steps 3–6 and have ancillaries supplied before late-activating sites have even begun their feasibility process. Regulatory delays can also push out site start-up timelines, and these delays may impact which ancillaries a site has available to run the study. These overlapping process steps underscore why clear responsibilities and cross-functional communication are so important to facilitate a smooth ancillary supply chain.

17.7 Key Takeaways

1) There are two categories of products supplied to a clinical trial, IMPs and non-IMPs. A comparator is defined as a medicinal product used in a clinical trial that is not the novel therapeutic being tested in the trial. There are placebo comparators and active comparators.
2) Non-IMPs include AMPs such as rescue medications and challenge agents, concomitant medications, and ancillary supplies. Ancillaries themselves are further subcategorized into items required to facilitate administration of the drug, lab kits and/or supplies, and other equipment.
3) From a regulatory perspective, all IMP comparators are categorized as investigational products, which has three implications, First, all IMP comparators must have a label applied to them. Second, any modification to a comparator, such as labeling the product, falls under the scope of GMP regulations. Third, the regulatory application of the trial needs to include any GMP manufacturing activities executed on the comparator after its acquisition.
4) Active IMP comparators have several unique supply chain considerations such as developing a sourcing and procurement strategy for the drug, incorporating the comparator into the demand and supply planning process, and managing supply availability challenges. The country list of the study drives the procurement strategy, which in turn impacts the SKUs that need to be created for the comparator. Cost, expiry period, and supply availability are also important considerations in developing the purchasing and packaging strategy for the comparator.
5) Blinding active comparators adds time, cost, and complexity to the entire supply chain. Significant resources are required to support manufacturing, supply chain, and regulatory activities for this category of comparators. A full analysis of the implications of blinding comparators should be prepared for the project team to consider and approve before going forward with this strategy.

6) Non-IMP comparator medications do not require a label application and does not fall under the scope of GMP. Using a local sourcing strategy, non-IMPs may be sourced by a clinical site directly from within their own country. Non-IMPs may also be globally sourced from a single country and then imported to clinical sites.
7) Ancillary supplies include all non-medication, non-IMP supplies that need to be provided to clinical sites during the course of a clinical trial. Ancillaries are a deceptively complex supply chain because of the diversity of materials that need to be provided, the differing site requirements, and the amount of coordination required to identify the site requirements and sourcing strategy. An effective, efficient ancillary supply business process is needed to manage these materials. The cornerstone of this business process is the ancillary supply list which documents all the ancillary supply requirements by site and indicates the sourcing strategy and the status of each ancillary material.

Notes

1 https://clinicaltrials.gov/ct2/about-studies/glossary.
2 https://www.cancer.gov/publications/dictionaries/cancer-terms/def/standard-of-care.
3 https://english.ccmo.nl/investigators/clinical-trials-with-medicinal-products-ctr/preparation-ctr/research-dossier-part-i/auxiliary-medicinal-product-dossier.
4 https://www.cancer.gov/publications/dictionaries/cancer-terms/def/concomitant.
5 https://www.accessdata.fda.gov/cder/cmc/topic7/topic7/index07.htm.
6 https://www.ecfr.gov/current/title-21/chapter-I/subchapter-D/part-312
7 https://eur-lex.europa.eu/legal-content/EN/TXT/HTML/?uri=CELEX%3A32014R0536&from=NL%23d1e32%2D72%2D1#d1e32-72-1
8 https://english.ccmo.nl/investigators/clinical-trials-with-medicinal-products-ctr/preparation-ctr/research-dossier-part-i/labeling-imp-and-axmp
9 https://www.fda.gov/drugs/fdas-labeling-resources-human-prescription-drugs/prescribing-information-resources
10 https://www.ema.europa.eu/en/glossary/summary-product-characteristics

18

Make-to-Order and Just-in-Time Supply Chains in Clinical Trials

18.1 Make-to-Stock and Make-to-Order Supply Chain Management

There are two supply chain manufacturing strategies primarily used across the biopharmaceutical industry, make-to-stock (MTS) and make-to-order (MTO). Recall from Chapter 1 – Supply Chain Management that under an MTS approach, batches of drug substance (DS), drug product (DP), and finished product (FP) inventory are made before receiving a demand signal from the clinic. Put another way, material is manufactured, released, and shipped to the clinical site in advance of a subject screening and enrolling in the trial. Figure 18.1 illustrates the MTS strategy for clinical trial supply chains. At each node of the supply chain, material is manufactured and then shipped to the subsequent downstream node in advance of an independent demand signal from a participating subject.

Historically, clinical trial supply chains have largely utilized the MTS approach because the clinical trial supply chain profile fits the criteria where MTS is the superior production and inventory management strategy. On the demand side, patient convenience is a high priority in clinical trial execution, and having the drug available everywhere it's needed at all times has been an expectation for subjects. An MTS approach guarantees FP availability when a subject is ready to receive the drug. There are also supply-related advantages when utilizing an MTS strategy. Most importantly, upstream manufacturing of traditional modalities like small molecules and biologics takes weeks or months to complete. There simply isn't time for the upstream supply chain to respond to a demand signal and have the product ready for a patient in a matter of days. Additionally, batch-based production reduces the cost of goods of each FP unit, improves the efficiency of the manufacturing process, and drives consistency from unit to unit. All of these traits are important in the biopharmaceutical supply chain. DS and DP manufacturing is complex and has high fixed costs, so reducing the number of batches as much as possible is desirable. We've also learned that achieving regulatory compliance means producing DP that meets a predefined Specification that guarantees product quality. Implementing an MTS supply chain helps achieve all of these goals.

However, MTS supply chains have their disadvantages as well, the most glaring of which is their tendency to generate excess inventory. We've covered waste and numerous demand and supply planning strategies to combat this phenomenon in earlier chapters. Reducing waste isn't the same thing as eliminating it entirely though. There are many drivers that add waste to clinical supply chains, but two in particular are exacerbated by using an MTS supply chain strategy. First, clinical trial supply chains have a high degree of demand uncertainty, and, second, DPs have a predefined

Supply Chain Planning for Clinical Trials – A Practical Guide, First Edition. Ryan Mills.

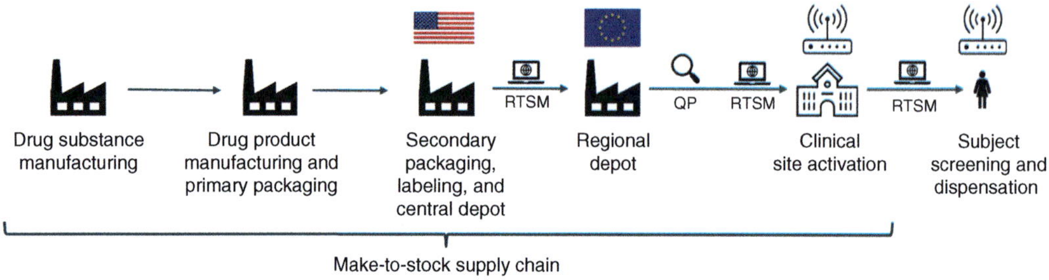

Figure 18.1 Make-to-Stock Clinical Trial Supply Chain Management

shelf life. These two elements combine to increase the risk of excess waste because when inventory is built to meet demand and that demand fails to materialize, the product expires and has to be replaced with even more inventory.

In an effort to reduce waste and expense, some biopharmaceutical companies have looked to MTO strategies as an alternative approach that theoretically reduces the total amount of inventory needed to supply the clinical supply chain. With an MTO strategy, the demand signal arrives before production begins and then a quantity of FP inventory sufficient to meet demand is manufactured and sent to the clinical site. This approach not only reduces FP inventory but also lowers the total amount of DP and DS that needs to be made for the study. For this strategy to work, labor, capacity, and materials all need to be available and ready to act as soon as the demand signal is received. MTO supply chains have other advantages over their MTS counterparts beyond reducing inventory and costs. Managing an MTO supply chain means building close relationships with upstream suppliers for improved responsiveness and execution. While MTO does offer some theoretical benefits, the unfortunate reality is that MTO production in the clinical trial supply chain space brings unique challenges that most consumer goods don't face. These include operational roadblocks such as component readiness, dealing with the international logistics process, and overcoming regulatory hurdles.

Even with these challenges, many biopharmaceutical companies have explored implementing MTO-like supply chains via inventory pooling and just-in-time manufacturing (JITM) at the FP level. And, new modalities like cell and gene therapies are forcing biopharmaceutical companies to adopt MTO strategies. Autologous cell therapy manufacturing in particular most closely resembles a complete MTO supply chain. Blood is taken directly from a clinical trial patient, sent to a manufacturing facility for immediate processing, and then returned to the patient for administration. The DS, DP, and FP manufacturing process takes place in a matter of days to weeks, and all aspects of the supply chain need to be carefully coordinated. Autologous cell therapy manufacturing is the most extreme MTO-like supply chain in the industry, and so it offers the most complete contrast to the advantages and disadvantages of MTO versus MTS supply chains.

We'll begin this chapter by further exploring the concept of MTS and MTO supply chains and lean manufacturing principles such as JITM. Through the remainder of the chapter, we'll examine a variety of clinical trial supply chains that incorporate MTO supply chain strategies. We'll begin with traditional modalities that incorporate inventory pooling and JITM at the FP level. Then, we'll move to allogenic cell therapy supply chains which also incorporate many of the inventory pooling and JITM strategies. We'll conclude the chapter with a review of autologous cell therapy supply chains and how these products utilize an MTO approach from DS all the way to FP. Within each application, we'll review supply chain planning and execution considerations.

18.2 MTS, MTO, Lean, and JIT Supply Chain Principles

Two synonyms used to describe MTS and MTO production strategies are "push" and "pull." In an MTS supply chain, inventory is "pushed" by the manufacturer to the consumer as in Figure 18.1. In an MTO supply chain, inventory is "pulled" from the manufacturer based on a demand signal from the consumer. The concept of push-versus-pull has several implications for how biopharmaceutical supply chains are managed from both a planning and execution perspective. With respect to planning, "push"-based supply chain systems rely on an independent FP demand forecast, like the ones built in Chapter 8 – Finished Product Demand Planning: The Basics. The demand forecast is then translated into a production plan at the FP level which then creates dependent demand for DP, DS, and raw materials needed to make the active substance. Safety stocks are incorporated at each level of the supply chain to guarantee desired service levels. Unfortunately, building up safety stocks to achieve a high service level then leads to increased waste.

However, there is a significant advantage with MTS supply chains, which is the predictability of upstream production planning. Production orders for FP, DP, and DS materials are communicated months in advance of the target delivery date. This is critical for biopharmaceutical products because of the long lead times associated with production. DS manufacturing often takes several weeks to produce a batch, and that's before analytical testing is conducted to ensure the product meets its specification. The total lead time from manufacturing start to batch release is often measured in months. DP manufacturing also has notoriously long lead times, although in many cases, DP manufacturing can proceed while analytical release testing occurs on the DS. Long lead times for sourcing raw materials and plant capacity constraints also greatly reduce the agility and flexibility of DP and DS manufacturing. Even FP batches take several weeks to execute because of label printing lead times, pre- and postproduction paperwork, and organizing labor to complete the job. A "push" system gives the biopharmaceutical supply chain the time it needs to prepare for and complete its highly complex upstream manufacturing processes.

"Pull"-based systems operate in a radically different fashion. From a planning perspective, the demand forecast takes on even greater importance because it drives raw material and production capacity requirements. But production doesn't depend on the forecast like it does in an MTS supply chain. Instead, when an order for the FP is received, components are assigned to the production order, and capacity is booked to complete the job. When upstream production (e.g. a subassembly) is required, these steps must be incorporated into the planning process as well.

MTO supply plans extend into these subassemblies, and each production or purchasing event must be carefully linked and scheduled as part of the planning process to ensure component availability for the job. Pure MTO supply chains rely on accurate bills of material (BOM) and capacity estimates to ensure all the upstream activities, materials, labor, and capacity are in place to produce the good when the demand signal is received. For this reason, pure "pull"-based supply chains are typically limited to highly customized items that consumers are willing to order and then wait a long time to receive. Designer furniture and art installations fit the bill. Clearly, it's not possible for traditional DS and DP biopharmaceutical supply chains to adhere to "pull" supply chain principles. The lead times are simply too long, and clinical trial subjects want the FP supply immediately upon enrolling in the study.

Another massively influential approach to supply chain management, lean manufacturing, emerged from Japan in the 1960s. Lean principles, including just-in-time (JIT) manufacturing and continuous production, were popularized and implemented around the world in the 1980s and 1990s. The objective of lean is to eliminate every aspect of waste in the manufacturing process. JIT accomplishes this by delivering only the materials, capacity, and labor required to produce a good when those materials, capacity, and labor are needed. Taking this approach reduces raw material and FP inventory levels throughout the supply chain. Continuous production literally means making goods through an uninterrupted sequence of activities, and its hallmarks include elimination of bottlenecks, minimal wait times between activities, and an optimized sequence of activities. Lean manufacturing depends heavily on close coordination and collaboration with upstream suppliers to ensure that components are available and ready to meet the needs of the supply chain.

When successfully implemented, lean manufacturing speeds up the cycle time of the supply chain, reduces inventory levels, reduces waste, and ultimately reduces the total cost to manufacture a good. However, lean manufacturing is often difficult to successfully implement, especially when demand is highly uncertain, or raw material supply chains are unreliable. When demand varies from the forecast, the JIT supply chains struggle to adjust because the requisite components, labor, and/or capacity aren't available which leads to delayed production times and increased costs. Similarly, if upstream suppliers can't deliver the components needed for production on time, this leads to delays. Lean manufacturing works best in environments with highly predictable demand and reliable suppliers who can quickly respond to the downstream demand signal. The COVID-19 pandemic exposed many lean supply chains due to a surge in demand from people staying home and shopping and raw material shortages.[1] As a result, companies are shifting their inventory management strategies to increase resiliency against demand and supply shocks.

Lean manufacturing principles complement both MTS and MTO supply chains. Maintaining component inventory on an as-needed basis works for MTS batch production as well as it does for MTO batch production. Working closely with suppliers to understand lead times, constraints, and potential issues in the supply chain is a best practice no matter the industry or FP output. In some situations, it simply doesn't make sense to apply lean manufacturing principles to the supply chain. For example, delivering glass vials to biologic manufacturing campaigns the day before they're needed is a highly risky strategy because if the vials fail to arrive and production is delayed, the underlying DS itself is put at risk. Determining when, where, and how to utilize MTS and MTO production strategies and to apply lean manufacturing principles is one key to effective supply chain management.

Many high-volume consumer goods manufacturers typically rely on a combination of MTS, MTO, and lean principles to optimize supply chain performance. Independent demand forecasts at the FP level are created, but not necessarily to pre-build final inventory. Instead, the forecast provides visibility to upstream component requirements and the amount of labor and capacity needed to make FP inventory. Then, components are preordered and "pushed" to the FP manufacturing site where labor and capacity standby to complete FP orders as they are "pulled" by the final consumer. This combination works well when consumers are willing to wait at least some time for their good. The attractiveness of this combination is that it greatly reduces FP inventory and excess waste while also reducing the lead time to produce the FP to the amount of time it takes to complete that final manufacturing step and then ship the good.

Taking the best attributes of "push" and "pull" systems and lean manufacturing principles has long appealed to clinical trial supply chain leaders to reduce excess FP waste. For the small molecule and large molecule modalities, DS and DP manufacturing always needs to be made using an MTS strategy because the lead times are so long, and clinical trial subjects don't have time to wait. The FP level is different. If the supply chain is agile enough to receive the demand signal, make and release a batch, and then ship it to the clinical site, this could theoretically reduce waste and save money. Using an MTO approach for FP theoretically translates to the elimination of excess waste, and this can be accomplished by using lean manufacturing principles to drastically reduce lead times and deliver the product in an acceptable lead time to the clinical trial subject.

There are common themes to successful MTO supply chain management that pop up throughout this chapter. Developing a detailed and accurate supply plan, building and maintaining close relationships with component suppliers, and setting appropriate expectations with stakeholders surfaces over and over again. In the next section, we'll examine the practical application of "inventory pooling" and "just-in-time labeling" in the clinical trial space.

18.3 Inventory Pooling and Just-in-Time Manufacturing

Inventory pooling is the practice of holding unfinished inventory at a FP depot for use across multiple studies or arms within a study. Unfinished in this case means that the final FP operation has not yet been completed. Secondary packaging, full labeling, or printing variable text on an existing label are all examples of "finishing" operations that could take place on pooled inventory. In isolation, inventory pooling accomplishes little if a traditional MTS strategy is used for FP manufacturing. Instead, the power of inventory pooling is unlocked when combined with JITM.

JITM is the process of making FP inventory after a demand signal (e.g. a subject entering screening) has been received by a clinical site. There are two points to note with this definition. First, JITM is a true MTO strategy because production occurs after the demand signal has been received from a clinical site. Second, JITM is limited to FP manufacturing for traditional modalities like small molecules and large molecules because the lead times to manufacture and release DS and DP are too long to wait for the demand signal. Figure 18.2 illustrates an inventory pooling and JITM supply chain for a clinical trial product.

As Figure 18.2 shows, inventory pooling and JITM work together by supplying DP to a central or regional depot prior to receiving the demand signal. After the signal is received, the FP manufacturing step begins. The FP batch pulls inventory from the available pool of DP, which is then labeled, released, and made available for shipment to the site within the span of a few days. The potential advantages of a JITM supply chain are numerous. These include reduced waste from elimination of FP safety stocks, increased flexibility to respond to demand as it occurs, and potentially reduced costs by making only what's needed when it's needed.

While this combination strategy holds promise as a superior approach to an end-to-end MTS supply chain, there are several strategic, operational, and cost-related considerations to assess. From a strategic perspective, inventory pooling works best when there are multiple clinical trials or arms within a single trial that pull from the same DP lot. Open label trial designs also eliminate some of the operational complexities that blinded trials must navigate. Additionally, the more active clinical trials or arms within a trial there are to support, the more advantageous inventory pooling becomes. An excellent use case for inventory pooling is an open-label basket trial that tests a single drug against multiple disease types. There are two other strategic considerations when considering a JITM strategy, the first of which is the screening window. If the screening window is very short (e.g. <14 days), then JITM may not be feasible because the supply chain can't react

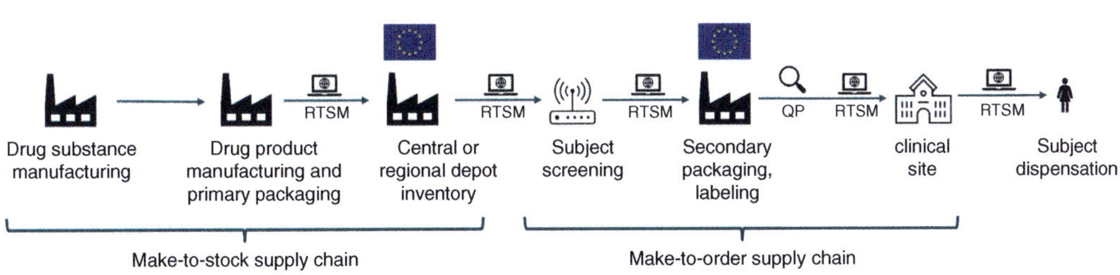

Figure 18.2 Inventory Pooling and JITM Manufacturing Supply Chain

quickly enough to supply kits before randomization and dosing of a subject. The second consideration is the expiry of the drug. Products that have a relatively short expiry period and planned extensions, JITM is advantageous because the label application occurs at a later date than in an MTS supply chain. This provides more opportunity to place a longer shelf-life date on the label and prevent unnecessary expiry extension labeling jobs.

The operational considerations for implementing inventory pooling and JITM are more daunting, and this starts with demand and supply planning. For inventory pooling to work, all demand sources against that inventory need to be included against the appropriate DP to ensure the plan drives sufficient supply to meet demand. Maintaining a demand plan at the FP level is also important for component availability such as labels, cartons, and tamper seals. Determining the FP batch size also becomes more onerous in a JITM strategy because there are so many more batches produced than in an MTS strategy, and it's still possible to generate waste when batches have an expiry date printed on them. Thought needs to be put into the quantity of each batch to avoid overproduction in the event a subject drops out of the study.

In order to accurately capture demand, the stockkeeping unit (SKU) strategy for JITM also needs to be determined. Recall from Chapter 8 – Finished Product Demand Planning: Basic Concepts that one driver for creating unique FP SKUs is the clinical label. Clinical labels are either single panels that are fit for use in one or two countries or booklet labels that support many or all countries in a study. In some situations, it makes more sense to use single panel labels because they're cheaper than booklets and because the demand signal comes from a specific country. However, that means keeping extra label stock on hand of each country and using a single panel limits flexibility to ship that product to that country. If the subject drop s out, then that labeled stock may not be usable. In other situations, it makes sense to continue using booklet labels. That's because booklets provide the flexibility to ship a JITM batch to any country on the label and because the variable text (e.g. the protocol number, the kit identification number, the batch number, the expiry date) can be printed on the label during the JITM manufacturing event. Using a booklet strategy has the added advantage of simplifying the SKU strategy and the number of forecasts that need to be maintained across the portfolio of studies. Note that the protocol number is also listed as variable text. This is an important detail that enables the pool of DP inventory to be used across any study.

Successful execution of JITM relies on the depot's flexibility to schedule and execute manufacturing jobs and on a responsive quality unit. In typical MTS manufacturing, it can take weeks to schedule a packaging job to ensure operator availability, complete the pre-production paperwork, and gather the required components. In the JITM space, the supply chain needs to react much more quickly. Labeling suites need to be ready to execute small jobs within a day or two of production order receipt, and pre- and postproduction batch records need to be templatized and limited to the variable pieces of information that come with each job.

From a functional perspective, quality plays a huge role in making the JITM strategy feasible because they hold responsibility for the review and approval of each generated batch record. Negotiating reduced lead times and service level agreements with the FP manufacturing site and with the applicable quality lead is necessary for this. Note that in Figure 18.2, the regional depot is located in the European Union. In that scenario, the qualified person certification process also needs to be integrated into the JITM process. Clinical operations is another key stakeholder. Communicating JITM production and shipment lead times back to clinical operations and to the contract research organization (CRO) keeps these stakeholders aligned on expectations and the supply chain's ability to deliver supply within a tight timeframe. These groups can then communicate the supply chain's capability to the clinical sites and subjects so there is no confusion during the process.

Another operational consideration is the country list and associated distribution network of the trial. There may be some countries that are supplied by depots that don't have FP good manufacturing practice (GMP) manufacturing capabilities to support JITM. This is especially common in Latin America and the Asia-Pacific region. Where that capability is missing, JITM may not be possible at all because the distribution lead time to import IMP precludes JITM as a feasible strategy. Let's examine a scenario at a clinical trial site in Brazil. A clinical subject enters screening, which sends a demand signal to the JITM depot in the United States. Even if the batch is made and released in three business days, it may still take several weeks to export the product into Brazil and transport it to the clinical trial site. If the screening window is less than this importation lead time, the drug may not arrive at the site in time to meet demand. As a result, for some studies, it may make sense to use JITM in some regions like the United States and the European Union and MTS in other regions like Latin America and Asia-Pacific. Combination approaches like this add complexity though, which further increases the importance of stakeholder alignment and communication before and during the study.

From a technology perspective, the randomization and trial supply management (RTSM) platform needs to be programmed to support inventory pooling and JITM where applicable. For example, in Figure 18.2, the RTSM begins with entry of the pooled DP inventory, which is another SKU to manage in the RTSM system. Notice too that customized integrations with the secondary packaging site are needed to pass order requirement information from the demand signal and to receive the batch upon its release. This is unique from standard MTS supply chain RTSM configurations where a direct linkage to the upstream manufacturing site only covers inventory at the depot, not the actual manufacturing events. Other items like drug ordering, batch release, order cancellations, and reporting also require specialized programming so the RTSM enables the JITM process rather than hinders it. This is especially true for blended trials where some depots use JITM, and others rely on a traditional MTS supply chain. When JITM supply chains are planned for a trial, extra time should be given to the RTSM design and user acceptance testing (UAT) phases to work through these details with the RTSM supplier.

If the strategic context makes sense and the operational considerations are deemed manageable, there's a final piece to analyze which is the total cost of running a JITM supply chain versus an MTS supply chain. MTS involves manufacturing fewer batches of FP with larger quantities while JITM involves executing many small FP manufacturing campaigns. There are cost trade-offs with each approach as summarized in Table 18.1.

There are five potential cost drivers to analyze between MTS and JITM strategies: packaging, excess waste, components, site shipments, and relabeling. Packaging costs for MTS are typically lower than JITM because there are many fewer jobs and fixed production costs like pre- and post-production record generation and approval are reduced. In other words, MTS has the advantage of

Table 18.1 Comparison of Relative MTS and JITM Costs

Cost Driver	MTS	JITM
Packaging	Lower	Higher
Excess waste	Higher	Lower
Components	Higher	Lower
Site shipments	Lower	Higher
Relabeling	Higher	Lower

economies of scale. However, this is offset by the fact that MTS supply chains produce more kits than JITM supply chains. This creates more excess waste at both the FP level and at the upstream level of the supply chain. The difference between MTS and JITM supply chains can be calculated through the overage multiplier that's used in MTS demand forecasting. Overage is used to protect against demand uncertainty which is eliminated in JITM supply chains. As a result of packaging more kits, component costs are also higher for MTS supply chains than for JITM supply chains. On the distribution side, JITM supply chains generate more clinical site shipments because kits are labeled and sent for a single subject where MTS site shipments send enough material to cover multiple subjects for weeks or months at a time. Finally, when JITM is implemented to mitigate poor expiry, there may be reduced shelf-life extension relabeling jobs when compared to the MTS approach. Putting together this analysis takes time but is a necessary input to consider before moving forward with implementing a JITM supply chain.

Let's review the themes that drive MTO success from the introductory section of this chapter. Detailed planning is needed to identify all the SKUs that an inventory pooling strategy supports. Accurate batch sizes drive component availability and reduce potential waste. Close relationships with the manufacturing depot, the qualified person, and the RTSM supplier enable JITM manufacturing process agility. Stakeholder engagement with quality, clinical operations, and the CRO creates alignment and reduces confusion during supply chain execution. Implementing an MTO-style JITM supply chain reshapes the entire FP supply chain, and everyone needs to understand their role in supporting its execution.

18.4 Allogenic Cell Therapy Supply Chain Considerations

Cell and gene therapy supply chains were introduced in Chapter 3 – Introduction to the Clinical Trial Supply Chain. Here, we'll take a deeper look at the allogenic and autologous supply chain structures and how MTO supply chain principles are applied within each of them. Recall that allogenic cell therapy products take cells from healthy human donors and modify them for distribution to broad patient populations. Conversely, autologous cell therapy products take cells from a single patient and modify them for explicit use in that same patient. This difference translates to two significantly different supply chain structures. We'll begin with allogenic products because their supply chains combine elements of MTS production and JITM manufacturing, but they also have their own unique considerations. Then, we'll move to autologous products where the supply chain behaves in a purer MTO fashion.

The current manufacturing process for many allogenic products begins with collecting cells from healthy donors via apheresis. Apheresis is the process of removing blood from a person and then separating cells from the blood plasma for genetic modification. After the cells have been extracted from the donors, they are placed into leukopaks which store and protect the extracted cells as they travel to the next node in the supply chain. It's typical for cell therapy manufacturers to ship the leukopaks under a refrigerated storage condition at 2–8°C for immediate processing within 48 hours of the apheresis step at the downstream manufacturing facility. However, scheduling these activities is very challenging because there is so little available time to collect, ship, and manufacture. As an alternative, some processes allow for leukopaks to be cryopreserved and shipped directly to the manufacturer or to a temporary warehouse where they are kept in specialized cryogenic frozen storage units at temperature levels below −150°C. The leukopaks are then used as starting materials for manufacturing, which is often performed in a single step. Immediately after DS production, the resulting product is typically filled into its final primary container, labeled, and cryopreserved. Secondary labeling may also happen prior to cryopreservation or later, closer to clinical distribution, depending on the packaging configuration. In this respect, allogenic supply chains behave similar to traditional large molecule MTS production strategies where large quantities of bulk DP are manufactured, though use for various studies may be limited by the pre-cryopreservation labeling strategy.

Once the DP is made, it's shipped to a regional depot where it's held for use in a clinical trial. Note here that the distribution network for cell therapy products also looks different than standard modalities. In-country depots are used far less frequently for allogenic (and autologous) cell therapy products to avoid complex customs clearance processes. Instead, product is shipped directly from the central depot to the clinical sites. This difference emphasizes the importance of performing the distribution network mapping exercise outlined in Chapter 11 – Finished Product Supply Planning: The Distribution Plan.

At this point, the allogenic supply chain starts to resemble the inventory pooling and JITM supply chain outlined in the previous section. That's because allogenic products must be stored frozen, and though most large academic institutions that conduct cell therapy clinical trials have this capacity, it's often limited. As cell therapy moves beyond oncology and toward community center and outpatient dosing, fewer sites will have this capability. As a result, the product needs to arrive at the clinical site just prior to administration to the patient. Depending on the configuration, it may be possible to delay secondary labeling with local label requirements so this activity can be executed closer to subject dosing. As inventory is often kept in a single or limited number of depot(s), it may be possible to implement a JIT labeling strategy within the desired lead time, executing clinical labeling after receipt of the demand signal from the clinical site.

The FP segment of the allogenic supply chain must therefore contend with the same issues a small molecule or biologic needs to manage when operationalizing JITM manufacturing. The label strategy, production scheduling strategy, batch release process, qualified person (QP) certification process, RTSM programming, and shipment coordination activities all need to be worked out. But allogenic supply chains have the added complexity of needing to be stored cryogenically throughout the supply chain. Cryogenic temperature makes relabeling of the primary vial extra challenging, and any secondary packaging and/or handling must use specialize equipment to minimize exposure to warmer temperatures. These products consist of living cells, where warming or thawing for any duration may impact the viability and potency of the product. That also means specialty storage containers need to be provided at each node of the supply chain and during shipment of the product through to the clinical site. These containers must be properly placed at ideally geographically close to the depot to be preconditioned and on hand at a moment's notice when the demand signal is received from the clinical site.

Though JITM is possible with allogenic cell therapies, the value of these therapies over their autologous counterparts is that they can largely be treated as a MTS product, readily available for patients who need treatment quickly. Cryopreservation provides an added challenge over traditional biologics and small molecules, limiting the labeling flexibility and requiring quick, reliable, distribution with specialized shippers. Additionally, allogenic products are a relatively new modality so there's often long lead times, inconsistency in upstream manufacturing yields, and raw material suppliers are not always reliable in delivering what's needed for upcoming production campaigns. Thus, having an accurate FP demand forecast becomes even more important for allogenic supply chains because the FP supply plan drives demand for manufacturing and the required raw materials, as well as the distribution components such as specialty shippers. While the demand and supply planning processes don't differ when compared to MTS supply chain management, the scope needs to increase to include raw materials for upstream manufacturing and components for downstream distribution.

Another unique characteristic of allogenic products is that the FP dose is measured in the number of cells delivered, which may vary batch to batch. The demand planning process needs to convert the number of cells a patient is expected to receive into the number of FP units that contain those cells. These conversions extend into the upstream supply planning process. If they're not performed correctly, this could lead to insufficient DS and DP production to support clinical trial demand.

The operational clinical demand and operations planning (CD&OP) process also takes on added importance with allogenic supply chains for two reasons. First, getting an early read on demand signals reduces the potential that raw materials aren't available and demand and supply of these critical, custom raw material demand and supply may be the primary focus of this process. Second, upstream manufacturing has more variability in cell therapy manufacturing than in traditional biologics, especially in early phases as the process is being developed. Yields can swing dramatically from one batch to the next, and raw material shortages can extend all the way back to upstream manufacturing. It's necessary for the clinical planner and the chemistry, manufacturing, and controls (CMC) team to closely scrutinize the output of each batch to determine whether increases or decreases in yields are part of donor variability, a trend or simply one-off situations. Strategic CD&OP cycles should also take place regularly to assess the manufacturing strategy and to relay potential production strategy changes to the manufacturing sites and their suppliers in a timely fashion.

Like the inventory pooling and JITM strategies, accurate demand and supply planning and consistent stakeholder engagement are key to effectively managing allogenic supply chains. Supplier collaboration deserves special attention here because the second-degree supply chain comes under intense pressure at every level. Critical raw and/or starting material, component, and shipper suppliers are integral parts of keeping the supply chain moving, so the sponsor needs to keep a close eye on where potential shortages pose a risk. The best way to do this is to establish a partnership-style relationship with every primary manufacturing site (critical raw materials, cell therapy, and final product) and with the preferred logistics partners that arrange for the shipment of goods. Often the critical raw material suppliers are providing custom products and may require oversight more similar to a CMO than a direct material supplier. Note that logistics providers are explicitly called out here. With traditional modalities, logistics has often been thought of as a purely transactional task, but in cell therapy supply chains, they are a critical supplier for the reasons outlined above. Establishing a partnership-style relationship enables the sponsor to share their project plan and priorities and to draw attention to risk mitigation strategies they want to see their suppliers implement within their own supply chains. Chapter 20 – Contract Organization Management and Oversight covers strategies for establishing partnerships with contract manufacturers and distributors in greater depth. Of course, setting up these relationships takes time, resources, and energy, but that's simply a part of managing an MTO supply chain.

18.5 Autologous Cell Therapy Supply Chain Considerations

Unlike the allogenic cell therapy manufacturing process which uses cells from healthy donors as the basis of the DS, autologous cell therapy products are patient specific. In other words, each batch of DS is produced for an individual patient. This characteristic makes autologous cell

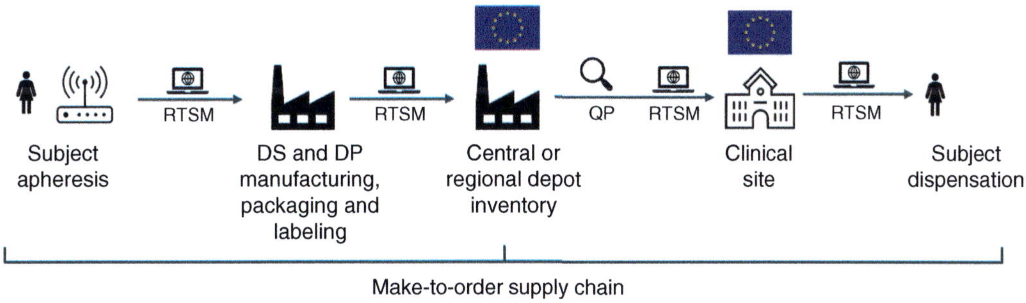

Figure 18.3 Autologous Cell Therapy Supply Chain

therapy supply chains unique among all modalities because they are true end-to-end MTO supply chains. Not only is FP manufacturing responding to a demand signal like with JITM, but the manufacturing and release also need to occur after receipt of the demand signal. Autologous supply chains must be responsive and agile because the patients who receive these treatments are often extraordinarily sick, and the cell therapy treatment represents their last or only chance of survival. Every day counts.

There are many subtypes of autologous therapies including chimeric antigen receptor T-cell (CAR-T) therapies, T-cell receptor (TCR) therapies, and autologous natural killer (NK) cell therapies. While each category has unique considerations, they are each supported by the same high level supply chain illustrated in Figure 18.3. Donor cells are taken directly from the patient via apheresis and then transported to the manufacturing site. Here, upstream manufacturing and analytical release take place on an extremely condensed timeline, usually in a matter of 2–3 weeks.[2] This is incredible given the number of steps involved including cell thaw, cell wash, cell selection, cell activation, cell engineering, cell expansion, cell harvest, and fill/finish and labeling/packaging activities. As much as possible, analytical release testing takes place concurrently with manufacturing. Release testing must still be performed on the final DP as well. The quality unit works side-by-side with the manufacturing teams to assess the quality of the product as rapidly as possible through each step. The product is then shipped back to the patient for infusion. Like allogenic products, autologous therapies also need to be cryopreserved throughout the supply chain and stored at $\leq -150°C$.

Given the MTO structure of the supply chain and the combined requirements of speed and quality throughout, autologous cell therapies must surmount numerous supply chain risks and challenges:

- *Manufacturing*: The manufacturing process for autologous cell therapies is highly complex; yields are uncertain from batch to batch; and process variability leads to a desire for greater control over each batch. Cell therapy products also need a mechanism to deliver the modified cellular material to its desired target. The most widely used of these mechanisms are viral vectors which also require an end-to-end GMP manufacturing process for production and release. Viral vectors are neither easy to make nor acquire at scale, and this presents a significant risk to the supply chain because the product can't be made without sufficient vector inventory on hand.
- *Sourcing*: Raw material, component, and shipper availability also present issues for cell therapy companies. Viral vectors are made from plasmids which have been in short supply since the COVID-19 pandemic, and the same cryogenic shipping containers used for allogenic cell therapy products are required for autologous cell therapy products.

- *Logistics*: Coordinating the pickup and delivery of cell therapy material between nodes of the supply chain is another challenge. Similar to allogenic products, close collaboration with transportation service providers is needed to ship material in a timely and compliant fashion.
- *Technology*: Traditional RTSM systems don't have the functionality to adequately manage the delivery of IMP from the apheresis stage through manufacturing and back to the patient. More advanced systems with additional capabilities that both schedule and control activities within the supply chain are needed in this space.
- *Cost of Goods*: Autologous cell therapy products are extraordinarily expensive to manufacture and distribute. Looking at the challenges listed above, that's no surprise. Multiple processes, scarce components, insufficient capacity, frozen storage conditions, and manufacturing for a single patient all contribute to this.

Let's take a deeper look at each of these challenges beginning with the manufacturing process. Variability in patient donor cells leads to mixed results during the cell expansion and harvest phases of the process because different cells don't interact with the cell culture media in the same way.[3] As the process scales up to produce the desired dose, in-process monitoring is required to ensure that critical process parameters are in line with expectations. Adjustments may need to be made during cell expansion in order to obtain the desired yields. Another challenge with DS manufacturing is that the plant needs to be ready to perform as soon as the cells have been received at the site. All aspects of production readiness need to be tightly coordinated from the moment the patient's apheresis procedure is scheduled. Capacity, raw materials, and the people who make it happen all need to be ready, and ready to readjust if a patient must reschedule. These characteristics translate to an environment where a high degree of control is needed over DS and DP manufacturing. Thus, many cell therapy companies have opted to build their own upstream manufacturing sites rather than outsource this to contract manufacturers. While it's a costly investment, this strategy ensures the sponsor has full control over process development, scheduling, and executing manufacturing.

Sourcing compliant raw materials and components in adequate volumes has also proven to be a risk for cell therapy products. The most important of these are the viral vectors that deliver the genetically modified material to the targeted cells.[4] Viral vectors come in many varieties including lentiviruses, retroviruses, adenoviruses, and adeno-associated viruses (AAVs). Viruses are made via the same process as antibodies or enzymes with an upstream and downstream DS manufacturing process, a DP manufacturing process, and a quality release of the final batch. However, unlike traditional large molecules, viral vector manufacturing does not have a standardized approach across industry. Different companies use different processes, different materials, and different specification parameters in their production.[5] There are also challenges that are unique to each type of viral vector.

This environment leaves autologous cell therapy manufacturers with a similar make-versus-buy decision on viral vector obtainment because of the difficulties and costs associated with outsourcing. Complicating matters further, the plasmids used as starting materials in viral vector manufacturing are customized to the product, so they have also been scarce and costly to source.[6] Alternative delivery mechanisms have been explored such as transposon platforms, but these struggle to deliver cellular material to the desired target at anywhere near the efficiency levels of viral vectors.[7] Another common gene editing approach uses clustered regularly interspaced short palindromic repeats (CRISPR) technology to knock out or in genes of interest. This requires a Cas enzyme and custom pieces of RNA that act as guides for the enzyme. Both these material categories face similar sourcing challenges, with long lead times, high costs, and inconsistent suppliers.[8] Beyond gene editing reagents, some cell therapy products rely on other unique reagents and supplements during cell expansion. This presents two potential issues. First, all raw materials need to be manufactured under GMP, and this isn't always the case for materials that aren't standard across the industry. Second, raw materials also need to be adequately characterized to ensure consistent results across manufacturing campaigns.

Third-party logistics (3PL) suppliers and technology platforms are both integral pieces of the autologous supply chain. From a logistics perspective, autologous cell therapies are stored in a frozen condition throughout the supply chain. Unlike their allogenic counterparts though, autologous products move much more quickly, and this puts even more pressure on transportation service providers to have appropriate shippers and storage containers at each node and to facilitate shipments as efficiently as possible. Technology partners offer systems typically referred to as orchestration platforms that help in two areas. First, they capture the product's chain of identity and custody through its journey, and, second, they create a detailed plan of the sequence of activities from apheresis to infusion. Tracking and ensuring chain of identity is critical to autologous cell therapies to ensure the cells collected are returned to the same patient. Any mix up in this process presents a severe threat to patient safety. The critical identifiers must be correctly associated, tracked, and verified throughout the process. For example, each clinical trial patient has a patient number, and each apheresis collection is assigned a unique number. Each manufacturing instance also has a unique batch number, and the apheresis and final product shipments each have their own shipment tracking numbers. The orchestration system marries all of these identifiers together to allow chain of identity verification at each step.

The scheduling piece is fundamental to ensure manufacturing capacity availability and that resources are on hand to execute analytical testing and batch release. Customized systems combine traditional elements of RTSM with some elements that are reminiscent of enterprise resource planning (ERP) systems. Once a patient is ready to enroll, the system offers apheresis slots that correspond to a manufacturing slot. Once an apheresis slot is booked by the site, the end-to-end supply chain is visible, and the key activities are scheduled. Unlike a traditional clinical supply chain where an investigator enrolls and doses a patient whenever convenient, this approach engages all stakeholders – patient, clinical site, manufacturing, quality, logistics – to agree on an achievable dosing schedule. This setup has the added benefit of creating a chain of custody where GMP documents are associated to each activity as they occur.

The story of autologous cell therapy supply chain management is nothing short of a miracle. All the traditional difficulties of running a MTO supply chain are present and made exponentially more difficult by the compressed timelines, the complexity of the manufacturing process, the number and availability of the materials required, and the wide swatch of stakeholders and suppliers involved. It's not a stretch to say autologous cell therapies are among the most challenging supply chains of any product in the world. Fortunately, these difficulties have been overcome, and the incredible potential of these products have been realized.

This has happened for numerous reasons. First, the traditional keys to successful MTO supply chain management have been applied. These include forecasting and planning at all levels of the supply chain including critical raw materials; engaging internal and clinical site stakeholders to fundamentally redesign the biopharmaceutical supply chain; and establishing close relationships with key suppliers to ensure availability of compliant material. Cell therapy companies have added another ingredient to the mix as well, innovation. It's taken a fresh approach to traditional clinical and commercial biopharmaceutical supply chain management to build and execute an autologous cell therapy supply chain. More advancements are on the horizon as the manufacturing process and materials become more standardized.

18.6 Key Takeaways

1) MTS and MTO are the two primary supply chain strategies used in clinical supply chain management. In an MTS strategy, all inventory including the FP is made in advance of receiving a demand signal from a clinical site. In an MTO strategy, some or all inventory is made after receipt of the demand signal.
2) MTS supply chains have been primarily used in traditional upstream clinical trial manufacturing because of long lead times and low agility within DS and DP supply chains. MTO supply chains offer potential advantages when compared to MTS supply chains because they reduce excess inventory.
3) Lean manufacturing complements MTS and MTO supply chain strategies. The goal of lean is to eliminate waste at every level of the supply chain by delivering only the materials, capacity, and labor when needed and in the quantity needed to execute a manufacturing event. Biopharmaceutical companies have been exploring ways to optimize their clinical trial supply chains by combining elements of MTS, MTO, and lean principles.
4) Inventory pooling is the practice of holding unfinished inventory at a FP depot for use across multiple studies or arms. JITM is the process of making FP inventory after a demand signal has been received by a clinical site. When combined, inventory pooling and JITM create significant potential value as an MTO-style supply chain by reducing FP waste through the supply chain.
5) There are numerous strategic and operational considerations to implement inventory pooling and JITM. These include the type of clinical trials, the distribution network of the trials, the planning process, the manufacturing and batch release process, and the RTSM system. Cost is also a fundamental input for consideration because in some cases, JITM drives higher costs than traditional MTS supply chains do.
6) Allogenic cell therapy products take donor cells from healthy volunteers and edit them for distribution to a broad patient population. Allogenic supply chains behave similarly to the inventory pooling-JITM strategy but with added operational complexity. Manufacturing yields are more uncertain, and the product needs to be frozen throughout the supply chain. This places even more pressure on the planning process to drive raw material and component availability. Close collaboration with contract manufacturing organizations and third-party logistics providers is also necessary for supply chain execution.
7) Autologous cell therapy products take cells directly from the patient and edit them for use in that single patient. This translates to a true end-to-end MTO supply chain where all aspects of the manufacturing process from DS to FP take place after the demand signal is received. Autologous cell therapy supply chains rely even more heavily on the principles that enable MTO success because of the compressed lead times, complex manufacturing process, and sourcing concerns throughout the supply chain.

Notes

1 https://www.ncbi.nlm.nih.gov/pmc/articles/PMC8776498/.

2 https://www.cellandgene.com/doc/will-point-of-care-manufacturing-unlock-the-value-of-autologous-cell-therapies-0001.

3 https://www.insights.bio/cell-and-gene-therapy-insights/journal/article/422/Challenges-opportunities-for-manufacturing-autologous-cellular-therapies-such-as-CAR-T-cells.
4 https://www.nature.com/articles/s41392-021-00487-6.
5 https://www.mckinsey.com/industries/life-sciences/our-insights/viral-vector-therapies-at-scale-todays-challenges-and-future-opportunities.
6 https://www.ncbi.nlm.nih.gov/pmc/articles/PMC7564888/.
7 https://www.ncbi.nlm.nih.gov/pmc/articles/PMC8019376/.
8 https://www.ncbi.nlm.nih.gov/pmc/articles/PMC10120314/.

19

Direct-to-Patient Supply Chain Management

19.1 Introduction to Decentralized Clinical Trials

Over the past decade, clinical trial designs have become increasingly complex in their design and execution. Trial subjects often need to travel long distances to clinical sites, take part in a battery of tests, and receive their prescribed treatment. All these activities combine to place a tremendous burden on study participants, and intense travel requirements are a large deterrent from trial participation. Sponsors are always keen to make travel easier on participants, and one attractive solution is to decentralize the study. Decentralized clinical trials (DCT) aim to reduce patient burden by bringing the clinical trial directly to the subject's home or to a nearby healthcare facility. The FDA defines a DCT as "a clinical trial where some or all of the trial-related activities occur at locations other than traditional clinical trial sites.[1]" These activities include taking samples, performing assessments, and administering investigational medicinal product (IMP) to the subject.

In 2020, the outbreak of the COVID-19 pandemic had a seismic impact on the ability of sponsors to continue the conduct of standard clinical trials. Shelter-in-place policies forced many subjects to avoid travel to their designated site, and almost overnight sponsors had to pivot to a DCT model to continue the conduct of their studies. In one 2020 report, more than 67% of clinical trials were impacted by shelter-in-place policies.[2] In another study, "researchers noted a 50% increase in global drug trials with a DCT or virtual element from 2020 to 2021 and a 28% increase from 2021 to 2022.[3]"

The challenge for sponsors new to decentralized trials is that remote study visits do not have a one-size-fits-all approach. Every patient population has unique needs because study objectives and safety considerations differ from one trial to another. Some studies require relatively simple assessments while others are dizzyingly complex and require highly trained specialists and equipment to complete. Therefore, the operational format of DCT has to be customized to accommodate the needs of the trial and patient population to ensure successful conduct of the trial.

Examples of DCT models include fully decentralized trials, hybrid trial designs, and mobile clinical trials. In fully decentralized trials, all aspects of study conduct take place away from the clinical trial site, while hybrid trials blend some aspects of traditional trial conduct with decentralized principles. One hybrid trial may require a subject to attend their first visit of the trial at a clinical site with subsequent visits conducted remotely. A different hybrid trial may require subjects to receive IMP administration at a site, but other visits may be conducted remotely. Mobile clinical trials incorporate technology and mobile applications to collect clinical trial data. Wearable technology such as activity trackers, heart rate monitors, and glucose monitors have all been incorporated into trial designs.

Supply Chain Planning for Clinical Trials – A Practical Guide, First Edition. Ryan Mills.

Customization in decentralized trials extends to IMP management. Most small molecules are relatively straightforward to administer in a decentralized model because they are consumed orally, and they have explicit directions for use (e.g. take two tablets once per day) listed on the label. Small-molecule products have been sent home with clinical trial participants for years and are the ideal modality for use in a DCT model. On the other hand, large molecules are more complex because they require administration intravenously or intrathecally. Large-molecule administration may involve reconstituting a lyophilized product, compounding the IMP with other materials (e.g. saline for infusion), and flushing lines following administration of the drug. Each of these steps needs to be performed by a trained professional to avoid risk to patient safety.

Before the drug can be administered to a subject though, it first needs to get from the supplying depot to the patient's home. Shipping IMP to a subject participating in a DCT is colloquially known as a direct-to-patient (D2P) supply chain. Like the overall decentralized trial design, D2P models come in a variety of flavors such as depot-to-patient, site-to-patient, and pharmacy patient. Factors such as the patient population, trial geography, drug modality, and clinical site capabilities all influence the design of a compliant and efficient D2P supply chain. Executing a D2P supply chain also requires incorporating numerous stakeholders including the subject, clinical trial site, randomization and trial supply management (RTSM) system, the supplying depot, and the logistics companies responsible for physical delivery of the good. Using a certified nurse to prepare and administer large-molecule IMPs is also necessary because these can't be prepared by the subject or an unqualified caretaker.

D2P models add significant complexity to the clinical trial supply chain and need to be carefully designed and managed. Over the course of this chapter, we'll break the D2P supply chain down beginning with an analysis of the different D2P models, and the various stakeholders involved with the D2P supply chain. Then, we'll explore the ramifications of D2P on the demand and supply planning process. Next, we'll introduce a comprehensive business process that supports the most complex D2P supply chains involving nurses and pharmacies alongside the traditional clinical trial supply chain stakeholders. The example business process map provides context for several D2P-specific execution considerations such as the handling of sensitive patient health information, managing non-IMP ancillaries, and blinding large-molecule products. Decentralized trials offer an opportunity to reduce patient burden and improve subject retention in the trials. Designing and executing an optimal D2P supply chain strategy is a fundamental component to achieving this goal.

19.2 Direct-to-Patient Supply Chain Models

There are three primary D2P models: depot-to-patient, pharmacy-to-patient, and site-to-patient. These descriptors succinctly capture the origin location of the shipment before it reaches the subject. In the depot-to-patient model, IMP is shipped from a central or regional depot directly to the subject's home. In the pharmacy-to-patient model, IMP is sent from a central or regional depot to a licensed pharmacy. The pharmacy then receives a prescription order from the clinical site where the subject is registered as a participant in the trial, and the IMP is labeled and then shipped to the patient's home. In the site-to-patient model, the IMP is repackaged for shipment by the clinical site and then sent directly to the subject's home.

There are three elements to consider when designing a compliant D2P model: trial geography, regulatory considerations, and the drug's modality and storage condition. From a geographic perspective, the United States has a unique set of constraints because of state-level regulations. Many states have a dual license requirement for the pharmacist-in-charge both to export a drug across state lines and to receive drug from another state.[4,5] Many clinical sites in America do not have the necessary licenses to prescribe and ship medicines to subjects who live outside the state where the site operates. Practically speaking, this means that if a study conducted in the United States wants to enable a D2P supply chain, a licensed depot or pharmacy is necessary to legally get the IMP to the patient. One exception to this is if all the sites participating in the trial have pharmacies that are licensed to distribute drug to all 50 states. The 50-state license is important because it's impossible to predict where subjects who enroll in the trial live, so each clinical site needs to be prepared to send drug to any state in the country.

The situation is different in the European Union where clinical sites are generally able to ship IMP directly to a subject's home no matter where in the country the subject lives. However, countries are generally not able to ship IMP to subjects who live in another country. This restriction has led sponsors and suppliers to push for an allowance to create a central pharmacy-like setup where IMP could be shipped from a central location to subjects wherever they live in the EU; however, this has not yet been codified into a law or regulation. Always keep in mind that the regulatory landscape around D2P is evolving. Some countries allow sites to ship product directly, while others restrict D2P entirely and require subjects to visit their clinical site to receive IMP. Before selecting a D2P model for the clinical trial, it's important to research the latest regulations and restrictions in each country that will participate in the trial.

The final factor to consider when selecting a compliant D2P model is the product's storage condition. Small-molecule products with an ambient storage condition can be safely shipped from any location, but large-molecule products with refrigerated or frozen storage condition pose a special challenge. If a pharmacy does not have appropriately qualified storage units on its premises, the drug may need to be sent directly from the site or depot.

Geography, regulatory considerations, and drug modality all potentially restrict which D2P model can be implemented for a particular trial. There are also benefits to consider with each of the models as well. Depot-to-patient models work well in the United States if site-to-patient models are restricted by state licensing requirements and the drug is relatively simple to consume by the patient (e.g. an oral small molecule or an auto-injector). Depots are also the most efficient supply chain because the drug does not need to travel to a second site (e.g. a pharmacy) before making its way to the subject.

On the other hand, pharmacies offer a host of benefits that depots do not. First, pharmacies are required to add a pharmacy label to the outer packaging that often contains more specific preparation and consumption information related to the drug. This is beneficial when the drug is complex to prepare and administer. Pharmacies also re-verify the drug and document which patient is receiving the drug, which is helpful for IMP accountability and reporting at the end of the trial. Finally, using a pharmacy increases patient safety because trained pharmacists also check for patient contraindications and allergies which depots are not trained to do. Using a pharmacy does have drawbacks though. It increases supply chain complexity because another depot (e.g. the pharmacy) is needed to distribute the IMP, and those extra services typically cost more than depot shipments do.

From a pure supply chain perspective, site-to-patient models may be the most attractive D2P model. In a site-to-patient model, there are no additional depots or associated costs, and the sites often have pharmacies on site that can provide all the benefits of the pharmacy-to-patient model. That said, site-to-patient models retain many of the general challenges involved with any D2P supply chain such as correctly setting up the RTSM solution, managing temperature excursions, and facilitating return shipments back to the appropriate site or depot. These activities are covered extensively later in this chapter.

Before moving on, there's another way to categorize D2P supply chains between those requiring a nurse to support IMP administration and/or assessments and those that do not. Trained nurses are necessary to support decentralized visits when the IMP is a biologic or complex to administer, assessments need to be performed by a trained person, or samples need to be taken from the subject. When nurses are involved in a decentralized visit, their availability needs to be coordinated with the subject's availability and with the arrival of the IMP. This adds another layer of coordination to make the D2P supply chain function. Home nursing suppliers who provide a central source of trained nurses to support specific therapeutic areas have grown rapidly over the past decade to service decentralized trials.

19.3 Direct-to-Patient Supply Chain Stakeholders and Costs

D2P supply chains rely on myriad stakeholders to function. From clinical sites to pharmacies to logistics providers, let's take a closer look at the role each party plays in the D2P supply chain:

- *Subject*: The subject is the single most important stakeholder in the process because they are the ultimate recipient of the drug. In clinical trials that test a drug with a complex or lengthy administration procedure, the subject may also determine when during the week or month a visit takes place. That visit date then becomes the goal for delivery of the drug to the patient's home to complete that visit.
- *Clinical Site*: The clinical site is also central to the D2P supply chain even though IMP administration takes place elsewhere. The site investigator and/or staff identifies each subject as a participant in at-home dosing, and they're also responsible for executing the RTSM transaction that assigns specific kits to each visit. In depot-to-patient and pharmacy-to-patient models, the site writes a prescription order that records the IMP and ancillary supplies that need to be shipped to the subject's home to support the visit. When a nurse is involved to facilitate administration of the drug or study assessments, the site also needs to coordinate the timing of the visit with the nurse and the subject.
- *Depot*: In depot-to-patient supply chains, the depot receives the order from the clinical site and prepares the IMP and any supporting ancillaries for shipment. Depots also typically receive return shipments of IMP at the completion of each visit which is necessary for IMP accountability and reconciliation at the completion of the trial.
- *Pharmacy*: In pharmacy-to-patient supply chains, the pharmacy receives the order from the clinical site and prepares the IMP and any supporting ancillaries for shipment. The pharmacy acts as a regional depot within the supply chain and stores IMP and ancillaries as needed to support site shipments.
- *Home Nursing Project Manager*: For trials that require nurses, a home nursing supplier is selected, and a project manager is assigned. The project manager identifies and schedules nurses to support each visit, and they play a role coordinating the exact timing of the visit with the subject, clinical site, and the depot or pharmacy who is supplying the IMP. The project manager also troubleshoots visit issues with the nurse and serves as a primary point of contact with the sponsor.
- *Nurse*: As part of each visit, the nurse is typically responsible for receiving the IMP from the courier, checking for temperature excursions, and returning the IMP at the completion of the visit.
- *Third-party Logistics Provider (3PL)*: The 3PL schedules transportation of the IMP from the depot, site, or pharmacy to the patient's home. The 3PL also typically identifies and schedules

the courier who picks up the IMP from the origin site or airline and delivers it to the subject's home.

- *Courier*: The courier is responsible for physical delivery of the IMP from the depot, pharmacy, or airline to the subject's home. If a temperature excursion occurs or a subject has to unexpectedly reschedule a visit, the courier may also be responsible for returning the IMP to the origin depot or pharmacy. Explicit instructions on the handling of the IMP and the window within which the IMP should be delivered are given to both the 3PL and the courier. Even with these precautions, it's common for problems to arise during the physical delivery of the IMP.
- *RTSM Supplier*: When a decentralized trial utilizes an RTSM system, the solution needs to be programmed to capture D2P visits, dispensation of the drug, shipment and receipt of the drug, and return of the drug back to the origin depot or pharmacy for reconciliation.
- *Contract Research Organization (CRO)*: As the main point of contact with the clinical site, the CRO often serves as a liaison between the clinical site, home nursing supplier project manager, and the sponsor. This role is important early in the process to help identify which subjects intend to participate in a D2P option and for troubleshooting issues within the D2P supply chain when they're identified.
- *Sponsor, Clinical Operations*: From a sponsor perspective, both clinical operations and clinical supply chain are heavily engaged in the D2P supply chain. Clinical operations typically owns selection of the D2P supply chain model for the study; activities that take place at the clinical site to support D2P; identification, selection, and oversight of the home nursing supplier; and updates to the RTSM solution if they have primary oversight responsibilities with the supplier.
- *Sponsor, Clinical Supply Chain*: Clinical supply chain typically defines the end-to-end D2P supply chain process once the model has been determined; manages the relationships with the depot and/or the pharmacy to monitor inventory levels and shipment execution; identification and selection of the 3PL and/or courier services as needed; and is the responsible function for triaging physical supply chain issues that may arise during D2P supply chain execution.

The responsibilities outlined above for clinical operations and clinical supply chain are fluid between the two functions. For example, in some organizations, clinical operations takes more ownership of defining the end-to-end D2P process, while in others, clinical supply chain owns more of the RTSM design for D2P. The key to success is for the roles and responsibilities to be clearly defined to avoid confusion. Given the need for alignment between clinical operations, clinical supply chain, and the large number of external stakeholders, creating a business process outlining each step within the D2P supply chain is strongly recommended, and an example business process map is provided later in this chapter.

Before moving on to more detailed supply chain considerations, let's also consider the total cost of the D2P supply chain which can run upward of several million dollars per year. Setting up a pharmacy, modifying the RTSM system, hiring a home nursing supplier, and sourcing ancillary supplies needed to support IMP administration are all expensive activities to undertake. That doesn't include the logistics costs which are much greater than in a standard clinical trial because there are so many more shipments in a D2P study. In a standard model, each site shipment covers multiple dosing for all the subjects enrolled at that site. Contrast that to the D2P supply chain where each shipment only supports a single dose.

Ironically, the costs of D2P are the highest in a hybrid model because all the start-up costs associated with activating clinical sites are accrued along with all the decentralized trial costs. Completely decentralized trials and traditional trial models avoid one set of costs or the other. Building a holistic budget for the D2P supply chain helps the sponsor assign responsibility for each component of the D2P supply chain and supports an analysis of whether or not D2P's cost is ultimately worth the benefits.

19.4 Direct-to-Patient Supply Chain Planning Considerations

Demand and supply planning for a D2P study begins with selecting the D2P model and then mapping the D2P distribution network, similar to the process outlined in Chapter 11 – Finished Product Supply Planning: The Distribution Plan. Fortunately, in depot-to-patient and site-to-patient models, no additional depots are required. However, in the pharmacy-to-patient model, the pharmacy acts as a separate regional depot. IMP is shipped from a central or regional depot to the pharmacy where it is stored before getting packaged and labeled for delivery to the subject's home. Consider the hybrid D2P stockkeeping unit (SKU)-depot example in Figure 19.1 that uses a pharmacy-to-patient model in the United States and a site-to-patient model in the European Union.

In this example, the finished product central depot is located in the United States. The central depot ships drug to US-based clinical sites who then administer the drug to subjects on the premises. Notice that the central depot also ships drug to a US-based pharmacy who ships the drug directly to subjects who have opted into receiving treatments in the home. The immediate ramification of setting up the pharmacy as a depot is to ensure the depot has sufficient inventory of IMP to meet demand from home-based subjects, which leads us to the demand planning implications of the pharmacy-to-patient model.

When a subject receives IMP from a site or from a pharmacy, it's necessary to understand which dispensation visits take place at the clinical site and which take place at home. In a standard clinical trial model, the individual subject demand forecast is continuous and placed against a single SKU–depot location that supplies the relevant clinical site. In a D2P clinical trial model, the individual subject demand forecast is calculated using the same methodology outlined in Chapter 8 – Finished Product Demand Planning: Basic Concepts, but the forecast is split against the SKU–depot location that supplies the clinical site visits and the SKU–pharmacy depot location that supplies at-home visits. Figure 19.2 illustrates this concept.

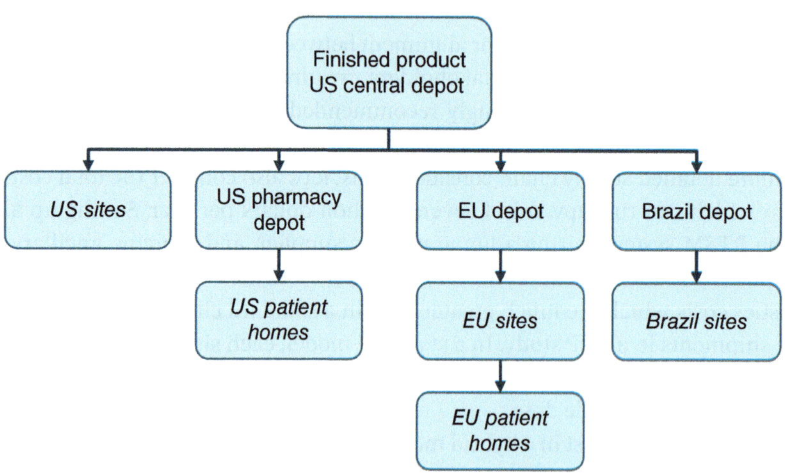

Figure 19.1 Direct-to-Patient Hybrid SKU–Depot Supply Chain Map

Subject	W1	W2	W3	W4	W5	W6	W7	W8	W9	W10	W11	W12	W13	W14	W15	W16	W17	W18	W19	W20	W21	W22
1			2	2	2	2	2	2	2	2								2				
2					2	2	2	2	2	2	2	2								2		
3							2	2	2	2	2	2	2	2	2	2	2	2	2	2	2	2
4									2	2	2	2	2	2	2	2						
5											2	2	2	2	2	2	2	2	2	2	2	2
Total subject demand	**0**	**0**	**2**	**2**	**4**	**4**	**6**	**6**	**8**	**8**	**8**	**8**	**6**	**6**	**6**	**6**	**4**	**6**	**4**	**6**	**4**	**4**

SKU description: Wonder drug placebo, 30-ct bottle, global label
Depot location: US central depot

Subject	W1	W2	W3	W4	W5	W6	W7	W8	W9	W10	W11	W12	W13	W14	W15	W16	W17	W18	W19	W20	W21	W22
1											2	2	2	2	2	2	2		2	2	2	2
2													2	2	2	2	2	2	2		2	2
3																						
4																	2	2	2	2	2	2
5																						
Total subject demand	**0**	**0**	**0**	**0**	**0**	**0**	**0**	**0**	**0**	**0**	**2**	**2**	**4**	**4**	**4**	**4**	**6**	**4**	**6**	**4**	**6**	**6**

SKU description: Wonder drug placebo, 30-ct bottle, global label
Depot location: US pharmacy

Figure 19.2 Direct-to-Patient Individual Subject Demand Forecast Tables

In this example, each subject is administered IMP in their Weeks 1–8 visits. Following those visits, the subject chooses between receiving at-home dosing and going into the clinic from Weeks 9 to 15. Then, the subject must come into the clinic for their Week 16 visit before resuming at-home or in-clinic dosing for the remainder of the study. Notice how Subjects 1, 2, and 4 have opted into the at-home dosing option so their forecast is split between the two depot locations. Meanwhile, Subjects 3 and 5 have opted to receive all of their treatment at the clinical site, so there is no forecast for these subjects out of the pharmacy depot.

There are two takeaways from this example for the clinical planner to heed as they create the forecast. First, the planner needs to work closely with clinical operations to understand which visits are eligible for at-home dosing so they can create an accurate individual subject demand forecast for those subjects who opt into D2P. Second, in most D2P models, the subject has a choice between receiving drug at the clinic or at home. This creates uncertainty in the forecast because the clinical planner needs to stock the pharmacy depot with sufficient IMP to supply subjects rolling into the D2P program which is challenging to do when the number of participating subjects is unknown.

Fortunately, we've already discussed strategies the clinical planner should take to manage this uncertainty. The first step is to collaborate with clinical operations on early identification of subjects that intend to take advantage of at-home visits. Adding D2P enrollment projections to the demand review meeting agenda in the operational clinical demand and operations planning process is a great start. Next, the same principles that apply to initially stocking other regional depots can be applied to developing an initial stocking strategy for the pharmacy. The estimated number of subjects, the confidence level in that estimate, the cost of the drug to manufacture or purchase, and the shelf life of the finished product batch should all feed into an analysis of how many kits are appropriate to send to the pharmacy before the first subjects have enrolled in the program. Another factor for the clinical planner to consider is how much flexibility patients have to return to the clinical site for administration of IMP after they've opted into D2P shipments. Monitoring in-clinic versus at-home visit trends through the demand actualization process gives the planner data they can use to improve the forecast over time and prevent excess waste at the pharmacy.

From a supply planning perspective, the pharmacy functions as any other regional depot does, but special consideration should be given to safety stock and order target levels. With respect to safety stock, carrying at least 4–6 weeks-of-supply is recommended to hedge against demand uncertainty when it's unclear how many subjects may enroll in D2P or if there is significant flexibility for patients to switch from in-clinic to at-home visits. When determining order targets and distribution order quantities to the pharmacy, keep in mind that some pharmacies may not have sufficient physical space to hold months of inventory at a time. The clinical planner needs to understand potential storage capacity limitations and work to incorporate those into their initial and resupply distribution orders to the pharmacy.

19.5 The Direct-to-Patient Supply Chain Execution Process

Following D2P model selection for the study, the clinical planner and clinical operations lead should begin developing the business process for executing the D2P supply chain. Implementing a D2P model impacts the entire downstream clinical trial supply chain and given the large number of internal and external stakeholders involved, a business process provides structure and clarity to all involved in executing the supply chain. The objective of the business process is to ensure delivery and return of the IMP required to support at-home visits during the trial, and the scope includes all activities required to make this happen from subject identification through the return of used IMP back to the depot.

At the outset of business process design, clinical operations and the clinical planner must decide who will be the process owner accountable for the D2P supply chain. The business process owner is responsible for maintaining the business process map and ensuring that each stakeholder knows their responsibilities within the process. If changes are made to the process, the business process owner documents and socializes those changes to the impacted stakeholders to ensure everyone is trained and knows how to complete their assigned tasks. Without a clear process owner, there may be confusion and misalignment if changes to the supply chain need to be made or when new team members roll onto the project.

Creating a detailed, step-by-step business process map for the D2P supply chain is essential. This is the document that connects the myriad activities, stakeholders, and systems together and provides the foundation for process success. Figure 19.3 illustrates a complex pharmacy-to-patient model that incorporates a home health nurse to facilitate administration of the IMP.

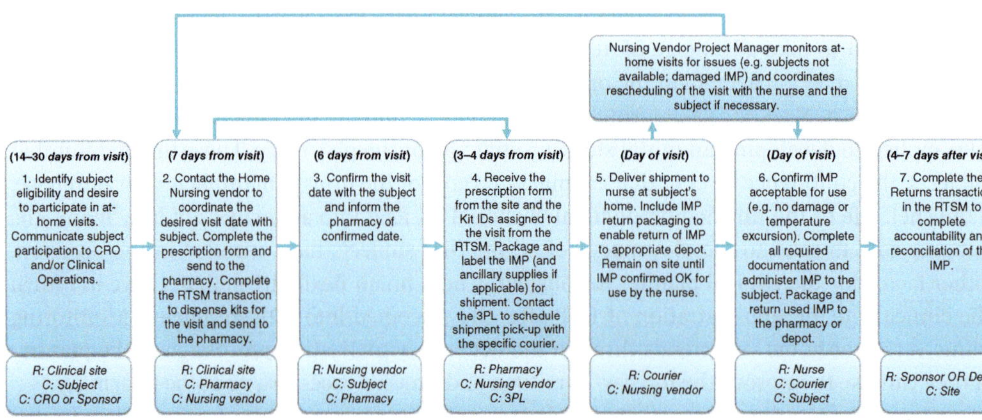

Figure 19.3 Pharmacy-to-Patient D2P Supply Chain with Home Health Nurse Support

Even this example is streamlined compared to a detailed process map, yet each step is packed with information. Notice that each description contains the time relative to the day of the visit when those activities must take place. That's because each step needs to be carefully sequenced to enable preparation and delivery of the IMP with subject and nurse availability. If the delivery window is too narrow, the physical supply chain won't have time to support the desired visit date because it usually takes at least three to four business days for the pharmacy to receive and prepare the order and for the courier to deliver the drug to the subject's home.

The process begins when the site identifies a subject who wants to participate in at-home visits. At least 2 weeks before the subject's visit, the site communicates to the CRO or clinical operations that the subject intends to switch from in-clinic to at-home visits. The subject's upcoming participation is then shared with the clinical planner so the demand forecast for the study accurately reflects which depot the subject will receive their IMP from. The 2-week window also gives the clinical planner time to ensure sufficient supply of IMP and ancillaries are available at the pharmacy. This timeframe holds extra importance when the protocol allows for flexibility between in-clinic and at-home dosing.

During the second process step, the site contacts the home nursing supplier project manager to identify a nurse who can support the subject's visit at least 1 week prior to the intended visit date. Nurses are not always readily available depending on geography, disease area, and training requirements. Performing this step at least a week ahead of the visit gives time to ensure a nurse can support the desired visit date. Then, the clinical site representative completes the prescription form which contains information about the subject, the visit, and the IMP and supporting ancillaries that will be administered at that visit. After the prescription form is complete, the clinical site executes the RTSM transaction for the visit, and the specific kit IDs are allocated from the pharmacy's inventory to the subject's visit. The prescription order form and an extract of the RTSM visit transaction are sent to the pharmacy for IMP shipment preparation. Note that the prescription order form and the RTSM transaction data are kept separate to protect the private health information of the subject, a topic we'll return to in the next section.

In the third step, the site sends the prescription form and the RTSM extract to the pharmacy, the nursing supplier also confirms the visit will take place with the pharmacy in the third step of the process. This communication ensures the pharmacy does not package and label material for a visit that can't take place due to lack of nurse availability.

After the pharmacy receives the documents from the site and confirmation of the visit from the nursing supplier, they begin the packaging and labeling process for the IMP and any applicable non-IMP ancillaries to enable subject dosing. The fourth step takes place about 3–4 days before the visit to allow at least 1–2 days for transit from the pharmacy to the subject's home. The pharmacy also contacts the 3PL provider to schedule pick-up and delivery of the shipment. When multiple hand-offs are required during transit of the shipment (e.g. from pharmacy to origin airport; origin airport to destination airport; destination airport to subject's home), the 3PL works with courier services for pickup and delivery of the shipment to and from the respective airports. This last component of the process step is important to distinguish because the courier plays an important role in the fifth process step.

On the day of the visit, the courier arrives at the subject's home, and they hand the shipment off to the nurse at the site. At the point of IMP drop-off, the courier also provides a second shipper for the nurse to use at the completion of the visit to return the used IMP kits to the depot. But the courier doesn't leave right away. Instead, they wait until the nurse has an opportunity to check the IMP for damage or for a temperature excursion which occurs when the product goes above or falls below the pre-specified storage condition (e.g. allowable temperature range).

At the beginning of the sixth process step, the nurse completes their quality check. If the vials are damaged or there is an excursion, the nurse immediately notifies the home nursing project manager who then informs the sponsor of what's happened so they can provide guidance on next steps. In the event the vials are damaged, the visit needs to be rescheduled to a later date, and the process begins again at the second step of coordinating the rescheduled visit. In the event of a temperature excursion, the sponsor needs to assess whether or not the IMP is still allowable for use. Performing excursion assessments are logistically challenging because the sponsor needs to have the appropriate personnel (e.g. analytical operations and/or quality) available to immediately address the excursion. If those team members aren't available, delayed responses can lead to a missed dose. To work around potential staffing limitations, the sponsor may provide an "allowable" excursion range to the 3PL/couriers and home nursing project manager, and if the shipment falls out of the allowable range, the shipment is automatically ruled not suitable for use.

In the event the IMP is delayed, damaged, experiences an irrevocable excursion, or the subject or nurse has to cancel the visit at the last minute, the visit needs to be rescheduled as soon as possible. The home nursing project manager has the responsibility in this process to monitor each visit for potential issues and work with the subject and the nurse to reschedule the visit if necessary. Airline delays, temperature excursions, and human factors are all part of managing the D2P supply chain so it's unavoidable that problems will happen. Having a clear process for rescheduling visits and managing stakeholder expectations on the limitations of the supply chain are both keys to success.

The sixth process step continues with the nurse administering the IMP to the subject. After the visit completes, the nurse returns the used IMP back to the supplying pharmacy using the return packaging provided by the courier. In that return shipment, the nurse provides documentation of the kits used in the visit so this can be reconciled against the kits assigned in the RTSM dispensation transaction from earlier in the process. The seventh and final step in the process is performance of accountability and reconciliation of the kits in the RTSM following return to the pharmacy or depot. If the pharmacy or depot is not willing to perform this transaction, then either clinical operations or clinical supply chain needs to take on this responsibility.

Figure 19.3 only captures a single example of a D2P supply chain, but it provides a flavor for the various steps that need to be defined from identification of the visit through IMP accountability and returns. The business process map is unique to each study and is based on the selected D2P model, the geographic region of the study, and whether or not a home nurse is used. The most important takeaway from this example is the sheer number of activities and details that need to be addressed no matter the D2P supply chain design. In the next section, we'll cover several additional execution-related elements not addressed in this particular example.

19.6 Other Direct-to-Patient Supply Chain Execution Considerations

Once the D2P model has been selected and the end-to-end supply chain process has been mapped, there are several execution-related details to finalize. These include:

- Ensuring safe handling of the private health information of each clinical trial subject.
- Ensuring the visit schedule enables the D2P supply chain.
- Managing non-IMP ancillary supplies, if applicable.
- Programming the RTSM to enable the physical D2P supply chain.

- Blinding large-molecule products in the home setting, if applicable.
- Administering sterile large-molecule products in the home setting.
- Shipping to locations outside the subject's home.

The first and most important item on this list is the protection of each clinical trial subject's private health information. The Health Insurance Portability and Accountability Act (HIPAA) was passed in 1996, and it established the privacy rule which created a category of protected health information (PHI) with special confidentiality protections in the context of clinical research.[6] The privacy rule protects all "individually identifiable health information" such as the name, age, address, and contact information of the clinical trial participant.[7] The sponsor of the trial has an obligation to work with the clinical trial site and the other stakeholders in the trial to keep this information limited only to the parties absolutely required to execute the supply chain. In practice, this means the only stakeholders who should receive PHI are the investigator and designated staff at the clinical site, the pharmacy receiving the prescription order, the nurse who prepares and administers the IMP (if applicable), and the courier who delivers the IMP. Not included in that list are the sponsor stakeholders including clinical operations and clinical supply chain and the RTSM supplier which controls assignment of IMP to each subject visit. This is the reason why the prescription order form is completed outside of the RTSM solution. If PHI were included in the RTSM, it could unnecessarily expose PHI.

A second aspect of the D2P supply chain that's easily overlooked is the visit schedule. The visit schedule needs to allow enough time between in-clinic visits and at-home visits to allow the D2P supply chain to work. For example, if the protocol calls for IMP administration at randomization, but the next at-home visit takes place 3 days later, there isn't sufficient time to coordinate and execute the D2P shipment. Clinical operations and clinical supply chain should review the visit schedule against the D2P process map and lead times to ensure the at-home visit timing is practical against the constraints of the supply chain.

Another element of the D2P supply chain to manage is the non-IMP ancillary supply strategy. Small molecules that require reconstitution, oral solutions, and large-molecule products for subcutaneous injection are likely to require ancillary supplies to prepare and/or administer each dose. For example, a small molecule that needs to be reconstituted in liquid may require a mixing container indicating the precise volume of liquid and a tool for stirring the mixture together. In cases where non-IMP ancillary supplies are required, the sponsor should:

- Determine the sourcing and warehouse strategy of the ancillaries based on the D2P model. If depot or pharmacy-to-patient models are used, the best path is to globally source each ancillary and store at the respective warehouse. If the site-to-patient model is used, local sourcing is preferred, but global sourcing may be required.
- Identify if rescue medications are required. Rescue medications are a special subtype of non-IMP ancillaries because they require a prescription from the investigator, so they need to be included on the prescription form sent in by the investigator or designated site staff.
- Determine whether to include non-IMP ancillaries in the RTSM alongside the IMP. The advantage of including non-IMPs in the RTSM is that it gives visibility and control over them from a supply chain perspective to the clinical planner. There is a significant disadvantage though, which is that adding non-IMPs adds material and batch complexity to the RTSM, and this can become overwhelming if several non-IMPS are used. Non-IMPs are often purchased in smaller quantities and batches, and this batch data needs to be migrated into the RTSM for the system to function appropriately. For this reason, including non-IMPs in the RTSM is discouraged unless deemed absolutely necessary.

With respect to the RTSM, appropriately programming the system to support the selected D2P model is another significant component of the D2P supply chain. The RTSM needs to:

- Enable the clinical site to select on-site or at-home visits where D2P is allowed in the protocol.
- Capture the correct supplying depot or site for at-home visits whether the depot, pharmacy, or clinical site.
- Maintain the ability to account for each used kit and whether that kit has been returned to the depot and/or destroyed for IMP reconciliation purposes at the end of the trial.

To accomplish this, the RTSM supplier should be included in the design of the D2P model as early as possible so the system can be programmed to accurately reflect the physical supply chain. Assigning the appropriate role for each transaction in the RTSM is also important, especially for accountability and returns which are typically done by the clinical site.

There are three more execution considerations that specifically apply to large-molecule products. First, in blinded trials that compare the active product to a commercially available standard-of-care product, it's not always possible to completely disguise the IMP and the comparator. In some cases, measures can be taken through additional packaging to achieve the blind, but in other cases, this simply isn't possible. The sponsor needs to weigh unblinding risk within the D2P set up, and if the unblinding risk is too great then D2P may not be a viable option for the trial.

The second large-molecule-specific consideration is how to ensure safe administration of a sterile compound in a non-sterile environment. The American Society of Health System Pharmacists has issued a guideline on Compounding Sterile Preparations (CSPs) that categorizes CSPs by risk level and by time of use.[8] The key category with respect to D2P is "Immediate-Use CSPs." Quoting from the guidance: "The immediate-use category should be reserved for emergent use or situations in which adhering to low-risk compounding procedures would add additional risk due to delays in patient care. Examples of such situations may include cardiopulmonary resuscitation, diagnostic procedures, or short-stability medications that must be prepared immediately before administration outside health care facilities (e.g. in home infusion or emergency care at the accident site or in an ambulance). Immediate-use CSPs do not need to be compounded in an ISO Class 5 environment and garbing and gowning are not required, as long as all of the following criteria are met:

- Hand hygiene per CDC recommendations.
- Aseptic technique is followed.
- No hazardous drugs are used.
- Only simple transfer of no more than three sterile, nonhazardous drugs in the manufacturer's original containers are involved in the compounding, and no more than two entries into any one container occur.
- No more than 1 hour elapses from the time compounding commences to the time administration to the patient begins (although best practice dictates that there are no intervening steps between compounding and administration).
- No batching or storage of CSPs occurs.
- The preparation is labeled with patient identification, names and amounts of all ingredients, name or initials of preparer, and exact 1-hour BUD and time."

If any of these conditions are not met, the product does not qualify for in-home use and D2P is not an option for that compound. The formulation lead for the CMC team needs to perform an assessment of these criteria to determine if the biologic meets the criteria before proceeding with implementing the D2P model.

Finally, there are situations where IMP may need to be shipped to a different location than a subject's home. One example of this is an infusion center where subjects can go to receive treatment. Setting up an infusion center is advantageous when the nursing supplier may not be able to identify a local nurse who can support infusions in the home. If another site is used, clinical operations should work closely with the good clinical practice (GCP) quality team to review the site's qualifications and storage capabilities. Specifically, cold chain storage capacity and temperature control are two categories where infusion centers may not meet the standards needed to appropriately control the IMP.

19.7 Parting Thoughts on the Direct-to-Patient Supply Chain

The D2P supply chain touches almost every aspect of supply chain planning and execution, and the level of coordination also increases the risk of missing patient doses. There are three strategies worth reemphasizing to achieve optimal performance within the D2P supply chain. First, recognize the amount of complexity in the D2P supply chain and spend time early on D2P model selection and detailed business process mapping. This early planning pays huge dividends when it's time to execute. There are too many steps where the supply chain can break down and too many stakeholders to coordinate to neglect outlining how the D2P supply chain should work.

Second, bringing third-party suppliers into the D2P design as early as possible is hugely beneficial. The depot, pharmacy (if applicable), RTSM supplier, and nursing supplier all have important roles to play in the process. Each supplier also has their own institutional expertise and their own ways of working that need to be incorporated into the D2P process. The only way to take advantage of that knowledge and to ensure the process works for them is to get them involved in the D2P process mapping exercise early.

Third, executing decentralized trials and D2P supply chains is a difficult endeavor, and there will inevitably be challenges along the way. Transparent communication with the clinical sites to set appropriate expectations about issue resolution is hugely important as is working closely with the respective suppliers to document issues and then troubleshoot them along the way. Process documentation and issue resolution is particularly useful for new team members as they come onto the program and support the supply chain.

Decentralized models and D2P supply chains have become a mainstay in the clinical trial landscape. Selecting the most appropriate model, incorporating the D2P supply chain into the demand and supply planning process, creating a detailed business process map for D2P supply chain execution, and minding all the execution-related considerations are all necessary steps to building an effective, efficient, and compliant D2P supply chain.

19.8 Key Takeaways

1) DCT are clinical trials where some or all of the trial-related activities take place outside of the traditional clinical trial site. There are three types of DCT: fully decentralized trials, hybrid trial designs, and mobile clinical trials. When drug is administered at home, the process of shipping IMP from the supplying depot to the subject is known as the D2P supply chain.
2) There are three D2P models: depot-to-patient, pharmacy-to-patient, and site-to-patient. The first word in each model describes the point of origin of the IMP before it reaches the subject. Geography, regulatory considerations, and the drug's modality should all be considered when

selecting an optimal D2P model. In the United States, it's common to use the pharmacy-to-patient model because of state licensure requirements, while site-to-patient models are more popular in other areas of the world that allow D2P shipments.

3) Myriad stakeholders are involved in the D2P supply chain including the subject, clinical site, supplying depot, pharmacy (if applicable), home nursing project manager, nurse, third-party logistics provider, courier, RTSM supplier, CRO, clinical operations, and clinical supply chain.
4) The pharmacy-to-patient model has the largest impact on the demand and supply planning process. In that model, the pharmacy acts as another regional depot that supplies at-home visits. A demand forecast needs to be created against the regional depot to generate a supply plan. Demand for each subject participating in D2P needs to be split against visits that take place in the clinic and visits that take place at home.
5) Developing a detailed business process map for D2P supply chain execution is important because it provides structure and clarity to all the stakeholders in the process. Clinical operations and the clinical planner need to decide who is ultimately accountable for the process to keep the process current and all stakeholders trained on their responsibilities. The business process map should outline each activity and how far in advance of the visit the activity should take place. The temporal element of the map is important to ensure the supply chain has time to react and deliver IMP for the visit.
6) There are several execution-related details that need to be reviewed before putting the D2P supply chain in motion. These include ensuring the safe handling of private health information, reviewing the visit schedule for practicality, managing non-IMP ancillary supplies, programming the RTSM, blinding considerations, administration of sterile molecules in a non-sterile setting, and setting up shipments to locations outside the subject's home. All decisions should be captured in the business process map as needed.
7) Three strategies are especially noteworthy when designing and implementing a D2P supply chain. First, recognize the high degree of complexity in the D2P supply chain and spend time on business process mapping. Second, bring the third-party suppliers into the process early to take advantage of their institutional expertise. Third, communicate transparently about the supply chain risks with the clinical sites.

Notes

1 https://www.fda.gov/regulatory-information/search-fda-guidance-documents/decentralized-clinical-trials-drugs-biological-products-and-devices.

2 https://www.fiercebiotech.com/biotech/more-than-two-thirds-trials-hit-by-covid-enrollment-halts-midstage-tests-worst-affected.

3 https://www.clinicaltrialsarena.com/features/2022-forecast-decentralised-trials-to-reach-new-heights-with-28-jump/.

4 https://sdlegislature.gov/Rules/Administrative/20:67:02.

5 https://law.lis.virginia.gov/vacode/title54.1/chapter34/section54.1-3420.2/.

6 https://privacyruleandresearch.nih.gov/clin_research.asp.

7 https://www.hhs.gov/hipaa/for-professionals/privacy/laws-regulations/index.html.

8 https://www.ashp.org/-/media/assets/policy-guidelines/docs/guidelines/compounding-sterile-preparations.ashx.

20

Contract Organization Management and Oversight

20.1 Strategic Supplier Management Within the Biopharmaceutical Industry

Contract development and manufacturing organizations (CDMOs), distributors (CDOs), research organizations (CROs), and technology providers (CTOs) are vital partners to sponsor companies that either don't have an internal capability or find it more cost effective to outsource the work to a third party. As of 2022, more than 80% of biomanufacturers reported outsourcing at least one activity in the manufacturing and batch release space.[1] Such a high percentage makes sense because building or purchasing drug substance (DS), drug product (DP), and finished product (FP) manufacturing sites is highly capital intensive, and most companies don't have the resources to build or purchase these capabilities. This is especially true for clinical-phase sponsors who commonly outsource the end-to-end supply chain to suppliers that specialize in their respective spaces.

In one applicable supplier management framework presented by Jonathan O'Brien in *Supplier Relationship Management: Unlocking the Hidden Value in Your Supply Base*, there are three relationship styles that a sponsor could choose from when engaging with contract organizations.[2] The simplest style is a transactional relationship which is characterized by low engagement from the buyer and is focused on obtaining the lowest price for the good. Transactional relationships are most appropriate for the purchase of simple goods or services. In a biopharmaceutical company, examples of transactional relationships include the purchase of lab supplies and office supplies. Operational relationships are the second style, and these are characterized by recurring purchases of goods and basic evaluations of supplier performance. An example of an operational relationship is one with a domestic transportation service provider responsible for shipping non-good manufacturing practice (GMP) materials between labs. These shipments are routine, yet they are important enough to warrant a deeper review than the transactional model offers. The third style is a strategic partnership relationship. This style is characterized by joint meetings between the buyer and the seller, information sharing, and taking a longer-term perspective with the supplier. In the clinical supply chain space, the strategic partnership style is most appropriate for DS, DP, and FP manufacturing and distribution site relationships.

Establishing strategic partnership relationships with clinical DS, DP, and FP manufacturing sites is essential because the activities undertaken by these suppliers are highly complex, time consuming, and expensive to complete. In clinical development, upstream CDMOs typically support manufacturing process development, analytical development, and manufacturing. At the FP level, the production and distribution activities are more straightforward than in upstream manufacturing, but they're also higher in frequency. A single clinical trial may require one or two DP batches,

Supply Chain Planning for Clinical Trials – A Practical Guide, First Edition. Ryan Mills.

several packaging jobs, and dozens of distribution orders. FP manufacturing also involves many sub-steps including label development, packaging design, and pre-/post-production batch record review.

Treating any of these engagements as a transactional or operational relationship can also have negative consequences. Successful process development, manufacturing, and distribution depends on having a clear set of goals and objectives for the supplier to work toward and consistent engagement from the sponsor to ensure the work stays on track. Transactional models don't offer this. Continuous change is also a hallmark of clinical development where radical changes in program and chemistry, manufacturing, and controls (CMC) strategy are common. Operational relationships also don't provide a sufficient amount of engagement to react to these strategic changes and implement solutions that work for both the sponsor and the CDMO. Even at the FP level where the work is more routine, relying on an operational model can leave the sponsor vulnerable when a batch needs to be moved quickly through the supply chain and there's not a complete understanding of the supplier's processes and capabilities.

Over the course of this chapter, we'll focus on the supplier relationship lifecycle introduced in Chapter 1 – Supply Chain Management. We'll begin by assessing the decision to insource or outsource that a sponsor with internal manufacturing and process development capabilities faces. Next is a review of the supplier identification and selection process where potential CDMOs and/or CDOs are evaluated against predefined criteria to make the best possible choice for a given program's needs. Following that, we'll turn to a discussion of the qualification and contracting process that enables the suppliers to begin working with the sponsor. Then, we'll examine how to engage CDMOs and CDOs through a partnership-style relationship. We'll also zoom out from analyzing individual supplier relationships to review single- versus multisourced strategies for the clinical supply chain. Finally, we'll end the chapter by focusing on an underappreciated partnership, the CRO, and all the touchpoints between the clinical supply chain function and the CRO.

20.2 Insourcing, Make-Versus-Buy, and Outsourcing

When a new drug candidate moves through the preclinical development process, the initial CMC team forms, and one of their first tasks is to define the DS and DP supply chain. For small molecule products entering a Phase I clinical trial, a separate FP manufacturer may not be required because the drug can be administered as active pharmaceutical ingredient (API)-in-capsule, and many Phase I research sites offer this capability. Large molecule products on the other hand typically require supplier selections for the entire GMP supply chain because they require both secondary packaging and labeling activities, and these must happen in a refrigerated environment. Most early-phase research units don't offer this refrigerated capability. As the product moves through development from Phase I to later phase trials, the supply chain also needs to change to accommodate increasing demand and to prepare for commercial readiness. In earlier phases, the CMC team's priorities are manufacturing and analytical process development, flexibility to respond to changes, and speed to produce and release material. In later phases, the priorities shift to scaling up the process to support additional clinical demand, increased reliability in execution, and suitability for a potential commercial launch.

At biopharmaceutical companies with internal manufacturing and process development capabilities, the first question the team needs to ask is whether to insource or outsource the work. The answer to this is usually obvious, but not always. Where there is a capability and capacity fit within the internal network, it makes sense to keep the work in-house because there is a greater degree of visibility into and control over what's happening. There's also more flexibility to make changes to the production plan. These are the same visibility, control, and flexibility (VCF) principles discussed in Chapter 7 – Supply Chain Planning for Clinical Trials. The role responsible for managing plant capacity is the master production planner or scheduler.

For new programs, the CMC lead and relevant DS and/or DP leads meet with the master production planner or scheduler to review the plant's upcoming production schedule and determine whether or not the product should be placed there. Within the largest sponsors, there may be multiple plants in the manufacturing network capable of executing the project. In this situation, there may also be a network planner involved who works with the CMC team and the master production planner(s) to find the best site for the new product.

When there isn't a strong capability fit or sufficient capacity to keep the product in-house, an alternative home for the product needs to be considered. In some cases, the sponsor may want to avoid outsourcing the product to a third-party supplier and instead purchase or build an internal manufacturing site. Building or purchasing an internal capability is attractive in two scenarios. The first is when a sponsor has a large-enough portfolio to justify investing millions of dollars into an internal site. Contract manufacturing sites charge large premiums to conduct work at their facilities, and with enough products and batches, there is an economic case to build or buy capacity. The second scenario is when the product requires a highly specialized and/or flexible manufacturing process. For this reason it's become common for cell and gene therapy companies to build their own manufacturing sites. This gives those companies maximum control over the plant and not risk losing needed capacity to serve their clinical trial needs.

Executing a make-versus-buy analysis to determine if a new plant should be built or purchased is relatively straightforward:

1) Estimate the fixed cost of building the new facility. Inputs include the building, equipment, systems, and other items required to start running the plant.
2) Estimate the cost of goods for manufacturing a batch at the new facility. Inputs include materials, labor, and maintenance for each batch.
3) Estimate the cost of manufacturing a batch at a third-party CDMO. This can be done by reviewing previous work or adding a mark-up to the internal estimate.
4) Execute a breakeven analysis showing how many years and/or batches it takes to recoup the cost of building the plant by plotting cumulative dollars spent internally (e.g. make) and externally (e.g. buy).

In Figure 20.1, the "make" data assumes a fixed cost to build the plant of $40,000,000, that each batch costs $2,000,000 to make at the plant, and that there are six batches manufactured per year. The "buy" data assumes each batch costs $3,000,000 to purchase from a CDMO, and the same six batches manufactured per year. The breakeven point is reached by Year 7 where the buy line becomes cumulatively cheaper than the make line. Based on these assumptions, the decision to move forward with investing in the new plant seems obvious. One trap to avoid though is overconfidence in the underlying assumptions of a breakeven model. This is especially true when building a new plant where unforeseen issues can add significant costs. Clinical portfolios are also notoriously fickle, and if programs are cancelled leading to fewer batches per year for the plant, the financial justification quickly becomes muddled. Running Monte Carlo-style simulations under a variety of fixed cost, variable cost, and batch frequency assumptions is recommended to quantify the potential risks of moving forward with a "buy" decision. For many biopharma companies though, the decision to outsource is obvious. Without internal capacity or a strong justification to

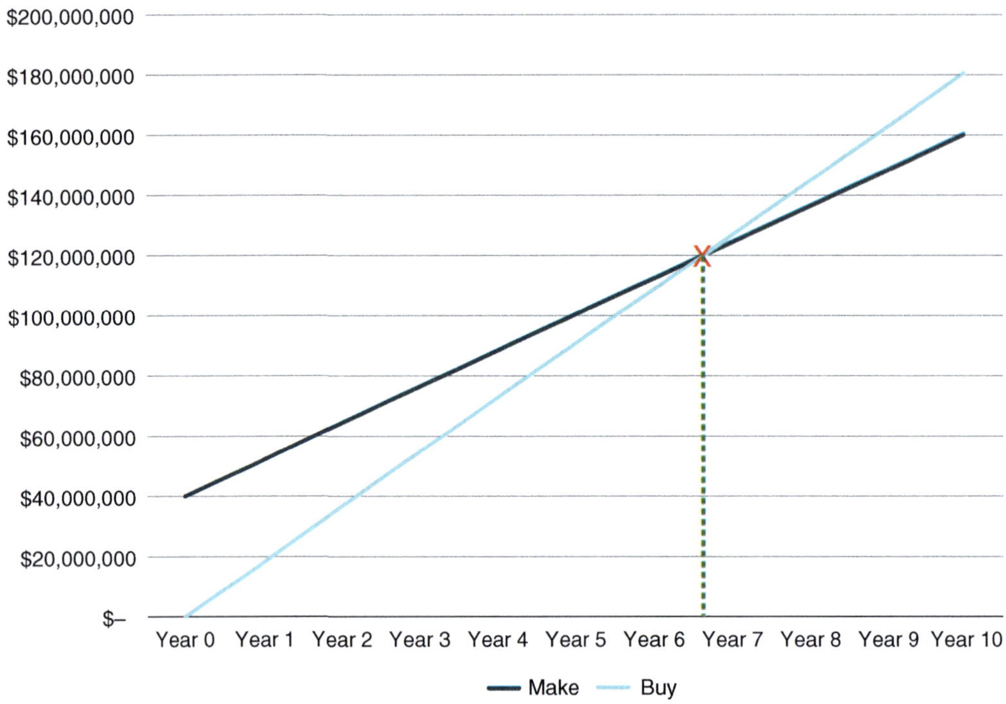

Figure 20.1 Example Breakeven Analysis

build or purchase a new site, the natural conclusion is to find a CDMO who can perform the required activities on behalf of the sponsor.

20.3 Supplier Identification and Selection

Selecting a new GMP manufacturing or distribution partner is a major decision for the sponsor. Relationships with DS, DP, and FP manufacturers last for years, and total spend with a supplier often runs into millions of dollars. Given the weight of the decision, it's sensical to lean on a methodical business process that involves the appropriate stakeholders and drives the best possible choice. The objective of the supplier selection process is to pick third-party suppliers that reliably and economically produce and distribute material on the sponsor's behalf. Usually, the business process owner for the supplier selection process is the procurement function, but technical expertise resides within the applicable CMC team members. Executing the process is therefore a partnership between procurement, the CMC team, finance, and other applicable functions. Figure 20.2 outlines the business process for supplier identification and selection:

In the first step, the CMC team leader or applicable DS, DP, or supply chain lead works to form a supplier selection team. There are many potential internal stakeholders who should be considered for inclusion. Adding a team member from procurement is obvious, but it should also be clear which responsibilities procurement and the technical leads will fulfill during the selection process. Procurement typically facilitates the process steps including supplier identification, RFP solicitation, and contract negotiation, but technical expertise needs to come from the individual CMC team members who will engage with the CDMO. For example, if analytical development activities

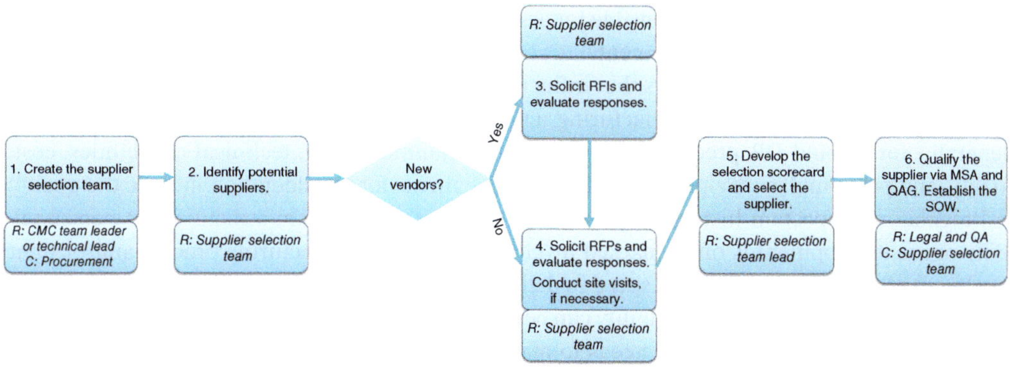

Figure 20.2 Supplier Selection Process Map

are in scope, then the relevant DS or DP analytical lead needs to be included. Likewise, if an external manufacturing function exists that oversees business operations at the supplier, they should participate in the process. The quality unit is also a required participant in every supplier selection because they need to evaluate whether potential suppliers meet internal and external qualification standards.

The second step of the process is to identify potential suppliers for inclusion in the selection process. Contract manufacturers and CDOs are identified through numerous sources. Direct marketing, market research reports, conferences, and networking with the salespeople who represent CDMOs are all mechanisms for learning about suppliers and their capabilities. The most valuable reference point though is prior experience working with a CDMO on a previous project or obtaining a trusted reference from someone else who has. However way the contacts are obtained, it's helpful to create a database outlining potential suppliers and key background information such as their location, size, regulatory history, and manufacturing capabilities. This information then becomes a starting point for supplier identification in future selection processes.

When new suppliers are considered in the selection process or if the supplier selection pool is large (e.g. more than five potential suppliers), the supplier selection team may consider soliciting a request for Information (RFI). An RFI contains a series of questions organized by various categories such as technical capabilities, personnel, and quality/regulatory history. RFIs are a useful tool for learning more about each supplier, and they provide valuable feedback for comparing different suppliers. RFIs are most useful for winnowing a field of potential suppliers down to a more manageable number (e.g. less than four potential suppliers) for the next step in the process, the request for proposal (RFP). Executing an RFI can take several weeks or even months, so if time is relatively limited, this step may also be skipped in favor of going straight to the RFP. The CDMO typically assigns a business development (BD) contact who is then responsible for coordinating the RFI response. The BD representative is also usually accountable for managing the relationship between the CDMO and the sponsor both during the supplier selection process and later on during the engagement phase of the relationship.

In the fourth step, the supplier selection team issues a request for a detailed proposal covering the intended scope of work. The CMC team members should also engage in writing and reviewing the RFP to ensure the content is a complete representation of the potential project. While the potential suppliers complete their responses to the RFP, the supplier selection team should also begin development of the selection scorecard. The selection scorecard is a list of the most important categories weighted by importance against which each supplier will be evaluated.

The supplier selection team is responsible for assigning both the decision criteria elements and the relative weights of each of those elements.

Two of the most important factors that influence the decision criteria are the phase of the clinical program and the type of manufacturing activity (e.g. DS, DP, and FP) under consideration. Generally, for DS and DP, process development and manufacturing, technical capabilities, cost, quality, and personnel are weighted heavily. Other elements for consideration include the quality and regulatory inspection history of the CDMO, and the project team structure and personnel. Early-phase development is often marked by changes in the manufacturing process and/or tight timelines to the clinic, so a CDMO's speed in delivering the project and flexibility to react to change may take precedence. Note that flexibility can be measured via capacity utilization within the plant and how much lead time a CDMO needs to make changes to a production order. Sites that have >80% utilization don't have the space to easily accommodate changes, and sites that require several months to prepare for a production order don't offer much flexibility. For late-phase projects, the focus shifts to the CDMO's potential as a long-term commercial partner and to performance reliability as the process settles and clinical demand increases. Reliability is measured through on time in full (OTIF) production orders and the average number of deviations associated with each batch. Another important element is the geographic location of the CDMO which has tax and geopolitical risk implications.

FP CDMOs have a different set of criteria than their upstream counterparts because of the nature of the work. Reliability takes on more weight because the packaging and labeling processes are so much simpler than DP and DS manufacturing. Other potential criteria at the FP level include

- Label development, translation, and printing capabilities.
- Primary packaging capabilities.
- Access to qualified persons who are required to release batches into the United Kingdom, European Union, and Israel.
- Reach of the global distribution network into Latin America and Asia Pacific.
- Access to a central pharmacy for direct-to-patient trial designs.

When the suppliers under consideration are new CDMOs to the organization, it's also a best practice to conduct site visits. Touring the facilities and meeting the CDMO team offers numerous benefits because they provide an opportunity for the sponsor to evaluate the physical space where the work takes place, to observe the facility and equipment firsthand, and to confirm the technical capabilities of the site. During a site visit, the supplier selection team can also meet and interact with the CDMO personnel they'd be working with on a regular basis. These impressions are invaluable because the sponsor can assess both the technical fit and the cultural fit with a potential partner. In lieu of a site visit, the sponsor may choose to invite the CDMO to their organization to present on their facility, capabilities, and team. These visits are colloquially known as proposal or bid defenses. While they don't offer quite the same impact as a site visit, bid defenses may be more appropriate when there is some past familiarity with the CMDO or the final decision between two CDMOs is particularly close.

After the RFPs have been provided and the site visits have been completed, the fifth step occurs which is to populate the selection scorecard and make the final decision. A sliding scale from 1 to 5 or 1 to 10 with the higher score indicating a better ranking typically works for this kind of analysis. Table 20.1 presents an example supplier selection scorecard for a late-phase upstream CDMO.

One other trend worth monitoring and potentially incorporating into the supplier identification and selection process is CDMO acquisition and integration. The largest CDMOs have built or acquired numerous capabilities to present end-to-end supply chain solutions to their clients. Integrated CDMOs

Table 20.1 Upstream Supplier Selection Scorecard Example

		Scored 1–5 (5 best; 1 worst)		
Element	**Weight**	**CDMO 1**	**CDMO 2 (Winner)**	**CDMO 3**
Manufacturing and process development capabilities	20%	4	5	4
Manufacturing reliability and commercial potential	20%	5	4	4
Cost	15%	4	5	2
People	15%	5	5	5
Analytical development capabilities	10%	4	4	4
Quality	10%	5	5	5
Geographic location	10%	3	2	3
Total	100%	4.35	4.40	3.85

like these offer theoretical advantages over selecting multiple unassociated sites such as an ability to move product more quickly through the supply chain because of harmonized business processes and systems. That said, capability acquisitions and integrated offerings may sound intriguing, but the reality doesn't match the promise. It can take years to fully incorporate a newly acquired site into a broader manufacturing and distribution network, so it's fair to take claims of integration with a measure of skepticism. Asking a potential supplier about the degree of integration between people, business processes, and technology during the selection process is also recommended.

Supplier identification and selection doesn't always need to follow this path. When a sponsor has a stable of qualified third-party manufacturers they are comfortable with, selection can be as simple as finding the CDMO with the best capabilities or the most available capacity. In this case, the CMC team may not need to be involved at all. Instead, the external manufacturing teams who manage those CDMOs can take the team's requirements and make the selection decision on their behalf. For new manufacturing sites though, it takes the combined expertise of the supplier selection team and a thorough process that provides as much information as possible to the sponsor so they can make the best possible decision for the program and the company. After a new supplier has been notified of their selection, the next step in the process begins, qualification and contracting.

20.4 Supplier Qualification and Contracting

Before work begins at the new supplier, a master services agreement (MSA), a statement of work (SOW), and a quality agreement (QAG) must be established between the sponsor and the CDMO. The MSA, SOW, and QAG are legally binding documents that commit the sponsor and CDMO to a series of obligations to each other during their partnership. Each agreement covers numerous categories and clauses, and each side wants the best possible terms for their own organization. Therefore, the process to draft, negotiate, and execute both the MSA and the QAG can take several months to complete. Supplier qualification, audits, and QAGs were covered extensively in Chapter 4 – Quality Considerations so the remainder of this section focuses on the MSA and SOW.

MSAs are a framework document commonly used in long-term business partnerships that outline the legal terms and conditions governing the engagement. An MSA documents the scope of services rendered, liability and indemnification terms, how intellectual property will be managed between the two parties, payment terms, and dispute resolution. Establishing an MSA with third-party CDMOs is commonplace in the clinical biopharmaceutical industry for three primary reasons. The services provided are complex, the duration of services rendered is lengthy, and those services are subject to change as the sponsor's clinical portfolio evolves over time. When changes do happen, an MSA allows both parties to quickly pivot without the need for a lengthy renegotiation over contract terms. MSAs are typically negotiated by teams of lawyers, but technical experts may also need to participate and provide input.

After an MSA is established, the next step is to define the exact work the CDMO will perform on behalf of the sponsor. This is captured in the SOW which outlines the specific services and deliverables the CDMO will provide. At the upstream level, the SOW identifies the manufacturing process development, analytical development, batch manufacturing, and stability studies the CDMO will conduct. At the FP level, secondary packaging, labeling, distribution, QP certification, and destruction activities may all be included. The SOW states the charges and costs associated with each activity listed. This information is the basis of the budget forecasting process outlined in Chapter 14 – Clinical Supply Chain Reports, Budgets, and Metrics.

Management and oversight of the activities listed within the SOW requires active management from both the sponsor and the CDMO to ensure services are billed according to the terms of the contract. Most sponsor organizations load SOWs into a procurement management system that captures each invoice provided by the CDMO for services performed. The invoices are associated to specific activities within the SOW, and this allows the sponsor to track how much remaining budget is left on the SOW. Invoice tracking may be performed by the technical lead (e.g. clinical supply chain for FP CDMOs), finance, procurement, or a combination of all three functions.

In theory, all goes according to the terms laid out in the SOW, and the CDMO invoices for services rendered until the contract is completed. In practice, changes in program strategy and study design happen all the time, and these often result in a need to make changes to the underlying SOW. It's therefore important to understand the CDMO's process for managing changes to their SOWs. When activities are added or removed, a CDMO may prefer to update the existing SOW or they may prefer to create an entirely new SOW. The sponsor has the responsibility to work the CDMO to understand the preferred strategy and ensure both the procurement system and the program's budget accurately reflect the activities and costs associated with the SOWs.

Before moving on to supplier engagement, note that the end-to-end supplier selection, qualification, and contracting process is a lengthy one. It's not uncommon for the entire process to take – 6–9 months (or more in extreme cases). This is especially true for late-phase supplier selections where the scope of activities, costs, and desire to find a long-term commercial supplier place enormous pressure on the decision. Sponsors must factor this time into account when determining the timing for kicking off the supplier selection process. Allowing at 9 months to bring an entirely new supplier on board before work begins is recommended to avoid compressing process development and manufacturing timelines.

20.5 Strategic Supplier Management

Supplier management encompasses all the actions a sponsor takes to oversee work performed on their behalf at a contract manufacturer or CDO. In transactional relationships, supplier management is limited to confirming receipt of goods and maintaining current SOWs. In operational relationships, supplier management activities expand to include

- Establishment of a joint project team with representatives from the sponsor and CDMO.
- Establishment of project deliverables and timeline.
- Identification of sponsor priorities within the project deliverables.
- Holding regularly occurring team meetings to facilitate completion of the project.

With a project team structure in place, clear deliverables and timelines outlined, and priorities clarified, the core elements are in place to successfully execute a project. Notice though that these elements establish what needs to get done, when it needs to get done by, and how the CDMO should oversee and manage completion of the work. The operational supplier management framework exists to complete the project, but there is more the sponsor can and should do to not only guarantee the success of a single project, but also maximize the value of the partnership and successfully deliver a portfolio of deliverables. These activities give both parties a broader understanding of each other so they can align on the project's goals and how best to reach them. Strategic supplier management elements include

- Establishment of escalation pathways for issue resolution.
- Discussion and documentation of ways of working between the two organizations.
- Measurement of supplier performance against the sponsor's priorities.
- Establishment of a broader steering committee with project team members and senior leaders from both organizations.

Through the remainder of this section, we'll discuss each of these actions and how they apply to managing clinical CDMOs beginning with the creation of a joint project team. While the supplier onboarding, qualification, and contracting process is ongoing, the sponsor and CDMO identify the team members assigned to participate in the project. For upstream process development and manufacturing engagements, sponsor team members typically include the respective DS, DP, and analytical lead(s); the quality lead(s); and the CMC project manager. For organizations that have an external manufacturing function, they are also included on the project team. With FP engagements, the clinical supply chain function, quality lead, and CMC project manager may be included. On the CDMO side, they assign their respective technical leads and a project manager who develops and manages the detailed project plans for the deliverables outlined in the SOW. The CDMO project manager often drives the success or failure of meeting the project timelines because they are often tasked with coordinating work internally at the CDMO and with flagging potential timeline risks to the sponsor. The more experience a project manager has, the better equipped they are to perform these tasks.

Supplier engagements typically begin with a kick-off meeting where the team members introduce each other and then review the project deliverables and timelines. All activities contained within the SOW are discussed, and the CDMO project manager should provide feedback on the feasibility of the project's goals and potential risks to meeting those. At this point, the sponsor should also highlight the project's top priorities such as meeting analytical development milestones, improving manufacturing yields, or simply meeting the deadline by which the material must be ready for the clinic. It's often the case that project deliverables conflict with each other, so it's especially important for the sponsor to guide the CDMO on how to prioritize between the deliverables. For example, a sponsor may want to see specific process improvement targets and to deliver material by a certain deadline. But, if the priority is to meet that deadline, there may not be enough time to improve the process to the desired targets.

The joint team also needs to define the meeting cadence and frequency to review progress against the project's goals. Most clinical-phase CDMO engagements require a weekly meeting to check in and provide updates and feedback on what's happening in the project. This is especially true of upstream manufacturers, but it's also a best practice for FP CDMOs. It's always easier to end a meeting early or cancel an unnecessary meeting than to scramble and schedule an ad hoc meeting at the last minute.

With the operational framework in place, strategic supplier management elements should be layered into the relationship. The first element to establish is an escalation pathway between the two companies. An escalation pathway is a structured approach to resolving issues, conflicts, or disputes in the relationship. Defining that pathway involves identifying specific stakeholders within both the sponsor and the CDMO who are flagged when a problem can't be solved within the project team. Examples of situations that may need to be escalated include poor performance from a project team member or a significant risk to achieving the deliverables outlined in the SOW. Typically, senior leaders within each organization are identified who have the authority to make personnel or financial-related decisions and have the power to bring issues to a close. That said, escalation pathways should only be used as a last resort. Conflict should be resolved within the project team wherever possible to maintain a productive working relationship. Involving senior leaders may result in negative outcomes for project team members and this can easily breed ill will within the partnership.

The second strategic supplier management element is to discuss and document ways of working between the two organizations. Every sponsor and CDMO has a unique organizational structure, business process framework, and culture they use to accomplish their goals and objectives. When the sponsor and CDMO are located in different regions of the world, local regulatory requirements may also influence the project plan and deliverables. Regional cultural differences may also need to be navigated when teams work globally. Developing a strong understanding of each other helps the sponsor and CDMO work better together in pursuit of the project's goals. Documenting ways of working can take many forms such as

- Outlining the role and responsibilities of each team member and the communication pathways within the project team. Some CDMOs prefer the project manager act as a single point of contact who filters information to the appropriate technical leads or departments. Other CDMOs prefer the technical leads liaise directly with each other to solve problems. Other CDMOs prefer a combination of the two approaches depending on the topic. Responsibilities and expectations for communication need to be clear across all parties.
- Reviewing the detailed project plan and highlighting critical path activities. It's important for the sponsor to understand the project plan structure, predecessors, and lead times of the CDMO. If the goals or deliverables of the engagement change, the sponsor can better assess the potential impact of those changes on the timeline and explain that impact to other stakeholders within the sponsor organization.
- Conduct business process mapping sessions with the CDMO. This is particularly useful for FP CDMOs who perform regularly recurring work on secondary packaging, labeling, and distribution orders. Documenting each process and where handoffs occur between the sponsor and the CDMO ensures the right information gets to the right stakeholder at the right time to keep the work flowing smoothly.

Measuring supplier performance via key performance indicators (KPIs) is the third element of strategic supplier management. A KPI measures the performance of an organization in meeting a predefined objective. The first step in establishing a supplier measurement program is to identify the KPIs that are most important to the sponsor. KPIs vary by the type of services offered and by the phase of the program. There are countless KPIs to choose from when measuring CDMO performance. Manufacturing process performance, production yields, manufacturing and batch release lead times, batch failure rates, and cost of goods are all important at the DS and DP level. OTIF, distribution order lead times, and right first-time rates matter more at the FP level.

In every case though, KPIs accomplish two goals. First, they allow the sponsor to objectively measure CDMO performance over time. KPIs illustrate a clear picture of where the CDMO is meeting or exceeding expectations and where there is opportunity for improvement. This data forms the basis for a regular supplier performance review. Second, KPIs indicate to the CDMO what the sponsor prioritizes in the project. This is another avenue for the sponsor and CDMO to align not only on what needs to be achieved, but also on how to achieve it. For example, imagine a sponsor who is running a study with a high-value comparator and a relatively short screening window. That sponsor may emphasize minimizing distribution order lead times to conserve how much product they're sending to each depot. By highlighting that KPI, the sponsor is reinforcing to the FP CDMO not only what the goal is (e.g. minimizing spend on the comparator), but also how they intend to achieve that goal (e.g. reducing distribution lead times).

The final strategic supplier management element is to establish a joint steering committee between the sponsor and the CDMO that conducts regular business reviews. The joint committee includes the project teams from both organizations and the BD lead from the CDMO and the senior leaders identified in the escalation pathway process. Topics discussed at the joint committee meeting include a review of all the projects underway between the two companies, ways of working between the sponsor and CDMO, KPIs and recommended actions to improve performance, innovation opportunities, and potential future projects. Joint committee meetings take place annually, semiannually, or even quarterly depending on the volume of work and the need to bring the project teams and respective executives together. Where possible, these meetings should take place face-to-face to further strengthen the partnership. Nothing replaces the value of meeting someone in person and learning more about who they are, what they do, and what they need from the other organization to make their job easier. Investing in these meetings strengthens the partnership on both sides and is a powerful mechanism to facilitate accomplishing the goals of the sponsor and the CDMO.

Even with a highly effective supplier management program, supplier performance ebbs and flows over time. Changes to personnel, leadership, and processes can all impact how well a CDMO delivers on the project. No CDMO is perfect. But that's exactly the point of strategic supplier management and viewing CDMOs as true partners and not transactional vendors. When the transactional approach is applied to a poorly performing CDMO, the background and context don't exist to understand why the problems are happening or how to appropriately address them.

20.6 Clinical Supply Chain Sourcing Strategies

Once the initial clinical supply chain for a program is established, the next question to consider is whether or not to use a single DS and/or DP CDMO at each level of the supply chain (e.g. a single sourcing strategy) or to engage with multiple CDMOs (e.g. a dual-sourcing or multisourcing strategy). The answer of whether to single source, dual-source, or multisource lies in the supply needs of the program, the costs of bringing on a new CDMO partner, and the risk tolerance of the sponsor. There are clinical supply chains where demand alone drives the need to dual- or multisource DS and/or DP production. Molecules under study in multiple late-phase clinical trials sometimes have enough clinical demand to exceed capacity at a single CDMO. If a molecule generates strong data in one trial, a decision may be taken to expand the program into other areas of study. This change can outstrip a single supplier's capacity over time, and bringing another CDMO online becomes necessary to meet that extra demand.

Those situations are relatively uncommon though. In most cases, a single DS and/or DP CDMO has sufficient capacity to meet clinical demand, so the question turns to an assessment of cost versus risk. Bringing a new CDMO online to manufacture is expensive in terms of dollars and resource burden. Even if the CDMO has already been qualified, the manufacturing process and analytical methods need to be transferred to the CDMO and non-GMP engineering batches need to be executed to demonstrate the CDMO is ready to begin production of GMP batches. Costs for these activities can run into millions of dollars, and an internal team of resources needs to oversee that work. For this reason, most clinical supply chains utilize a single sourcing strategy because those costs outweigh the risk of a plant failure or other supply interruption at the primary CDMO. As a program moves towards a commercial launch, the question of single and multisourcing takes on greater importance because the demand profile typically increases substantially, and revenue generated by the product offsets the cost of bringing on multiple suppliers. We'll return to this topic in Chapter 23 – Integrating the Clinical Supply Chain into a Commercial Launch.

At the FP level, the analysis changes because FP CDMOs have the capacity to handle hundreds of clinical trials at once. Therefore, individual clinical trials almost always utilize a single CDMO to handle a specific trial. There are other strategies a sponsor may utilize to de-risk the FP supply chain. First, the sponsor has the ability to list multiple FP CDMOs and sites in the regulatory application for the trial. This gives the sponsor flexibility to switch from one CDMO to another in the middle of a trial should disaster strike. The costs of transferring product to a new CDMO at the FP level are also much lower, so this strategy makes sense as long as the CDMOs listed in the application are audited on a regular basis. The second strategy is to work with different FP CDMOs on different trials or programs across the portfolio. This approach makes sense when there are multiple products with enough ongoing trials to justify engaging with two or more CDMOs. Taking this portfolio-level approach to dual- or multisourcing provides incentive for each supplier to offer their best pricing and service because they know there will be competition for business in future studies.

20.7 Contract Research Organization Engagement

Before closing, there is another third-party vendor that has a profound impact on the clinical supply chain, the CRO. They are typically selected, onboarded, and managed by the clinical operations function, but the CRO engages in many aspects of the clinical supply chain. These include

- *Demand Planning*: The CRO is usually the primary source for enrollment rate projections, site activation schedules, and potential country additions through the course of the clinical trial.
- *Clinical Label Development*: CROs often have local experts who provide country-specific requirements for what information needs to be included on the clinical labels.
- *Non-IMP Requirements and Sourcing Strategy*: The CRO typically performs clinical site readiness visits to identify which non-IMP materials or equipment need to be provided to the site in order to conduct the study.
- *Importer of Record (IoR) Services*: The IoR is listed on import documentation for shipments of IMP and non-IMP supplies to clinical trial sites. The IoR is legally responsible for those imports,

and in many countries, the IoR must have an entity established in the country of import. Many CROs have global reach and can offer this service when a sponsor does not have a legal entity in place. See Chapter 21 – International Logistics for Clinical Biopharmaceutical Products for more.
- *Decentralized Trial Strategy and Execution*: The CRO also plays a key role in defining decentralized clinical trial strategy and execution. This includes setting up and providing oversight for the direct-to-patient supply chain where applicable.

CROs also play a central role in IMP management at the clinical site, and they often act as a liaison between the site and the sponsor for IMP-related issues. The clinical site is the most downstream node of the supply chain because it's the final location IMP is stored before dispensation to a clinical subject. Shipments of IMP to the site represent the handoff between the good manufacturing/distribution practice (GM/DP) supply chain and the good clinical practice (GCP) supply chain. As soon as the IMP reaches the clinical site, oversight of the IMP shifts from the clinical supply chain function (e.g. the GMP supply chain) to the clinical site (e.g. the GCP supply chain). Clinical operations and/or the CRO have oversight responsibility of all clinical site activity including the handling, preparation, and accountability of IMP. While this handoff sounds straightforward on paper, there are several points where the clinical supply chain function may be engaged by the CRO:

- *IMP Site Shipments*: The CRO typically trains the site on how to ensure IMP is well received and properly stored. When kits are damaged or suffer temperature excursions, the CRO may notify the clinical supply chain function directly to rectify the issue. The clinical supply chain team may need to work with both GMP quality to disposition the material and the GCP quality team to prevent future excursions at the site.
- *IMP Storage at the Site*: Similarly, if kits are damaged or suffer an excursion during storage at the site, the CRO may notify clinical supply chain to address the issue.
- *IMP Accountability*: Every kit of IMP received by each site must be accounted for at the conclusion of the trial whether it was dispensed, expired, damaged, returned, or destroyed. CROs often work with the clinical site to review handwritten accountability logs and ensure that individual kits match the record in the RTSM. Where there are discrepancies between the accountability logs and the RTSM, the CRO may engage clinical supply chain to clarify and correct differences between the two.
- *IMP Returns and Destruction*: As product expires at the clinical site and/or when the study completes, the clinical site has a choice to either return the IMP to the supplying depot or destroy the unused IMP on site. The CRO then notifies the sponsor of the site's preference. For depot returns, clinical supply chain needs to provide clear direction to the CRO and the sites on how to execute depot returns.

Given these myriad touchpoints, it's important for the clinical supply chain function to engage with clinical operations and/or the CRO during the CRO onboarding process. Each of the elements listed above needs to be discussed so the scope of the CRO's role, the business processes they'll engage in, and the roles and responsibilities between clinical operations, the CRO, and the clinical supply chain function are well understood by all parties.

The final area of engagement between clinical supply chain, clinical operations, and the CRO is on the trial master file (TMF). The TMF contains study-related documents which "individually and collectively permit evaluation of the conduct of a trial and the quality of the data produced.[3]" Documents that need to be captured in the TMF encompass several aspects of IMP management including the labels, instructions for handling of the IMP at the site, shipping records of the IMP to the sites, certificates of analysis and QP certifications, accountability records, and documentation of

IMP destruction. To summarize, the entire lifecycle of the IMP at the clinical site needs to be recorded within the TMF. Before beginning the trial, clinical supply chain should meet with clinical operations and the CRO to identify the specific documents that need to be uploaded in the TMF as well as potential documents that are out of scope. Responsibilities for collecting, uploading, and reviewing documents all need to be assigned and understood because once the trial starts, the volume of documents produced is extensive and can overwhelm an organization if they're not proactively prepared to manage it.

20.8 Key Takeaways

1) From O'Brien, there are three potential supplier management relationship styles: transactional, operational, and strategic partnership. The strategic partnership style is most appropriate for managing clinical DS, DP, and FP contract organizations because the work is highly complex, time consuming, and expensive to complete, and only the strategic partnership style enables the sponsor and CDMO to best understand how they can achieve a project's goals together.
2) The first decision a sponsor with internal DS, DP, and/or FP capabilities must make is whether to use an internal plant or to outsource a program's supply chain to a CDMO. Insourcing is usually preferred where there is a capability and capacity fit, but where that doesn't exist, outsourcing is required. Sponsors may also consider building or purchasing internal capabilities where there is a strong financial or strategic rationale to do so.
3) Supplier identification and selection is a multistep process that's usually led by procurement but requires participation and input from key CMC team members including the technical leads, quality lead, and finance. When new suppliers are considered or a new capability needs to be brought online, a vendor selection team should be formed, potential vendors identified, and then an RFI should be solicited to collect basic background information on each CDMO. Then, an RFP can be made to the final few CDMOs, and those responses can be evaluated against a supplier selection scorecard.
4) The MSA, SOW, and QAG are three legally binding documents that must be in place before work can begin at a CDMO. These documents are complex and can often take months of lead time to negotiate and finalize, so it's important to build this lead time into the overall supplier selection process.
5) Supplier management encompasses all the actions a sponsor takes to oversee work performed at a contract manufacturer or CDO. Operations elements include establishment of joint project teams, project deliverables and timelines, identification of sponsor priorities, and holding regular meetings. Strategic supplier management expands on this to establishing escalation pathways, documenting ways of working, establishing KPIs, and establishing a joint committee.
6) KPIs are particularly important in strategic supplier management because they allow the sponsor to objectively measure CDMO performance over time and because they illustrate what priorities the sponsor values within the CDMO.
7) Clinical trial supply chains are most commonly single-sourced meaning there is only a single DS, DP, and FP CDMO supporting a given program. Exceptions to this occur when program demand outstrips a single CDMO's capacity to support the program or when the desire to de-risk the supply chain exceeds the costs of bringing multiple CDMOs online. At the FP level, some sponsors choose to utilize multiple CDMOs across different programs and studies within the overall portfolio.

8) The CRO is another important vendor for the clinical supply chain function to engage with before and during the clinical trial. CROs support many aspects of the clinical supply chain including demand planning, label development, and IoR services. CROs also play a role in overseeing IMP management at the clinical sites and often serve as a point of contact for the clinical supply chain function in resolving IMP-related issues.

Notes

1 https://www.outsourcedpharma.com/doc/outsourcing-trends-in-biopharmaceutical-manufacturing-0001.

2 O'Brien J. (2018). *Supplier Relationship Management: Unlocking the Hidden Value in Your Supply Base*. London: Kogan Page.

3 https://database.ich.org/sites/default/files/E6_R2_Addendum.pdf.

21

International Logistics for Clinical Biopharmaceutical Products

21.1 International Logistics for Clinical Biopharmaceutical Products

In Chapter 1 – Supply Chain Management, we introduced logistics as the strategy and execution of shipping goods from an origin location to a destination location. International logistics expands that definition to the shipment of goods from one country (e.g. exports) into another country (e.g. imports). Executing international shipments requires careful coordination and collaboration between numerous parties including the seller of the goods, the buyer of the goods, the importer of record (IoR) who has legal responsibility for the shipment, the customs brokers who facilitate document review and approval of the shipments, and the transportation service providers (TSP) and couriers who arrange for the physical movement of the goods. There are so many stakeholders in this process because logistics is not simply about the physical movement of goods; it's also about the financial implications of each shipment and the risks and liabilities borne through the movement of goods if something goes wrong during the delivery process. International logistics is also complex because individual countries have their own legal and tax frameworks that govern imports and exports.

Clinical biopharmaceutical products are particularly challenging to manage for the three reasons outlined in Chapter 1:

1) They are often shipped internationally, subjecting them to the process outlined in Figure 1.4.
2) They are high-value products and thus subject to significant taxes and duties, but they don't have an obvious value because they are not saleable products.
3) They require constant temperature control and monitoring through the shipping process which creates additional pressure to deliver the product on time to its destination or risk losing the shipment.

International logistics for biopharmaceutical products is typically not well understood outside of the experts who coordinate shipments on a regular basis. In large pharmaceutical companies, whole departments are dedicated to the practice. But, in many clinical-stage companies, the task usually falls on the clinical supply chain function where that expertise may not exist within the team members. As a result, each shipment becomes a high-stakes exercise fraught with peril from a number of potential sources. Initiating a shipment too late doesn't leave enough time to prepare the requisite shipment documents and shipping containers. Incorrect or incomplete documentation leads to delays at the point of customs review which in turn jeopardizes manufacturing timelines. Or, the physical shipment itself gets lost in the transportation process.

Supply Chain Planning for Clinical Trials – A Practical Guide, First Edition. Ryan Mills.

Over the course of this chapter, we'll review the basics of international logistics and the special considerations for clinical biopharmaceutical products. We'll begin that discussion with a deeper look at the taxes levied against imports and exports, how to properly value clinical products, and the economic registration activities required in the European Union to import goods into that region. In order to determine how much tax is owed, every product must have a value associated with it. Typically, valuations are based on the sales price of the good, but clinical products aren't for sale so valuing a product requires a different approach. Then, we'll cover the many roles involved in setting up and executing an international shipment. In that section, we'll also introduce IncoTerms and how they dictate the responsibilities of the seller and the buyer of the shipment. IncoTerms outline when the transfer of ownership of the goods takes place, who bears the risk of the shipment during transit, and who prepares any applicable export or import licenses.

With that context in mind, we'll detail a step-by-step business process outlining how to set up and execute shipments for clinical biopharmaceutical products. This process begins with shipment identification and includes preparation of shipment documents, arranging the appropriate shipment materials, customs entry, and delivery of the goods. Finally, we'll review two common scenarios. The first scenario is the application of the import-for-export (IFE) process to clinical biopharmaceutical goods where an upstream (e.g. Drug Substance or Drug Product) material is imported into the United States for further processing before getting exported to another country for final consumption. The second scenario is the use of free trade zones (FTZs) and special economic zones (SEZs) to enable storage and further processing of goods in countries like China and India.

21.2 Duties, Taxes, Valuation, and Registration

Civilizations have engaged in global trade for centuries, and as long as organized governments have existed, they've treated the movement of goods as a source of income through tariffs, which are taxes on imported goods. In the United States, the Tariff Act was passed in 1789, and it codified a 5% tax on all imports into the country.[1] Since then, international trade has evolved into an extraordinarily complex web of taxes that vary by country and by the type of good under import.

Taxes on imports are loosely categorized into two buckets: tariffs and taxes. Tariffs are taxes levied on goods when they are imported into a country. The tariff rate is determined by the Harmonized Tariff Schedule (HTS) code. HTS codes are an internationally standardized system used to classify and categorize goods. The US HTS library has 22 general classifications of goods, and pharmaceutical products are explicitly called out in Section VI, "Products of the Chemical or Allied Industries.[2]" The HTS coding system categorizes materials through a numeric system where the first six digits are common across every country, but the last two or four digits are country specific. This allows each country to subclassify materials and apply their own tariffs. Each country then assigns tariffs and taxes to the classification of goods, and these vary for clinical biopharmaceutical products from country to country. But, there are also opportunities to classify clinical biopharmaceutical products into research and development HTS categories that are subject to lower tariffs depending on the country. For example, the United States categorizes placebos and blinded clinical trial kits as free for entry.[2]

There are several other taxes that can potentially be levied on imports. The most impactful of these are typically consumption taxes such as the value-added tax (VAT) or the goods and services tax (GST). Consumption taxes work by assessing the value added in each production stage of the supply chain and then providing tax credits back to the companies who perform the production operations.[3] The end consumer then pays the penultimate tax on the final good, which is why these taxes are referred to as consumption taxes. Nearly every country in the world except for the United States charges VAT or GST on imports, and rates average between 15% and 20%. This is a high rate when compared to tariffs on biopharmaceutical products that usually range from 0% to 5%. Fortunately, VAT can usually be recovered on clinical pharmaceutical imports, but the reclamation process is unique to each country, often administratively complex, and usually lengthy to complete. For example, in the European Union, to qualify for a VAT refund, the business "must not have been based in the refunding EU country or supplied goods or service there – except for exempted transport and ancillary services or supplies to customers liable for payment of the related VAT under the reverse charge mechanism.[4]" That's a mouthful to say that a sponsor who contracts with a third party in the EU to manufacture clinical products potentially qualifies for a VAT refund.

The amount of duty and tax paid on imports is taken by multiplying the customs value of the import against applicable duty and tax rates. According to the World Trade Organization (WTO), customs valuation is "a customs procedure applied to determine the customs value of imported goods.[5]" The first principle of customs valuation is that the value of a product is equivalent to the transaction (e.g. the sale) price of the good. When a transaction value is not available, alternative methods are used to calculate the customs value of a good. This is relevant for the products used in clinical trials because there is no sale involved in the provision of clinical drugs to participating subjects. In the absence of a saleable value, it's tempting to assign a minimal value (e.g. $1/unit) and move on, but this isn't a fair representation of the costs that go into manufacturing clinical products. Entering a value of $1/unit on an import may alert a customs authority and cause the shipment to be held until it is properly valued.

To avoid this risk, it's recommended to value the clinical product using one of the five alternative methods provided by the WTO if the transaction value is unavailable. The most applicable method for valuing clinical products is the computed method which takes the production costs of the goods plus an amount for profits and general expenses. Applying this to the clinical supply chain involves taking the cost of production for the drug substance, the cost of production for the drug product, and secondary packaging and labeling costs and deriving a total per unit or per kit cost of production. Table 21.1 contains an example valuation for a large molecule product.

Table 21.1 Example Valuation of a Large Molecule Finished Product

Description	Value
Drug Substance Total Cost ($ per Batch)	$2,000,000
Drug Substance Yield (grams)	2000
Drug Substance Cost ($ per gram)	$1000
Drug Product Formulation (g per vial)	0.1 (100 mg/vial)
Drug Substance Cost ($ per vial)	$100
Drug Product Total Cost ($ per batch)	$500,000
Drug Product Yield (vials)	15,000
Total Drug Product Cost ($ per vial)	$33
Finished Product Total Cost ($)	$75,000
Finished Product Yield (kits)	15,000
Total Finished Product Cost ($ per kit)	$5
TOTAL CUSTOMS VALUATION ($ per kit)	$138 per vial

In this example, the manufacturing costs for each production stage are taken and then converted into a per unit value. The drug substance, drug product, and finished product values are totaled to arrive at the total customs valuation. Note that costs associated with process and analytical development are not included in the total customs valuation, and no profit margin is added because clinical material is not saleable. Also, keep in mind that the total customs valuation for clinical products will change over time as processes are scaled up to larger volumes and the total cost to manufacture a batch improves. Reviewing and updating the total customs valuations on an annual or semi-annual basis is recommended to avoid overexposure to import duties as the cost of goods to manufacture a product comes down.

To monitor import and export activities within their borders, some countries require importers to register themselves within a national database. While the United States does not have such a requirement, the European Union mandates all importers obtain an economic operators registration and identification (EORI) number.[6] A EORI number may be granted by any of the EU member states, and an EORI number issued by one state is typically accepted for use by all other member states. However, obtaining an EORI number does not imply that a shipment will clear customs. The IoR must have a physical entity within the European Union in order to gain shipment clearance.[7] Other countries, such as Canada which requires importing entities to obtain a business number, also require registration prior to importing products.[8] Economic registration regulations are also subject to change, so it always makes sense to understand what is required with the exporting and/or importing sites.

21.3 Logistics Stakeholders and IncoTerms

There are many stakeholders involved in requesting, coordinating, and executing international shipments of clinical biopharmaceutical goods. At their center, international shipments are a transaction between two parties, a seller who makes the good available for sale and a buyer who acquires the good. It's often the case for shipments of consumer goods that the seller also serves as the exporter (e.g. the shipper) of the good and the buyer also serves as the consignee (e.g. the recipient) of the good. These roles are fluid though. It's not uncommon for a buyer to purchase a good from a seller and then organize a shipment between an upstream exporter and a downstream consignee where all four parties are different entities.

For shipments of clinical biopharmaceutical products, who plays the role of seller, buyer, exporter, and consignee is also determined by the supply chain network of the sponsor. When the sponsor has internal, wholly owned manufacturing and distribution capabilities, the sponsor simultaneously plays the role of buyer and seller, with each of the respective manufacturing and/or distribution sites playing the role of exporter and consignee. When the sponsor outsources manufacturing and distribution capabilities to third parties, the sponsor plays the role of the buyer, the upstream manufacturing/distribution site plays the role of seller/exporter, and the downstream manufacturing/distribution site plays the role of consignee. In both cases, the sponsor, acting as the buyer of the good, is responsible for coordinating the shipment between the exporter and the consignee.

Below is a summary of the other stakeholders involved in international logistics and their respective responsibilities in the shipping process:

- *Transportation Service Provider (TSP)/Freight Forwarder (FF)*: The TSP, also known as a freight forwarder, organizes the physical logistics of the shipment and may also help coordinate and review shipment documents. These activities include scheduling transport of the shipment to and from airports, ensuring qualified containers and temperature monitors are available to appropriately store and monitor the shipment, and tracking the shipment from pickup to delivery. Some specialty TSPs and FFs also offer customs brokerage and indirect representation capabilities to provide an end-to-end service offering for each shipment.

- *Carrier*: The carrier picks up the shipment and transports it from the exporter to the consignee. In some cases, the TSP and carrier are synonymous. In other cases, the TSP outsources carrier responsibilities to third-party logistics providers that are located in the exporter or importer's location.
- *Airline*: Air travel is the preferred mode of transport for international clinical biopharmaceutical shipments. Expiry period limitations mean there's no time to waste with sea-based modes. For air-bound shipments, the carrier gives the shipment to an airline for transport from airport to airport. The airline has custody from receipt of the shipment until it's delivered to the carrier at the destination airport.
- *Customs Broker*: Customs brokers facilitate the movement of goods out of the exporting country and into the importing country. Customs brokers review shipment documentation to ensure the information provided is acceptable to gain customs clearance. This includes reviewing the HTS classification of the good, the valuation of the good, the IoR, and other applicable local regulatory requirements.
- *The Importer of Record (IoR)*: Listing an IoR is required for all international shipments. The IoR is legally responsible for the imported goods and, for this reason, many countries mandate the IoR to have a physical entity in the country or region of import. The European Union, Latin America, and China all share this requirement. The IoR dictates who is responsible for: communicating with customs authorities (in partnership with the customs broker, if applicable), paying duties and taxes on the shipment, collecting applicable import licenses for the shipment, and bearing the financial risks of the shipment. If the sponsor does not have a physical entity in the country of import, they can find a third party or an indirect representative to play this role on their behalf. Many contract research organizations (CROs) offer IoR services because they have local entities where clinical trials are conducted. When a CRO acts as the IoR, it's important that the clinical supply chain function overseeing distribution should have visibility to this if issues arise with getting customs documents prepared and approved.
- *Indirect Representative*: An indirect representative plays the role of the IoR on behalf of the sponsor and takes on the responsibility of paying duties and taxes, collecting import licenses, and general regulatory compliance. However, the IoR remains financially and legally responsible for the shipment. A power of attorney agreement is necessary to pass these responsibilities from the IoR to the indirect representative.

Ultimately, it's up to the sponsor to understand who plays each role in the shipment process. Multinational biopharmaceuticals with internal supply chain networks may have different sites and functions that play the role of buyer, seller, exporter, consignee, customs broker, and IoR because they have physical entities established in the countries of import and export. These companies also typically have internal logistics departments that coordinate shipments with TSPs as needed and global trade compliance departments that specialize in international customs. Global trade compliance is typically responsible for ensuring all local customs regulations are adhered to with each shipment, so they play a critical role in facilitating shipment clearance.

However, for smaller sponsors that rely on third-party contract manufacturing and distribution networks, there might be a unique supplier each playing the role of seller, exporter, consignee, customs broker, and IoR. Managing all of these stakeholders can be difficult to sort out without an internal logistics and/or global trade compliance department. For this reason, it's strongly recommended to find a TSP that specializes in facilitating international biopharmaceutical shipments from initiation to delivery who can provide resources and expertise. Typically, contract manufacturers have internal logistics departments, and these groups can also provide insight on import documentation requirements and their shipment processes.

The responsibilities of the buyer and seller are determined by the international commercial term (IncoTerm) selected for the shipment. IncoTerm assignment is negotiated between buyer and seller, and they lay out the point of ownership transfer between buyer and seller, the cost obligations, risk obligations, and insurance obligations of each party. IncoTerms are "a set of 11 internationally recognized rules which define the responsibilities of sellers and buyers ... including who is responsible for paying for and managing the shipment, insurance, documentation, customs clearance, and other logistical activities.[9]" The 11 IncoTerms are organized on a sliding scale of responsibility from the ExWorks (EXW) IncoTerm which places maximal obligations on the buyer to the Delivered Duty Paid (DDP) IncoTerm which places maximal obligations on the seller. The most common IncoTerms used in biopharmaceutical shipments are

- *ExWorks (EXW)*: Under EXW, the seller makes the goods available for shipment. The transfer of ownership occurs when the buyer's carrier places the goods on the mode of transport. All other shipment obligations fall on the buyer including arranging the logistics, gaining import/export clearance, and carrying the risk of the shipment through delivery.
- *Free Carrier (FCA)*: Under FCA, the point of transfer occurs when the seller loads the goods onto the means of transport. The seller is also responsible for clearing the goods for export, but all other shipment obligations fall on the buyer.
- *Delivery at Place (DAP)*: Under DAP, the point of transfer occurs when the seller delivers the goods, ready for unloading, at a predetermined location. The seller is responsible for arranging the logistics, gaining the export clearance, and carrying the risk of the shipment through delivery. The buyer is responsible for gaining import clearance.
- *Delivery Duty Paid (DDP)*: Under DDP, the point of transfer occurs when the seller delivers the goods at the disposal of the buyer ready for unloading and transport to the final destination. The seller is responsible for the same activities under DAP and for gaining import clearance of the goods.

The stakeholder that determines the IncoTerm also depends on the sponsor's supply chain network. For sponsors with an internal supply chain, they decide the IncoTerms that help the shipment process function smoothly. If an exporter has a logistics department able to facilitate shipments, then DDP may be optimal to place the responsibility of obtaining import and export licenses on that exporter. Conversely, if an importer has a customs broker on site, then EXW makes sense to place the shipment responsibility on that importer. For sponsors with a virtually outsourced network of manufacturers, the exporter typically dictates the IncoTerm to limit the amount of risk they're exposed to during the shipment. It's always the sponsor's responsibility to review and understand the assigned IncoTerm and the implications it has for arranging the shipment and insuring the shipment.

21.4 The International Shipping Business Process

Like many of the supply chain processes described earlier, the complexity of international shipping warrants development of a business process to ensure the steps, roles, and responsibilities are clear to all stakeholders. The objective of the international shipping process is to successfully prepare

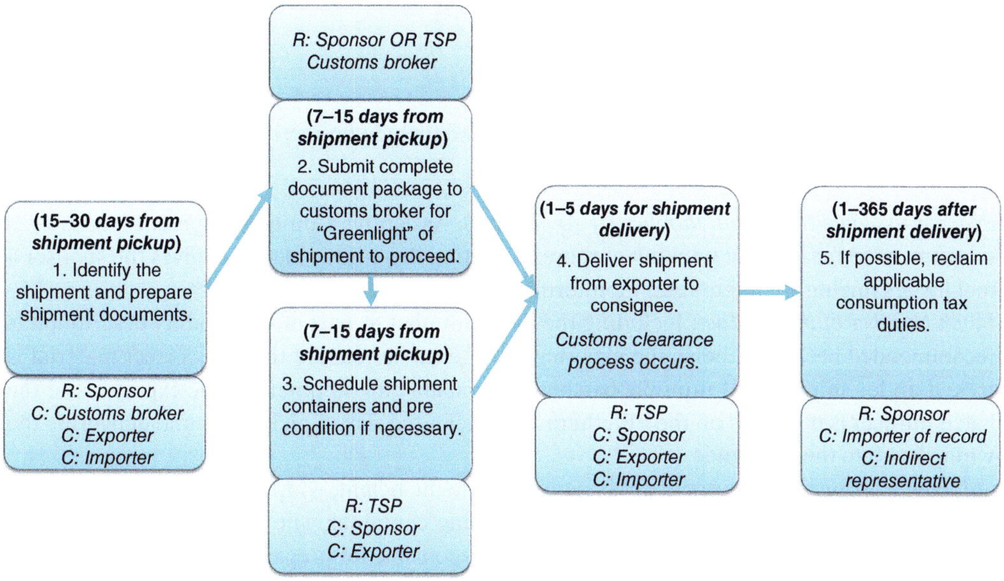

Figure 21.1 International Shipment Process for Biopharmaceutical Products

and deliver each shipment in a compliant manner while simultaneously limiting cost and tax exposure to the greatest extent possible. In global organizations, the logistics department typically owns this process, but for emerging biopharmaceuticals, this responsibility may fall on the clinical supply chain team. Either way, the process owner has the accountability to develop a process map and document which party plays which role in supporting the sponsor's international shipments. Similar to direct-to-patient supply chains, the timing of each process step is critical in international shipping. For this reason, the business process map outlined in Figure 21.1 captures the steps in the shipment process and when those activities should take place relative to pickup and delivery of the shipment:

The shipping process begins with identification of a batch of material that needs to move from an upstream node of the supply chain to a downstream node. Shipment identification needs to happen several weeks before the batch's desired delivery date because of the time required to draft and finalize the shipment documents, obtain the approval to ship from the customs broker, prepare the physical elements of the shipment, and deliver the product. Within the sponsor organization, it's usually the clinical planner who is responsible for shipment identification. If the planner also plays the role of logistician, then they should initiate the next steps in the process. If logistics is a different function from clinical planning, the planner should notify the appropriate party in logistics to initiate the next step.

Following shipment identification, the second step is the preparation of the shipment documents. The single most important document is the proforma invoice (PFI), which is common to all international shipments. The PFI contains, at minimum, the following elements:

- Exporter and consignee names, locations, and points of contact.
- IoR name, location, and point of contact.
- TSP and/or the carrier responsible for delivering the shipment and point of contact.
- A description of the goods being shipped including the molecule name and phase of production (e.g. Molecule ABC Drug Substance).

- The quantity of goods contained in the shipment.
- The total valuation of the shipment.
- HTS code and IncoTerm assignments.

Below is a list of additional points that are helpful to include on the PFI for biopharmaceutical shipments:

- *Sponsor-assigned Shipment Number (If Applicable)*: Assigning a unique shipment identifier is recommended for internal tracking purposes. Creating the unique identifier gives the sponsor a tool for organizing shipments and monitoring them through to delivery.
- *Batch Number of the Product*: Including the batch number as part of the product description is recommended because it helps the exporter and the carrier ensure that the correct material is picked up for shipment. If multiple batches are available for shipment at the exporter and the batch number is not listed on the PFI, there is an increased risk of picking up and delivering the wrong batch to the consignee.
- *Shipment Dimensions*: Shipment dimensions include the height, width, length, and weight, and the TSP needs these to identify a workable shipping container. The dimensions are typically provided by the exporter of the shipment who is responsible for preparing the goods.
- *Product Storage Condition*: The product storage condition also helps the TSP select an appropriate shipping container that is qualified to store the product at the desired condition.
- *Hazardous Goods Declaration*: If the product has been declared a hazardous good, this should also be listed on the PFI so it can be managed appropriately through the customs clearance process.
- *Other Information*: Any other information that's pertinent to customs clearance or to the importer's receipt process should also be included in the PFI. Examples include the EORI number, applicable import license numbers, and purchase order numbers.

The PFI is only the first of many shipment documents required to obtain a customs preclearance. Other documents include a certificate of analysis for drug substance and drug product batches to indicate the material has passed specifications and the BSE/TSE statement. BSE stands for bovine spongiform encephalopathy, and TSE stands for transmissible spongiform encephalopathy. BSE and TSE are degenerative brain disorders that impact cattle livestock. The BSE/TSE statement declares the product to be free from material of animal origin and therefore a minimal risk to passing BSE/TSE to people.

Imports into the United States require additional documents that must accompany all biopharmaceutical shipments. The first document is the end use letter which requires the sponsor to declare whether or not the import will be consumed by human subjects in the United States. If the product is intended for human use, an approved investigational new drug (IND) application number must be provided within the end use letter for reference. Without an approved IND, the shipment will be put on hold by US Customs until an approved IND is made available. Another document that's required if the product contains any material of animal origin is a US Department of Agriculture (USDA) permit.

In the European Union, a summary declaration and the single administrative document (SAD) are required when importing products from outside the EU into the EU. Use of these documents is harmonized across all EU member states, which simplifies the process. Finally, all applicable export and import licenses must be available. While import licenses were covered extensively in Chapter 11 – Finished Product Supply Planning: The Distribution Plan, they also apply when moving drug substance or drug product into a country that requires a license. Some countries also require an export license to move biopharmaceuticals out of their nation.

Once all the documents required for customs clearance are ready, they should be made available for review to a customs broker or to a global trade compliance expert located in the country of import. This is colloquially known as the "green light" step because the shipment should not be initiated (e.g. it has a "red light") until this review has been completed. If the shipment arrives at the border and the customs agent detects a problem with the shipment documentation, they will block importation of the shipment until the documents are corrected. Typically, the IoR facilitates the green light process with a customs broker or internal experts, but the sponsor has the ultimate responsibility to ensure this step has occurred.

Meanwhile, as soon as a draft version of the PFI is ready, this is sent to the TSP to begin the third step of the process, preparing for shipment pickup. The first element is selecting a shipping container to hold the product through transport. Biopharmaceuticals must be held at their specified storage condition through the duration of the shipment, and for cold chain products, this means the shipping container may need to be preconditioned for several hours or days before the pickup date. Second, temperature monitors must be included with each shipment. Temperature monitors are typically provided by the TSP, but they're placed into the container by the carrier at pickup. It's important that explicit instructions are given to the carrier and to the exporter for how the temperature monitors are to be handled so accurate temperature data is recorded and unnecessary excursions are avoided. Combination devices exist that have GPS and temperature monitoring capabilities combined into a single unit. These generate real-time tracking and monitoring for shipments and are recommended for high-value shipments of drug substance and drug product. Finally, scheduling the shipment from pickup through delivery also requires careful planning. Shipment pickup should be timed to occur after the green light process has completed, and potential logistical impediments such as federal holidays and plant shutdowns at the consignee need to be avoided. While the TSP prepares for shipment pickup, they also issue the house airway bill (HAWB). The HAWB document captures the terms and conditions of the shipment as specified by the TSP, and it has a unique airway bill number associated to it that allows the TSP and the sponsor to track the shipment if a real-time GPS monitor is not used in the shipment.

Delivering the shipment is the fourth step, and it usually takes several days or even weeks to complete. First, the shipment needs to be picked up from the exporter and then transported to the airport. The shipment may sit at that airport for several hours before it makes it onto an outbound flight toward the consignee's home country. Multiple flights may also be necessary depending on the distance between the exporter and the consignee. Imports into the United States are categorized into two entry types, consumption entries and "other" entries.[10] Biopharmaceutical products are considered formal entries and therefore require a bond which is posted by the IoR, the TSP servicing the shipment, or the customs broker. Other entry types are for products not ultimately intended for consumption in the United States. A common shipment type categorized as "other" is IFE, which is detailed in the next section.

After the shipment arrives in the consignee's country, it's held for customs review and clearance. A fast clearance occurs within 24–48 hours, but this process may take several days depending on the country and the volume of imports into that country. In the EU, some countries employ random customs documentation checks where every detail on each document is scrubbed top to bottom. Once customs clearance is granted, the shipment is then delivered to the consignee by the carrier.

Assuming the shipment arrives safely at the destination location, the fifth and last step in the process is to collect the proof of delivery (POD) document and to pay outstanding duties and taxes. When the sponsor is listed as the IoR, they are billed directly, and when another party acts as IoR, the duties and taxes are typically passed through to the sponsor for repayment. At this point, the recovery process begins for applicable consumption taxes. It's a best practice to archive the PFI, HAWB, and POD of every shipment to aid in the capture of recoverable duties.

To summarize, success in international logistics largely depends on preparation. Initiate the process at least 3–4 weeks ahead of the desired arrival date. Research and prepare the required customs documents. Work closely with the TSP to guarantee the proper containers are available and the shipment schedule is feasible. Track the shipment through delivery and archive all applicable documentation. But, even with the most carefully laid plans in place, random disruptive events happen, and there are so many ways a shipment can go wrong. Inclement weather delays a flight. A carrier forgets to put a temperature monitor in the container. An airline mislabels a shipment for the wrong destination. A temperature excursion renders the product not fit for use. A customs agent discovers an error in the documents and puts the shipment on hold. When something bad happens to a shipment, it's important to have a contingency plan in place. Contingencies include having stability data available to justify temperature excursions or having a retest strategy to check the quality of the product at the receiving location.

The worst outcome though is for a shipment to be lost altogether. When this happens, the first question asked is whether or not there is sufficient available inventory to avoid a stockout situation. It's never wise to place all available drug substance or drug product inventory into a single shipment unless it's absolutely necessary. The only time it doesn't make sense to split a batch up into two shipments is if the quantities are so small that they can't be processed into a downstream batch or if there is sufficient inventory in the supply chain to absorb the lost shipment without interrupting supply. These decisions are left in the hands of the clinical planner responsible for initiating the shipment.

21.5 Import for Export, Free Trade Zones, and Special Economic Zones

A subcategory of international shipments commonly used in the biopharmaceutical industry is IFE. The Food and Drug Administration (FDA) defines IFE as "importation of a product that is unapproved or otherwise does not comply with FDA laws and regulations if it is coming into the United States for further processing and ultimately exported out of the United States.[11]" Put another way, IFE allows a sponsor to import drug substance or drug product into the United States for further processing without an approved IND application in place. There are strict documentation requirements for IFE shipment including a statement of intent, chain of possession, and certificates of analysis. The statement of intent acts similarly to the end use letter in that it declares the product will be exported from the United States prior to use. The chain of possession documents show that the sponsor has complete visibility into the end-to-end supply chain of the product. Finally, IFE shipments must also be set up as a temporary importation under bond (TIB) entry type. The TIB entry allows goods to enter duty free, but they must be exported within 1 year or destroyed at the end of this period.

Also note that finished products do not generally qualify for IFE because no further processing takes place on these materials. That means both IMP and non-IMP finished products manufactured in another country cannot pass through the United States on the way to their final destination. The exception to this rule is if the finished product materials pass through an FTZ or a bonded area.[12] FTZs are fenced in, duty-free areas where storage and reexport distribution is facilitated. Bonded areas are similar to FTZs, but they may also allow for further processing of goods. Before attempting to import biopharmaceutical products to an FTZ or bonded area, confirm the shipment strategy, allowable activities, and documentation required to stay compliant within those zones through the green light process.

As a customs union, the EU enables the free movement of goods between member states, and therefore IFE is allowed when goods are moving from one member state to another. The EU also permits imports for other countries to be further processed without an approved regulatory application for those goods. However, all imports into the EU are subject to VAT, and the reclamation process becomes increasingly complex when the product is imported for upstream processing (e.g. drug substance or drug product) or exported to a country outside the EU for downstream processing (e.g. finished product) and then reimported to the European Union. VAT reclamation processes require tracing the good to its final point of consumption within the EU, and this is particularly challenging for global clinical trials where the finished product is distributed to multiple sites and countries within the region.

China and India each have significant upstream biopharmaceutical manufacturing industries and complex import/export regulations that are continuously evolving. Both countries also leverage the concept of SEZs to facilitate economic activity within their borders. SEZs are geographically delineated areas that offer benefits to investors and participants and have a separate, typically duty-free customs area.[12] FTZs and bonded areas are a subtype of SEZs. The rules and regulations around SEZs are complex and can vary dramatically from country to country. If the clinical supply chain relies on manufacturing in China, India, or other countries outside the United States and the European Union, consult a reputable TSP, customs broker, or other global trade compliance expert to understand the import/export requirements and regulations.

21.6 Key Takeaways

1) International logistics is the discipline of shipping goods from one country into another country. Logistics for clinical biopharmaceutical products is complex for three reasons: they are often shipped internationally subjecting them to an array of tax and compliance regulations; they are high-value products without an obvious valuation because they aren't for sale; they require constant temperature monitoring and control which increases the pressure to deliver them on time or risk losing the shipment.
2) Clinical biopharmaceutical products are potentially exposed to two types of import taxes, tariffs and consumption taxes. Tariff rates are country specific and determined by the HTS code. Consumption taxes such as the VAT and the GST are charged in almost every country except the United States. Consumption taxes are typically recoverable, but this process is lengthy and complex. The European Union also requires an importer to obtain an EORI number to enable imports to proceed.
3) The four primary roles in international logistics are the buyer, the seller, the exporter, and the consignee. Other stakeholders include the TSP, carrier, airline, customs broker, IoR, and indirect representative. Determining which parties play which role is the sponsor's responsibility for each shipment.
4) IncoTerms assign responsibilities and obligations between the buyer and seller of each shipment. These obligations include who bears the cost of the shipment, who pays for insurance, and who arranges for export/import clearances. There are 11 IncoTerms that are organized on a sliding scale of responsibility between the seller and the buyer, and the four most commonly used in clinical biopharmaceutical shipments are EXW, FCA, DAP, and DDP.
5) Developing a business process for international shipments is helpful to clarify the sequence and timing of activities and to identify the applicable stakeholders responsible for each step. The business process has at least five distinct steps: shipment identification, preparation of the

shipment documents, preparation for the shipment pickup, customs clearance and delivery, and archiving all applicable documents at shipment completion.

6) The most important shipment document is the PFI which is a common document to all international shipments. The PFI outlines basic shipment information including the exporter, consignee, IoR, description of goods, quantity of goods, shipment valuation, and the HTS/IncoTerm assignments.
7) Never place all available drug product or drug substance inventory into a single shipment unless absolutely necessary. Shipments sometimes go wrong, and the clinical planner never wants to find themselves in a position where all inventory is lost because of a weather event or a temperature excursion.
8) IFE is a subcategory of international shipments that are commonly used for biopharmaceutical imports into the United States. IFE allows for importation of a product that is otherwise unapproved if that product will ultimately be exported out of the United States without getting consumed. The European Union allows IFE for economic activity taking place within its borders. China and India leverage SEZs to facilitate manufacturing and distribution within their borders. Customs regulations change frequently though, so it's always recommended to work with reputable TSPs and customs brokers and to use a thorough green light process for every shipment.

Notes

1 https://www.cbp.gov/about/history/1789-first-congress-provides-customs-administration.
2 https://hts.usitc.gov/.
3 https://taxfoundation.org/taxedu/glossary/value-added-tax-vat/.
4 https://taxatione-customs.ec.europa.eu/vat-refunds_en.
5 https://www.wto.org/english/tratop_e/cusval_e/cusval_info_e.htm.
6 https://taxation-customs.ec.europa.eu/customs-4/customs-procedures-import-and-export-0/customs-procedures/economic-operators-registration-and-identification-number-eori_en.
7 https://trade.ec.europa.eu/access-to-markets/en/content/guide-import-goods.
8 https://www.cbsa-asfc.gc.ca/import/guide-eng.html.
9 https://www.trade.gov/know-your-incoterms.
10 https://www.fda.gov/industry/import-basics/common-entry-types.
11 https://www.fda.gov/industry/import-basics/import-export.
12 https://pedl.cepr.org/content/i-concept-and-definitions-sezs.

22

Clinical Supply Chain Planning Tools and Technology

22.1 Data, Technology, and Clinical Supply Chain Planning

Clinical trial supply chain planning and execution relies on data and information technology (IT) platforms. From a supply chain perspective, the most important technology platform is the one that supports the demand and supply planning process. In clinical-stage biopharmaceutical companies with relatively small supply chain organizations, the default planning tool is almost always a spreadsheet-based software program. When a sponsor commercializes a product and their clinical portfolio expands, they sometimes move away from offline spreadsheets to more advanced planning systems that enable greater integration and collaboration across the organization.

There are other technology platforms that impact clinical supply chain management. One of these, the randomization and trial supply management (RTSM) system, had an entire section, Chapter 6 – RTSM and Its Role in the Clinical Supply Chain, dedicated to it. The RTSM system captures demand-related data like subject enrollment, subject weights, and site stocking requirements. Most commercial-stage companies also have an enterprise resource planning (ERP) system as part of their IT landscape. ERP systems are a data warehouse for business operations including financial transactions, manufacturing, and inventory management. ERP systems typically contain supply-related data like product descriptions, batch history, and on-hand inventory (OHI) levels. Demand and supply planning tools rely on inputs from RTSM systems and, for companies that have them installed, ERP systems.

Technology solutions don't exist in a vacuum though. One trap a supply chain leader can fall into is to view technology as a panacea for supply chain optimization. The thought process goes, "A clinical supply chain technology that seamlessly integrates demand and supply data to automatically generate optimal production plans could simultaneously solve for stockout and excess waste risk." This viewpoint ignores three realities. First, supply chain solutions rely on timely, accurate data inputs to generate timely, accurate data outputs. When incorrect or outdated data is used in the tool, the resulting outputs don't make sense. This depletes trust in the system and reduces user adoption of the solution. Second, technology platforms are operated by people, and anyone interacting with the solution needs to understand how it works to maximize its potential benefits. When people don't understand how to use the technology, they either use it incorrectly and generate poor outputs or become frustrated and turn away from it. Third, the technology needs to fit alongside and complement the business process framework of the company. When business processes are ignored, technology becomes an impediment to business process execution.

Supply Chain Planning for Clinical Trials – A Practical Guide, First Edition. Ryan Mills.

Clinical supply chain leaders need to consider all three of these elements: the data that goes into and comes out of their planning system, the people who interact with the system on a regular basis, and the business processes the system is designed to support. If any of these areas are neglected, it's a recipe for disaster. A holistic supply chain planning technology strategy is a smart way to ensure user adoption and get the best results from the tool. A comprehensive strategy has each of the following components:

- For pre-commercial companies or companies without the resources to invest in a more advanced solution, development of an initial spreadsheet-based demand and supply planning model.
- A data governance strategy to ensure the planning model has accurate data inputs from the RTSM and, where applicable, the ERP system.
- A training strategy for teaching users how the planning model works, how to input data, and how to interpret the model's outputs.
- A business process integration strategy that connects the planning model to the supply chain planning business processes including demand and supply planning, operational clinical demand and operations planning (CD&OP), and strategic CD&OP.
- For companies looking to make a transition out of spreadsheets, a selection and implementation strategy for moving the organization into a more advanced planning solution.

Figure 22.1 provides a visual illustration of this strategy. The first step in a pre-commercial company is to build out a demand and supply planning model using a spreadsheet-based program. This model should then be supported by a simple data governance strategy, a training strategy, and a business process integration strategy. As the company matures and considers transitioning to a more advanced planning solution, the technology strategy shifts to selection and implementation of that new tool. When this happens, the data governance, training, and business process integration strategies also need to evolve to enable the new solution.

Over the course of this chapter, we'll cover these elements beginning with development of a spreadsheet-based model. In this section, we'll explain why spreadsheets are such a powerful tool and present a framework for designing an end-to-end demand and supply planning engine in spreadsheets based on the concepts introduced earlier in this book. Then, we'll turn our attention to creating a data governance strategy. No planning solution, including a spreadsheet-based model, can be used effectively without a data governance strategy in place to ensure the quality and integrity of the data used in supply chain planning. Following that, we'll discuss training strategies for the selected tool and offer pointers for increasing user adoption and maintaining

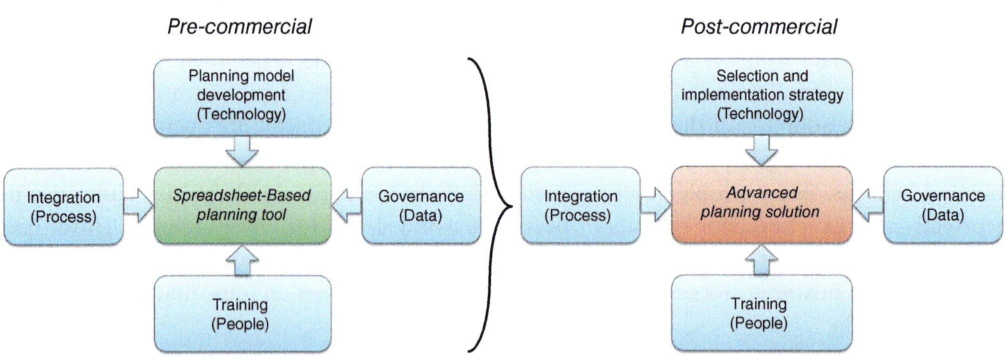

Figure 22.1 Holistic Technology Strategy for Clinical Supply Chain Planning

version control of the planning solution. Then, we'll discuss the business processes that the planning solution should support. Finally, we'll review selection and implementation strategies for replacing spreadsheets with a more advanced planning tool that offers improved collaboration, integration, and scenario-modeling capabilities. We'll conclude the chapter with a look at future technology trends that may impact and improve how planning systems are developed and used across the industry.

22.2 Demand and Supply Planning in Spreadsheets

Spreadsheet-based software programs enable data entry, storage, analysis, and visualization, all of which are necessary for supply chain modeling. Spreadsheets are also widely used for supply chain management in small businesses because of their relative simplicity to program, their flexibility to accommodate a wide variety of business use cases, their powerful data analytics tool sets, and their cheap cost to use. When a user opens a new spreadsheet file, they typically see a ribbon of functionality at the top and an array of empty cells to work with. A spreadsheet's simplicity as a database starts with the column/row structure. The intersection of each column and row is a cell where data is entered or formulas are written to manipulate data that's contained in other cells. A spreadsheet's flexibility stems from the user's ability to exert complete control over this data entry and formula-building process to suit the user's own purposes. A spreadsheet's power comes from the available formulas and analytical tools the software program offers. The most popular spreadsheet applications have dozens of standard formulas that are commonly used in business operations management. These include

- *SUM*: Adds the quantity of a group of cells together.
- *SUMIF/SUMIFS*: Adds the quantity of a group of cells but only selects cells that fit predefined criteria.
- *AVERAGE*: Averages a group of cells together.
- *MIN*: Takes the minimum value of a group of cells.
- *MAX*: Takes the maximum value of a group of cells.
- *CEILING*: Places a maximum value on the output of a formula.
- *FLOOR*: Places a minimum value on the output of a formula.
- *IF*: Makes logical true or false comparisons between values.
- *OR*: Tests multiple conditions to determine if any of them are true.
- *AND*: Tests multiple conditions to determine if all of them are true.
- *INDEX*: Returns a value or a reference to a value from within a range of cells.
- *MATCH*: Searches for a value within a range and then returns the relative position of that value.
- *VLOOKUP and XLOOKUP*: Searches a range or array and returns an item corresponding to its match.

When a user opens a new spreadsheet (e.g. a new document), there are tabs at the bottom, each of which represents a worksheet within that document. New sheets can be added as needed. The initial challenge with spreadsheets is to transform an empty workbook into a functional demand and supply planning model for clinical trials. Fortunately, the components of a workable planning model have been covered extensively in previous chapters. The logical place to begin is with demand plan creation. Recall, from Chapter 8 – Finished Product Demand Planning: Basic Concepts, the demand plan elements: the stockkeeping unit (SKU), the depot location, the demand forecast period (e.g. weeks or months), the subject demand forecast, the enrollment curve, overages, and site stocking demand requirements. All of these are needed to arrive at the total demand forecast, and spreadsheets naturally enable each of them:

- *SKU–depot Management*: Individual tabs within a workbook can be created to represent unique SKU–depot locations at the finished product level. Note that additional tabs can also be created to represent unique SKUs at the upstream planning level.
- *Demand Forecast Period*: Each column can represent a unique week or month depending on which demand forecast period the model is designed to support.
- *Subject Demand Forecast*: Each row can represent an individual subject enrolled in the trial. The individual subject demand curve can be plotted across an entire row with each cell representing a single dispensing visit or multiple dispensing visits depending on the schedule of assessments. Multiple subjects can also be aggregated into a single row.
- *Enrollment Curve*: The enrollment date of each subject can be timed by placing the initial dispensing visit in the appropriate week or month, which creates a natural enrollment curve.
- *Overage*: Additional demand planning inputs like overage and average subject weight can be added in unique cells and then referenced in formulas to multiply overages against total subject enrollment.
- *Site Stocking Demand and Total Demand*: Additional rows can be used to plot site stocking demand and then add this to subject demand to arrive at total demand.

Similarly, spreadsheets also enable the key principles of finished product supply planning including creation of a min–max-based predictive resupply algorithm based on weeks- or months-of-supply, the ability to toggle safety stock and order target values as needed, and the ability to segregate suggested orders from actual orders. After the demand forecast has been built, additional rows can be added for each of the attributes needed to create a supply plan at that SKU–depot combination. These rows include OHI, actual production orders, planned production orders, weeks-of-supply, safety stock values, and order target values.

There are two concepts to note here. First, the planned production order formula should reference future demand over the desired safety stock and order target values. Utilizing an INDEX formula that also references the safety stock and order target values to create ranges of time over future demand returns a suggested production order quantity. Second, the model should always take an actual production order value over a planned production order value. This can be accomplished using a conditional IF formula which gives the user control to enter production order quantities that supersede the suggested order values from the production plan formula.

Spreadsheets are also powerful enough to incorporate expiry planning and distribution planning concepts into the same SKU–depot tabs. Using nested IF statements, it's possible to disaggregate the production plan into the individual lots that compose total OHI. This capability enables a user to input and update expiry dates for each lot and to zero out inventory of a specific lot when its expiration date is reached. Building a lot-specific inventory section also enables more accurate finished product distribution planning. Similar to the production planning section, the model can also be programmed to take actual lot-specific values instead of projected lot-specific values. This gives the user control to enter an inventory quantity for a specific batch on a specific date which in turn enables the user to actualize inventory within the model. This section can also be used to make distribution allocation decisions at the finished product level and production allocation decisions at the upstream level. Refer to Figure 10.3 – Production Plan Table with Batch Data Populated for an example of this model structure. The only element missing from Figure 10.3 is a table to manually overwrite the lot-specific inventory values.

With respect to upstream planning, the same concepts used in the finished product SKU–depot tabs can be applied to build out the planning model for an upstream SKU. Instead of placing subject demand in individual rows, unique downstream SKUs that consume upstream material are entered into individual rows. Spreadsheets can also handle all of the various small molecule, large molecule, and cell therapy supply planning inputs that influence yields and bill of material (BOM) conversion quantities. These can be adjusted as needed over specific time periods (e.g. ranges of columns) to account for manufacturing process improvements. The same approach that's used to create a finished product production plan can be applied to a drug substance or drug product production plan. This includes building lot-specific inventory rows for each batch of upstream material to enable allocation of upstream batches to downstream production. Refer to Figure 12.8 – Large Molecule Drug Substance Upstream Production Plan Table for an example of this model structure.

Finally, basic formulas can be written to summarize production and distribution plans in a concise way. To do this, create a summary section at the bottom of each SKU–depot tab that indexes the actual production or distribution plan. Each of the key data points in the production/distribution plan summaries is available including the SKU description (from the SKU–depot description), manufacturing site (from the SKU–depot description), batch numbers (from the lot-specific inventory section), quantities (from the production/distribution actual row), and date of manufacture or shipment (from the column header). Other inputs to the production and distribution summary table such as the unit of measure, allocation decisions, and the process version can be added manually to the summary tables. Budget summaries can also be built directly off the summary reports using VLOOKUP or XLOOKUP formulas to pull packaging and distribution orders into the tables.

Given its ability to effectively model the clinical trial supply chain, it's no surprise that spreadsheets are a widely used planning tool in the biopharmaceutical industry. However, there are many potential drawbacks to consider when using spreadsheets. The first potential issue is that a planning model needs to be created before it can be used by a clinical planner. Building all the formulas that go into creating an end-to-end planning model takes time and expertise, and that's not always available. Most clinical planners in clinical-stage companies are responsible for a wide variety of activities beyond planning, and they may have come from organizations that used an advanced planning solution so they may not have a lot of experience creating models in spreadsheets. Coding an entire model is a daunting task because there are so many formulas, and misplacing even one character can lead to an incorrect output.

The second drawback is that spreadsheets do not integrate well with other systems like RTSMs, ERPs, and other planning software solutions. These systems warehouse the inputs for planning tools and, without systematic integration, data has to be downloaded from them and then manually uploaded into the spreadsheet-based planning model. This process is slow and risky. It takes time to download data from one system, manipulate that data so it fits into a spreadsheet-based format, and then map it into the spreadsheet-based solution. It's easy to make formatting or copy–paste errors that produce an incorrect output or break the model's formula structure.

Third, as the clinical portfolio and the team of clinical planners expand, different versions of a spreadsheet-based model tend to pop up over time, and control over how the model is used tends to get lost. When each planner has their own version of a spreadsheet, ensuring the integrity of each model becomes more difficult. Each planner also has to figure out how to import data from the RTSM and/or ERP system on their own, and each planner ends up providing summary reports, production plans, and distribution plans in a different way. All of this places an extra burden on the clinical supply chain function and makes the clinical planner's job more difficult. In the most extreme cases, a poorly designed spreadsheet can force a clinical planner to build an entirely new model from the ground up. Most people don't have the background in clinical supply chain and sufficient expertise to do this. The final problem with this is that when a planner has designed their own model and they leave the organization, nobody knows how to use that version of the model. It can take weeks of frustration just to figure out what somebody else was trying to do in their plan.

Most small and pre-commercial planning companies are going to use spreadsheets for their clinical planning solution. To mitigate the issues outlined above, there are a few rules of thumb to follow:

- Build a planning model that encompasses the basic functional needs of managing the clinical supply chain including demand planning, supply planning, expiry planning, distribution planning, upstream planning, and summary reporting. Without these elements, clinical planners will either make suboptimal plan, attempt to create a model of their own that fills the gaps, or both.
- Assign someone on the team who is an expert in coding to build the model and become its super-user. Don't ask someone to learn clinical supply chain and spreadsheet coding at the same time or the model will be designed poorly. The super-user can also help the team efficiently troubleshoot formula errors and other mechanical problems within the model.
- Document how to use the model with an emphasis on how to provide data inputs and how to interpret the data outputs. This is where clinical planners need the most guidance, so they understand how to use the model and how to communicate the outputs.
- Limit the number of planning model versions as much as possible. When a clinical planner needs additional functionality, enlist the super-user to provide a recommended solution for that particular model and determine whether all other versions of the model need to be updated with that same functionality. Standardizing the models ensures that everyone works from the same base and that each clinical planner can cover for the others.

Even with these strategies, large biopharmaceutical organizations eventually outgrow the use of spreadsheets as a planning tool. At a certain size, it's no longer practical or efficient to maintain dozens or hundreds of clinical trial plans in a desktop-based software application. Version control becomes impossible, the lack of integration with other systems causes too much inefficiency, and there are more powerful applications available to handle a diverse portfolio of products. Before moving to the selection strategy, we'll turn our focus to data governance and training which are applicable to spreadsheet-based planning models as well as more advanced technology solutions.

22.3 Data Governance

All demand and supply planning systems, including spreadsheet-based applications, rely on current and accurate data inputs to function properly and produce reliable outputs. It's wise to have a **data governance** strategy in place to ensure the data that's loaded into the planning tool meets this standard. Data governance is a data management framework designed to ensure that data is current, accurate, and actively managed in a company's technology platforms. More specifically, data governance is a structured approach to data management that includes data definitions, data rules, and data owners. Most importantly, implementing a data governance program establishes trust between the technology and the user community. This increases user adoption and helps organizations capture the full value of what technology-based solutions offer.

Manufacturing companies, and biopharmaceutical companies in particular, generate a massive amount of data through their business operations. To illustrate this, think about the production of a batch of GMP material. There is data about the material itself: the product name, the product level (e.g. drug substance, drug product, and finished product), the dose for drug product and finished product materials, the packaging configuration, and the unit of measure. There is data about the manufacturing site: the address, the country of manufacture, and the supplier qualification status. There is data about the manufacturing process: the BOMs, the manufacturing steps, and the equipment used. And, there is data about each batch: the quantity produced, the date of manufacture, and the disposition status of the batch. All of these data points are relevant to the demand and supply planning process, and if any of these data points are incorrectly captured or entered into a planning tool, the entire plan may be thrown off course.

A strong data governance strategy establishes which data elements need to be managed, assigns roles and responsibilities for creating and maintaining data, and establishes data management business processes within an organization. A robust data governance strategy has many elements to it. Below are the five elements most applicable to clinical supply chain planning models and systems:

1) *Data Element Definition*: Clearly defining the data elements that are used in a system is the starting point for a data governance strategy. Organizations first need to know what data they need to manage before they can do anything else. Within clinical planning systems, there are numerous data elements that have been identified, and these are summarized in the Appendix.
2) *Source Document or System Record*: The source document or system of each data element is important to note because the same data element often lives in many different documents and systems. For example, the source document for the batch yield is the executed batch record. In companies with an ERP system, that yield is then entered into the ERP which becomes the source system of record. Then, the yield is exported into the planning solution. When there is a conflict between documents and/or systems, it's helpful to understand what the true source of the data point is so these discrepancies can be resolved with the accurate data point.
3) *Data Ownership*: After the data elements and source systems are captured, the next step is to identify the data owners who are accountable for generating and maintaining that data element in the system of record. For example, the quality function owns the batch disposition status data element. As the data owner of this element, quality is responsible for defining the batch disposition options and for maintaining the batch disposition status in the appropriate system of record.
4) *Data Rules*: Data rules define how data is structured within a system, and their primary value is to keep data consistent across users, roles, and functions. This is important when there are multiple users who generate data within a technology platform. If two users enter the same piece of data differently, this is problematic because the data could be interpreted incorrectly by another user or the data may not integrate properly with another system. For example, material descriptions should have clearly defined data rules to ensure they provide sufficient information to clearly describe what the material is without ambiguity or question.
5) *Data Quality Management*: The final segment of a data governance strategy is to establish a data quality management program for reviewing data on a regular cadence. This encompasses both the actual data entered into a technology solution and the data rules and the data owners who are responsible for data maintenance. Data quality management provides a check on the entire data governance strategy and ensures that data governance principles are adhered to across the organization.

A data governance strategy does not need to be complex to be effective. In smaller organizations, the clinical supply chain function should at minimum identify the data elements used in a spreadsheet-based planning model and how that data should be entered and updated. This information is excellent background for training materials on how to operate the model from actualization through the entire demand and supply planning business process. Creating this documentation is also an ideal task for the planning model super-user who understands how the model was built and should be used and who has the ability to teach other team members about these topics.

As biopharmaceutical companies grow and their technology strategy moves toward implementation of an ERP system, data governance becomes even more important. Typically, the responsibility for data governance shifts from individual functions to the IT team responsible for corporate-wide technology implementations and oversight. The clinical supply chain function needs to stay engaged with the data governance policies, procedures, and rulebook because they influence how data from external systems like an ERP will be made available for input into either a spreadsheet-based model or a more advanced planning tool.

22.4 Training

The next element of a holistic technology strategy is training. Even a well-designed planning model supported by a robust data governance strategy will fail if people don't understand how to use it. Training gives clinical planners the information they need to engage comfortably with the planning model on an everyday basis. A training strategy includes development of training documents to educate new team members on how to use the model; establishing a system super-user who has built the planning model and/or understands the mechanics of how the planning model works; and implementation of a feedback mechanism to improve the planning model over time so changes can be instituted and rolled out to users in an organized fashion.

Training documentation is the first component, and training content should cover both clinical supply chain principles and how those principles are applied in the planning solution. There are multiple ways to approach training document creation. One approach focuses first on theoretical supply chain concepts and then shifts to the application of those concepts in the model. An alternative approach focuses on one supply chain concept at a time and then illustrates how that particular concept is applied in the model. The first approach works best for team members who are relatively new to supply chain management or clinical supply chain and need a broad overview of the space. The second approach works best for more experienced clinical planners who need to learn how supply chain principles they already know are incorporated into the planning model. When new team members are onboarded, their past experiences and skill sets should be evaluated, and training documentation should be customized to meet their needs.

Training should also cover how data migrates from other systems like the RTSM or ERP into the planning model and how the planning model is incorporated into the business process framework of the clinical supply chain function. Demand and supply planning is the most obvious process to cover, but linkages to other processes like operational CD&OP and supply plan handover should also be included. Conversely, general training should not focus on troubleshooting formula or programming issues within the planning model. That level of detail should be limited to the assigned super-user who can step in and fix problems with the planning model as they come up.

The second component of the training strategy is to assign that super-user role. We've previously referenced appointing a super-user to build a spreadsheet-based model, but there's more this role can do. The expertise acquired through mastering model mechanics can also be shared with other team members during the training process. It's particularly valuable to have super-users collaborate directly with clinical planners as they're performing demand and supply planning activities in the model. When this happens, it provides an opportunity for clinical planners to ask the super-user questions in real time and to identify areas where planners may need additional training and where improvements to the planning model could be made.

Instituting a formal feedback mechanism is the third component of the training strategy. Feedback should be sought during the training period and during routine day-to-day operations. Two areas where feedback can be collected is through super-user engagement and through the business process framework. For example, the plan review step in operational CD&OP is an excellent opportunity for clinical supply chain leaders not only to evaluate how well clinical planners are using the planning model but also to identify gaps in the model itself. These gaps can then be passed back to a super-user who can work on updating the planning models to reflect the proposed updates. Communication throughout this process is key, especially when making updates to spreadsheet-based planning models. That's because spreadsheets are not a cloud-based solution, so if models need repair or updates, only one person at a time can make the desired changes. Without strong communication and alignment between the super-user and the clinical planners, planning model updates can be made in one version while the clinical planner uses a different version for demand and supply planning. Reconciling separate versions is a frustrating result because work has to be duplicated to get back to a single version of the model.

22.5 Business Process Integration

Integrating the planning model with the business process framework used to manage the clinical supply chain is the third pillar of an effective technology strategy. Here, integration means that the planning model is designed and constructed to effectively support and enable the business processes which rely on its outputs. In Chapter 15 – Clinical Supply Chain Planning Processes, five business processes were introduced that are fundamental to clinical supply chain management. These were demand and supply plan creation, demand and supply planning, operational CD&OP, strategic CD&OP, and supply plan handover.

Demand and supply plan creation and the demand and supply planning cycle obviously require a planning tool to execute. These two processes are entirely dependent on the model so the design directly impacts a clinical planner's ability to execute these processes in an efficient and effective manner. The planning model needs to provide all the functionality required to appropriately plan the clinical trials that are run by the sponsor. If the planning model doesn't have these capabilities, then clinical planners lose productivity and make suboptimal decisions with respect to preventing stockouts and reducing excess waste. This is also why having the feedback mechanism introduced in the last section is so important. If the clinical portfolio changes or expands to new modalities, the planning model needs to evolve too or planners will face these struggles.

The CD&OP processes also deserve attention when considering the planning model design. Operational CD&OP provides the inputs that enable the demand and supply planning process to happen. The planning model should be designed to capture these inputs in a simple, intuitive manner. Meanwhile, strategic CD&OP relies more heavily on planning model outputs. Where calculations or conversions need to happen, the model should also be programmed to perform these and provide easy-to-understand outputs. The charts and graphs covered in Chapter 14 – Clinical Supply Chain Reports, Budgets, and Metrics that illustrate inventory levels through the supply chain and the total cost of inventory should be available as part of the planning model summary reports. Alternatively, planning model outputs should be compatible with a data visualization tool that enables the reports that feed the strategic CD&OP process. Like the strategic CD&OP, supply plan handover also requires outputs from the planning model. The production plan summaries and distribution plan summaries covered in Chapter 9 – Finished Product Supply Planning: The Production Plan and Chapter 11 – Finished Product Supply Planning: The Distribution Plan should also be programmed into the model.

For companies that use spreadsheets as the platform for their demand and supply planning model, implementing these strategies should not be too difficult. Spreadsheets are infinitely flexible, and adding data points or capabilities is only limited by the creativity and ingenuity of those responsible for building the model. The challenge of business process integration increases when more advanced platforms are considered as a replacement for spreadsheets. Customization opportunities in these tools may be limited, and not every data field or capability may be available to support existing business process requirements. As a result, the business processes may need to be modified to fit what the planning solution can deliver.

22.6 Beyond Spreadsheets: Selection and Implementation Strategies for Advanced Planning Systems

Eventually, biopharmaceutical companies reach a point where the breadth and depth of the clinical supply chain stretches spreadsheets to a breaking point as an efficient tool for demand and supply planning. There are a few potential triggers for this breaking point. The first trigger is the total number of clinical trials a sponsor is actively conducting. While there isn't an exact number of studies where spreadsheets stop working, once a sponsor has more than 50–100 active clinical trials, clinical supply chain leaders should monitor their team's ability to maintain optimal supply plans in spreadsheets. Second, as the team grows in size, version control becomes more difficult. Once a clinical supply chain organization has more than five clinical planners, leaders should also look for different versions of the model cropping up to handle the variety of trials the organization is undertaking. Third, the breadth and depth of the clinical portfolio has an impact on spreadsheets' long-term viability. In the previous section, we noted that when a sponsor diversifies their portfolio into new therapeutic areas or new modalities, the planning models may need to adapt as well. This additional complexity often leads to multiple versions of the model and a loss of organizational efficiency. Finally, the organization's IT infrastructure is another factor. When there is no ERP system for production and inventory management, spreadsheets are usually a natural solution to act as a production and inventory management data warehouse. Once an ERP system is installed though, this opens up a new avenue for data integration and migration that makes an advanced planning tool a more attractive alternative.

To summarize, the more organizations grow, the more evident the flaws in spreadsheets become. Maintaining a diverse set of planning models becomes more resource intensive, the planning process becomes more inefficient, and it becomes more challenging to transition clinical planners from one program to another because they have to learn the nuances of a new program and the nuances of how a different version of the planning model works. Fortunately, there are numerous cloud-based planning applications that attempt to solve for these drawbacks. These solutions can be categorized into two general buckets. In the first category are planning solutions that are extensions of existing software applications like the RTSM or ERP system. In the second category are planning tools that exist as stand-alone platforms but have the ability to integrate directly with systems like the RTSM and ERP to improve data management efficiency.

The first category can be further subdivided into planning tools that are extensions of the RTSM and planning tools that are extensions of the ERP. RTSM-based planning tools are excellent for finished product-level forecasting because of the direct access to demand-related data that RTSMs enjoy. Site activations, projected screening and enrollment, and the individual subject demand curve can all be programmed into the RTSM, and this data can then be transformed into suggested site stocking and depot stocking strategies. The most advanced RTSM solutions are now leveraging artificial intelligence and machine learning programs to simulate enrollment scenarios and provide site and depot demand forecasts to guide inventory management strategies for the finished product supply chain. One drawback of RTSM-based planning tools is that they are not yet an end-to-end planning solution. Upstream supply planning strategies are absent from RTSM-based planning engines, so this needs to be managed through a different planning tool and having multiple planning models and/or systems for one program is less than ideal.

Meanwhile, ERP-based planning systems are an extension of the ERP system. Where RTSM-based solutions have direct access to demand-related data, the ERP system has direct access to supply-related data. Manufacturing records, yields, lead times, and current inventory levels are typically available in an ERP system, and these data elements enable upstream supply planning. Interestingly, ERP-based planning systems also have the ability to integrate directly with RTSM solutions to import finished product demand data. This end-to-end connection between downstream finished product demand and finished product inventory levels via the RTSM and upstream drug substance and drug product supply-related data via the ERP offers a best-of-both-worlds approach. The inefficiency of having to manually export data from one system and import into another system is eliminated, and everything needed for demand and supply planning is available in one system. However, ERP-based planning systems have their own drawbacks. They are typically configured for commercial products and may not have the required fields or capabilities to manage the nuances of the clinical supply chain. Customization may be possible, but making changes to preexisting system architecture is typically expensive to design and difficult to maintain over time.

The second category of advanced planning systems are stand-alone platforms that exist entirely outside the RTSM and ERP. These tools are typically designed specifically with clinical trial needs in mind, and they can be further categorized into demand planning systems, supply planning systems, and end-to-end systems. Demand planning systems are designed to generate study-level forecasts which can then be extracted and loaded into a separate supply planning system. Similarly, supply planning systems are able to generate finished product and/or upstream supply plans for a program. The most efficient type of system though is end-to-end with both demand and supply planning capabilities embedded in the system design.

End-to-end advanced planning systems can also integrate directly with both RTSM and ERP systems to pull demand and supply-related data, so they offer the same supply chain visibility and control advantages that ERP-based solutions do. Additionally, these solutions sometimes offer more advanced stochastic planning capabilities to quickly model a range of clinical trial supply chain scenarios that ERP-based solutions may not have. One disadvantage to using this type of solution is that it's an entirely different system to manage in addition to the RTSM and ERP solutions, so additional human and financial capital may be needed to maximize the potential value of the tool.

When the decision is taken to move from spreadsheets to a more advanced planning tool, developing selection criteria that fit the needs of the organization is paramount. We've already discussed several selection considerations, and they are summarized again below:

- *Capability*: Does the planning solution have the necessary capabilities to execute demand and supply planning comprehensively and efficiently for the portfolio's product modalities and therapeutic areas? Is customization of the planning solution required to add capabilities? Does the planning solution generate the outputs required to support the business process framework? Can the planning solution run stochastic simulations for larger, more complex clinical trials to optimize supply strategies?

- *Integration*: Is the planning solution already a part of an existing system? Can the planning solution integrate with other systems? Will data in another system need to be manipulated in order to successfully integrate with the planning solution?
- *User Interface*: Is the planning solution intuitive to use for demand and supply planning? How simple and efficient is it to upload data for scenario planning if this capability exists? How simple and efficient is it to interpret planning model outputs?
- *Financial and Human Cost*: How expensive is the system to implement? How much does customization cost if customization is required? How much human capital will be required to import or export data? How much human capital will be required to train the team on the system?

Before initiating the selection process, it's wise to treat planning solution providers like RTSM solution suppliers and contract manufacturing organizations. A thorough supplier selection process should be prepared and executed. This means soliciting proposals from multiple potential technology providers, conducting demonstrations of the planning solution, and developing a supplier selection scorecard that captures input from the relevant stakeholders (e.g. clinical planners, finance, and IT) prior to choosing a winner.

After the selection decision has been made, clinical supply chain leaders should revisit the technology strategy. All three areas – data, training, and business process integration – are likely to be impacted by the new planning solution. From a data governance perspective, there may be new data elements to manage, data ownership may need to change hands, and new data rules may need to be written to ensure that inputs and outputs are appropriately structured for the new planning tool. From a training perspective, new training documents will need to be developed that teach the clinical planning team how to use the new tool. One or more super-users should also be identified as soon as the selection decision is made to engage with the solution provider to begin mastering the tool and documenting how it works to support the clinical planning team. Business processes may also need to be modified or updated based on the solution's planning and reporting capabilities. If any of these pillars are neglected, successful implementation and user adoption of the new solution will be put at risk.

From the beginning of the implementation process through the end of the engagement, the solution provider should be treated as a long-term partner by the sponsor. Approaching the relationship through this framework accomplishes multiple objectives. First, there will be an intense amount of knowledge-sharing from solution provider to sponsor about how to set up and execute the demand and supply planning process within the tool. The assigned super-users should quiz the solution provider on the mechanics of the tool and on best practices for implementing the new tool. Second, the planning solution may need customization prior to implementation. Designing custom solutions requires a deep understanding of the solution provider's capabilities and limitations and of the sponsor's data structures and business process framework. Establishing a collaborative working relationship is the only way for each side to get to know each other so compromises can be found. Finally, the solution provider is likely to release updated versions of the planning solution over time. When there is a strong relationship, the sponsor can quickly capitalize on these improvements and more easily integrate them into their business process framework.

22.7 The Future of Technology in Clinical Supply Chain Management

We've kept the focus of this chapter on demand and supply planning systems, but there are several other exciting technological advances that are already having an impact on clinical trial supply chain planning and execution. The first such technology is the digital twin concept. A digital twin is a virtual representation of a physical object, person, or process. Creating a digital twin enables a person to simulate real-world conditions under a variety of different scenarios to identify risks and guide decision-making. The application to clinical trial supply chain planning is obvious. Using a digital twin that mimics the clinical supply chain (e.g. the production sites, predicted yields, and lead times) and the clinical trial design parameters potentially enables a clinical planner to quantify the risk of a lost batch or to identify an optimal site and depot stocking strategy under a variety of enrollment scenarios. Digital twins are even more exciting when combined with stochastic planning principles that assess a wide range of inputs and outcomes. With that vision in mind, today's digital twins are costly to build and maintain. This is especially true for complex system representations like clinical trial supply chains. There are technology solution providers actively working to reduce costs and make digital twins available as an analytical tool for sponsors with highly expensive clinical trial supply chains that would make the up-front investment worthwhile.

Digitalization is another emerging trend, particularly at global biopharmaceutical companies that have advanced information technology infrastructure and sufficient capital to invest in broad system integrations. Digitalization is the process of converting information to a digital or computer-based format. Supply chain digitalization is about more than simply capturing supply chain data in electronic systems though. It's also about harmonizing and integrating that data across systems and converting that data into recommendations. Consider a planning solution that not only takes source data from RTSM and ERP solutions but also translates those inputs into actionable outputs. Imagine a tool that analyzes enrollment trends and site and depot inventory levels to make suggested regional depot resupply orders. Imagine a tool that performs trending on production campaigns and automatically estimates future yield improvements. Imagine a tool that proactively suggests optimal safety stock and order target levels based on real-time demand and supply data. The applications to clinical planning are virtually limitless.

Further advances in machine learning and quantum computing are necessary to making a planning tool like this not only possible but also cost effective. Machine learning is a subfield of artificial intelligence where computers are trained to learn from advanced data sets and make decisions based on that information. Machine learning's potential application to demand and supply planning is clear. However, clinical trial supply chain management involves enormous data sets, so powerful computers are necessary to run these programs. That's where quantum computing enters the equation. Quantum computers process information differently than classic computers by changing how computers process information from classical bits (e.g. 0s and 1s) to qubits (e.g. 0s, 1s, and any superposition of these states). While quantum computing is still unproven, it has the potential to exponentially improve complex modeling and decision-making. In other words, it's tailor-made to complement machine learning and supply chain optimization.

Generative artificial intelligence (GAI) could further empower this type of planning solution. Like machine learning, GAI is a subset of the broader artificial intelligence field that is designed to create new content out of existing data sets. What makes GAI unique is that it's a computer model that presents a user interface with human-like communication features. Enter a question into a GAI tool and it provides a response, usually instantaneously. Combine the features of GAI with a machine learning model programmed to address clinical supply chain management that's powered by quantum computers, and an ultra-powerful planning solution that combines digitalization, machine learning, and GAI could become a reality for clinical supply chain planners and leaders. Somewhere in the future, a clinical planner may show up to work, open their planning tool and have a conversation with it about the latest data, trends, and recommended strategies for each trial in their portfolio. Gone are the days of downloading data from an RTSM, fixing spreadsheet formulas, or trying to interpret clunky planning systems. That is an exciting future to consider!

Another emerging technology with the potential to impact the clinical trial supply chain is the blockchain. A blockchain is a distributed and decentralized digital ledger that creates a system in which a transaction record is maintained across computers that are linked in a peer-to-peer network. Blockchains are exciting in the world supply chain management because they can be used to securely track the movement of inventory through the supply chain. This is an acute need in clinical trial supply chain management because the movement of drug, especially at the finished product level, is managed in manual processes that are rife with the potential for errors. RTSM systems are designed to capture depot-to-depot and depot-to-site shipments, but the RTSM is at the mercy of humans remembering to enter manual transactions. In a future world where finished product kits are managed through a blockchain, each shipment could be scanned through a barcode upon departure and then again upon receipt. Information about the inventory status could then be automatically captured on the blockchain to provide a more real-time view of where those kits are at any time.

22.8 Key Takeaways

1) Clinical trial supply chains rely on data and information technology, but focusing exclusively on technology leads to suboptimal results. A holistic technology strategy includes building a data governance strategy, a training strategy, and a business process integration strategy to support user adoption and to extract maximum value from the solution.
2) Most pre-commercial biopharmaceutical companies rely on spreadsheets for demand and supply planning. As companies grow in size, they sometimes decide to move away from spreadsheets because of its limitations and toward a more advanced planning tool. For companies looking to make this transition, there should also be a selection and implementation strategy for the new system.
3) Spreadsheets are a powerful, flexible, and cheap spreadsheet-based software that enable data entry, storage, analysis, and visualization. Using a spreadsheet's formula set, it's possible to build a completely end-to-end demand and supply planning model that encompasses demand planning, Finished product supply planning, distribution planning, expiry planning, and upstream supply planning capabilities. Summary reporting and analytics can also be programmed into the planning tool.
4) Spreadsheets also have significant drawbacks including the need to manually program a model from scratch and its inability to directly integrate with other technology platforms. Spreadsheets are also not a cloud-based solution so it's easy for different versions of a planning model to pop up over time, and this creates challenges with training and program transitions when clinical planners leave the group. To mitigate these drawbacks, a spreadsheet-based planning model should be end-to-end, a super-user from the team should be assigned who understands how to troubleshoot issues in the model, and training documentation should be written.
5) Data governance is a data management framework designed to ensure that data is current, accurate, and is actively managed in a company's technology platforms. There are five elements of a data governance strategy that are particularly applicable to clinical planning systems: data element definition, source document/system definition, data ownership, data rules, and data quality management. A simple data governance strategy begins with identification of the data elements used in a spreadsheet-based planning model and how that data should be entered and updated.

6) Training is another foundational element of a holistic technology strategy. The training strategy should have at least three components: creation of training documents; assignment of a planning solution super-user; and establishment of a feedback mechanism to incorporate suggested improvements to the planning solution.
7) Business process integration is yet another element of a holistic technology strategy. The demand and supply planning solution should be designed to support and enable the core clinical supply chain management business processes: Demand and supply plan creation, demand and supply planning, operational CD&OP, strategic CD&OP, and supply plan handover.
8) Some biopharmaceutical companies outgrow spreadsheets as a technology platform for demand and supply planning. Advanced planning tools fall into two categories, those that exist within an RTSM or ERP solution and those that exist as stand-alone platforms. When going through a selection process, the sponsor should consider system capabilities, system integrations, the user interface, and the costs of implementing and maintaining the new solution.
9) There are several technological advancements with the potential to significantly impact clinical trial supply chain planning. These include the concepts of digital twins, supply chain digitalization, machine learning, quantum computing, and GAI. It's possible to imagine far more powerful planning tools in the future that automatically synthesize data and provide recommended supply chain strategies.

23

Integrating the Clinical Supply Chain into a Commercial Launch

23.1 Transitioning from Clinical Supply Chain Management to Global Supply Chain Management

Launching an approved drug is the primary goal of the clinical development process. At the end of registrational clinical trials, the data collected from study participants is analyzed to determine whether the endpoints have been met and the drug is both safe and efficacious. If the data is positive, the sponsor then completes and submits a marketing application for the drug to the competent health authority of each country they wish to market the product in. For small molecules, a new drug application (NDA) must be submitted and for large molecules and cell therapy products a biologics license application (BLA) must be submitted. The NDA/BLA review process takes up to a year or more depending on the country and the filing process before an approval is potentially given. Only after that approval is obtained can the drug finally be made available for sale in that country.

Preparing to launch a commercially approved biopharmaceutical product is a massive undertaking that takes years of preparation and involves almost every functional area within the sponsor. During this process, the chemistry, manufacturing, and controls (CMC) team is responsible for developing, manufacturing, and releasing the first commercial batches of material and for authoring the quality sections of the commercial application. The global supply chain function has several responsibilities of their own within this effort. These include supporting the commercial supply chain strategy, integrating clinical and commercial demand and supply planning, and enabling commercial supply chain execution.

The commercial supply chain strategy is composed of several elements including identification and selection of the commercial supply chain sites, determining the manufacturing scale and, if necessary, executing technical transfer activities to support a larger commercial-scale process within the commercial supply chain network. Another important piece of the commercial supply strategy is to decide whether ongoing clinical studies will be supplied from the commercial supply chain or will continue to be supplied from the existing clinical supply chain. There are advantages to consolidating production within a single supply chain, but there are other considerations that also complicate this decision.

On demand and supply planning, the clinical and commercial supply chains of a single molecule are often naturally intertwined with each other. In many cases, when a product launches commercially, there are active clinical trials happening during the NDA/BLA application process and after

Supply Chain Planning for Clinical Trials – A Practical Guide, First Edition. Ryan Mills.

the product has received approval. All active clinical demand streams must be accounted for in the demand plan, or the supply plan will be inadequate to support them. Additionally, new studies like Phase IV post-marketing trials, investigator-sponsored trials (ISTs), and expanded access programs (EAPs) may begin during the commercial application process or after an approval has been received, creating even more demand on the commercial product. Integrating the demand and supply planning process isn't limited to demand source identification.

Integrating the clinical and commercial supply chains is key to a successful launch. To accomplish this, a decision must be made on whether or not to integrate the clinical and commercial supply chain functions under one global supply chain organization or to split them out into two separate groups. There are advantages and disadvantages to both approaches. The business process framework that supports global supply chain management also deserves attention. Demand and supply planning itself remains a core business process for commercial products, but a formal sales and operations planning (S&OP) process also needs to be implemented to bring together commercial supply chain stakeholders and align on an optimal supply chain strategy for a commercial product. S&OP also facilitates collaboration between clinical and commercial planners to identify demand streams and ensure they are captured in the appropriate supply plans.

Lastly, managing commercial supply chain execution involves many now-familiar elements of managing clinical supply chain execution. New good manufacturing practice (GMP) production and distribution sites need to be selected, onboarded, and managed; the quality management system (QMS) needs to be updated to meet international regulatory standards for commercial products; the regulatory framework needs to account for increased SKU complexity and more onerous change control requirements; and the information Technology strategy typically includes implementation of an enterprise resource planning (ERP) solution to manage financial transactions and inventory. While the global supply chain function is typically not solely responsible for driving these pieces, it does need to be engaged so the organizational impacts on the clinical supply chain are well understood.

In this chapter, we'll discuss the responsibilities of the global supply chain function during commercial launch preparation and how they impact the clinical supply chain in greater detail. We'll begin with a look at commercial supply chain strategy development and then move to demand and supply planning integration, the S&OP process, and commercial supply chain execution. Finally, note that we've replaced the "clinical" precursor with "global" in our description of the global supply chain function in this chapter. That's because as a biopharmaceutical organization moves toward its first commercial launch, the scope of the supply chain function expands to cover both clinical and commercial products around the world. We'll conclude the chapter with a discussion of what that means for the department's organizational structure and measures global supply chain leaders can take to foster integration between the clinical and commercial supply chain teams.

23.2 Commercial Supply Chain Strategy

The physical supply chain for commercial products often differs from that used to manufacture and release clinical products. While the manufacturing sites sometimes remain the same, commercial drug substance and drug product manufacturing often takes place on a much larger scale than clinical manufacturing. Scaling up a manufacturing process usually requires different equipment and occasionally requires an entirely new manufacturing process altogether. The analytical control strategy for commercial products is also more rigorous and the DS and DP specifications are tighter than in clinical development. The finished product labeling strategy is also different.

Instead of using a single booklet label with multiple languages, commercial products are typically labeled with a single country's language, and this approach creates additional finished product SKU complexity. From a distribution perspective, the commercial supply chain is typically centralized through third-party logistics (3PL) providers, and the product is shipped around the world to other regional distribution hubs and then on to hospitals, pharmacies, and other points of sale. That's in contrast to clinical supply chains which operate through a fixed number of regional depots and only distribute directly to clinical sites or to patients. There are also stricter regulatory requirements for commercial products than clinical products. As part of the NDA/BLA application, the sponsor needs to show regulators a consistent commercial-scale manufacturing and batch release process for each of the supply chain.

Given this landscape and the unique requirements of commercial pharmaceutical products, the commercial supply chain strategy needs to include

- Development of a commercial demand plan.
- Identification of the intended commercial manufacturing scale.
- Identification and selection of commercial manufacturing sites.
- Demonstration of the commercial drug substance, drug product, and finished product manufacturing processes.

Building a commercial supply chain strategy starts with understanding the demand profile of the commercial product. Typically, the commercial function within a biopharmaceutical company is responsible for developing the launch strategy of a product. The primary outputs of this strategy that influence the demand profile are the timing of potential approval and launch within each country and the estimated market size within each country. Global launches often occur in a staggered fashion where major markets like the United States, Germany, France, Spain, Italy, the United Kingdom, and Japan are prioritized. This strategy is due to both the market size of these countries and the well-established regulatory frameworks for review and approval of marketing applications. As a result, demand for commercial products typically ramps up over several years before peak demand is reached. As part of the launch planning process, the commercial function provides a forecast for the lifetime of the product from initial launch up to peak demand and beyond. That forecast also includes a breakdown of how many patients are expected to go on the drug by year and by country. Upside and downside demand scenarios are also helpful to understand the widest range of outcomes the supply chain may need to support if those aren't already included in the analysis.

The global supply chain function is typically responsible for analyzing this demand forecast and providing a recommended commercial manufacturing batch size and scale. With this information, the CMC team then works with the late-phase clinical manufacturing sites to assess whether or not they can support commercial launch supply. For example, consider a small-molecule drug substance manufacturing site that produced 2×100 kg clinical batches per year to meet trial demand. If peak commercial demand is 25× clinical demand, then 5000 kg of commercial drug substance is required to meet that demand. The CMC team has two levers to explore with the manufacturing site to meet this requirement: the batch size and the number of batches that need to be produced. In this example, let's assume the maximum batch size the DS manufacturing site can accommodate is 250 kg. That means the site would also need to be able to produce 20 batches per year to meet the 5000 kg requirement. If the site doesn't have sufficient capacity to support this, then another DS supplier would need to be qualified and onboarded as part of the supply chain strategy. Large-molecule product launches go through similar analyses, but the available bioreactor sizes limit the batch size options. For example, if the current clinical manufacturing site only offers

2k and 4k bioreactor sizes and the ideal batch size to meet commercial demand requires a 12k bioreactor size, then executing a technical transfer to a new manufacturing site with a 12k bioreactor makes sense as part of the commercialization strategy.

Performing this analysis and deciding on whether to transfer to a new manufacturing site(s) should take place at least 3 years ahead of the proposed launch date. That's because as part of the NDA/BLA application, the commercial manufacturing sites need to be qualified, and the manufacturing process needs to be performed and validated at the intended launch sites. This validation process involves executing a process performance qualification (PPQ) manufacturing campaign. While Food and Drug Administration (FDA) guidance doesn't require a minimum number of batches, typically at least three batches of drug substance and three batches of drug product are manufactured with the intended commercial process at the intended commercial manufacturing sites to prove the process performs consistently.[1] Given this requirement, technical transfer to new commercial manufacturing sites needs to begin at least ~ 2–3 years in advance of the NDA/BLA application submission for the drug substance and ~ 1–2 years in advance for drug products. This allows sufficient time to perform site qualification activities, transfer the manufacturing process, and execute the PPQ batch campaign in time for the NDA/BLA filing. Another strategy to consider is postponing technical transfer activities until after the initial NDA/BLA approval is received. This approach works when the existing clinical supply chain has sufficient capacity to support commercial demand through the first few years of launch.

Standing up a new commercial supply chain also leaves the CMC team with a choice between using the existing upstream (e.g. drug substance and drug product) clinical supply chain to supply ongoing or newly planned clinical trials or switching to the new commercial supply chain to service that demand. The factors that influence this decision include timing, supply chain complexity, cost, and regulatory complexity. With respect to timing, the commercial supply chain needs to be qualified and running in order to supply clinical trials. If there is risk that technical transfer activities won't be completed, then supplying studies with clinical material is the only option. From a supply chain complexity perspective, it's always simpler to plan and execute a single supply chain, and this argument favors consolidating the two supply chains as soon as possible. However, manufacturing with the clinical supply chain may be cheaper to execute, and switching to commercial material requires filing a substantial regulatory amendment to introduce the new supply chain. For ongoing global clinical trials, this is a time-consuming and costly process of its own.

Figure 23.1 captures the considerations that go into the commercial supply chain strategy. In this example, the clinical development plan (CDP) calls for a Phase III trial to complete by the end of Year 3. Approximately 18 months later, the NDA submission takes place in that indication. Additionally, a Phase II and subsequent Phase III study is planned in a second indication, and an EAP begins in Year 5. The clinical supply chain produces drug substance and drug product to support those trials through the end of Year 4. Meanwhile, the CMC team has outlined a commercial supply chain transition that begins halfway through Year 2 where drug substance site selection and qualification begins. Drug product site selection and qualification begins about 6 months later at the beginning of Year 3. This technical transfer process culminates in Year 4 where both DS and DP PPQ batches are expected to complete. This strategy accomplishes two objectives. First, it provides PPQ data to support the NDA application in Year 5, and second it gives the CMC team optionality to supply the second Phase III and the EAP with either the clinical material or the commercial material.

Ultimately, the global supply chain function needs to work closely with the CMC team on the commercial supply chain strategy as soon as that planning process begins. The CMC team needs to understand ongoing clinical demand requirements from the global supply chain function, and the

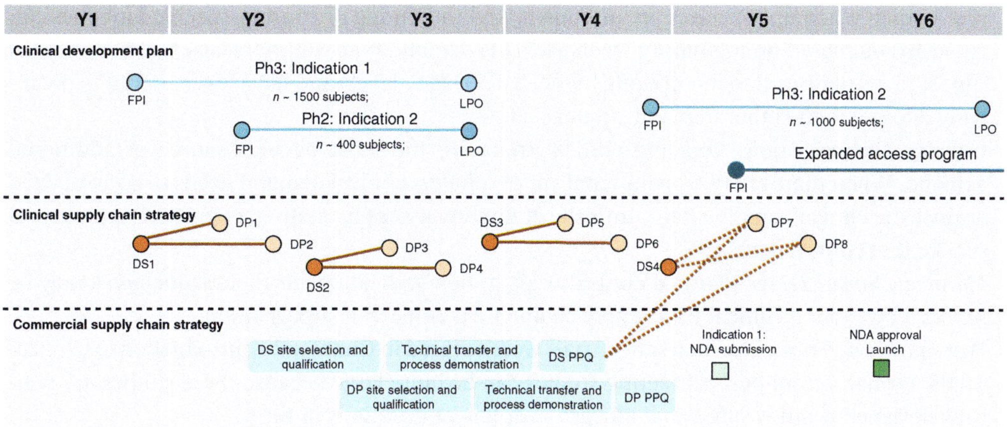

Figure 23.1 Example Commercial Supply Chain Strategy

global supply chain function needs to understand when commercial material will be available and how that material needs to be allocated across both commercial and clinical demand streams.

23.3 Integrating Clinical and Commercial Demand and Supply Planning

During commercial launch planning, it's tempting to think of commercial demand as completely separate from clinical demand, and to an extent, this is true. Patients who access a commercially approved drug through their doctor or health care provider represent a fundamentally different population than clinical trial subjects. But, it's important to remember that demand against commercial supply has many potential sources including EAPs and ISTs. It's a terrible feeling as planner to learn of a demand stream that was not included in the supply plan. At best, a situation like this leads to reduced safety stock levels which make the supply chain vulnerable to further demand or supply shocks. At worst, this leads to a potentially significant stockout event where difficult allocation decisions need to be made across demand streams.

To ensure the supply plan is sufficient, every potential demand source needs to be identified and integrated into the applicable clinical or commercial supply plan. In commercial biopharmaceutical companies with multiple active supply chains (e.g. a clinical supply chain and a commercial supply chain), this isn't as straightforward as it appears for a number of reasons. First, demand for commercial material comes from a variety of places, and the functional oversight of these demand streams is often spread across the sponsor. Second, ownership of the demand and supply planning processes is often split up across the clinical and commercial supply chain functions. Third, the planning systems used by the clinical and commercial functions may differ. For example, clinical planning may still be performed in spreadsheets while commercial planning takes place in a more advanced planning system.

The first potential issue to address is properly identifying all potential demand sources for clinical and commercial material post-launch. These sources include

1) *Commercial Patients*: Suffice to say this demand is self-explanatory. But, it's also worth mentioning that commercial demand is highly variable, especially during the first few years

post-launch when uptake rates are unknown, and the timing of country-specific launches can move around based on regulatory feedback. This demand uncertainty means that commercial supply plans may need to incorporate higher safety stock levels and have less material to potentially allocate to the other demand streams below.

2) *Ongoing Clinical Trials*: Ongoing clinical trials are the most obvious source of additional demand. When there are two operational supply chains, clinical demand needs to be associated against the clinical supply chain until the commercial supply chain is ready and approved to service that clinical demand.
3) *Planned Clinical Trials*: Planned clinical trials in new indications or therapeutic areas may be serviced by either a clinical trial supply chain or a commercial supply chain.
4) *Post-marketing Trials*: Post-marketing trials were covered in Chapter 2 – Introduction to Clinical Trials. Demand from post-marketing studies can be quite large because these studies are typically designed to study safety and efficacy over longer durations of time.
5) *Investigator-sponsored Trials (ISTs)*: ISTs are a sub-type of post-marketing study that is planned by a third-party investigator. Demand from ISTs can be dispersed across many small trials, so getting a complete picture of the demand profile is a challenging task.

There is a fifth and final category of potential demand that deserves extra attention, EAPs. EAPs are unique because they are not a clinical trial. Instead, they are designed for patients living with serious or immediately life-threatening diseases or conditions, who have limited or no access to therapeutic alternatives and are not eligible for or able to participate in a clinical trial. EAPs go by many different names including compassionate use, named patient access, and post-trial access, but they all attempt to get a potentially life-saving or life-extending drug into the hands of someone who wouldn't otherwise have access through a clinical trial or from commercially available material.

Initiating an EAP is never taken lightly by a sponsor. That's because providing drug to a patient population before that drug has been made available to other populations raises ethical and operational questions that must be adequately addressed. These include ensuring fair access to the drug, managing potential resource trade-offs, and abiding by all country-specific regulations. A cross-functional group including representatives from development, commercial, and CMC needs to work together on this analysis. During this effort, the global supply chain function typically needs to provide input on the feasibility and cost of manufacturing supply to meet EAP demand. The cross-functional team needs to provide the assumptions that impact a demand forecast including identification of the countries where the EAP will be made available and the patient population characteristics within each country. With this information, the global supply chain function can craft a demand and supply plan and provide feedback to support a decision on whether to open an EAP.

Identifying demand streams across the clinical landscape, post-marketing studies, and EAPs is also complicated by the different stakeholders who manage these categories. Clinical operations is almost always the source for trial-related demand information, but ISTs and EAPs are typically overseen by different functions. Medical affairs is the function responsible for communicating with external medical professionals and health care providers about the sponsor's drugs. Given their relationships with doctors and investigators, medical affairs is typically responsible for investigator-sponsored trial identification and oversight. Responsibility for EAPs also typically falls on the medical affairs team, although other functional representatives from commercial operations, regulatory affairs, and clinical operations are also heavily engaged. It's imperative for the global supply chain function to understand the entire demand landscape and is accomplished by working closely with these functional stakeholders.

At this point, all demand sources and applicable functional stakeholders should be identified. The next issue to tackle is the division of responsibilities within the respective supply chain functions. In commercial biopharmaceutical companies, it's not uncommon for demand and supply planning to take place in two different organizations, clinical supply chain and commercial supply chain. These two teams may use entirely different processes for creating demand and supply plans, and they may also use two different technology systems for planning. When the clinical and commercial supply chain functions are divided organizationally, by technology, or by both, there needs to be a mechanism or process that connects them together.

Once this integration takes place, the planning process itself is straightforward. For post-marketing trials and ISTs, demand planning largely follows the process outlined in Chapter 8 – Finished Product Demand Planning: Basic Concepts. These studies have protocols and the intrinsic and extrinsic characteristics of the study mimic those of standard clinical trials. EAPs behave differently though. With EAPs, it's often difficult to predict both the total number of requests and from which countries those requests will originate. Medical affairs or the responsible functional area needs to provide as much information as possible about the scope of the EAP, and the supply chain planner responsible for the program needs to provide the maximum number of EAP patients the supply chain can support back to medical affairs.

On the commercial side, demand and supply planning also functions similarly to its clinical counterpart, but the stakeholders and feedback mechanisms for collecting demand and supply inputs are different. Instead of working with clinical operations, commercial planners work closely with their counterparts in the commercial organization who are responsible for country-specific sales forecasts. Supply planning principles are also largely similar, but there are notable differences based on the physical supply chain. On the upstream side, there may be multiple production sites available so assigning planned batches across the network is an extra step in the process. Commercial DS and DP manufacturing is typically more consistent with respect to yields than clinical manufacturing, but there still needs to be an active CMC team monitoring this and feeding outputs back to the clinical supply chain team when the allocation strategy changes. On the finished product side, the commercial planner also needs to deal with the added SKU complexity brought on by country-specific labels, and the depot network also differs from its clinical counterpart.

The inputs and outputs that affect the clinical and commercial supply plans need to be made visible to the responsible clinical and commercial planners. Whether it's identification and integration of demand streams or allocation of manufactured batches, sharing information among the right stakeholders at the right time is critical to effective clinical and commercial supply chain management. In the next section, we'll review the S&OP process and how useful it is when applied to this space.

23.4 Sales and Operations Planning as an Integrator of Clinical and Commercial Supply Chain Management

S&OP is a well-established business process designed to align key functional stakeholders around a production strategy and financial plan for the entire company. The primary objective of S&OP is to balance projected demand, which is overseen by the sales and marketing functions, with supply, which is managed by the supply chain and manufacturing functions. The manufacturing strategy also dovetails into financial planning for the company as sales forecasts and revenue projections are married with the cost of goods and other organizational expenditures. The scope of S&OP is intended to cover all revenue-producing goods and encompasses the functional areas responsible for demand, supply, and financials. This group typically includes sales, marketing, manufacturing, supply chain, operations, and finance. Given the breadth of the process and the need to align all of these functions, S&OP is typically sponsored by the executive leadership team of a company, and business process ownership is assigned to one or more of the functional areas listed.

S&OP is traditionally run on a monthly cadence and involves a series of meetings that culminate in executive endorsement of the plan. The meeting sequence begins with demand planning where the previous period's demand forecast is reviewed by sales and marketing, and adjustments are made to the next period's demand plan. Next is the supply planning meeting where that forecast is assessed against the supply chain's ability to meet that forecast. The manufacturing and operations functions work to schedule production across the manufacturing network and to highlight potential risks and financial opportunities along the way. Pre-S&OP is the third meeting and here all the functions come together to review the sales projections, demand plan, supply plan, and cost of goods. Key operational and financial metrics are prepared, and potential risks are consolidated. The process culminates with the executive S&OP meeting where company leadership reviews the entire strategy, resolves any disputes between the functions, and endorses the final demand forecast, production plan, and financial analysis.

The potential ability of S&OP to integrate clinical and commercial demand and supply planning is relatively clear. In Chapter 15 – Clinical Supply Chain Planning Processes, we covered the clinical demand and operations planning (CD&OP) processes and how they mimic aspects of the S&OP process. The same approach can be used to map S&OP principles to commercial supply planning. As part of demand planning, the commercial planner should include the clinical planner to identify all active and planned clinical trials that may require commercial material to support them. The clinical planner should also call out which trials will remain on the clinical supply chain and when a potential cutover to commercial material is planned to take place. The commercial and clinical planner should also align on who is responsible for creating a demand forecast for post-marketing studies and EAPs that don't naturally fall under sales and marketing or clinical operations. If the planning systems differ between the clinical and commercial functions, the planners should agree on which system should be the source of truth for each plan. The demand planning meeting also serves as a forum to call out how much demand uncertainty there is in a particular forecast. EAPs and newly planned clinical trials have the greatest amount of demand uncertainty, and highlighting this risk gives the planners the visibility they need to develop an appropriate production and inventory management strategy.

Clinical and commercial planners should also come together at the supply planning meeting. Here, the team checks that the demand streams are associated to the correct supply chains and that the supply plan is producing enough material to cover potential demand. For demand streams with high uncertainty, the clinical and commercial planner should align on the upstream allocation strategy and where there is flexibility to allocate material across demand streams. Potential supply risks should also be discussed when technical transfers to new sites are planned or when there are upcoming production campaigns. And, the total cost of the production and inventory management strategy should be summarized by program and across the entire portfolio for executive input.

S&OP isn't a one-size-fits-all approach, and every biopharmaceutical company is structured differently with respect to their physical supply chain and to the functional areas that manage or influence demand and supply. Designing and implementing a suitable S&OP process takes commitment from the executive leadership team and engagement from the respective functions. No matter where the clinical and commercial supply chain planners sit in the organization, global supply chain functional leaders should either lead or, at minimum, participate in establishing

S&OP within the broader organization. Without the connections that a strong S&OP process provides, the integration of clinical and commercial planning may be put at risk.

23.5 Commercial Supply Chain Execution

Beyond planning, the global supply chain function also needs visibility to production and distribution activities within the commercial supply chain. Similar to the clinical supply chain, at the drug substance and drug product level, oversight and engagement depend on whether the material is manufactured internally or outsourced to a contract manufacturing organization (CMO). When internal sites are leveraged, the sponsor has much more flexibility and control over manufacturing campaigns. The S&OP process should incorporate functional representatives from the sites to review capacity and resource requirements and endorse the proposed supply plan. At the completion of each manufacturing campaign, the site feeds actuals back to the CMC team and/or the commercial planner for reconciliation with the supply plan.

When CMOs are used for upstream production, the same strategic supplier management activities used with clinical CMOs should be used. Establishing a joint project team, holding regular meetings, establishing escalation pathways, and documenting ways of working are all best practices. The difference between commercial and clinical upstream production is that the focus shifts from process development to more routine manufacturing. Consistency is paramount and utilizing metrics and key performance indicators becomes a higher priority. One tool that's often used to align on desired production targets is the supply agreement. A supply agreement is another legally binding document that's meant to complement the master services agreement and the quality agreement. The supply agreement documents the sponsor's expectations with respect to manufacturing and batch release lead times, yields, quality targets, and on time, in full delivery of manufacturing campaigns. It's also not uncommon for a single CMO to perform both clinical and commercial manufacturing on behalf of a sponsor. In this situation, the functions responsible for CMO oversight need to align on how best to manage the overall relationship with the CMO while also keeping the objectives of each project on track.

Downstream from drug substance and drug product, commercial secondary packaging and labeling also shares similarities with its clinical counterpart. Recall that the clinical supply chain function typically has direct responsibility for this step, and the same is usually true for the commercial supply chain function. This means commercial supply chain drives the packaging site selection process and directly manages work performed there. When internal packaging and labeling capabilities exist, the selection process is usually skipped, but when these activities need to be outsourced, it's best to follow the process outlined in Chapter 20 – Contract Organization Management and Oversight. One additional consideration specific to secondary packaging and labeling is whether to select the same supplier used for clinical packaging and labeling or to choose a new supplier. Leveraging the same supplier has advantages because of the sponsor's familiarity with their network and processes, but some clinical partners may not have expertise in the commercial space. On the other hand, choosing a new supplier takes more up-front work for qualification, but many commercial packaging CMOs also have clinical capabilities and have the potential to become backup suppliers for new clinical programs.

The scope of secondary packaging and labeling activities in the commercial space also differs from its clinical counterparts. In the clinical space, packaging and labeling campaigns do not need to be validated. In the commercial space, validation of the packaging and labeling process is required as part of the NDA/BLA submission. This means the commercial supply chain function needs to have a strategy for which batches of material will be used in the validation process. In the clinical space, kit identification numbers are used to uniquely identify each kit. In the commercial space, recent regulations like the Drug Supply Chain Security Act (DSCSA) now require commercial finished product material to have unique product identifiers attached to them so the product can be traced through every level of the supply chain.[2] The process of developing these identifiers and establishing a track and trace system for finished product inventory is colloquially known as serialization throughout the industry. There are technology providers who specialize in serialization solutions and partner with secondary packaging and labeling partners to implement this for biopharmaceutical products. Commercial packaging and labeling operations are also typically automated because of increased volume where clinical packaging and labeling is usually done manually. This reduces the cost for commercial operations when compared to clinical operations.

Distribution of finished product works differently for commercial products than for clinical products. Where clinical product is stored and distributed through a depot network and shipments are controlled by randomization and trial supply management (RTSM) systems, commercial product is stored in warehouses by 3PL providers who arrange for shipments through a wide variety of channels such as hospitals, pharmacies, and/or specialty health care providers. There are a variety of distribution models that commercial products may take, and these are influenced by factors like the disease indication, the patient population, and the countries where the product is marketed. Typically, the commercial function dictates the distribution strategy, and the global supply chain function focuses on inventory management through the distribution network. Inventory-related information is usually captured in an ERP system which is often integrated with the 3PL's ERP system for real-time data transfer and visibility.

While the commercial supply chain operates separately from the clinical supply chain, it's useful for both sides to understand the structure, limitations, and risks of the other. This information is helpful for planning purposes and for deciding which supply chain is best suited to support specific demand streams. With respect to planning, the demand profile of a commercial product may vary widely post-launch, and this may limit the amount of commercial material available to support clinical demand even when upstream production is more consistent. Commercial manufacturing also takes better advantage of economies of scale in upstream production and with automated packaging and labeling. This is a major factor when deciding how quickly to consolidate into a single supply chain for all demand streams.

23.6 The Commercial Quality Management System and Regulatory Considerations

When companies prepare to launch a commercial product for the first time, they review their QMS and evaluate its ability to ensure the quality of a commercial product and to withstand the scrutiny of a preapproval inspection (PAI). During NDA/BLA application review in the United States, the FDA may conduct a preapproval facility evaluation or a PAI of the sponsor,[3] and successfully passing a PAI is a prerequisite to obtaining NDA/BLA approval. According to FDA guidance, the purpose of the PAI is to verify "the methods used in, and the facilities and controls used for, the manufacture, processing, packing, and testing of the drug are found adequate to ensure and preserve its identity, strength, quality, and purity." During a PAI, the FDA may visit either the sponsor or any of the GMP manufacturing and analytical testing sites within the supply chain. The FDA has the right to examine all aspects of the QMS to assess whether or not GMP regulations have been satisfactorily met by the QMS. While the European Medicines Agency (EMA) does not conduct PAIs, competent health authorities are empowered to perform inspections of sites located within their countries.[4] Chapter 4 – Quality Considerations covered the scope of a PAI and emphasized the importance of the clinical supply chain function and the quality function working together toward a successful PAI outcome. Here, we'll focus more on the impact PAI readiness has on clinical and commercial supply chain management.

To prepare for a PAI, the sponsor should assess every aspect of the QMS including supplier qualification, quality agreements, specifications, manufacturing control (e.g. batch records), deviations, corrective action and preventive actions (CAPAs), and investigational medicinal product (IMP) handling. There is one process in particular that has potential implications for both commercial and clinical supply chain management, change control. In Chapter 4 – Quality Considerations, we discussed the differences between change management, which is more appropriate for clinical supply chain management, and change control, which is more appropriate for commercial supply chain management. Recall that the clinical change management process allows for technical implementation of a change prior to regulatory assessment, while the commercial change control process requires a complete impact assessment of the change prior to implementation. The increased rigidity of commercial change control exists because making a change to a commercial manufacturing process often has more wide-ranging regulatory and cost impacts than a change to a clinical product. The NDA/BLA for a commercial product typically contains much more detail about the manufacturing process and also has tighter specifications and controls around the manufacturing process. Therefore, a change made to the commercial process is more likely to have a regulatory impact. Changes to a commercial manufacturing process or supply chain are further complicated by the number of individual marketing applications for a product. For many products, there are a greater number of commercial marketing applications to manage than there are clinical trial applications.

As a sponsor prepares for commercial launch, the change management process needs to be robust enough to effectively control commercial changes but flexible enough to process clinical changes. To accomplish this, there should be one parent change management process for all changes. The first step in this process should be to filter changes into early-phase clinical changes, late-phase clinical changes, and commercial change. Then, separate workflows can be used to process the different change types, so the right levels of flexibility and control are used to assess and implement each change. Effectively implementing a nuanced change management process like this on an organization-wide scale is potentially challenging because the global supply chain function and the quality function are often siloed from each other. But, the potential benefit is worth the effort to keep the clinical supply chain agile through change and to make sure that commercial changes have the right cost–benefit analysis prior to implementation.

Managing increased regulatory complexity in a post-commercial world is another hurdle to clear, particularly when commercial material is used to supply clinical trials and EAPs. Recall from Chapter 5 – Regulatory Considerations that IMP lifecycle management is the process through which batches of investigational medicinal product are manufactured, released, filed, approved, and distributed while accounting for supply chain, manufacturing, and analytical changes to the product. Commercial products have the same requirement where the manufacturing process, specifications, and supply chain used in the production of a batch need to exactly match the approved NDA/BLA content in order for that batch to be compliantly distributed within each country. The commercial change management process therefore needs to include a holistic regulatory impact assessment of the NDA/BLA and any clinical trial applications that reference NDA/BLA content.

Process version management (PVM) becomes even more important in a company with both clinical and commercial products because the number and type of regulatory applications increases. The PVM process creates a process version code that summarizes the combination of regulatory-relevant attributes of each DS and DP batch into a single data element. The process version code is then associated to the batch and to the regulatory application which creates a link between the batch release and regulatory status tracking processes. This approach works just as well for commercial products as it does for clinical products and makes the batch release process more compliant and efficient when there are multiple active supply chains within a program. Many sponsors choose to invest in a regulatory document management system (RDMS) technology platform to manage their regulatory applications more efficiently. RDMSs contain not only the applications themselves, but also the data elements necessary to implement PVM.

23.7 The Evolution of the Global Supply Chain Organization

The organizational growth that's necessary to bring a commercial product to market directly impacts the global supply chain function. Prior to its first launch, supply chain management is limited to the clinical space. In smaller organizations, there may only be one or two people responsible for all planning and execution activities, while in larger companies, there may be explicit subfunctions. During the launch readiness process, a separate commercial supply chain function needs to be established to manage planning and execution activities for the commercial product. With this development, the global supply chain function evolves from focusing exclusively on clinical to a more global focus on all clinical and commercial products; hence, the change to global supply chain as the functional title.

At the point of a company's first commercial launch, it's typical for the clinical and commercial supply chain functions to report through the overall global supply chain function. This creates a natural environment for the clinical and commercial team members to work together and collaborate. However, as the number of commercial products and clinical trials increases, the commercial supply chain function often grows into a completely separate entity from the clinical supply chain function. This separation takes multiple forms. In some instances, the clinical and commercial functions are split up from a reporting perspective. In other instances, the teams physically sit in different areas of the company. Without physical proximity or leadership that brings the teams together, they lose touch with each other, and this creates potential information-sharing gaps that lead to poor planning.

On one level, this differentiation makes sense. The stakeholders that commercial supply chain works with to collect demand and supply planning inputs are completely different from the clinical supply chain team. The physical supply chain, business processes, and systems are also usually different. Constructing an organization dedicated to commercial supply chain management is an entirely understandable approach. But, the one consistent theme through this chapter is the need to keep the clinical and commercial supply chains integrated where they intersect with each other.

There are a few strategies global supply chain leaders should consider to increase communication and collaboration between the two functions. These strategies are grouped by the People–Process–Technology framework introduced in Chapter 1 – Supply Chain Management. From a people perspective, the supply chain functions should be physically proximate to each other. Where that's not possible, leadership should consider establishing a rotational program where clinical planners and commercial planners have an opportunity to spend either a 6- or 12-month period working in the other's functional area. A rotational program builds skills and awareness between the two functions, and it creates a development opportunity for people to work toward outside of their current role. If a rotational program isn't possible, offering lunch-and-learn-style networking events that cover specific topics within clinical and commercial supply chain management is a potential alternative. Finally, onboarding and training documentation needs to explicitly call out the links between the clinical and commercial functions. Dedicated training sessions that focus on the connection between clinical and commercial should be held on a regular basis.

We've already touched on the most important business process to consider earlier in this chapter. Where an S&OP process exists, the commercial planner should invite their clinical counterpart to relevant demand and supply reviews. If an S&OP process doesn't exist, the commercial planner should take the initiative to set up a monthly review of activity with the clinical planner. Clear roles and responsibilities should also be established between the two planners so there's no confusion about which demand sources will be covered by which planner. Business process adherence checks should also be established that document whether or not these conversations are happening, and adherence metrics should be provided to global supply chain leadership.

With respect to data and technology, the first step is to document the end-to-end system architecture used by the clinical and commercial supply chain functions. Understanding who's using which systems for what work is necessary to identify potential optimization opportunities. Optimization does not necessarily mean integration though. Think back to Chapter 22 – Clinical Supply Chain Planning Tools and Technology. Considering systems in isolation or as the sole solution for planning activities is a recipe for disaster. It may be tempting for global supply chain leaders to think that if both clinical and commercial planners used a single planning system, there would be increased efficiency and better results. This might be true in some instances, but not necessarily. As we've covered extensively, clinical trial supply chain management has so many nuances that a planning system has to support, and those nuances simply don't exist in the commercial space. In some cases, it may make more sense to keep planning in two different systems and to find other areas where efficiency may be captured. For example, master data standardization facilitates faster and more accurate import/export activities between two systems, so this might be a better starting point than shooting for integration right away.

No matter the functional landscape, thinking holistically about managing clinical and commercial supply chain integration should be a priority for global supply chain leaders. There will always be fundamental differences between the two supply chains, but they're also interconnected with each other. Successful integration enables the sponsor to effectively manage their supply chain and to best serve the stakeholder group that matters the most, the patients and clinical trial participants who rely on sponsors to deliver drug in a timely and cost-effective manner.

23.8 Key Takeaways

1) The primary goal of the clinical development process is to launch a commercially approved drug. Filing an NDA or BLA is required to obtain commercial marketing approval. The global supply chain function has many responsibilities during the launch readiness and post-approval phases. Successful integration of the clinical and commercial supply chains is key to a successful launch, and this requires a holistic strategy that's focused on the organization, business process framework, and planning systems.
2) A commercial supply chain strategy is composed of the commercial demand plan, the commercial manufacturing scale, the commercial supply chain sites, and demonstration of the commercial manufacturing process. The strategy should be assembled at least 3 years from

the intended launch date to allow sufficient time for technical transfer of the manufacturing process and provision of demonstration batch information to inform the NDA/BLA.

3) Integrating clinical and commercial demand and supply planning is necessary to avoid missing potential demand streams or allocating demand to the wrong supply chain. There are many potential demand streams on commercial material post-launch including commercial patients, ongoing and planned clinical trials, post-marketing trials, ISTs, and EAPs.
4) S&OP is a business process that's designed to align key functional stakeholders around a production strategy and financial plan for a company. The application of S&OP to commercial supply chain management also provides a pathway to integrate the clinical and commercial supply chain functions. Clinical planners should attend demand review and supply review to provide inputs and identify risks within the clinical supply chain for the commercial planner to use in the commercial planning process.
5) Commercial supply chain execution shares many similarities with its clinical counterpart at the drug substance, drug product, and packaging levels, but there are a few differences. It's useful for the clinical supply chain team to understand the capabilities and limitations of the commercial supply chain when commercial product is supplied to clinical trials.
6) Reviewing and updating the QMS to meet regulatory requirements is another aspect of commercial launch readiness. The introduction of a commercial change control process has the potential to impact clinical supply chain management the most. It's important to have a change management process that's flexible enough to support the needs of the clinical supply chain and rigorous enough to thoroughly assess the impact of commercial changes prior to implementation. This can be accomplished by filtering all changes into clinical or commercial buckets for further processing.
7) As the sponsor becomes a commercial organization, it's natural for the clinical and commercial supply chain functions to grow apart from each other. It's the job of global supply chain leaders to foster communication and collaboration where necessary. There are many strategies to do this including finding ways to bring the teams together, focusing on business process integration, and minding the systems landscape the functions are using.

Notes

1 https://www.fda.gov/drugs/guidances-drugs/questions-and-answers-current-good-manufacturing-practice-regulations-production-and-process#5.
2 https://www.fda.gov/drugs/drug-supply-chain-security-act-dscsa/what-do-i-need-know-about-supply-chain-security-requirements-under-drug-supply-chain-security-act#Product%20Identifiers.
3 www.fda.gov/media/121512/download.
4 https://www.ema.europa.eu/en/human-regulatory/research-development/compliance/good-manufacturing-practice/guidance-good-manufacturing-practice-good-distribution-practice-questions-answers#inspection-coordination-section.

Appendix

Clinical Supply Chain Planning Data Elements and Formulas

a RTSM Data Elements

- Kit identification number (expressed as a numerical value)
- Treatment group number (expressed as a numerical value)
- *SRLT*: Site replenishment lead time (expressed in number of days)
- *Subject Demand Over SRLT*: (expressed as a numerical value)
- *QKD*: Quantity of kits dispensed (expressed as a numerical value)
- *NESV*: Number of enrolled subject visits (expressed as a numerical value)
- *SOHI*: Site on-hand inventory (expressed as a numerical value)
- *F*: Floor quantity (expressed as a numerical value)
- *C*: Ceiling quantity (expressed as a numerical value)
- *SW*: Short window (expressed in number of days)
- *LW*: Long window (expressed in number of days)
- $NESV_{SW}$: Number of enrolled subject visits over the short window (expressed as a numerical value)
- $NESV_{LW}$: Number of enrolled subject visits over the long window (expressed as a numerical value)
- *SWD*: Short window demand (expressed as a numerical value)
- *LWD*: Long window demand (expressed as a numerical value)
- *ROD*: Resupply order decision (expressed as a yes/no statement)
- *ROQ*: Reorder quantity (expressed as a numerical value)
- *DND*: Do not dispense period (expressed in number of days)
- *DNS*: Do not ship period (expressed in number of days)
- *DNC*: Do not count period (expressed in number of days)
- *DNI*: Do not include period (expressed in number of days)

b RTSM Predictive Resupply Threshold Methodology Calculations

$$\mathrm{SWD}(\mathrm{units}) = \mathrm{QKD}(\mathrm{units}) \times \mathrm{NESV}_{\mathrm{SW}}$$

$$\mathrm{LWD}(\mathrm{units}) = \mathrm{QKD}(\mathrm{units}) \times \mathrm{NESV}_{\mathrm{LW}}$$

ROD not required if

$$\mathrm{SOHI}(\mathrm{units}) > \mathrm{SWD} + \mathrm{F}$$

Supply Chain Planning for Clinical Trials – A Practical Guide, First Edition. Ryan Mills.

ROD required if

$$\text{SOHI}(\text{units}) < \text{SWD} + \text{F}$$

$$\text{ROQ}(\text{units}) = \text{LWD} + \text{C} - \text{SOHI}$$

c Individual Subject Demand Planning Data Elements

- *AWS*: Average weight of a subject (expressed in kg)
- TD_{wb}: Weight-based target dose (expressed in mg, g, cells, etc.)
- TD_{ab}: Absolute target dose (expressed in mg, g, cells, etc.)
- *TC*: Target concentration (expressed in mg/mL)
- *VS*: Vial size (expressed in mL)
- *TAD*: Total available dose (expressed in mg, g, cells, etc.)
- *KC*: Kit configuration (expressed as the number of units of drug contained in a single kit)
- *NDU*: Number of drug units per dose (expressed as a numerical value)
- *ADpK*: Available doses per kit (expressed as a numerical value)
- *DbA*: Doses between IMP administration events (expressed as a numerical value)
- Doses per day (expressed as QD, BID, TID)
- *QKD*: Quantity of kits dispensed (expressed as a numerical value)
- *FPI*: First patient in date (expressed as a date)
- *FSA*: First site activation date (expressed as a date)
- *SW*: Screening window (expressed as a numerical value)

d Small-Molecule Individual Subject Demand Planning Formulas

$$\text{NDU}(\text{capsules or tablets}) = \frac{\text{TD}_{\text{ab}}}{\text{TAD}}$$

$$\text{ADpK}(\text{doses}) = \frac{\text{KC}}{\text{NDU}}$$

$$\text{DbA}(\text{doses}) = \text{doses per day} \times \text{days between administration events}$$

$$\text{QKD}(\text{units}) = \text{ROUNDUP}\left(\frac{\text{DbA}}{\text{ADpK}}\right)$$

$$\text{FPI Date} = \text{FSA Date} + \text{SW}(\text{days})$$

e Large-Molecule Individual Subject Demand Planning Formulas

$$\text{TD}_{\text{ab}}(\text{mg}) = \text{AWS}(\text{kg}) \times \text{TD}_{\text{wb}}\left(\frac{\text{mg}}{\text{kg}}\right)$$

$$\text{TAD}(\text{mg}) = \text{TC}\left(\frac{\text{mg}}{\text{mL}}\right) \times \text{VS}(\text{mL})$$

$$\mathrm{NDU(vials)} = \mathrm{ROUNDUP}\left(\frac{\mathrm{TD_{ab}}}{\mathrm{TAD}}\right)$$

$$\mathrm{ADpK(doses)} = \frac{\mathrm{KC}}{\mathrm{NDU}}$$

$$\mathrm{QKD(units)} = \mathrm{ROUNDUP}\left(\frac{\mathrm{DbA}}{\mathrm{ADpK}}\right)$$

$$\mathrm{FPI\ Date} = \mathrm{FSA\ Date} + \mathrm{SW\ (days)}$$

f Enrollment Planning Data Elements

- *SR*: Screen rate (expressed as patients per site per month or p/s/m)
- *SFR*: Screen failure rate (expressed as a percentage)
- *ER*: Enrollment rate (expressed as p/s/m)
- Number of clinical sites (expressed as a numerical value)
- Site activation schedule (expressed as a schedule of events)
- QKD_{DFP}: Quantity of kits dispensed in the demand forecast period (expressed as a numerical value)
- *TSD*: Total subject demand (expressed as a numerical value)
- *DSi*: Discrete sites (expressed as a numerical value)
- *CSi*: Cumulative sites (expressed as a numerical value)
- *DSu*: Discrete subjects (expressed as a numerical value)
- *CSu*: Cumulative subjects (expressed as a numerical value)
- *OVG*: Overage (expressed as a percentage)
- *SSD*: Site stocking demand (expressed as a numerical value)
- *TD*: Total demand (expressed as a numerical value)
- TSD_A: Total subject demand with attrition (expressed as a numerical value)

g Enrollment Planning Formulas

$$\mathrm{ER(p/s/m)} = \mathrm{SR} \times \mathrm{SFR}$$

$$\mathrm{TSD(units)} = \Sigma \mathrm{QKD_{DFP}}$$

$$\mathrm{DSu(\#sub)} = \mathrm{Active\ sites} \times \mathrm{ER}$$

$$\mathrm{TD(units)} = \mathrm{TSD} \times \mathrm{OVG} + \mathrm{SSD}$$

$$\mathrm{TSD_A} = \mathrm{TSD} \times (1 - \mathrm{Attrition\ Rate})\ \hat{}\ \mathrm{Year}$$

h Finished Product Supply Planning Data Elements

- *SSQ*: Safety stock quantity (expressed in number of weeks or months)
- *MIN*: Minimum (expressed in number of weeks or months)
- *MAX*: Maximum or order target (expressed in number of weeks or months)
- *CW*: Current week (expressed as a date)

- *CW-1*: Previous week (expressed as a date)
- OHI_{CW}: On-hand inventory in current week (expressed as a numerical value)
- OHI_{CW-1}: On-hand inventory in the previous week (expressed as a numerical value)
- TD_{CW}: Total demand in current week (expressed as a numerical value)
- *OD*: Order decision (expressed as a yes/no statement)
- *FD*: Future demand (expressed as a numerical value)
- LT_R: Lead time to replenish a product (expressed in days, weeks, or months)
- FD_{LTR}: Future demand over the lead time to replenish
- *SSQ*: Safety stock quantity (expressed as a numerical value)
- *RORD*: Resupply order requirement date (expressed as a date)
- *ROSD*: Resupply order start date (expressed as a date)
- *POQ*: Production order quantity (expressed as a numerical value)
- *OT*: Order target (expressed as a numerical value)
- FD_{OT}: Future demand over order target
- *WOS*: Weeks-of-supply (expressed as a numerical value)
- *ED*: Expiry date (expressed as a date)
- *ESD*: Expiry short date (expressed as a date)
- $BOHI_{CW}$: Batch on-hand inventory (expressed as a numerical value)
- $TOHI_{CW}$: Total on-hand inventory (expressed as a numerical value)
- *AFPWOS*: Aggregate finished product weeks-of-supply (expressed as a numerical value)
- LT_{DTD}: Depot-to-depot lead time (expressed in number of days, weeks, or months)

i Finished Product Supply Planning and Expiry Planning Formulas

$$\mathrm{OHI_{CW}}(\text{units}) = \mathrm{OHI_{CW-1}} - \mathrm{TD_{CW}} + \mathrm{POQ_{CW}}$$

$$\mathrm{SSQ} = \Sigma(\mathrm{FD_{LTR}})$$

Order decision is yes if

$$(\mathrm{OHI_{CW}} \leq \mathrm{SSQ})$$

Order decision is no if

$$(\mathrm{OHI_{CW}} > \mathrm{SSQ})$$

$$\mathrm{ROSD} = \mathrm{RORD} - \mathrm{LT_R}$$

$$\mathrm{POQ} = \Sigma(\mathrm{FD_{OT}})$$

$$\mathrm{WOS} = \#\,\text{Weeks of future demand covered by on hand inventory}$$

$$\mathrm{ESD} = \mathrm{ED} - \mathrm{DND} - \mathrm{DNS} - \mathrm{DNC} - \mathrm{DNI}$$

$$\mathrm{BOHI_{CW}} = \mathrm{BOHI_{CW-1}} - \mathrm{Demand_{CW}} + \mathrm{POQ_{CW}}$$

$$\mathrm{TOHI_{CW}} = \Sigma(\mathrm{BOHI_{CW}})$$

j Upstream Supply Planning Data Elements

- LT_P: Production lead time (expressed in number of days, weeks, or months)
- *BOM quantity*: Bill of material quantity (expressed as a numerical value)
- Production yield (expressed as a numerical value or as a percentage)
- Upstream requirement conversion (expressed as a numerical value)
- Capacity constraint (expressed as a numerical value)
- Expiry window (expressed in number of days, weeks, or months)
- *ITRR*: Implicit total run rate (expressed as a numerical value)
- *Implicit WOS*: Implicit weeks-of-supply (expressed as a numerical value)

k Upstream Supply Planning Formulas

$$\text{Upstream requirement conversion} = \text{Qty of downstream order} \times \frac{\text{BOM Qty}}{\text{production yield}}$$

$$\text{DS BOM quantity}(\text{kg}) = \frac{\text{DP strength}(\text{kg})}{\text{DP Production yield}(\%)}$$

$$\text{DP BOM quantity}(\text{mg/unit}) = \text{DS conc.}(\text{mg/mL}) \times \frac{\text{DP fill volume}(\text{mL/unit})}{\text{DP yield}(\%)}$$

$$\text{DS yield}(\text{g}) = \text{reactor size}(\text{L}) \times \text{titer}(\text{g/L}) \times \text{harvest yield}(\%) \times \text{purification yield}(\%)$$

$$\text{DP yield}(\text{Number of units}) = \frac{\text{DS yield}(\text{g})}{\text{DP BOM quantity}(\text{g/unit})}$$

l Reporting Data Elements

- Protocol number
- Stockkeeping unit
- Manufacturing site
- Manufacturing type
- Supplying distribution depot
- Receiving distribution depot
- Batch number
- Quantity produced
- Unit of measure
- Production start date
- Release date
- Expiry date
- Shipment initiation date
- Shipment receipt date

- Process version number
- Target weeks-of-supply
- Cost of goods
- Cost of inventory

m Metrics Formulas

$$\text{On Time in Full} = \frac{\text{Number of Orders On Time In Full}}{\text{Total Number of Orders}}$$

$$\text{Order Lead Time} = \text{Average Lead Time of a Production or Distribution Activity}$$

$$\text{Stockout Rate} = \frac{\text{Number of Unavailable Orders}}{\text{Total Number of Orders}}$$

$$\text{Stockout Rate} = \frac{\text{Number of missed doses or visits}}{\text{Total Number of doses or visits}}$$

$$\text{Service level} = 1 - \left(\frac{\text{Number of missed doses or visits}}{\text{Total number of doses or visits}}\right)$$

$$\text{Waste} = \frac{\text{Number of Unused Finished Good Units}}{\text{Number of Finished Good Units Produced}}$$

$$\text{Total waste} = \Sigma\left(\frac{\text{Number of unused FP kits}}{\text{Number of FP kits produced}}\right) + \left(\frac{\text{Number of unused DP units}}{\text{Number of DP units produced}}\right) + \left(\frac{\text{amount of DS remaining}}{\text{amount of DS produced}}\right)$$

$$\text{Excess waste} = \text{total waste} - \text{inherent waste}$$

$$\text{Defect Rate} = \frac{\text{Number of Defective Orders}}{\text{Total Number of Orders}}$$

Glossary

Active pharmaceutical ingredient (API) See drug substance
Actualization The conversion of forecasted values to actual values
Adverse events Any unfavorable or unintended disease, sign, or symptom (including an abnormal laboratory finding) that is temporally associated with the use of a medical therapy or procedure[1]
Aggregate forecasting The process of placing all future subject demand at a single SKU–depot location
Allocation Selection of a specific batch that's designed to fulfill a downstream production or distribution order
Allogenic cell therapy Cell therapy product manufactured from healthy donor cells
Amendment A planned change to a regulatory application
Ancillary supply Any material required to conduct a clinical trial aside from the novel therapeutic and comparator products used in the study
Arm A group or subgroup of participants in a clinical trial who receive a specific intervention or no intervention, according to the trial's protocol[2]
Assessments A physical or cognitive test performed on a clinical trial subject
Attrition rate The rate at which subjects drop out of a clinical trial
Audit An official inspection of a contracted supplier
Autologous cell therapy Cell therapy product manufactured from an individual patient's cells
Basket protocol A type of master protocol designed to test a single investigational therapy or therapy combination in patients from different disease groups or subgroups in the same clinical trial[3]
Batch A quantity of goods produced at a single time
Batch lifecycle The process by which a specific batch of IMP is manufactured, released, and filed and approved in a country to enable that batch to ship to a clinical site
Batch record A description of a manufacturing process
Batch release The process of certifying that a batch meets its quality requirements and is good for use in a clinical trial
Batch release lead time The lead time between the date of manufacture and the formal batch release date
Bill of materials (BOMs) List of all the components and quantities of each component needed to produce a good
Biologic See large molecule
Bioprocessing Process of manufacturing large molecule drug substance compounds

Supply Chain Planning for Clinical Trials – A Practical Guide, First Edition. Ryan Mills.

Blinding A process used in clinical trials to prevent bias
Booklet Type of clinical label containing many panels that support multiple languages
Business process A sequence of tasks designed to accomplish a specific objective and to ensure that information is generated and passed to the stakeholders who need it to act
Buyer Party that has purchased a good
Cell therapy Drug made with genetically modified whole cells
Central depot A distribution depot that supplies other depots and/or clinical sites directly
Change control A sequential process where a potential change to a product's GMP supply chain, manufacturing process, or analytical strategy is identified and recommended for assessment
Change management The process of managing a change within an organization
Chemistry, manufacturing, and controls (CMC) The team responsible for developing the process to make the drug substance, drug product, and finished product, and providing oversight of the manufacture and release of the drug
Clinical development plan (CDP) A plan describing the clinical trials required to support registration of a drug
Clinical protocol A detailed plan of a scientific or medical experiment, treatment, or procedure[4]
Clinical/contract research organization (CRO) Third-party organization that supports conduct of a clinical trial
Clinical demand and operations planning (CD&OP) A business process that links demand and supply planning inputs to the production and distribution plan outputs
Clinical trial A research study in which one or more human subjects are prospectively assigned to one or more interventions (which may include placebo or other control) to evaluate the effects of those interventions on health-related biomedical or behavioral outcomes[5]
Clinical trial application The application/submission to a competent National Regulatory Authority for authorization to conduct a clinical trial in a specific country[6]
Cohort A group of subjects who share a particular characteristic such as age, gender, and pretreatment status entering the trial[7]
Comparator A medicinal product used in a clinical trial that is not the novel investigational medicinal product
Competent health authority (CHA) National regulatory body that reviews and approves or denies clinical trial applications
Component Material used in the manufacture of the finished product
Consignee The recipient of a good
Container closure See primary packaging
Contract development and manufacturing organizations (CDMO) Third-party organization that supports manufacturing process development and execution of drug substance, drug product, or finished product manufacturing
Control arm The control arm is the group of participants that does not receive the therapy being studied (investigational therapy). This control arm is compared to the group that receives the investigational therapy, to see if the investigational therapy works and is safe[8]
Courier Organization responsible for shipping raw materials, components, and goods through the supply chain
Current good manufacturing practice (cGMP) Current versions of GMP regulations. See GMP
Data governance A data management framework designed to ensure that data is current and accurate and is actively managed in a company's technology platforms

Date of manufacture Date upon which the shelf-life of a product is based

Decentralized clinical trial A clinical trial where some or all of a clinical trial's activities occur at locations other than a traditional clinical trial site[9]

Deliver The act of providing a good to a downstream node in the supply chain such as a manufacturer, retailer, or consumer

Demand The quantity of a good required by a consumer of that good

Demand-driven planning Plan that directly links demand to supply

Demand and supply planning The process of creating a demand forecast for a good and converting that forecast into a supply plan for production and distribution of that good

Demand forecast Prediction of how many units of a good are required within a specified timeframe

Dependent demand Demand on a component that is driven by a downstream production requirement

Depot A site within the clinical trial supply chain network that ships product to other depots or to clinical sites

Digitalization The process of converting information to a digital or computer-based format

Direct-to-patient A clinical trial supply chain where drug is delivered directly to a patient instead of the patient receiving drug at a clinical site

Distribution The process of sending finished product through a network of depots and on to the clinical site

Distribution order An instruction to ship and deliver a specific quantity of a good for delivery at a specific time and location

Distribution plan A series of distribution orders for finished product that dictate where finished product needs to be shipped, when material needs to arrive, and in what quantities

Dose reduction Decreasing the amount of drug delivered to a subject

Dose titration Increasing the amount of drug delivered to a subject

Downstream A temporally relative description of when a manufacturing or distribution step takes place after another step (e.g. finished product is downstream of drug product)

Dropout rate See attrition rate

Drug product The formulated state of the drug substance that is suitable for human consumption

Drug substance The active compound (e.g. the drug)

End-to-end plan A plan that captures all production and distribution activities within the supply chain of a good

Endpoint A targeted outcome of a clinical trial that is statistically analyzed to help determine the efficacy and safety of the therapy being studied[10]

Enrollment rate Rate at which subjects are projected to enroll in the clinical trial. Typically measured in patients per site per month

Enterprise resource planning (ERP) system Technology platform that serves as a centralized database for product, inventory, and financial information

Excipients Raw materials used in the manufacture of drug products

Exclusion criteria Features of the potential study participants who meet the inclusion criteria but present with additional characteristics that could interfere with the success of the study or increase their risk for an unfavorable outcome[11]

Executed batch record (EBR) Completed batch record documenting the manufacturing process of a batch

Expanded access program (EAP) Program designed for patients living with serious or immediately life-threatening diseases or conditions, who have limited or no access to therapeutic alternatives and are not eligible for or able to participate in a clinical trial

Expiry date Date upon which a batch of drug substance or drug product expires and is no longer suitable for use

Expiry period The period of time over which a batch of drug substance or drug product is suitable for use. Typically measured in number of months

Expiry short date Date upon which a batch of finished product can no longer be dispensed to a subject or shipped from a depot based on RTSM expiry parameters

Exporter The shipper of a good

Fill-finish Large-molecule drug product manufacturing

Finished product The final packaged, labeled unit of drug that is dispensed to a subject

First patient in (FPI) date Date the first subject randomizes. Typically, the first dose of IMP is administered on the date of randomization but this is not always the case

First site activation (FSA) date Date the first clinical site in a trial is activated for recruitment and enrollment activities

First subject screened (FSS) date Date the first subject enters screening for potential inclusion in the study

Fixed text Clinical label text found consistently on every batch of product made for a clinical trial (e.g. protocol number or storage condition)

Further processing lead time The lead time between date of manufacture and further processing date

Good Final output of the supply chain

Good distribution practice (GDP) Regulations governing the distribution of investigational medicinal product (IMP)

Good manufacturing practices (GMPs) A set of legal regulations designed to ensure the quality of a product and protect clinical trial subjects from harm

Importer of record (IOR) Legally responsible entity for the imported goods

Inclusion criteria The key features of the target population that the investigators will use to answer their research question[11]

Independent demand Demand that is independent of any other source beside the consumer of the good

Indication A sign, symptom, or medical condition that leads to the recommendation of a treatment, test, or procedure[12]

Insource Organization makes its own good or product

International Conference of Harmonisation (ICH) An organization whose mission it is to draft and publish a common set of guidelines around Quality and GMP for all biopharmaceutical companies to follow

Intervention A manipulation of the subject or subject's environment for the purpose of modifying one or more health-related biomedical or behavioral processes and/or endpoints. See investigational medicinal product (IMP)

Interventional clinical trial A clinical trial that studies an intervention such as a medicine or investigational medicinal product

Investigational medicinal product (IMP) A medicine being studied in a clinical trial[13]

Investigational Medicinal Product Dossier (IMPD) Document submitted as part of the clinical trial application that summarizes the quality, manufacture, and control of the IMP[14]

Investigational medicinal product lifecycle The process by which all batches of IMP are manufactured, released, filed, approved, and distributed while accounting for supply chain, manufacturing, and analytical changes to the product

Investigational New Drug Application (INDA) A request from a clinical study sponsor to obtain authorization from the Food and Drug Administration (FDA) to administer an investigational drug or biological product to humans[15]

Item Output of the manufacturing process. Synonymous with material and product

Just-in-time manufacturing The process of making finished product inventory after a demand signal (e.g. a subject entering screening) has been received by a clinical site

Key performance indicator (KPI) Metric of strategic importance

Kit Single unit of the finished product

Kit identification list A list of kit identification numbers that associates each kit identification number to a treatment group

Kit identification number A unique code assigned to a single kit of finished product

Label The text applied to the finished product that describes the product and directions for use

Labeling "At Risk" Labeling a finished product batch before the assigned expiry date has been approved

Large molecule Drug manufactured from biological sources

Last patient in (LPI) The date the last patient randomizes and receives their first dose of drug

Last patient out (LPO) The date the last patient completes their last visit

Lead time The time between the initiation and completion of a supply chain process

Logistics The discipline of shipping a product from one location to another location

Lot Synonymous with batch

Low case–high case planning A form of scenario planning that compares the supply plan under low and high demand forecast conditions

Make The act of producing a good that will be used in a downstream step of the supply chain or sold as the final product

Make-to-order (MTO) Production strategy where goods are manufactured after a sales order is received

Make-to-stock (MTS) Production strategy where goods are manufactured before a sales order is received

Make-Versus-buy analysis Analysis to determine if the manufacturing site for a good should be insourced (e.g. made by the company) or outsourced (e.g. made by a third party)

Manufacturer Organization responsible for producing components or products within the supply chain

Manufacturing The production of goods

Manufacturing "At Risk" Manufacturing a batch of a product when the upstream batch has not yet received a fully quality release

Material Output of the manufacturing. Synonymous with item and product

Material requirements planning (MRP) Process that translates a production order into component requirements

Metric Measurement of supply chain performance

Observational clinical trial A clinical trial that does not use an intervention such as a medicine or investigational medicinal product

On-time in-full (OTIF) Metric that measures a percentage of orders that are delivered on time and in full

Order quantity The total amount of a good that will be delivered as part of that order

Order target The desired amount of a good that should be produced or distributed based on future demand

Outsource Contract a third party to manufacture a good or product on the organization's behalf

Packaging "At Risk" Packaging a drug product batch prior to the full quality release of the underlying drug product batch

Participant See subject

Pharmacy manual Document that provides detailed instructions on how IMPs are to be handled, prepared, administered, and returned or destroyed over the life of the trial

Plan (noun) The result of the planning process

Plan (verb) The process of optimizing the manufacture and delivery of goods by balancing supply and demand

Planner Person responsible for executing the planning process and delivering the demand and supply plan

Planning horizon Amount of time prior to a desired production order release date that the production order needs to be shared with the manufacturing site to guarantee order readiness activities will be completed in time to support the release date

Planning level Refers to a specific node within the supply chain (e.g. the drug substance planning level)

Planning time fence The amount of time from the production order release date past which the quantity and timing of the order should not change

Preapproval inspection (PAI) An evaluation by the food and drug administration to ensure the supply chain meets GMP commercial requirements

Predictive threshold resupply model (PTRM) An RTSM forecasting methodology that uses predetermined look-ahead values to calculate the number of kits needed to resupply a clinical site

Primary packaging The container closure system of the drug product. Examples include bottles, blister strips, and vials

Pooling The practice of holding unfinished inventory at a depot for use across multiple studies or multiple arms within a study

Process version management Process which links the manufacturing process used to make a batch with the manufacturing process listed in the regulatory application

Product Output of the manufacturing process. Synonymous with item and material

Production order An instruction to produce a specific quantity of a good for delivery at a specific time and location

Production plan A series of production orders for drug substance, drug product, and/or finished product that dictate how much to produce, where to produce, and when to produce

Production yield See yield

Quality agreement (QAG) A legally binding document that spells out the roles and responsibilities that each party must uphold with respect to GMP compliance

Quality management system (QMS) A formalized system that documents processes, procedures, and responsibilities for achieving quality policies and objectives[16]

Qualified person (QP) Legal entity in the European Union, United Kingdom, and Israel who is responsible for two activities within the supply chain, declaring the GMP supply chain meets EU regulatory requirements and certifying each batch of product for release to a specific trial and country

Qualified person certificate Document issued by a qualified person certifying a specific batch for use in a specific country

Qualified person declaration Document issued by a qualified person declaring the supply chain of a product meets EU GMP regulatory requirements
Randomization The process of assigning clinical trial participants to therapy or control arms of the trial by chance rather than by choice[17]
Randomization schema Predefined ratio of randomization into the arms of a study
Randomization and trial supply management (RTSM) Technology platform that controls subject randomization, IMP dispensation, accountability, and reconciliation during a clinical trial
Randomization list See kit identification list
Raw material Materials used in the manufacture of the drug substance or drug product
Regional depot A distribution depot that only supplies clinical sites
Regulatory commitment A condition the sponsor is held to as part of the approval of a clinical trial application
Regulatory restriction A limitation or restriction on a component of a clinical trial application
Release See batch release
Replenishment lead time (RLT) Duration of time to order and receive a material
Requirement A quantity of a good needed to support the next manufacturing or distribution step in the supply chain at a specific time
Return The act of returning a good to an upstream node of the supply chain
Route of administration Mechanism by the drug is delivered to the subject's body
Safety stock Additional inventory that's held to mitigate the risk of a stockout
Sales and operations planning (S&OP) A business process that brings cross-functional stakeholders together to match a manufacturer's supply with demand
Sample size ("n") Number of subjects included in a clinical trial
Screen failure Disqualification from participation in the trial during the screening process
Screening The process of determining subject eligibility for a clinical trial
Screening window Period of time over which the screening process takes place
Secondary packaging An outer layer of packaging the primary packaging is placed into prior to distribution to a subject
Seller Party that has manufactured and sold a good
Serialization The assignment and application of a unique code to the packaging of a drug
Shelf life Synonymous with expiry date
Shelf-life extension Finished product manufacturing event that extends the shelf life date printed on the clinical label
Single panel Type of clinical label with only one panel containing the text for one or two countries
Sites Locations at which clinical trials take place
Small molecule Chemical compound
Source The act of purchasing or acquiring a good (e.g. a raw material or component) that is needed to facilitate a downstream manufacturing step
Sponsor An individual, company, institution, or organization that takes responsibility for the initiation, management, and/or financing of a clinical study or trial[18]
Standard operating procedure (SOP) Established or prescribed methods to be followed routinely for the performance of designated operations or in designated situations
Stockkeeping unit (SKU) A unique type of product that's made available to a consumer
Stockout Event where supply is not available to meet demand

Storage condition Temperature range at which the drug substance or drug product must be kept
Subject A person taking part in a clinical trial
Supplier, first-degree Provides components needed to make a good
Supplier, second degree Provides raw materials or components to first-degree suppliers for use in their component manufacturing
Supply A good that a manufacturer must produce to meet demand
Supply chain A network of suppliers, manufacturers, and distributors that collaborate to manufacture a product and make it available for consumption
Supply chain agility How quickly the supply chain can react to a change in the plan
Supply chain management The discipline of planning, coordinating, and executing supply chain activities
Supply chain planning The process of optimizing the manufacture and delivery of goods by balancing supply with demand
Supply chain reliability How consistently a manufacturing site delivers a quality batch within the estimated lead time
Supply chain resilience The ability of the supply chain to withstand the realization of a supply chain risk
Supply chain risk Anything that can disrupt the supply chain from delivering sufficient supply where and when it's needed
Supply plan Indicates how many unites of a good that need to be produced and/or distributed to a specific location within a specified timeframe
Supply planning parameters Safety stock and order target values
Tariff Tax on imported goods
Temperature excursion Event where a batch of drug substance, drug product, or finished product exceeds or falls below its intended storage condition
Transportation service provider (TSP) Organization responsible for shipping raw materials, components, and goods through the supply chain
Treatment See investigational medicinal product
Treatment arm See treatment group
Treatment group Another description for treatment arm which kit identification numbers are associated to within a kit identification list
Upstream A temporally relative description of when a manufacturing or distribution step takes place before another step (e.g. drug product is upstream of finished product)
Upstream supply planning Colloquially refers to drug substance and drug product supply planning
Value-added tax (VAT) Tax levied on imported goods. VAT is often but not always recoverable for clinical trial material
Variable text Clinical label text that varies from one batch to another in a clinical trial (e.g. batch number or expiry date)
Visit Predetermined timepoints where a subject has assessments performed on them, receives IMP for administration, or both
Visit schedule Frequency at which visits take place over the course of the study
Visit window Allotted range of days or weeks within which a visit must take place
Waste Amount of unconsumed goods leftover when a supply chain is discontinued
Waste (inherent) Waste that is structurally imbedded in the clinical protocol design elements and the physical supply chain of the study

Waste (excess) Waste that occurs when material expires or simply goes unused by the end of the study because supply has exceeded demand

Weeks-of-supply Describes how much on-hand inventory is available within a specific forecast period

Yield The percentage of component productivity expressed as the number of units produced/number of units entered in the manufacturing process

Notes

1 https://toolkit.ncats.nih.gov/glossary/adverse-event/
2 https://toolkit.ncats.nih.gov/glossary/arm/
3 https://toolkit.ncats.nih.gov/glossary/basket-protocol/
4 https://www.cancer.gov/publications/dictionaries/cancer-terms/def/protocol
5 https://grants.nih.gov/policy/clinical-trials/definition.htm
6 https://rsc.niaid.nih.gov/sites/default/files/CTASubmissionGuidance.pdf
7 https://www.cancer.gov/publications/dictionaries/cancer-terms/search/cohort/?searchMode=Begins
8 https://toolkit.ncats.nih.gov/glossary/control-arm/
9 https://www.fda.gov/drugs/news-events-human-drugs/evolving-role-decentralized-clinical-trials-and-digital-health-technologies
10 https://toolkit.ncats.nih.gov/glossary/endpoint/
11 https://www.ncbi.nlm.nih.gov/pmc/articles/PMC6044655/
12 https://www.cancer.gov/publications/dictionaries/cancer-terms/def/indication
13 https://www.ema.europa.eu/en/glossary/investigational-medicinal-product
14 https://www.ema.europa.eu/en/documents/presentation/eu-gmp-requirements-investigational-medicinal-products-bernd-boedecker_en.pdf
15 https://www.fda.gov/vaccines-blood-biologics/development-approval-process-cber/investigational-new-drug-applications-inds-cber-regulated-products
16 https://asq.org/quality-resources/quality-management-system
17 https://toolkit.ncats.nih.gov/glossary/randomization/
18 https://toolkit.ncats.nih.gov/glossary/clinical-study-sponsor/

Index

Note: Page numbers in *italic* and **bold** refers to figures and tables, respectively.

Supply Chain Planning for Clinical Trials – A Practical Guide, First Edition. Ryan Mills.

d

q

Printed and bound by CPI Group (UK) Ltd, Croydon, CR0 4YY

16/03/2026

14843324-0005